Innovative Renewable Energy

Series Editor
Ali Sayigh
World Renewable Energy Congress
Brighton, UK

The primary objective of this book series is to highlight the best-implemented worldwide policies, projects and research dealing with renewable energy and the environment. The books are developed in partnership with the World Renewable Energy Network (WREN). WREN is one of the most effective organizations in supporting and enhancing the utilisation and implementation of renewable energy sources that are both environmentally safe and economically sustainable. Contributors to books in this series come from a worldwide network of agencies, laboratories, institutions, companies and individuals, all working together towards an international diffusion of renewable energy technologies and applications. With contributions from most countries in the world, books in this series promote the communication and technical education of scientists, engineers, technicians and managers in this field and address the energy needs of both developing and developed countries.

Each book in the series contains contributions from WREN members and cover the most-up-to-date research developments, government policies, business models, best practices, and innovations from countries all over the globe. Additionally, the series publishes a collection of best papers presented during the annual and bi-annual World Renewable Energy Congress and Forum each year.

Ali Sayigh
Editor

Achieving Building Comfort by Natural Means

Springer

Editor
Ali Sayigh
World Renewable Energy Congress
Brighton, UK

ISSN 2522-8927 ISSN 2522-8935 (electronic)
Innovative Renewable Energy
ISBN 978-3-031-04713-8 ISBN 978-3-031-04714-5 (eBook)
https://doi.org/10.1007/978-3-031-04714-5

This Springer imprint is published by the registered company Springer Nature Switzerland AG
The registered company address is: Gewerbestrasse 11, 6330 Cham, Switzerland

Introduction

This book sets out to explore the important issue of achieving building comfort through natural means. First of all, it has to be established why this is considered such an important issue. It is important because in order to reduce carbon emissions which lead to climate change it is crucial to minimize the use of energy derived from fossil fuels.

If we look at the construction and everyday operation of buildings, we find that buildings consume approximately 45% of total energy. Thus, it is imperative that this consumption is not only substantially decreased but also that such energy that is necessary is obtained from environmentally friendly and sustainable sources.

Building comfort can be described from the viewpoint of the inhabitants of the building. This encompasses such factors as heat levels, ventilation, space, lighting, acoustic, and levels of activities.

The aim of this book is to investigate the reduction of the use of fossil fuels–based energy and excessive use of electricity to achieve comfort in buildings. To do this, many factors have to be considered: global locality, thermal mass, daylighting, ventilation, solar gain, shading, earth shelter, night radiation, building orientation, humidity and climate, choice of building material, insulation, noise, energy management and control, the use of vegetation, the creation of micro-climate, and the colour of the building.

A first step towards achieving comfort is to look at the American Society of Heating, Refrigerating and Air-Conditioning Engineers (ASHRAE) directive (55-1992) which sets out two comfort zones as plotted on the psychometric chart of Fig. 1.

ET lines refer to constant enthalpy lines.

Many investigators have specified that the ideal comfort in domestic and commercial buildings is to maintain a dry bulb temperature of 25 °C and relative humidity of 50%; however, a tolerable level of comfort is acceptable when the dry bulb temperature is 25–31 °C with the relative humidity range of 20–60%. It is worth noting that building comfort can be achieved only when the following two factors are considered, the environmental factor, which involves air temperature, radiant temperature, air velocity, and humidity, and the personal factor, which consists of clothing insulation and metabolic heat.

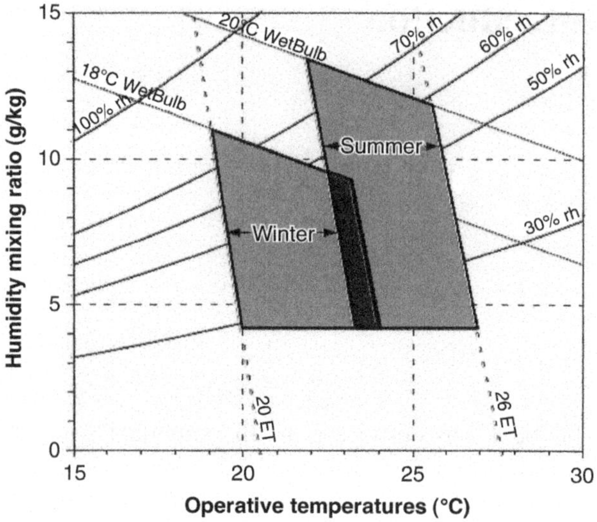

Fig. 1 ASHRAE psychometric chart for comfort

From 1930 to 2000 several outstanding physicists and architects, such as Corbusier, Givoni, Szokolay, Olgyay, and Balcomb, have theorized about achieving comfort levels. However, some architects attributed comfort as being mainly based on the individual and derived their criteria (for example, from Ole Fanger) from the percentage of people dissatisfied (PPD) and a function of predicted mean vote (PMV); see Fig. 2.

The electricity consumption of most households is devoted to space and water heating, space cooling, cooking, lighting, laundry, electrical appliances, and outside applications such as gardening and outdoor cleaning. According to Eurostat, in 2018, this represented 26.1% in Europe.

In 2015 some researchers have indicated that typical buildings and their construction account for a massive 36% of energy use and 39% of carbon dioxide (CO_2) emissions. However, progress towards sustainable buildings and modern construction methods have failed to keep up with a growing demand for energy. It is of paramount importance to improve the energy intensity per metre square of the building industry by an average of 30% by 2030 compared with the 2015 figure set forth in the Paris Agreement.

The energy consumption, in kWh/m^2, of the construction industry according to an IEA source (IEA, Energy Technology Perspectives 2017, IEA/OECD, Paris www.iea.org/etp), in 2000, was 200 kWh/m^2 or 200 EJ (exajoule). In 2015 it was 150 EJ. The UN in Paris proposed to reduce this to 100 EJ by 2030, that is, by 30%.

If one judges domestic electricity use in different parts of the world then one realizes the diversity and uncontrolled reason for it; see Table 1, taken from Electricity—BP Statistical Review of World Energy (2020).

The disparity in domestic electricity consumption between different parts of the world is highlighted in the table, which indicates that the wealthiest countries are profligate in their level of consumption and give scant consideration to the environment and climate change.

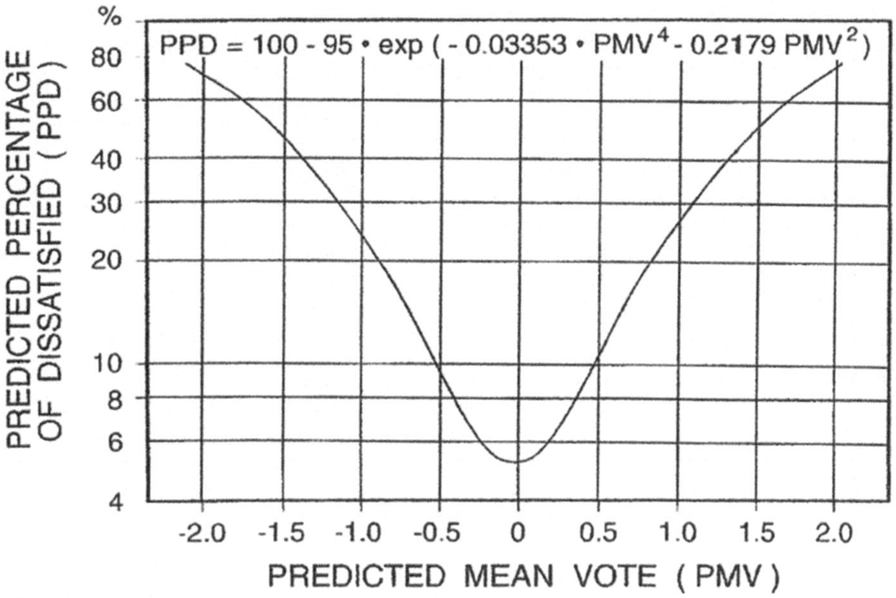

Fig. 2 The ideal comfort is at zero predicted mean vote (PMV). (Ref: http://ceae.colorado.edu/~brandem/aren3050/docs/ThermalComfort.pdf)

The UK government (Health and Safety Executive, HSE) issued a check list to determine comfort in working premises; see Table 2. More than two yes answers trigger an investigation into the causes of discomfort.

Natural ventilation is an important tool for achieving comfort and reducing both CO_2 emissions and electricity demand. Ventilation not only improves comfort levels but removes damaging toxins and CO_2 from the air. The inhabitants are not the sole beneficiaries of good natural ventilation; the 'health' of the building itself is boosted by the reduction of humidity and condensation, which are associated with the growth of mould, which gives rise to health issues and the deterioration of the building's fabric. Impurities such as dust, pollen, radon, and variable organic compounds are also reduced. Effective natural ventilation must be incorporated at the early stage of building design.

One of the forgotten issues in comfort is noise pollution. Acoustic isolation is an important issue for comfort. In the UK, there were 313,000 complaints in England and Wales of domestic noise disturbance in 2004. Buildings have improved acoustic insulation requirements since 2002. One of the main recommendations regarding acoustics is to get the building shell right because it is difficult to retrofit an acoustic material properly afterward.

In this book there are 20 chapters written by authors from various parts of the world covering multiple buildings and complexes achieving comfort by considering: the variability of climate temperature; shading; using the retrofitting technique; improving efficiency and temperature control; using natural cooling and ventilation; using ventilative

Table 1 Electricity consumption per capita in selected countries

Country	Population × 10^3	Electricity Generation – TWh	Electricity Consumption per Capita, kWh
Kuwait	4207	75.0	17,827
UAE	9771	138.1	14,134
USA	329,065	4401.3	13,375
Saudi Arabia	34,269	357.4	10,430
France	65,130	555.4	8528
Germany	83,517	612.4	7333
China	1,433,794	7503.4	5233
UK	67,530	323.7	4795
Brazil	211,050	625.6	2964
Algeria	43,053	81.3	1888
India	1,366,418	1558.7	1141
Egypt	100388	14.7	146.4

Table 2 Comfort check list questionnaire

Factor	Description	Yes
Air temperature	Does the air feel warm or hot?	
	Does the temperature in the workplace fluctuate during a normal working day?	
	Does the hot or cold season affect the working place temperature?	
Radiant temperature	Is there a heat source in the environment?	
	Is there any equipment that produces steam?	
	Is the workplace affected by external weather conditions?	
Humidity	Are your employees wearing (personal protection equipment) PPE that is vapour impermeable?	
	Do your employees complain that the air is too dry?	
	Do your employees complain that the air is humid?	
Air movement	Is cold or warm air blowing directly into the workspace?	
	Are employees complaining of draught?	
Metabolic rate	Is work rate moderate to intensive in warm or hot conditions?	
	Are employees sedentary in cool or cold environments?	
PPE (Personal Protection Equipment)	Is PPE being worn that protects against harmful toxins, chemicals, asbestos, flames, extreme heat, etc.?	
	Can employees make individual alterations to their clothing in response to the thermal environment?	
	Is respiratory protection being worn?	
What your employees think	Do your employees think that there is a thermal comfort problem?	

cooling systems; diachronic analysis of daylight design and management technique in Mediterranean constructions; enhancing the microclimate towards outdoor thermal comfort in urban isles in the Mediterranean region; implementing passive solar design using the building's geometry and orientation; creating thermal and visual adaptive comfort in buildings; considering the well-being of the end-users; keeping design in mind of sustainability; neuroscience and architecture; using external shading in humid-to-moderate climates; building with living bricks to generate energy and wean humanity off fossil fuels; creating natural cooling in the very hot Arewa sun of Nigeria in traditional Hausa buildings; is still 'ornament' a 'crime'; using orientation and courtyards in arid zones of Arabia; integrating thermal mass in sustainable buildings; creating visual comfort in UAE heritage buildings such as museums.

The authors come from the Netherlands, Argentina, Egypt, Cuba, the UK, Canada, Italy, Cyprus, Greece, Malaysia, Portugal, Iran, Nigeria, the US, and Bahrain.

WREN/WREC
Brighton, UK

Ali Sayigh

Contents

About the Editor

Ali Sayigh, UK citizen, graduated from London University and Imperial College in 1966 and holds BSC AWP, DIC, and PhD. He is a fellow of the Institute of Energy, fellow of the Institution of Engineering & Technology, chartered engineer, chairman of Iraq Energy Institute, and Fellow of the Royal Society of Arts. Prof. Sayigh taught in Iraq, Saudi Arabia, and Kuwait and at Reading University and the University of Hertfordshire from 1966 to 2004. He was head of the Energy Department at Kuwait Institute for Scientific Research (KISR) and an expert in renewable energy at AOPEC, Kuwait, from 1981 to 1985. He started working in the field of solar energy in September 1969. In 1972, he established with some colleagues in Saudi Arabia *The Journal of Engineering Sciences* in Riyadh, Saudi Arabia, and in 1984, he established the *International Journal for Solar and Wind Technology*, as editor-in-chief. This has changed its name in 1990 to *Journal of Renewable Energy*. He is editor of several international journals published in Morocco, Iran, Bangladesh, Nigeria, and India. He established WREN and the World Renewable Energy Congress in 1990. Prof. Sayigh is a member of various societies related to climate change and renewable energy. He is chairman of Iraq Energy Institute since 2010. He was consultant to many national and international organizations, among them, the British Council, ISESCO, UNESCO, UNDP, ESCWA, UNIDO, and UN. He runs conferences and seminars in 54 different countries and has published more than 600 papers. He has edited, written, and associated with more than 100 books. He has supervised more than 82 MSc and 36 PhD students. He was editor-in-chief of the yearly *Renewable Energy Magazine*, 2000–2016. He is the founder of WREN and *Renewable Energy Journal* published by Elsevier and was the editor-in-chief for 30 years, from 1984 to 2014. Prof Sayigh is the editor-in-chief of *Comprehensive Renewable Energy* and has organized 154 top scientists, engineers, and researchers' contribution in eight volumes published in 2012 by Elsevier, which won 2013 PROSE award in the USA. He is the founder of Med

Green Buildings and Renewable Energy Forum since 2011. In 2016, he established the peer-reviewed international open access journal titled *Renewable Energy and Environmental Sustainability*—REES, which is published in English online by EDP publisher in Paris. He is winner of the Best Clean Energy Implementation Support NPO—UK. In 2018, WREN was rated globally as one of the best organizations in the UK promoting renewable energy. In November 2018, Prof. Sayigh was elected Fellow of the Royal Society of Art, (FRSA). Prof. Sayigh is working with Springer Nature in publishing books and proceedings since 2014.

Sustainable Schools: Their Passive Systems to Provide Comfort with Natural Means as an Educational Example for Pupils and Their Parents

Wim Zeiler

1 Introduction

> Architecture will, therefore, become more informed by the wind, by the sun, by the earth, by the water, and so on. This does not mean that we will not use technology. On the contrary, we will use technology even more because technology is the way to optimize and minimize the use of natural resources. (Richard Rogers)

As the results of the energy use of the built environment, which is around 40% of the total energy consumption [25], become more evident (depletion of fossil fuel and global warming), there is a demand for energy reduction. In order to reduce the high energy demand and pollution of greenhouse gasses, it is necessary to improve the performance of the buildings. In Europe, more than 64 million students and almost 4.5 million teachers work inside a school in preprimary, primary, and secondary schools [26]. School buildings represent a significant part of the building stock (17% of the nonresidential sector in Europe) and a noteworthy amount of total energy use (12%) [26].

Good education is an essential pillar for citizens' welfare and well-being. High-quality education infrastructure is one of the main pillars of economic prosperity and can put people on a path toward good health, empowerment, and employment. Evidence shows that, on average, each additional year of education boosts a person's income by 10% and increases a country's GDP by 18% [11]. Data from the Perry Pre-School Study, for example, show that every dollar spent on students when they were 4 and 5 years old had a benefit/cost ratio of more than 13 dollars by the time they were 40 [33]. Teacher recruitment, retention, absenteeism, and presenteeism are affected by poor indoor conditions [44], which also influences learning outcomes [12]. It is therefore of critical importance to ensure that schools enable

W. Zeiler (✉)
Department of the Built Environment, TU Eindhoven, Eindhoven, Netherlands
e-mail: w.zeiler@bwk.tue.nl

A. Sayigh (ed.), *Achieving Building Comfort by Natural Means*, Innovative Renewable Energy, https://doi.org/10.1007/978-3-031-04714-5_1

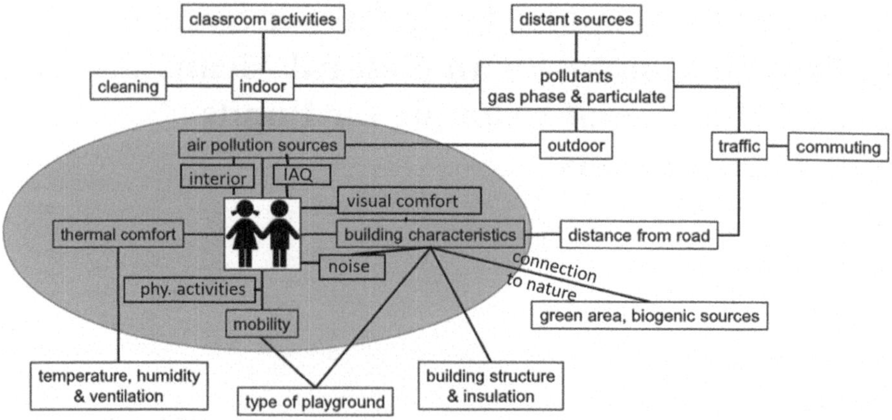

Fig. 1 Dynamic influence factors impacting the indoor climate in school classrooms [49]

children to thrive physically and mentally [49]. Despite this, inadequate ventilation is a common problem in school classrooms all over the world [2, 3, 36, 53]. A recent Dutch survey of August 2020 stated that from the more than 3386 schools that responded and performed a quick scan during summer conditions, 23% did not fulfil the minimum requirements for ventilation [40]. These outcomes are in line with the earlier outcome by van Dijken and Gelderblom [20], stating that in around 40% of the schools, classroom ventilation is seriously insufficient during the winter.

It is essential to create a healthy learning environment besides the sustainability aspects of schools [30] and to optimize the consistency of all dynamic influencing factors on the indoor environment (see Fig. 1).

2 Historical Perspective

It is important to learn from buildings of the past while still harnessing the opportunities presented by new technologies to think interdisciplinary and work collaboratively [19]. Since ancient times with the Greeks and Romans, the school plays an important role in the lives of children and their parents. But after the fall of Roman civilization, there was hardly any education in the middle centuries. Within a few years after the rise of Christianity, schools which were attached to churches and monasteries were built. These were intended for young people from the highest ranks to be formed into liturgy and church singers. Despite the fact that education was intended for the good, the situation in the big cities did not meet a healthy and comfortable climate at all. In rural areas, the condition of school buildings was even more dramatic, and school buildings were often compared with pigpens. The school from the beginning of the nineteenth century had only one classroom with sometimes hundreds of children. Old prints testify to these massive groups and give an idea of the housing. It was not until the nineteenth century that there was a

Fig. 2 Open-air school Dordrecht 1929 [13]

Fig. 3 Zuidwalschool Den Haag 1933 [13]

Fig. 4 Open-air school by Duiker and Bijvoet, Amsterdam 1930 [13]

turnaround; education laws proposed all kinds of measures to improve the situation. From the 1900s to the 1930s, due to all the health problems of pupils by the then prevailing tuberculosis, the concept of the open-air school was developed [13]. The phenomenon of the open-air school had a practical origin. Due to the significant health problems in industrial cities, such as tuberculosis, the solution was found in sanatoria. These were equipped with a lot of light and air as this was considered important for school-age children. The emphasis was put on ventilation and day-light entry intended for sick or weak children and usually situated outside the city or in a park-rich environment (see Figs. 2 and 3) [13].

The perhaps most well-known Dutch open-air school was designed by architect Jan Duiker in Amsterdam (1929–1930). This building is divided into the classrooms and terraces into the three-story structure in such a way that a maximum of light and air can be accessed here. The facades are located in one plane with the floor and a curtain wall with initially six large steel turn windows on the long sides of the rooms. The maximum openness, while minimizing construction, is in line with the principles of the new building style [6] (see Fig. 4). In 1994 the school was completely restored and adapted to the current educational requirements.

3 Use of Passive Strategies

With regard to passive strategies, architectural design elements, such as building shape, building façade, orientation, and window-to-wall ratio (WWR), can significantly influence the final energy use of a building. [32] made a complete overview of passive solar heating and cooling concepts and their effects on the performance of a building's thermal management. The concepts of Trombe wall, solarium, evaporative cooling, ventilation, radiative cooling, wind tower, earth air heat exchanger, roof pond, solar shading for buildings, and building-integrated photovoltaic and thermal panels are extensively covered in this review. Optimal combination of passive strategies can be analyzed, such as building shape, orientation, wall thickness and insulation, roof insulation, window orientation WWR, window-to-wall ratio, glazing, and overhang depth [31, 38, 42, 51] by applying building energy performance simulation (BEPS) such as [24]. These imply that the building's final energy can be significantly reduced through the use of passive and active strategies. However, excessive use of active strategies may be unproductive [46, 56]. Also, the effectiveness of passive strategies can vary depending on the individual thermal characteristics in a building [38].

4 Iconic Examples from the Past

Most architects dream about a school without installations, and many take St. George's County Secondary School in Wallasey as a great example (see Fig. 5).

The school designed by Emslie Morgan was built in 1961 and is considered the first English passive sun-controlled school to heat the school entirely with solar energy. However, the building has a heating system that only comes into operation under extreme outdoor conditions. Normally the sun, internal lighting, and the children themselves are a sufficient source of heat. The lighting plays a major role in heating the school, and lighting being reflected on the wall heats the room before the arrival of the children. The building has a two-story high transparent south façade,

Fig. 5 St. George's County Secondary School in Wallasey, Cheshire, England. (Photo: Jeremy Marshall [22])

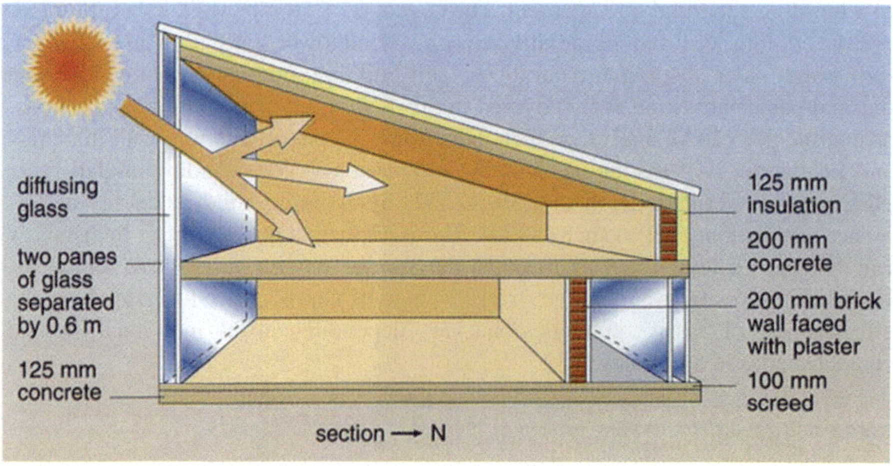

Fig. 6 Cross-section St. George's County Secondary School in Wallasey [22]

allowing the sunlight through a double skin façade consisting of two separate single glass windows with an intermediate distance of 60 cm. into space [23]. The walls and floors are made of heavy materials and thick to obtain a high buffering capacity of the structural construction (see Fig. 6).

A recent study into the current state of the building shows that due to new floor coverings on the concrete floors and the use of personal computers, the problems of overheating in the summer have increased sharply. Although the manually operated supply and discharge air valves function well in itself, it is not very user-friendly to operate in practice. As a result, the air valves are not always optimally adjusted, resulting in insufficient ventilation between 0.3 and 2.5 air changes per hour. The main problem of the building is the glare caused by the too much sunlight that enters through the all-glass south façade. This problem is very conflicting with twentieth century school equipment such as electronic whiteboards and computer screens. Although the school is an icon for the current designers of sustainable schools, one should also consider the critical aspects of the design. Nevertheless, the 1961 design with total energy use of 97 kWh/m^{2}a deserves praise and admiration, compared to the current English standard of 250 kWh/m^{2}a.

As the icon of a new generation of sustainable, low-energy, naturally ventilated sustainable educational buildings, the Queen's Building at De Montfort University, Leicester, was completed in late 1993. It gained a reputation with its stunning architecture using distinctive ventilation chimneys [18] (see Fig. 7). Architect Short Ford Associates worked on the building design alongside environmental engineer Max Fordham LLP, Cambridge Architectural Research (on the stack-effect chimneys), and Bristol University (on the physics of the airflow).

The building's operational performance was revealed in 1996 when a post-occupancy evaluation was carried out as part of the PROBE (Post-occupancy Review of Buildings and their Engineering) project. The assessment revealed a

Fig. 7 Queens Building, De Montfort University. [Photo Martine Hamilton Knight]

number of key shortcomings. Unresolved defects meant that for the first 2 years, the building operated with problems in critical mechanical and control systems [18]. The survey of its occupants showed dissatisfaction with high summertime temperatures and stuffiness in both winter and summer months [18].

The first energy-neutral sustainable elementary school of the Netherlands [58] made clear what the focus on sustainable school design should be [57]:

> A sustainable school is a sustainably designed, built, and used school building in which sustainable education is developed and provided. The building and its surroundings serve as an educational tool for illustration, demonstration, exploration, experimentation, and discovery. The concept of the sustainable school and its practical realization must be 'CONVINCING':
>
> **C** ONTEXTUAL, **O** RIGINAL, **N** EED, **V** ISIBLE, **I** NTEGRATED, **N** ATURAL, **C** LEVER, **I** NSTRUCTIVE, **N** ETWORKING, **G** LOBAL
>
> The whole concept of a sustainable school building is based on principles of sustainable development which deal with the limited availability of natural resources, the interdependence with nature, the fundamentals aspects of interdependence with nature, the fundamentals aspects of production and consumption, and the issue of equity within, between and among generations.

5 Methodology

Traditionally, the potential for energy reduction in the Netherlands was mainly determined by the applications of building energy reduction measures according to the Trias Energetica method (see Fig. 8). In the first design step, the energy demand is reduced as much as possible by avoiding waste of energy, applying high

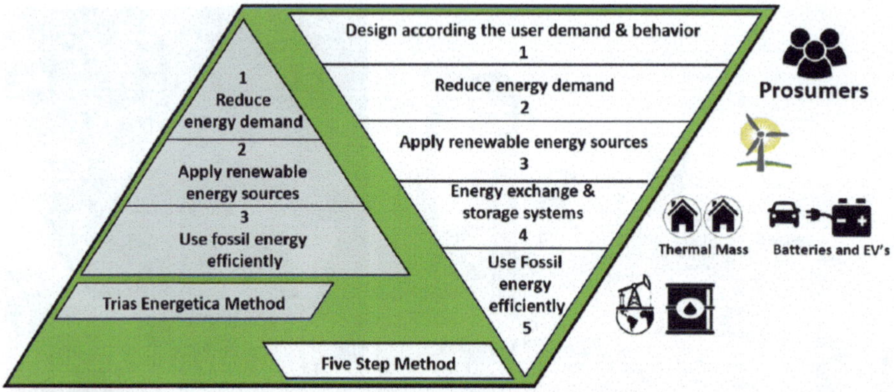

Fig. 8 The Trias Energetica method and the five-step method [55]

insulation levels, and implementing energy-saving measures, such as lowering the setpoint temperature in winter or increasing the setpoint temperature in summer slightly. In the second step, all possibilities to apply renewable energy sources are investigated. In the third step, the remaining energy demand is covered by using fossil energy solutions as efficiently as possible. An adapted version of the Trias Energetica method, the five-step method, has been developed [60] by adding the integration of user behavior, their capability to produce energy locally, as well as energy exchange and storage systems (buffer tanks, Aquifer Thermal Energy Storage) (see Fig. 8) [55].

More specifically, the first step to minimize the energy demand is to look at the actual behavior and occupancy in relation to the ventilation need, which is directly related to the number of pupils in a specific classroom. So this demand-driven ventilation is a good step to reduce the energy demand by the ventilators. A similar measure in this step is occupancy detection for the control of the lighting and daylight. This measure minimizes the use of artificial lighting, thus reducing the electricity demand. The second important step to reach nZEB is the application of the passive house strategy. This leads to energy savings on heating of 80% compared to conventional standards of new buildings. To reach these goals, the passive house concept focuses, first and foremost, on reducing the energy demand of the building through high insulation of walls, roofs, floors, and windows/doors, thermal bridge free construction, and airtightness.

In 2009, the Dutch government started their so-called UKP NESK program to stimulate innovation for energy-neutral buildings. UKP means unique chances projects, and NESK means "toward energy-neutral schools and offices" (Naar Energie neutrale Scholen en Kantoren). The published results by the Rijksdienst voor Ondernemend Nederland (RVO) showed the top 15 most energy-conscious schools in the Netherlands for 2014 and 2016 [47, 48]. The total energy consumption was determined by the energy use for heating, hot water, cooling, ventilation, and lighting. The number one school in the top 15 of the combined 2014 and 2016 list [62] is a primary school, where high thermal conductivity is obtained in the structural

Fig. 9 Overview energy generation by PV panels of school 1 [50]

elements to prevent large heat losses. The open façade parts consist of frames with three-layer glass. Solar screens are provided in order to avoid overheating. Besides, the school building generates a lot of energy due to a large number of solar panels on the roof, with an amount of 135 kWp in total: 41% of the generated energy is used by the school, the rest is delivered to the grid (see Fig. 9).

In Table 1, an overview is provided of the different HVAC concepts of the Top 15 schools. It is pretty remarkable that besides the five schools which are connected to a district heating system and the only one using a biomass pellet boiler, all other schools are using a heat pump. All Dutch projects are all applied with a balanced ventilation system and heat recovery [62].

Schools typically have a high internal heat load (heat generated by people, lighting, and buildings). As a result, the required amount of cooling is significantly higher than the average building. Nowadays, modern school buildings have reached the insulation quality at which the amount of cooling needed during the summer roughly equals the amount of heating required during the winter. Hypothetically, if all heat could be stored within the building, no external heat source would be needed throughout the year. An increasingly popular solution is energy storage in the groundwater below the building: aquifer thermal energy storage (ATES) in porous sand layers, called aquifers. These systems work in combination with a heat pump in order to ensure the required temperature difference for cooling and heating is reached. The principle of an ATES system is based on transferring groundwater between two separated storage wells. During summertime, water is extracted from the coldest well and used to cool the building (see Fig. 10).

During cooling, the water temperature increases from approximately 8 °C to 16 °C and is injected in the warmer well and stored until the winter season. During winter, the extraction/injection flow is reversed, and the heated water (which still has a temperature of approx. 14 °C) is pumped back to the building. With a heat

Table 1 Overview of different techniques used for heating/cooling and ventilation

School	HP	ATES	Pellet boiler	Borehole HE	District heating	HR boiler	Ventilation system	CO_2 control
1			X				MS+ME+HR	+
2	X	X					MS+ME+HR	+
3	X					X	MS+ME+HR	+
4	X						MS+ME+HR	+
5					X		MS+ME+HR	+
6	X	X	X				MS+ME+HR	+
7	X	X					MS+ME+HR	+
8	X	X					MS+ME+HR	+
9	X	X					MS+ME+HR	partly
10	X			X			MS+ME+HR	+
11	X		X				MS+ME+HR	partly
12	X						MS+ME+HR	partly
13					X		MS+ME+HR	+
14					X		MS+ME+HR	+
15					X		MS+ME+HR	-
16	X	X					MS+ME+HR	+
17					X		MS+ME+HR	partly
18	X	X					MS+ME+HR	partly
19	X	X					MS+ME+HR	-
20							MS+ME+HR	-
21	X	X					MS+ME+HR	-
22						X	MS+ME+HR	+
23	X	X					MS+ME+HR	-
Total	15	10	3	1	5	2	23	

pump, the heat is extracted and converted to higher temperatures to heat the building. The water is cooled to approx. 6 °C and is injected back into the cold well. A heat exchanger between the groundwater and the building system water is used to avoid water contamination. An optimal performing ATES system can deliver very efficient cooling: coefficient of performance (COP) of 12 compared to a regular (compression-based) cooling system which reaches a COP between 4 and 6 [8, 9]. Because of the favorable soil conditions, the use of ATES systems in the Netherlands has become increasingly popular since the first installations in 1990 and became an important sustainable alternative to conventional chillers. In the Netherlands, the number of ATES applications could rise from 2500 to 20,000 in the next decade, but it also has a high potential for other coastal areas throughout the world [10, 29]. The concept is suitable for residential complexes and small offices up to large-scale clusters of buildings like business neighborhoods or university campuses.

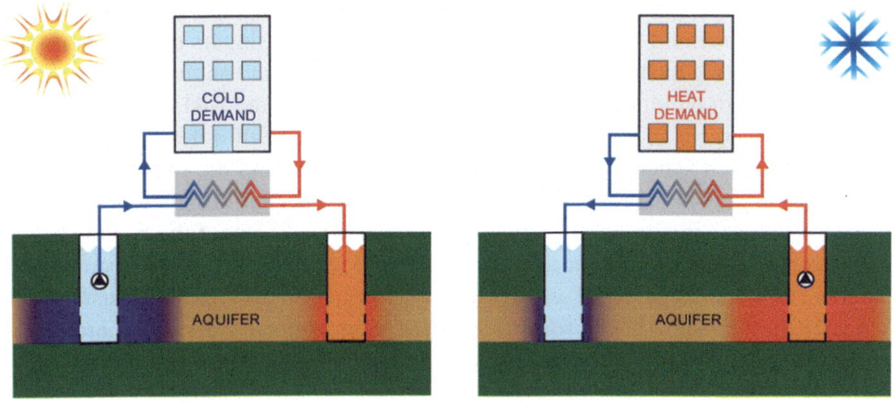

Fig. 10 Doublet ATES systems [21]

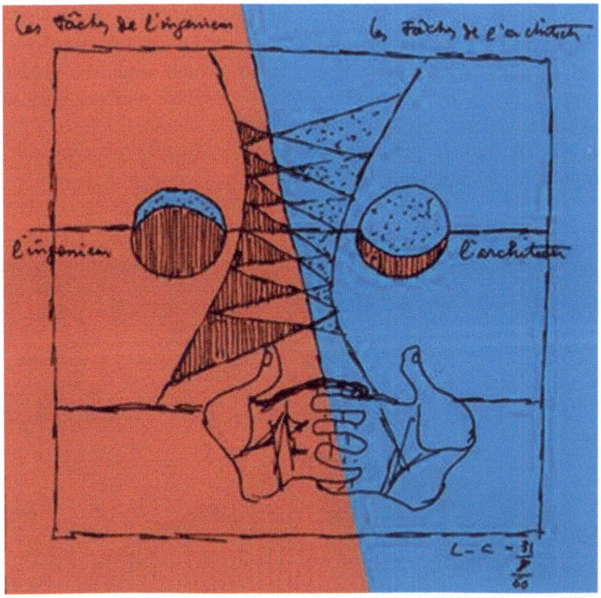

Fig. 11 Science et Vie [41]

Form and function should form a synergy instead of being sometimes in conflict with each other. Already many years ago, Le Corbusier showed how the interaction between architect and engineer should look like (see Fig. 11). Fortunately, more and more integral approaches are becoming common practice to find the optimal solution between passive and active energy elements and between architecture and building services.

Le Corbusier explains the roles of the architect and the engineer [45]: "Under the symbolic composition I have placed two clasped hands, the fingers enlaced horizontally, demonstrating the friendly solidarity of both architect and engineer engaged, on the same level, in building the civilization of the machine age" [41]. The architects and the engineers should work together from the very start of a design project and must aim to reach synergy by combining the knowledge and the experience of all disciplines already in the early stages of the conceptual design. An integral approach is needed to make this possible, which represents a broad view of the world around us that continuously needs to be adapted and developed from sound and documented experiences that emerge out of an interaction between practice, research, and education [61]. In the Netherlands, some very interesting examples show that architects and engineers can design as an integral team. These will be described next.

6 Dutch Iconic Examples from the Present and Near Future

> "Finally, it is not only a matter of methodology; it is much more. It is the urgent, and unconditional call for spirituality, and for a poetic mind, which would be able to recognize, respect, and support the beauty that nature offers for free. No methodology, no number, in fact, can be sufficient alone, as there is no technical layout for beauty." – Alessandra Scognamiglio, ENEA Italian National Agency for New Technologies, Energy and Sustainable Economic Development

Fig. 12 North facade of the EAE building with solar chimney visible [54]. (Photo: Egbert de Boer)

6.1 Energy Academy Europe- Groningen

The Energy Academy Europe (EAE) building (see Fig. 12), in Groningen, provides education, conducts research, and fosters innovation in the field of energy while working toward the transition to a sustainable energy future. The key energy themes of the EAE are the energy of tomorrow (such as wind and solar energy), energy efficiency, cost savings, and CO_2 reduction:

- Zero emissions (after 40 years, incl. construction)
- BREEAM-NL outstanding
- Energy performance coefficient = 0 or less
- 51 kWh/m^2 per year (which is extremely low for an education-related building)

The unique design demonstrates how a building can make optimum use of the natural elements of earth, water, air, and sunlight as a primary energy source.

The main external feature of interest on otherwise straightforward flat, triple-glazed facades is the wavy vertical *brise soleil* made from Siberian larch (from Dutch forests) [43]. They are the result of calculations on controlling solar gain to particular areas and keep the building as cool as possible on hot days. The ratio of glazing to a wall is 50% to the offices on the east- and west-facing facades but around 80% to the south winter garden side, and the north, which doesn't have direct sunlight.

A 200-meter-long air vent under the building uses the earth to cool and heat the air (see Fig. 13). The design connects outdoor air, collected on the lower south side, to the air demand on the buildings' higher northern side. Pushed through a "thermal labyrinth" and assisted by natural powers such as thermal buoyance caused by the sun and wind stream, the airflow due to occurring pressure differences is optimal for cooling the building in summer.

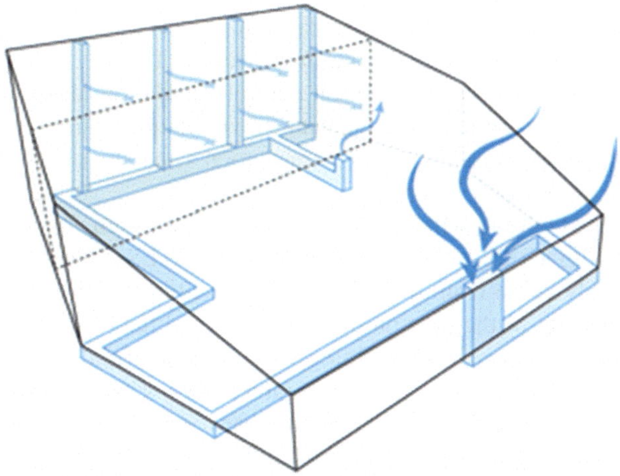

Fig. 13 Air supply through a "thermal labyrinth." [Source: Arch20]

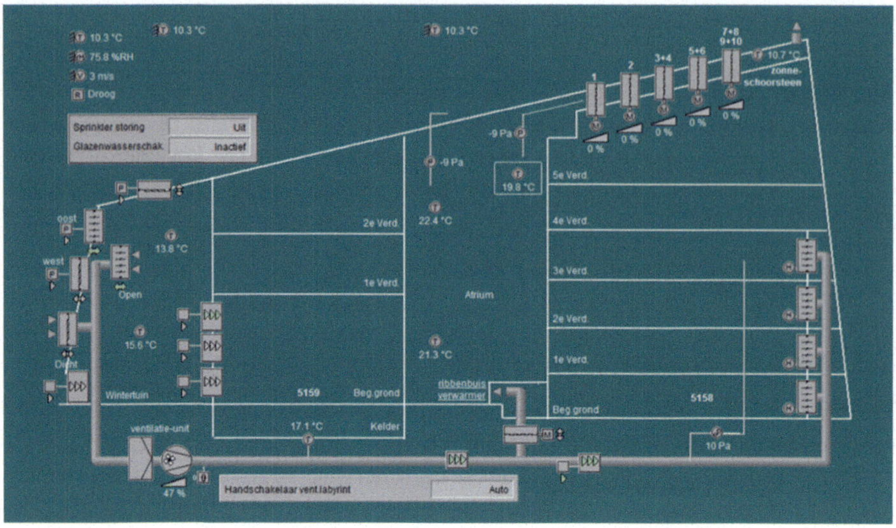

Fig. 14 Hybrid ventilation system with process control by building management system [54]

The north side (the highest point of the building) has a solar chimney, which stimulates the airflow using solar energy and controls the flow of the return air out of the building. The controls are quite complex as all weather conditions and internal conditions have to be taken into account (see Fig. 14). This natural ventilation saves approximately 20% in energy. On warm summer nights, there is additional cooling through a natural process that requires no power. Cool night air flows through the interior winter garden via the atrium and through the entire building so that the next day occupants can start their day in a cool working environment.

The atrium connects two sections; the north side is the research area with laboratories and related workspaces, and on the south side, there are workspaces, a winter garden, and the teaching areas. Via wide stair ramps, users will feel a natural preference to take the stairs rather than the elevator (see Fig. 15).

The building does not have to rely on natural ventilation alone; if the weather conditions are not good enough, the mechanical ventilation system will step in. The building has three air handling units with a total capacity of 81,000 m^3/h (see Fig. 16).

In addition, the building will be heated and cooled through concrete core activation combined with aquifer thermal energy storage (ATES). The heating and cooling sources mainly come from the stored energy within the ground. However, the total cooling capacity of the heat pump is 830 kW and the heating capacity is 550 kW. Two deep wells dug near the building will use geothermal energy of the ATES to keep the building at a comfortable temperature. Two water reservoirs are located at a depth of 100 m – one for heating and another for cooling. In summer, the cold water is pumped up and absorbs heat in the building. This now warmed water is then pumped back into the second reservoir. In winter, this process is reversed. The water

Fig. 15 The atrium as a central connection to the different building parts. [Source: Arch20]

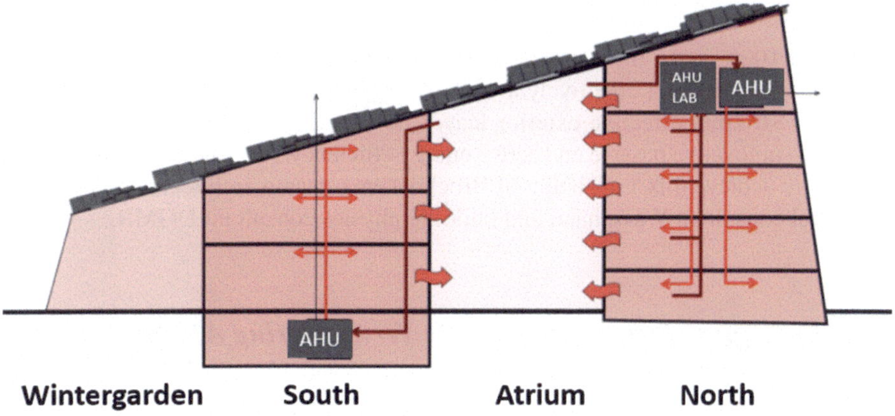

Fig. 16 The three different air handling units for the mechanical ventilation. [Source: Arch20]

is further heated by a heat pump which efficiently turns electricity into heat which is distributed throughout the building for underfloor heating (60% of the heating), to heat the airflow through climate ceilings, and to heat tap water. When needed, cold water can also be used from the nearby pond.

The building has an extraordinary energy roof (see Fig. 17). The energy roof has an optimal orientation toward the south with an unusual design of a large, slanted roof of 1600 solar panels. The panels are arranged in triangles (133 in total) so that

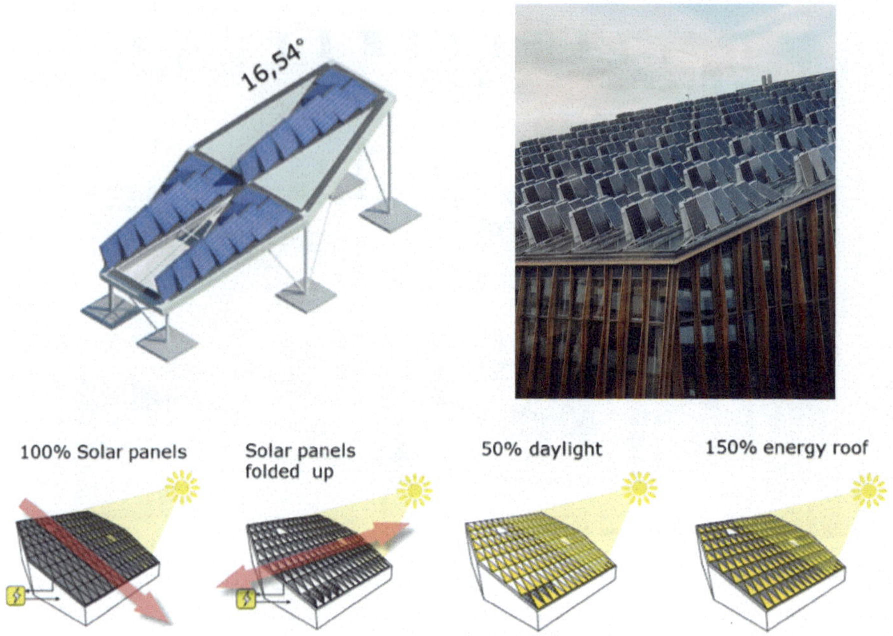

Fig. 17 Solar panel roof with windows for daylight. [Source: Arch20]

they cover 100% of the roof surface. At the same time, the design allows daylight to pass through, making it effectively a 150% energy roof.

Such a striking, effective exterior leaves no room for doubt: the building is a place with innovative passive and active energy efficiency. A team of experts worked together to achieve this, consisting of Broekbakema, pvanb architecten, ICS, Arup, Structural consultant Wassenaar, and building physics consultant DGMR.

6.2 Ecological Primary School De Verwondering Almere

The design of the primary school De Verwondering of 1900 square meters and 14 classrooms (see Fig. 18) is inspired by nature Architect Daan Bruggink, who suggested biophilic principles should be translated into all kinds of buildings levels and spatial context. The school consists of clusters of classrooms that form the biotopes for pupils of different age groups. These are so oriented toward each other that a natural gap is created that encompasses the central space. The school design is made as circular as possible in a dry and demountable system, with local materials and local wood. Natural and biobased materials are dominant, and wood is the primary building material for the school. In the classrooms, the interior walls are made of clay stones to get thermal mass into the building. Nowadays, airtight construction and good insulation are two key concepts that are becoming increasingly important

Fig. 18 Primary school De Verwondering Almere. [Photo: Ronald Auee]

in building design. However, this also means the need to cool a building increases as the indoor temperature can become too high during the warmer (summer) months. Therefore, each room is equipped with large ventilation hatches combined with large roof hatches in the roof of the central area to provide natural ventilation and night ventilation to cool the school in summer. Solar collectors on the roof (see Figs. 19 and 20) absorb the heat from the ambient air, and then the heat is stored in an underground ice buffer (see Fig. 21).

A heat pump transfers the water from the buffer to the desired temperature to heat the building. If it gets colder, that water in the ice buffer is cooled to below freezing. When cold water transitions to ice, energy is released, the crystallization heat. This heat can then be used to meet the heat needs. The ice can cool the building in the summer; the heat of the building is then used for heating again in winter. From the end of the heating season, the reverse process begins. The ice buffer is a form of thermal energy storage, and such can be an alternative for the earlier discussed ATES.

On the roof of the lowest cluster, there is an open-air classroom, partly covered and surrounded by the vegetation roof, on which even chickens walk. This outdoor room (see Fig. 22) is used all year round so pupils can experience the weather, the seasonal changes, the environment, and all other natural aspects that play a role in teaching. It is possible to teach within the covered classroom approximately 300 days a year. It was already "fully booked" during the first year of operation. Indeed, based on natural and biophilic considerations, this is a modern version of the old and earlier discussed open-air schools' principle.

Fig. 19 The green and energy roof of the school with solar thermal collectors and PV panels. [Photo; ORGA architect]

Fig. 20 The solar thermal collectors to feed the heat pump ice buffer system. [Photo: TDS Engineering]

Fig. 21 Ice buffer. [Source ORGA architect, Gemeente Almere, Heko Spanten]

Fig. 22 Open-air classroom on the top floor of the school. [Photo: Ronald Auee]

Design team: Architect Daan Bruggink-Orga architects, Structural engineer Lüning, Building Services consultant Bart Ruijs TDS Engineering, Building Physics consultant Nieman, Plant design Well Planted, Contractor Van Norel Bouwgroep. Special thanks to Daan Bruggink and Bart Ruijs for their collaboration.

6.3 Erasmus University Rotterdam: MFO II

The Erasmus University Rotterdam's (EUR) new multipurpose educational building MFO II (see Fig. 23), covering some 8500 m^2 (91,500 sq ft), is currently being built and will become one of the most sustainable university buildings in the Netherlands. It will house lecture halls, teaching rooms, a lounge area, and plenty of study space for around 3000 students from various faculties. Energy neutrality, circularity, and vitality are the main goals while simultaneously inspiring and stimulating, sustainable, and aesthetically appealing surroundings for students and staff. The following minimum requirements for its sustainability were formulated [27]: BREEAM opens external 1 rating as "outstanding," building circularity performance (BCP/CPG 2) score is 8, and energy performance coefficient (EPC) is 0. The design, development, and construction contract was granted to a consortium made up of Paul de Ruiter Architects, Halmos Building Services consultants, and LBP|SIGHT building physics consultants together with BAM as construction partner. Their design will actually generate more energy than it will use, thanks to roof-mounted solar panels, heat pumps, and ATES. One of its unique features will be a revolutionary new ventilation system using wind and solar energy. This "powered by nature" air-conditioning system was developed by Ben Bronsema [14–17], the Earth Wind and Fire (EWF) concept (see Fig. 24) in collaboration with the Universities of Technology of Delft and Eindhoven [[7, 34], Hooff et al. 2021] and first applied at the end of 2018 in hotel BREEZE in Amsterdam [15–17]. The design

Fig. 23 Erasmus University Rotterdam MFO II Building. [Photo: Paul de Ruiter Architects]

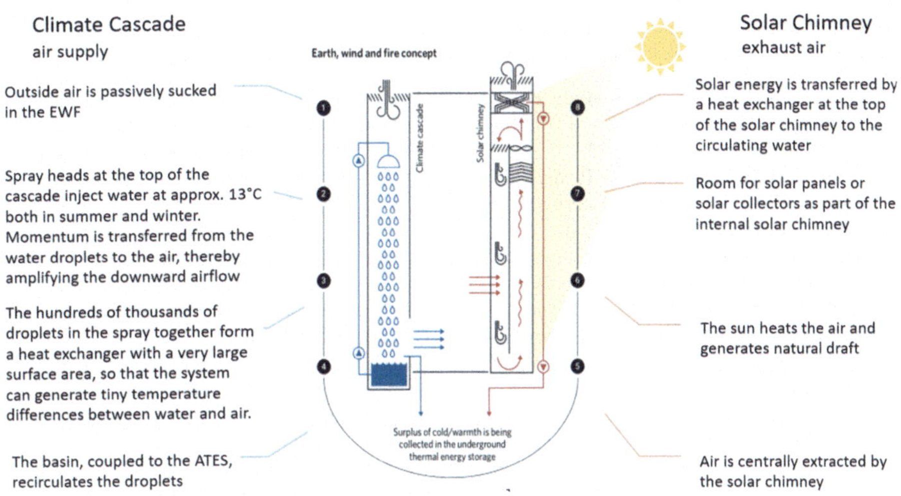

Fig. 24 The EWF concept [37]

requirements of the EWF concept have been meticulously applied, and multiple optimizations have been implemented on the system, such as energy recovery between the air from the solar chimney, and to the climate cascade, variable airflow by application variable volume controls with low resistance and application of multichannel heat exchangers (MCHE) as low resistance heating coils. Two EWF ventilation systems are applied with a total air quantity of 135,000m^3/h and a resistance of 140Pa in supply and in the exhaust of 80Pa (normal for air handling units is approximately 900 and 600Pa). With this system, energy savings of more than 50% are achieved compared to air treatment cabinets, of which 90% are on fan energy.

7 Conclusion

Building comfort by natural means is about finding the right balance between architecture and engineering, form and function. The goal is not to minimize the use of external or additional building services systems but to aim for the most optimal combination of natural means and other efficient systems, that is, the right balance between passive and active systems. In particular, when designing schools, this is of utmost importance to create a healthy and effective learning environment. There are a growing number of examples where the building itself becomes part of the education, not only to facilitate it alone. The focus of this chapter is especially placed on schools as they offer possibilities to include elements in the education of the pupils and involve their parents and therefore show the new generation what the possibilities to create a more sustainable built environment are. Some of the most inspiring Dutch examples are presented in this chapter. They are a good promise of even better things to come by a new generation more aware of the necessity and the possibilities that nature and technology offer.

References

1. Aa A. van der, 2002, Pilot study report: School building Waterland, International Energy Agency, Energy Conservation in Buildings and Community Systems, IEA ECBCS Annex 35 HybVent, report date 01-02-02.
2. Almeda R.M.S.F., Freitas V.P. de, 2014, Indoor environmental quality of classrooms in Southern European climate, Energy and Buildings 81: 127–140.
3. Almeda R.M.S.F., Freitas V.P. de, 2015, IEQ assessment of classrooms with an optimized demand controlled ventilation system, Proceedings 6th International Building Physics Conference, June 14–17, Turin, Italy
4. Almeida M.G., Ferreira M., Rodrigues A., 2015, CO2 zero schools School building renovation towards emissions neutrality REHVA Journal – December 2015: 34–38
5. Arch20, 2021, Energy Academy Europe (EAE) | Broekbakema + De Unie Architecten, https://www.arch2o.com/energy-academy-europe-eae-broekbakema-de-unie-architecten/
6. Architectureguide, 2021, http://www.architectureguide.nl/project/list_projects_of_tag/tag_id/15/prj_id/362
7. Blocken B., Hooff T. van, Aanen L., Bronsema B., 2011, Computational analysis of the performance of a venturis shaped roof for natural ventilation: venturi-effect versus wind blocking effect, Computers and Fluids 48(1): 202–213
8. Bloemendal, M., Hartog, N., 2018. Analysis of the impact of storage conditions on the thermal recovery efficiency of low-temperature ATES systems. Geothermics 17, 306–319.
9. Bloemendal, M., Olsthoorn, T., Boons, F., 2014. How to achieve optimal and sustainable use of the subsurface for Aquifer Thermal Energy Storage. Energy Policy 66, 104–114.
10. Bloemendal, M., Olsthoorn, T., van de Ven, F., 2015. Combining climatic and geo-hydrological preconditions as a method to determine world potential for aquifer thermal energy storage. Science of the Total Environment 538 621–633.
11. Brandt Y., 2015, Education: the most powerful investment in our future, https://blogs.unicef.org/blog/education-the-most-powerful-investment-in-our-future/
12. Brink H.W., Loomans M.G.L.C., Mobach M.P., Kort H.S.M., 2020, Classrooms' indoor environmental conditions affecting the academic achievement of students and teachers in higher education: a systematic literature review. Indoor Air. https://doi.org/10.1111/ina.12745
13. Broekhuizen D., 2005, Openluchtscholen in Nederland, Architectuur, onderwijs en gezondheidszorg 1905–2005, Uitgeverij 010, Rotterdam 2005.
14. Bronsema B., 2013, Earth, Wind & Fire – Natuurlijke Airconditioning, PhD thesis TU Delft, https://tudelft.on.worldcat.org/oclc/845637529.
15. Bronsema B., Bokel R., Bruggema H., Luijk R. van, Mook A. van, Potter M. de, Meerstadt O., Quist M., 2018c, Earth, Wind & Fire: The Evolution of an Innovation (2) – 'Fire': Natural ventilation and air-conditioning using the climate cascade, REHVA Journal October 2018: 65–73
16. Bronsema B., Bokel R., Bruggema H., Meerstadt O., Quist M., Spoel W. van der, Swier P., Veerman J., Vermeer J., 2018b, Earth, Wind & Fire: The Evolution of an Innovation (1) – 'Earth': Natural ventilation and air-conditioning using the climate cascade, REHVA Journal June 2018: 65–79
17. Bronsema B., Luijk R. van, Swier P., Veerman J., Vermeer J., 2018a, Natural air conditioning: what are we waiting for?, REHVA Journal April 2018
18. Bunn R., 2007, Live and learn: Queens Building, De Montfort University, Building ServicesJournal , April 2007
19. Clements-Croome D.J., 2020, Designing Building for People: Sustainable Liveable Architecture, Crowood
20. Dijken F. van, Gelderblom A., 2016, Performance of mechanical ventilation systems in primary schools in Rotterdam, the Netherlands, Proceeding Indoor Air, Gent, Belgium
21. Drijver B., Willemsen A., 2015, Groundwater as a heat source for geothermal heat pums, International summer school on direct application of geothermal energy, https://www.

researchgate.net/publication/267300619_GROUNDWATER_AS_A_HEAT_SOURCE_FOR_GEOTHERMAL_HEAT_PUMPS
22. Elsadig A.K., 2005, Energy Efficiency in Commercial Buildings, PhD-thesis Faculty of Engineering department of mechanical Engineering Energy Systems Research Unit, University of Strathclyde, June 2005, pp. 42–43.
23. Energie-Cités, 2006, Bringing sustainable energy into your school, A Handbook for teachers and school managers, produced by Energie-Cités as part of the School Buildings Integrated Energy development Operation (SchoolBIE-DO), pp. 36–40
24. EnergyPlus, 2021, EnergyPlus Manual; Department of Energy: Washington, DC, USA. https://energyplus.net/documentation
25. EU, 2020, https://ec.europa.eu/info/news/focus-energy-efficiency-buildings-2020-feb-17_en0
26. EU, 2021, EU Buildings Factsheets, https://ec.europa.eu/energy/eu-buildings-factsheets_en
27. EUR 2020, https://www.eur.nl/en/campus/tomorrows-campus/ongoing-projects/multipurpose-educational-building/sustainability
28. EWF-Lab, 2021, Earth, Wind and Fire, https://www.ewflab.nl/literatuur/
29. Fleuchaus, P., Godschalk, B., Stober, I., Blum, P., 2018. Worldwide application of aquifer thermal energy storage – A review. Renewable and Sustainable Energy Reviews 94, 861–876.
30. Golshan M., Thoen H., Zeiler W., 2018, Dutch Sustainable Schools towards Energy Positive, Journal of Building Engineering 19: 161–171
31. Gong X.Z., Akashi Y., Sumiyoshi D., 2012, Optimization of passive design measures for residential buildings in different Chinese areas. Build. Environment 58: 46–57
32. Gupta N., Tiwari G.N., 2016, Review of passive heating/cooling systems of buildings, Energy Science and Engineering 4(5): 305–333.
33. HighScope, 2021, Perry Preschool Project, https://highscope.org/perry-preschool-project/
34. Hooff T. van, Blocken B., Aanen L., Bronsema B., 2011, A venturi-shaped roof for wind induced natural ventilation of buildings: wind tunnel and CFD evaluation of different design configurations, Building and Environment 46: 1797–1807
35. Hooff T. van, Blocken B., Aanen L., Bronsema B., 2012, Numerical analysis of the performance of a venturi-shaped roof for natural ventilation: influence of building width. Journal of Wind Engineering and Industrial Aerodynamics 104–106: 419–427.
36. Hou Y., Liu J., Li J., 2015, Investigation of Indoor Air Quallity in Primary School Classrooms, Proceedings 9th ISHVAC & 3rd COBEE conference, Tianjin, China
37. Huijbers K., 2021, ABT en EWF-Lab, https://www.ewflab.nl/wp-content/uploads/2021/02/ABT-en-EWF-lab.pdf
38. Kang J.E., Ahn K.U., Park C.S., Schuetze T., 2015, Assessment of Passive vs. Active Strategies for a School Building Design Sustainability 7(11): 15136–15151
39. Kolokotroni M., Perera M.D.A.E.S., Azzi D., Virk G.S., 2001, An investigation of passive ventilation cooling and control strategies for an educational building, Applied Thermal Engineering 21 (2001) 183–199.
40. LCVS, Landelijk Coördinatieteam Ventilatie op Scholen, 2020, Eindrapportage, Beeld van ventilatie op scholen in het funderend onderwijs in Nederland, 1 oktober 2020.
41. Le Corbusier, 1960, Science et Vie, August.
42. Nguyen A.T., Reiter S., 2012, Passive designs and strategies for low-cost housing using simulation-based optimization and different thermal comfort criteria. J. Build. Perform. Simul. 7: 68–81
43. Parker J., 2017, Intellectual Energy, Energy Academy Europe Groningen, Netherlands, ADF December 2017, www.architrctsdatafile.co.uk
44. Plotka E., 2016, Better spaces for learning, RIBA.
45. Popovic Larsen O., Tyas A., 2004, Conceptual Structural Design: Bridging the Gap Between Architects and Engineers, American Society of Civil Engineers, Thomas Telford Ltd.
46. Ramesh T., Prakash R., Shukla K.K., 2010, Life cycle energy analysis of buildings: An overview. Energy Build 42: 1592–1600
47. RVO, 2015, Top 15 energiezuinige scholen onderzoeksrapport 2014, Rijksdienst Voor Ondernemend Nederland, Utrecht

48. RVO, 2016, Top 15 energiezuinige scholen onderzoeksrapport 2016, Rijksdienst voor Ondernemend Nederland
49. Salthammer T., Uhde E., Schripp T., Schieweck A., Morawska L., Mazaheri M., 2016, Children's well-being at schools: impact of climatic conditions and air pollution, Environ. Int., 94: 196–210
50. SCX Solar, 2014, https://www.scx-solar.eu/projecten/zonnestroominstallatie-op-mfaschool-hart-oijen/
51. Soussi M., Balghouthi M., Guizani A., 2013, Energy performance analysis of a solar-cooled building in Tunisia: Passive strategies impact and improvement techniques, Energy Build. 67: 374–386
52. Spring M., Dazzling achievement, Building connect, 8 June 2007, www.building.co.uk/story.asp?sectioncode=583&storycode=3088489&c=1
53. Vilecekova S., Kapalo P., Meciarova L., Burdova E.V., Imreczeova V., 2017, Investigation in Indoor Environment Quality in Classroom – Case study, Procedia Engineering 190: 496–503
54. Vos A., Groep W. van de, Türkcan O., 2021, Hybride ventilatie met natuurlijke bronnen, TVVL Magazine 01 Februari 2021
55. Walker S.S.W., 2021, Sustainable energy transition scenario analysis for buildings and neighborhoods: Data-driven optimization, PhD thesis TU Eindhoven.
56. Wang Q., Zhang Y., Ahuja S., Augenbroe G., 2014, Re-evaluation of passive design measures in the BASF house in recognition of uncertainty and model discrepancy, Proceedings of the 30th International PLEA Conference, Ahmedabad, India, 16–18 December 2014.
57. Weenen H. van, 2000, Passive Solar Schools International Expertise in Support of the First Sustainable Elementary School of The Netherlands International Workshop, 3–5 March 2000, Castricum, The Netherlands.
58. Weenen H. van, Dettmers W.J.M., Overtoom M.S.J., Poldermans H.G.J.M., 2000, Development of the First Sustainable Elementary School of The Netherlands, Proceedings International Conference Sustainable Building 2000, 22–25 October, Maastricht, the Netherlands.
59. Woodger H, 2017, Energy academy provides a low-tech lesson for the future, Future Constructor and Architect, fc&a, https://www.fca-magazine.com/features/greenhouse/941-breeam-energy-academy-provides-a-low-tech-lesson-for-the-future
60. Zeiler W., Gvozdenović K., Bont K. de, Maassen W., 2016, Toward cost-effective nearly zero energy buildings: The Dutch Situation, Science and Technology for the Built Environment 22(7): 911–927,
61. Zeiler W., 2017, Dutch academic Building Services design education: Integral Design to connect industry and university, essential for nZEB design, REHVA Journal: 01/2017:57–62
62. Zeiler W., 2018, Dutch efforts towards sustainable schools, Proceedings WREC, July 30th – August 3th, Kingston, London, UK

Wim Zeiler started in the building services industry in 1983 and is since 2001 Full Professor of Building Services at the Department of the Built Environment, Eindhoven University of Technology (TU/e), the Netherlands. In addition, Zeiler is still a part-time building services technology specialist at Kropman Building Services contracting. He was awarded in 2009 a gold medal by TVVL, the Dutch Technical Society for Building Services. In 2014 he received a professional science award from REHVA, the European Federation for Heating, Ventilation and Air-conditioning. His research is on integral design, intelligent buildings, and flexible smart grid integration.

Living Bricks Can Generate Energy in the Home and Wean Humanity Off Fossil Fuels

Rachel Armstrong

1 Introduction

As the world emerges from prolonged lockdowns associated with a viral pandemic, microbes probably feature very low down on most people's list of factors when thinking about potentially significant contributions to improving building comfort. Most of us will probably regard the fewer microbes there are in a space, then the better, but our understanding of the broader role of microbes in issues of health and well-being beyond the effects of pathogens is changing and is influencing how we design our living spaces. These factors not only affect human occupants but also the immediate, character, and environment of our buildings.

This chapter considers the nature of comfort within buildings beyond conventional parameters concerned with ventilation, heating and cooling systems, resource recycling, and energy conservation, which optimizes the livability of spaces and examines the implications of emerging microbial technologies that enable us to strategically generate environmental effects within the building. Taking a first step toward how we might imagine living spaces fueled by the "living" processes of metabolism, it also introduces the emerging concepts of the human microbiome and the microbiome of the built environment, with which the presence of microbes in our buildings are entangled. Owing to the cutting-edge nature of this field of research, we do not yet fully know the impacts of microbial distribution, within and beyond the human body, but we do know that these issues are likely to be important characteristics in the near future not only for issues related to building comfort but also with respect to their environmental impacts.

R. Armstrong (✉)
Experimental Architecture, School of Architecture, Planning & Landscape, Newcastle University, Newcastle upon Tyne, UK

Department of Architecture, Campus Sint-Lucas, KU Leuven, Ghent, Brussels, Belgium
e-mail: Rachel.armstrong3@newcastle.ac.uk; Rachel.armstrong@kuleuven.be

A. Sayigh (ed.), *Achieving Building Comfort by Natural Means*, Innovative Renewable Energy, https://doi.org/10.1007/978-3-031-04714-5_2

The presence of microbes as co-occupants of our living spaces raises important issues about our view of them, as modernity upholds a largely negative view that overlooks their role as the living world's regenerative technology. So to understand the issues at stake in the provision of building comfort, we must first examine how we obtained such present negative associations and consider why these ideas must change.

2 The Reign of Hygiene

First visualized in the seventeenth century as "animalcules" through Antonie van Leeuwenhoek's microscope—an inverse telescope that revealed an inverse cosmos—initially, microbes were a source of wonder. The only way we could relate to them was through their shapes, which were simple and, therefore, were associated with biological primitiveness. As visualization techniques improved, the long-held belief that life spontaneously appeared from nonliving substances during the process of spoilage was disproven, as microbial transmission through air and other invisible processes was demonstrated by Lazzaro Spallanzini in the eighteenth century, who found that boiling broth would sterilize it and kill any microorganisms in it.

Then, through the work of Louis Pasteur and Robert Koch in the late nineteenth century, the modern version of germ theory was formed that demonized and equated microbes with all kinds of illnesses. Faced with the human catastrophe of infectious disease—whereby, even in 1900, the three leading causes of death were pneumonia, tuberculosis, diarrhea, and enteritis (which together with diphtheria) caused one-third of all deaths and children under the age of five accounted for 40% of all those deaths from these infections—Western society went to "war" with microbes. It is no understatement to observe that the invention of the modern toilet was necessary for survival and by the 1920s, "sanitary imperialism," which freed people from the responsibility to personally dispose of their bodily excrements, was in full swing. Typified by their white, easy-to-wipe surfaces, modern indoor bathrooms were designed into new houses and plumbed into urban sewage systems. Regarded as contaminants that needed to be designed out of our lives—via indoor toilets, plumbing systems, sewage plants, white cleanable ceramics, and a host of environmentally noxious household products like bleach and detergents— their eradication shaped the microbially hostile *reign of hygiene*. Governed by the principles of germ theory, revolutionized by new rituals of cleanliness, and supported by epidemiology, this blunt instrument helped keep pathogens at bay.

Modernist architecture designed the fear of disease into its living spaces, where the liberal use of germicides became part of the *machine à habiter*'s defensive spaces. Le Corbusier's houses were a testimony to antimicrobial design being raised above the humid ground, where Villa Savoye's entrance hall sink could be considered an *altar to hygiene*. Hugo Alvar Henrik Aalto's Paimio Sanatorium functioned as a medical instrument designed to eliminate tuberculosis, while the clinical minimalism of Ludwig Mies van der Rohe or Marcel Breuer unambiguously boast their environmental sterility.

Within the course of a century, the reign of hygiene had taken its toll on the environment. With the increased use of water as part of the purification process, the volume of wastewater significantly increased but our handing capacity did not. In 1920, the world's population was just under 2 billion: today it is 7.5 billion and is served by almost the same infrastructures that were available a century ago. Now, wastewater is all around us—from the water flowing through the shower drain to the runoff from roads, where it seeps into underground water tables and contaminates local ecosystems. Compounded by frequent applications of chemical cleaners, phosphorous, nitrogen, and ammonia are tipped into toilets, sinks, drains, sluices, and dishwashers. Entering the wastewater system, these weak poisons gradually accumulate in the environment, particularly in soils, natural bodies of water, and the flesh of living things. The pure white layer of modern sterility therefore presents an ethical mask for critical issues of wastewater management, pollution, exploitation of nature, cheap materials, planned obsolescence, disposable consumption, and scarcity economics—amplified by modern consumer society. Having reduced the microbial loads in our kitchens and bathrooms, the reign of hygiene has simultaneously produced rivers, reservoirs, lakes, and seas that are drowning in chemicals, waste, plastic, microplastics, and pesticides. One of the consequences of these cumulative toxins is a drastic loss of biodiversity, which is powerfully described in Rachel Carson's *Silent Spring*.

> Why should we tolerate a diet of weak poisons, a home in insipid surroundings, a circle of acquaintances who are not quite our enemies, the noise of motors with just enough relief to prevent insanity? Who would want to live in a world which is just not quite fatal? [1]

Despite these antimicrobial designs, modernity has been unable to prevent us from being quarantined within our own domestic spaces. In fact, presently, our homes, and other interior spaces where we socialize, are the most likely places for the spread of contagion. The Pandora's Box that is the microbial era has been well and truly sprung, and the issues of infectious disease have not gone away, so a balance must be struck between the baselines for modernity's template for cleanliness and environmental health and well-being. Rather than pursuing relentless programs of environmental sterility, more agentized, probiotics habitats are needed to effectively negotiate a condition of mutual livability between human and microbial inhabitants alike. More than our inability to think and act ecologically, the present pandemic with its lockdowns, social distancing measures, new rituals of hygiene, and intrusions on our lifestyles spotlight the home as a tirelessly generative, experimental space capable of producing new design protocols that become part of our everyday routines. Potentially, here, the character of our living spaces can be reconceptualized and transformed into a grassroots response to the pandemic that actually changes our lives for the better.

3 The Environmental Role of Microbes

Less than one percent of all microbes are pathogens. Although this is a large number, to date, the role and function of more than 99% are still largely unknown and until now has been unknowable. Even at the time when microbes were being equated with pathogens in the nineteenth century, a parallel stream of scientific investigation looked at their largely beneficial environmental effects in agriculture, where their role in plant health and recycling life's primary elements, namely, carbon, hydrogen, oxygen, phosphate, and nitrogen, were firmly established. Working in partnership with the sun, these ancient beings were found to generate *a commons of biochemical goods all around us* that forms the "microbial commons" where their ongoing metabolic activity sustains the biosphere [2]. As primary producers of biomass, they can remove carbon dioxide from the atmosphere through photosynthesis and, using light energy, convert these ephemeral substances into organic (cellular) material, generating vital oxygen in the process. They also process the useful molecules within dead matter, so that it can (re)enter the food chain, linking the cycles of life and death through the production of compost. Such world-making narratives, however, were less anthropocentric and less relatable than the reign of hygiene, and so their critical importance in maintaining the productivity of the world was largely unrecognized until the late nineteenth century.

4 Microbial Diplomacy

The only living beings on the planets billions of years before humans arrived, microbes have effectively comprised the world's society and culture for most of the story of life on earth. While their life together is no utopia, it has nevertheless resulted in the optimization of site conditions that have produced the flourishing of other organisms, resulting in an overall increase environmental liveliness. Since no one microbe can perform all possible transactions alone, their systems of organization, modes of exchange, and forms of communication comprise an origin of sociable life [3], where diverse microbial species work together to produce livable spaces. Myra Hird develops these relational-material approaches within the biosphere as a *micro-ontology* that acknowledges the contributions of this unseen majority as being independent of human observation and action. The material forms that emerge from these negotiations are structures like biofilms that offer many advantages to their inhabitants and possess a high level of organization, where microbes are actively engaged in a community of life that is rich with social connections and countless acts of design. These "different ways of being that give valuable perspectives on life and the world" [4] may, at first glance, appear to be more chance than design, but the tools of molecular biology have recently enabled us to "see" how microbes work together using their genetics and metabolism, and researchers are starting to reveal their complex *inner life*. For example, using chemical languages, bacteria can observe their neighbors, understand how many microbes are present, and, if sufficiently large numbers of like-minded organisms are present, they may even agree to act in a coordinated manner through *quorum sensing.*[1] Such signaling systems enable bacterial cells to engage in diverse group behaviors that reach across different species—even to animals—to generate a host of outcomes including biofilm formation, swarming motility, justice systems to deal with cheaters [5], or produce specific products like extracellular proteases and iron-chelating siderophores. Such coordination enables bacterial communities to collectively perform the work of giants [6]—decomposing a whale carcass, fertilizing a forest floor, causing tooth decay, illuminating breaking waves with pulses of bioluminescence, or occasionally becoming toxin-producing pathogens, such as the bubonic plague which devastate their hosts [7]. None of these actions take place without coordination, or intent, which involve "collective sensing, distributed information processing, and gene regulation of individuals by the group" [8]. Since these material effects are produced over time and space, they can be considered as acts of microbial diplomacy or design, as all establish a discriminating primordial ruleset through which decisions are made. Navigating their world in ways that we barely understand they can, for example, communicate over long distances to strategically organize into complex patterns of multicellular configurations (fungi, archaea, and bacteria). It is also noteworthy that the human-centric tactics for dealing with microbes were identified from solutions produced by other microbes, namely, the identification of the antibiotic penicillin from mold. So, when it comes to the regulation of microbes, the 3.5 billion years of experience of dealing with other microbes position the microbial realm as the best negotiators when microbiomes are out of equilibrium.

5 Technology and Microbes

Since we cannot see microbes with the naked eye, our relationship with them is mediated by technology. Up until the twenty-first century, our understanding was relatively limited and rested largely on old ideas of morphology observed through light microscopy and a restricted view of their biotechnology compounded by issues of culturing them. Many microbial species actually fare very poorly in refined media [9], where usually only one dominant species responsible for a particular disease

[1] Quorum sensing is the regulation of gene expression in response to fluctuations in cell-population density. These bacteria produce and release chemical signal molecules called autoinducers that increase in concentration as a function of cell density. The detection of a minimal threshold stimulatory concentration of an autoinducer leads to an alteration in gene expression. Gram-positive and Gram-negative bacteria use quorum sensing communication circuits to regulate a diverse array of physiological activities. These processes include symbiosis, virulence, competence, conjugation, antibiotic production, motility, sporulation, and biofilm formation, see Miller, M.B., Bassler, B.L. (2001). Quorum sensing in bacteria. *Annual Review of Microbiology*, 55, pp. 165–199. https://doi.org/10.1146/annurev.micro.55.1.165.

can thrive [10]. In the case of laboratory-cultured microbes, typically those that survive under such abnormal conditions are not representative of the species but, by definition, "monsters" [11].

The molecular biology revolution catalyzed an important tipping point in microbial understanding, where cheaper, faster, sequencing methods in the emerging science of metagenomics enabled the insides of microbes to be "seen." This enabled a picture of their metabolic diversity, rather than morphology, to be mapped and the characteristics of entire microbial populations within specific "microbiomes" could be more easily understood as communities, rather than collections of individuals. Furthermore, new screening techniques using molecular techniques (e.g., 16S ribosomal RNA gene sequencing[2] based on polymerase chain reaction) enabled the low-cost broad surveys of microorganisms, which revealed just how diverse and important microbial metabolisms are in the fundamental transactions that make up the natural realm. While our view of microbial ecosystems is still piecemeal, their distribution and specific composition within our personal spaces, atmospheres, and habitats have implications for our health and environmental well-being. At the heart of this environmental enlivening a better relationship with microbes within our homes implies that world-making activities can take place within the domestic realm.

6 The Human Microbiome

> Microbiome … may be defined as a characteristic microbial community occupying a reasonably well-defined habitat which has distinct physio-chemical properties. The term thus not only refers to the microorganisms involved but also encompasses their theatre of activity [12].

Microbiomes are distinctive microbial communities that make up every ecological niche. Following advances in biotechnology it was suddenly possible not only to see the microbial ecosystems around us but also within us, deeply complicating what it means to be "human" [13]. The biggest revelation is just how microbial we

[2]The piece of DNA used for identifying bacteria is the region that codes for a small subunit of the ribosomal RNA (16S rRNA), or 16S rDNA. Different bacterial species have unique 16S rDNA sequences. 16S rRNA gene sequencing is commonly used for identification, classification, and quantitation of microbes within complex biological mixtures such as environmental samples. The 16S rRNA gene is a highly conserved component of the transcriptional machinery of all DNA-based life forms, see Cox, Michael J., Cookson, William O.C.M. and Moffatt, Miriam F. (2013). Sequencing the human microbiome in health and disease. *Human Molecular Genetics*, 22(R1), pp. R88–R94, and is highly suited as a target gene for sequencing DNA in samples containing up to thousands of different species. While the conserved region makes universal amplification possible, sequencing the variable regions allows discrimination between specific different microorganisms such as bacteria, archaea and microbial eukarya. The identification relies on matching the sequence from your sample against a database of all known 16S rDNA sequences. Identification of viruses requires metagenomic sequencing (the direct sequencing of the total DNA extracted from a microbial community) due to their lack of the phylogenetic marker gene 16S.

actually are. Metagenomics shows that ecosystems of microbes inhabit our bodies and living spaces, where around 50% by number of our own body cells are bacterial [14]. State-of-the-art understanding of the human microbiome is primarily salutary where microbes inhabit our guts and skin where, in these transactional spaces, they play a significant role in our welfare by acting as an "organ" that influences our moods, helping with digestion while releasing microbial "goods" that comprise essential vitamins, which we cannot make ourselves and acting as a first-line immune system against pathogens [15]. There is no point thinking that we can do without these microbial colonize as they are critical for our health. Consider, for example, how disordered intestinal flora can be following long courses of antibiotics and how sick we can become in their absence. These microbial populations, however, are not always benevolent and, when they become dysfunctional, are also responsible for switching on a wide range of diseases including cancer, cardiometabolic diseases, allergies, and obesity [16]. The discovery of this *human microbiome* [17] means that we are no longer "pure" bodies but "bodies-as-ecosystems," where we are irreducibly entangled with many nonhuman others [18], sharing entangled histories and situated narratives with them [19], where these microbes are not "other" than "us" but are kin. Requiring a more holistic understanding of who "we" are and the metabolic realms we inhabit, we must learn how to reconceive our relationship to our living spaces through an understanding of this expanded notion of "self" to identify as a *human holobiont* [20]. Thinking through the concept of holobiont helps us consider our well-being in relationship to multiple actors including our habitat, microbial "others," and the physical makeup of the domestic environment. Caution is needed, however, in assuming that all actors within the holobiont are either neutral or prioritize human needs, as it is a transactional system that depends on *how well* the constituent agents live together and is not predicated on the privileging of any specific entity—human or nonhuman.

Over the last century, the reign of hygiene has also endangered the integrity of the human microbiome through rapid changes in sanitation systems, medical advances, and modern lifestyles, which threaten essential microbial communities whose vital, metabolic networks are in decline and even face extinction. This is the dark side to clean living, where cleanliness rituals have resulted in changes in the way our bodies appropriately react to microbial "others," where children's immune systems are not trained appropriately through immunological encounters with dirt, resulting in a host of allergies [21]. Additionally, vaccination programs and antibiotics, which are conceived of addressing targeted organisms, also set up the conditions for resistance across the whole microbiome, resulting in the eradication of certain "silent" microbial species whose specific metabolic pathways are important to a specific individual's health. Such microbiome diversity losses are particularly important in immigrant populations, as by moving to places with new microbial ecosystems and health-care practices, people lose key microbial populations that are vital to their well-being. Such observations herald the advent of design and protocols for human inhabitation that engage the emerging microbial era, where appropriate design of probiotic activity can rearticulate our behavior and contributions to the maintenance of health and well-being—not only during a pandemic but also in untroubled times.

7 Microbiome of the Built Environment

As holobiontic[3] subjects, human and microbial ecology compel us to understand what kinds of microbes we live alongside, recognize their diversity, and pay attention to how they behave in different settings. Over the last decade, high-throughput molecular techniques became sufficiently affordable for microbial communities in indoor environments to be analyzed, which revealed that not only bodies but also all surfaces also possess unique microbial colonies. This *microbiome of the built environment* (MBE) is integral to inhabited spaces and is maintained by our everyday activities that disseminate, enrich, and even propagate microbes within our living spaces. Establishing a critical link between microbes, humans, and environment, interior spaces form a natural laboratory for shaping our lives and well-being. Present knowledge of the MBE is incomplete, and our understanding of its nature is therefore still evolving [22], so state-of-the-art includes its characterization. It is not possible to overemphasize the seismic shift in perspective for architectural practice with this "discovery," whose central tenets are the body and environment. Moreover, the *theater of activity that comprises indoor spaces* is where we also spend 90% of our lives, and certain organizational principles are emerging that impact our health and well-being [23], where the distribution of microbes within each microbiome can be variably influenced by how the space is designed and operated. A few architects have sampled the presence of microbes in buildings using modified laboratory apparatuses such as petri dishes to directly visualize them [24] and determine which materials provide the best substrates for microbial colonization [25]. Existing approaches for modifying the MBE are generic solutions working with established building technologies such as through the control of ventilation rates and use of air filtration systems [26], envelope tightness, disinfectant applications, moisture control [27], remediating damp [28], particle filtration [29], germicidal irradiation [30], UV irradiation [31], chemical disinfection [32], and natural ventilation to increase beneficial outdoor microbial biodiversity in buildings [33] and even by adding beneficial microbes to buildings [34]. Incorporating a longitudinal epidemiological study of exposures to chemical and biological agents within an indoor space over time, the exposome extends the relevance of the MBE into the realm of the kinds of metabolites its agents produce through an ecosystem of relational contracts that are now recognized as major drivers of human health and disease [35].

[3] Lynne Margulis coined the term *holobiont* to refer to any physical association between individuals of different species to form a coherent ecological unit, see Margulis, L. (1998). *Symbiotic Planet: A New Look at Evolution*. New York: Basic Books. This definition has since changed to incorporate more eco-systemic bodies like corals and trees with mycorrhiza, which is in keeping with the present usage of the term.

Changing the fundamental concepts of architecture, buildings are no longer inert spaces but environments that house complex ecosystems of trillions of diverse microorganisms [36]. All our design protocols are now entangled with a whole range of distinct yet entangled microbiomes including the human microbiome [37], the built environment microbiome (MBE) [38], and the soil/human microbiome [39]. Too important and pervasive for design to ignore, a way forward can be found in the intimate relationship between these complex "fabrics" which establishes a critical link between microbes, humans, and environment whereby *gardening the presence of microbes in a place* will inevitably alter the MBE. These associations are not formalized but are likely and so must be raised and considered at this point as a potential way of mediating the composition of an MBE potentially using microbe to regulate microbe, as was originally seen in the advent of antibiotics.

8 Microbial Technology

Even before we could "see" them, we have used microbes as workhorses from composting to fermentation and tanning, and the incredible metabolic diplomacy of microbes can be gardened and directed as technology, the fine control of which has become increasingly possible with the advent of advanced biotechnology.

In 1911, Michael Cressé Potter used microbes to produce (bio)electricity by designing a "living" battery, or microbial fuel cell (MFC), which used the vital processes of *Saccharomyces* bacteria to produce several hundred millivolts of energy [40]. Acting as biocatalysts, the microbes converted the chemical energy of organic matter from waste streams into electrons for as long as they were fed. Each "cell" consisted of two compartments, just like a chemical battery—the anode and the cathode—which were separated by a proton-exchange membrane. Bacteria anaerobically oxidized the organic matter in the anode chamber to release electrons that flowed through an external circuit to provide electrical power. Acidic protons were also produced that dissolved into solution and passed through the membrane into the cathode, where they reacted with oxygen to produce freshwater. This highly mediated relationship set up a power-sharing relationship across mechanical and natural bodies that was neither entirely biological nor exclusively mechanical. Blurring the relationship between organism and machine, the microbial fuel cell provided a technical environment that enabled a natural biofilm to perform a range of metabolic tasks at room temperature that in addition to the production of bioelectricity included cleaning wastewater and detoxifying pollutants.

However, the microbial fuel cell (MFC) could not compete with the sheer power of other electricity-generating systems that were demanded by industry and modern households and were therefore considered a curiosity rather than a potential building infrastructure. It was not until the late twentieth century when Bill Gates invested in research that explored how sanitation could be reinvented that the formal development of microbial fuel cells as implementable systems could begin.

9 Microbial Technology as Infrastructure for Building Comfort

Microbial technologies are not single-purpose systems but create a whole set of transformations that potentially change the livability of space. An example of all these relationships is provided in the next sections using the Living Architecture project and ALICE prototype as case study examples of microbial technologies. The protocols of space-making they engender, however, do not propose to be a soft equivalent for modern utilities but generate an alternative resource management platform with potential to change how we imagine, design, construct, and inhabit buildings.

10 Living Architecture

Aiming to establish a regenerative circular economy of the household, the Living Architecture project combined the bioelectrical properties of microbes with the principles of "animal economy," "metabolism," and "ecology"—all of which are centered on the every-day or "domestic" activities of daily living within the "oikos." Situating the capacity for transformation of building impacts within the homes of ordinary people, it established a circular economy of the household that was founded on the reutilization of organic waste. Applying the "living" metabolisms of bacteria, our domestic spaces are provided with an alternative resource and energy system than the modern dependency on fossil fuels through an alternative, regenerative technological platform and infrastructure that not only carries out housework (cleaning wastewater and producing small amounts of electrical energy) but also acts as site for transactions between humans and nonhumans. All of this takes place through vital, metabolic activities such as feeding, metabolizing, excreting, and reproducing. As both "vital" thing and inhabited structure, the constituent metabolic processes are not oblivious to us but respond to our multiple acts of care through the quality of housework performed and creating the possibility of interdependencies that enable environmentally engaged ways of living. Invoking the iterative, vital exchanges performed between human and soil, Living Architecture's synthetic metabolism was designed to perform regenerative transactions as a technical embodiment of the ancient relationship between a person and their land. Rather than the plot for tender being outside the home, it has been technologically internalized and micro-miniaturized (like an organic computer). The soil-inspired system of Living Architecture also indicates the same figure and ground that invokes the production of architecture—and the acknowledgment that people are not apart from but are integrated into the flows of natural (eco)systems. Funded by the Horizon 2020 Research and Innovation Programme, it ran from April 2016 to June 2019, as a transdisciplinary collaboration between the universities of Newcastle, UK; the West of England (UWE Bristol); Trento, Italy, and the Spanish National Research Council in Madrid and the Small-to-Medium Enterprises, LIQUIFER Systems Group, Vienna, Austria; and Explora, Venice, Italy (Fig. 1) [41].

The first step in producing the Living Architecture prototype was to consider the types of useable resources that could be extracted from household waste and then design the homes for microbes that were needed to form the foundations for its circular economy. Through the careful design of their surroundings, achieved by making special houses, or bioprocessors, for them, microbes could take cues from their specific environment that helped them carry out particular metabolic tasks.

Three basic types of "homes" were provided for microbes that were sequentially configured within an apparatus about the size of a large bookcase (Fig. 2).

Fig. 1 Fully inoculated Living Architecture "wall" and apparatus installed at the University of the West of England, Bristol, photograph courtesy of the Living Architecture project, 2019

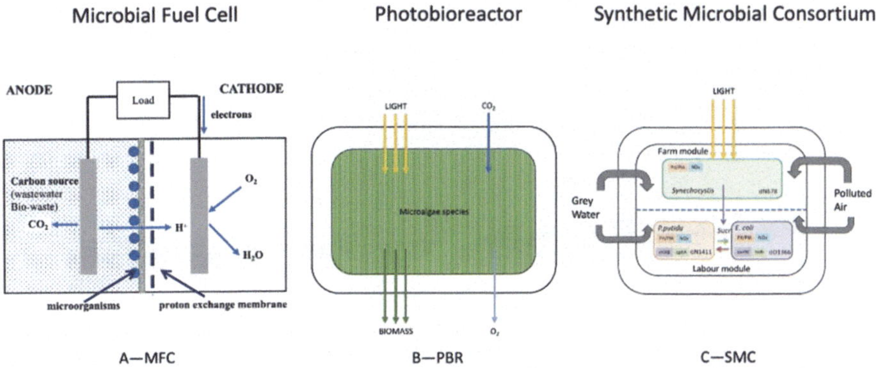

Fig. 2 Diagrams of the three basic types of bioreactor in Living Architecture; drawings courtesy of the Living Architecture project, 2019

End-stage metabolic products then become food for next-door microbial communities, whose own unwanted products go on to feed their neighbors and so on. Being self-contained homes, or "metabolic worlds," different bioprocessors can be sequenced in various ways as "metabolic apps" that can be combined to generate specific kinds of resources. Such an arrangement of multiple metabolisms, joined by one continuous stream of electron exchange, invokes a whole choreography of sustained transactions, where opportunities arise for microbes to respond and influence outcomes (Fig. 3).

The first type of microbial "home" was a photobioreactor. It cultured microalgae, which transformed sunlight and carbon dioxide into biomass (cellulose) and oxygen, which acted as an end-terminal acceptor for the electrons harvested from the metabolizing biofilm next door. This second type of bioprocessor was a microbial fuel cell, whose bioelectricity signal provided both an energy-generating and communications system that powered an artificial intelligence (AI) that observed the metabolic activity within the system and accordingly altered the external inputs using a system of valves to maintain a steady state. All the other metabolites produced by the biofilm in the microbial fuel cell passed into the cathode which contained cleaned and filtered water.

The final bioprocessor type was a modification of the microbial fuel cell, but this time, the two-chamber device (one chamber is transparent, one opaque) housed synthetically modified microbes to create an experimental environment where new microbial metabolism could be designed using synthetic biology techniques. The transparent chamber was likened to a "farm," where genetically modified, light-loving microbes produced high quantities of carbon-rich sugar that passed into the

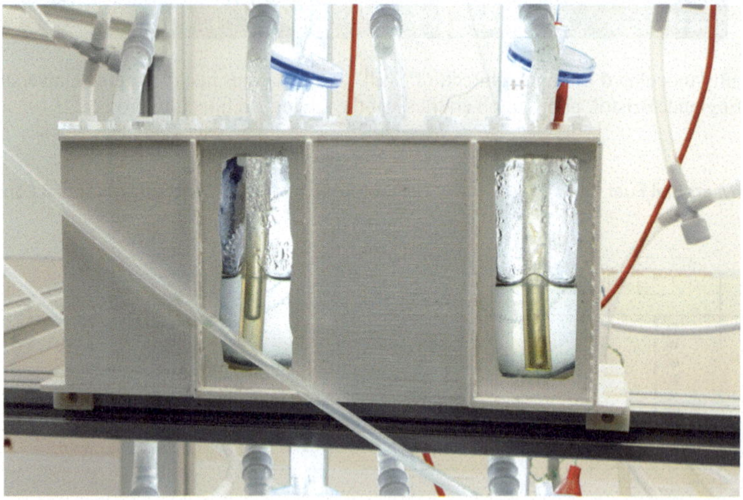

Fig. 3 Four-chambered "living" brick design for the Living Architecture project; photograph courtesy of the Living Architecture project, 2019

opaque "labor" module where a range of different "workhorse" organisms[4] lived. In nature, these microbes do not live together but could do so under energy-rich conditions, where constituent microbial metabolisms could even be designed to work across different species. This metabolic modularity greatly increased the available toolset for metabolic design with the potential to generate many different kinds of metabolic "apps." In the Living Architecture project, processes that reclaimed phosphate from liquid detergents and detoxified gaseous pollutants like nitrogen dioxide were designed and implemented as a result of these processes.

From its outset, Living Architecture established diplomatic relationships at the human/microbial interface and was central to the project to stop synthetically modified microbes from escaping into the environment. While a number of "kill" switches, such as splitting synthetic metabolisms across more than one microbe, so that an escapee could not survive on their own in the wild, were developed, the natural regulatory capabilities of biofilms in microbial fuel cells were also explored as an active interface for removing synthetic microbes from waste streams. Working with the pathogenic hepatitis B virus as a model system to test this principle [42], the findings were striking. When contaminated wastewater flow was looped back through the microbial fuel cell, the natural biofilm removed the pathogens. In other words, when flow through the bioprocessor was organized in a circular configuration, then the microbial fuel cells acted like an external microbial "immune system." With the potential to decontaminate domestic waste streams, this capability will be particularly useful at times of pandemic. Notably, the strategic use of natural microbial intelligence to regulate the composition and character of neighboring microbiomes could be a key technical principle for the ongoing principled investigation of technical systems to help us negotiate microbial environments.

11 Active Living Infrastructure: Controlled Environment (ALICE)

Microbes possess a very particular kind of intelligence that reveals a great deal about the environmental health of a place. To access this information for the purposes of building comfort requires formal appreciation of what they are telling us. This requires a unique, technologized understanding, since we cannot observe them

[4] "Workhorse" organism is a technical term that refers to microorganisms whose genes are well known and often have been completely sequenced. The term itself raises questions about the ethics of anthropocentric laboratory experiments—in synthetic biology and beyond—as it does not imply a partnership between human and microbe, but traditional command-and-control where creatures are the subjects for the execution of human will rather than co-constitutive partners in a mutual inquiry and is, therefore, open to exploitation (another term used in laboratory experiment specifically in relationship to results) and multiple abuses. I discuss the use of synthetic biology in the Living Architecture project in more detail in Hughes and Armstrong, 2021. *The Art of Experiment: Post-pandemic Knowledge Practices*. London: Routledge, pp.124–125)

with the naked eye—and when we can see them *en masse* as an organized biofilm, then quite frankly, they are deeply unrelatable. Typically, microbial performance is deciphered through their biochemistry—specifically through the expression of their nucleic acids and metabolites—but, in human terms, this process is quite slow. Tapping into the electron flows within biofilms, however, gives us a direct way of understanding the behavior of a microbial population at any given moment and, depending on the sensitivity of electron detectors, creates the possibility of developing a communications platform between human and microbe in the most basic manner.

Centered on the microbial fuel cell technology, the ALICE[5] prototype aimed to develop a readable communications interface between human and microbe to change our relationship with and expectations of healthy environmental microbes. It established a co-constitutive environment between human and microbe by using the direct connection between the electron flows generated by microbial metabolism and the digital realm [43]. Using conventional electronics systems to detect and respond to microbial "data"—i.e., electron flow—it directly correlated with microbial "comfort" in the biofilm.

The electrical data from the biofilm was translated by software into animations that conveyed the overall status of the biofilm in relatable terms (Fig. 4). Audiences could, therefore, respond to the microbial behavior—not by looking at unpleasant

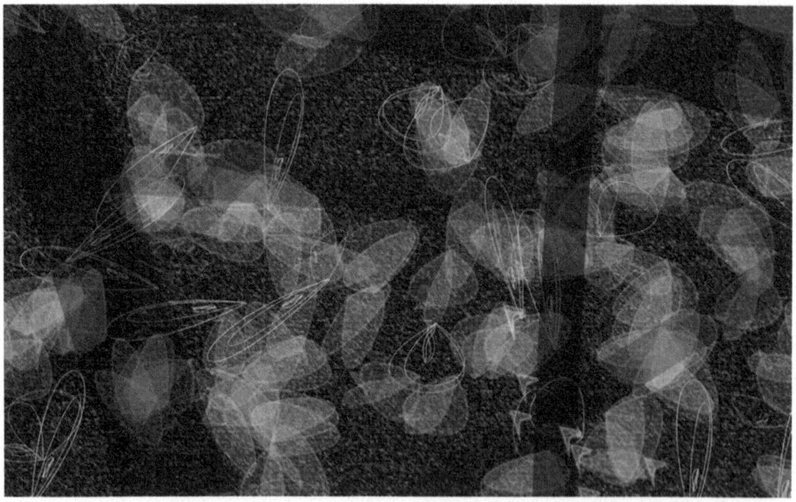

Fig. 4 "Mobes," from the ALICE website (http://alice-interface.eu) showing dynamic, interactive, graphical representations of microbes, courtesy of the ALICE consortium, 2021

[5] The *Active Living Infrastructure: Controlled Environment* (ALICE), is a collaboration between the University of Newcastle, University of the West of England and Translating Nature. This EU funded Innovation Award prototypes the construction of a novel bio-digital interface using Microbial Fuel Cells and augmented reality experience for "living" bricks developed in the Living Architecture project.

"slime" (the natural "face" of microbial colonies) but by interacting with appealing forms on a familiar screen-based interface. Avoiding having to deal with the strange appearance of biofilms, participants could play with resident microbes in an exploratory exchange through data and performance—as if they were a pot plant or even a pet. This world of "Mobes"—a characterful term coined for the data-based representations of microbes—helped human-microbial relations by revealing the incredible electron-rich microbial world inside the bioprocessor where they lived. Such a probiotic approach to interspecies communication provided access to the highly situated realm of microbes, in a relatable manner that could become part of our everyday routines. Being in conversation—rather than "exploiting" microbes—means we may start to learn-along-with them through their ability to generate clear and direct signals and data that relate to household resources so that we can manage them more strategically.

Due to the ongoing coronavirus pandemic, ALICE currently "lives" online as an interactive interface that can be remotely accessed through the many layers and options presented. Visitors can access real-time data to observe how a stable biofilm population is performing and interact with them remotely by selecting options from the project website. A person observing that microbes are "uncomfortable" may wish to feed them through a remote controlled valve or gently warm them with an LED to speed up their metabolism and generate more bioelectricity.

These simple exchanges between human and microbe create an environment where more equitable explorations can be conducted and further interaction invited that expands our understanding beyond the human realm. Seeking mutual enlivening, thriving, and continuity, the production of shared languages mediated through a digital interface that is co-composed between human and microbe enables us to observe and understand such responses to inform new value systems and suggest different kinds of (house)work and care for our domestic environments. With ALICE, there is no planned "end" to its experiment, as it marks the start of an ongoing conversation where the conditions for mutuality between microbe and human can continually be negotiated and developed.

12 Impact of Microbial Technologies on Comfort

Commercially, no domestic microbial technologies presently exist but they have a long informal tradition, having been used in farmhouse settings in rural areas where, for example, manure heaps are used to provide hot water. By formalizing the anaerobic processes, Living Architecture provides a prototype and platform for formally developing an implementable building technology whose protocols of servicing, care, and rituals of inhabitation enable a choreography of living that works with lower energy systems than typically powered by mains electricity, provides cleaned (not potable) water, and can generate the kind of heat typical of a body, as in medieval times when farmers brought their animals indoors during harsh winters partly as a heating system for their building.

Living Architecture does not provide all the power and resources needed to completely support for modern lifestyles. In fact, it establishes limits on resource consumption within our homes that makes us think about how we use energy and resources. In this way, it is a pedagogical system that helps us understand by seeing for ourselves, how we can work within broader resource limits set by the carrying capacity of the planet. Living Architecture is also the start of an innovation process that explores the potential for lower power devices and convergent energy systems to work smarter with household resources and increasingly empower end users to make strategic choices about their footprints working in concert with microbes.

To date, a version of this technology was developed for and installed at the Whitechapel Gallery's group exhibition "Is this Tomorrow?" as a collaboration entitled *999 years 13sqm (the future belongs to ghosts)* as a collaboration between Rachel Armstrong and Cecile B. Evans in 2019, which explored the possibility of a posthuman apartment powered by microbes and haunted by digital figures, or "ghosts," which promoted discussion of alternative futures for the household (Fig. 5).

Fig. 5 The installation 999 years 13 sqm (the future belongs to ghosts) is a collaboration between Cecile B. Evans and Rachel Armstrong for the "Is This Tomorrow?" exhibition at the Whitechapel Gallery, London, photograph courtesy Rolf Hughes, 2019

13 Beyond Building Comfort: Microbial Technologies as a "Living" Infrastructure

By establishing a communications platform, the relationship between microbes and humans is no longer passive but invites rituals of care for our living spaces, which deeply influence their performance and character. Synchronizing our habits through everyday actions (urination, washing hands, cleaning) and rituals, the nurturing work is achieved through a choreography of electron flows, which link the performative aspects of our personal habits with our needs. Living architectures will wake up with us, go to sleep when we do, will cope with our intimate habits, and will even be re-enlivened by our return after a holiday break—being "pleased" to see us on our return. In this sense, residents are more than inhabitants but prime these environments for further (metabolic) actions through their invisible (metabolic) traces, routines, and residues, generating a metabolic milieu within our haunts and domestic spaces.

Through the care and conscious design of their metabolisms, the day-to-day interfaces, and nuanced intuitions about the character of living architectures, residents will differently understand the character of their homes, what they can do, how they are part of them, and how they can establish responsibilities toward them. They can also expect nonhuman neighbors that form the MBE of their living spaces to reciprocate accordingly and, through choreographed engagements facilitated by technologies like Living Architecture, may provide valued materials like medicines, and edible biomass or generate livable amounts of electricity and heat. Subverting the modern expectations of what a "comfortable" home may be, designing with microbial metabolism is valued through the way it increases the liveliness of its communities and transforms our behavior into a performative ethics composed of invisible acts of environmental care, which is fundamental for sustaining worlds. No matter how living architectures are deployed, they will require our socialized engagement with them, and so they behave in convivial ways [44], for which we are accountable through the mutuality of everyday relationships. Through the conscious design of their metabolisms, the maintenance of day-to-day interfaces, and myriad acts of care, our homes may even acquire a unique character and inner "life"—and being truly haunted, these holobiontic spaces may even possess a spirit or soul.

14 Scaling Up

The next steps for Living Architecture and ALICE are to scale the microbial systems so they can provide public utilities by reactivating the commons to transform notions of "waste" into useable "goods" that are available to all through the electron flow economy—which can be accessed through bioelectrical systems. Imagine, for example, if public wastewater gardens were quite literally power "plants" where people can go to charge their mobile phones and have free access to online services and where LEDs provide street lighting—and all powered by community effluent.

ALICE is one among a range of bioelectrical system technologies that may be incorporated into our everyday lives by research conducted by the PHOENIX Cost Action network of scientists and urbanists. Applications range from remediating brownfield sites using "smart" self-powered sensors and robots to providing public utilities from urinals that charge your phone or enable you to play computer games—which is already available at Glastonbury Festival—and in the near-term whole settlements may also draw on waste to provide an alternative, mobile circular infrastructures that leave almost no footprint behind. Microbial technologies such as bioelectrical systems are poised to enable us in the near term to establish a truly circular (electron) economy for ecological lifestyles that provide an infrastructure for the founding of regenerative cities.

Dealing with a microbial world requires a re-envisioning of the built environment beyond the home and into the city, where our living spaces are sites for upholding environmental justice. This requires a revisioning of the very principles on which our buildings work including the comforts they afford—from their atmospheres to notions of hygiene, choice of materials, and modes of inhabitation where, like our own bodies, buildings breathe, sweat, leak around, and infiltrate us through shared microbial relations. A better understanding of microbes as versatile, world-making agents provides insights about how their actions can be scaled up—potentially to strengthen the holobiontic capacities of our living spaces. A multilayered approach to personal health, notions of hygiene and pandemic prevention can potentially be developed through a humbler, more care-oriented approach that is sensitive to a broad range of parameters from breath to touch, moisture, and airflow through the sensibility of the body. Applying the emerging knowledge of the microbiome enables us to think through our holobiontic entanglements with microbial "others" to establish unique principles of microbial design such as noninnocence, incompleteness, horizontality, and co-constitution to enable "permaculture" of microbial landscapes.

What these spaces might look like and what kinds of comfort they provide are not yet formalized but are characterized by principles of stasis and flow, the formation of micro-niches, metabolic opportunities, and life-promoting affordances. While our understanding of this space will always be incomplete—as we will never fully be able to anticipate the responses of our microbial "others"—our constant engagement with the more than human fabrics all around us provides a metabolic diplomacy that enables us to negotiate specific personal and environmental benefits. Working only at the human scale, however, will not generate the regenerative impact needed to overturn the impacts of the Anthropocene; so scaling and making visible the microbial realm amplify strategic microbial interventions while building co-constitutive relations between microbes and humans. Rendering visible our microbial neighbors helps us to relate with and care alongside them to design and landscape the invisible realm where "they" are no longer "othered" but become our kith and kin.

15 Conclusions

The concept of "living" microbial systems as a strategy for modulating building comfort is a new field of research, even with respect to the fundamental science that characterizes human/microbial relations, where findings are still emerging. Despite their novelty, the role of microbial technologies in buildings is of strategic importance in an ecologically stressed world, where gardened microbial systems in buildings will stop waste streams being tipped into the environment, reduce consumption, and provide probiotic living spaces with regulatable impacts on health and well-being. As we enter an age of increasing zoonotic epidemics and pandemics, their capacity to act as external immune systems by eliminating pathogens from waste streams is of particular interest. While microbial technologies are instrumented and not in themselves "nature," they nonetheless draw on the ancient processes of world-building that generated our fertile planet in the first place. Whether, owing to a whole host or socioeconomic-political factors, they areformally adopted in our future homes and cities or not, their ongoing interrogation and development provide critical environmental lessons with respect to our understanding of building comfort, alongside insights into our health and well-being that must be heeded.

Acknowledgments The *Living Architecture* project (2016–2019) was a collaboration between six different organizations including the Newcastle University, the University of the West of England, the University of Trento, the Spanish National Research Council, the Liquefier Systems Group, and Explora Biotech. It was funded by the European Union's Future Emerging Technologies Open Horizon 2020 program to the sum of 3.2 M euros under grant agreement no. 686585.

The *Active Living Infrastructure: Controlled Environment* (ALICE) is a collaboration between the University of Newcastle, University of the West of England, and Translating Nature. This European Union funded Innovation Award under grant agreement 851246 (2019–2021) prototyped the construction of a novel biodigital interface using microbial fuel cells and augmented reality experience for "living" bricks developed in the Living Architecture project.

References

1. Carson, Rachel. 1962. Silent Spring. Boston: Houghton Mifflin Harcourt, p.22.
2. Dedeurwaerdere, Tom (2010). Self-governance and international regulation of the global microbial commons: Introduction to the special issue on the microbial commons. *International Journal of the Commons*, 4(1), pp.390–403.
3. Hird, M. (2009). Origins of Sociable Life: Evolution After Science Studies. New York: Palgrave Macmillan.
4. Wendell, S., 2013. Unhealthy Disabled: Treating Chronic Illnesses as Disabilities. In: L. J. Davis, ed. The Disability Studies Reader. 4th ed. New York: Routledge, p. 161–73: p171.
5. Özkaya, Ö., Balbontin, R., Gordo, I. and Xavier, K.B. (2018). Cheating on cheaters stabilises cooperation in Pseudomonas aeruginosa. Current Biology, 28, pp. 2070-2080. https://doi.org/10.1016/j.cub.2018.04.093.
6. Gould, S. E. (2011). Bacteria with bodies – multicellular prokaryotes, Scientific American Blog, 16 November. [online]. Available at: https://blogs.scientificamerican.com/lab-rat/bacteria-with-bodies-multicellular-prokaryotes/. [Accessed 19 June 2020].

7. Angier, N. (2010). Listening to Bacteria, Smithsonian Magazine, August. [online]. Available at: https://www.smithsonianmag.com/science-nature/listening-to-bacteria-833979/.
8. Hird, M. (2009). Origins of Sociable Life: Evolution After Science Studies. New York: Palgrave Macmillan, p. 42.
9. Guidot, Alice, Jean-Claude Debaud, Roland Marmeisse, "Spatial distribution of the below-ground mycelia of an ectomycorrhizal fungus inferred from specific quantification of its DNA in soil samples", *FEMS Microbiology Ecology*, vol.42, no.3, 2002, p. 477–486.
10. Koch Robert. 1876. The etiology of anthrax, founded on the course of development of Bacillus anthracis. In: Carter KC; translator. Essays of Robert Koch. Connecticut: Greenwood Press; 1987. p. 14.
11. Woese, Carl R., "There must be a prokaryote somewhere: microbiology's search for itself", *Microbiology Reviews*, vol.58, no.1, 1994, p. 1–9, p.7.
12. Whipps, J.M., Lewis, K. and Cooke, R.C. (1988). *Mycoparasitism and plant disease control*. In: Burge, N.M. (ed). "Fungi in Biological Control Systems". Manchester: Manchester University Press, pp.161–187: p. 176.
13. Lederberg J. and McCray, A.T. (2001). "Ome Sweet'Omics—A Genealogical Treasury of Words". The Scientist, 15(7), p.8–9.
14. Sender, Ron, Fuchs, Shai, and Milo, Ron. (2016) Are we really vastly outnumbered? Revisiting the ratio of bacterial to host cells in humans. Cell 164: 337–340.
15. Dominguez-Bello, M.G., Godoy-Vitorino, F., Knight, R. and Blaser, M.J. (2018). Role of the microbiome in human development. *British Medical Journal*, 68(6): https://doi.org/10.1136/gutjnl-2018-317503.
16. Hadrich, D. (2018). Microbiome Research is Becoming the Key to Better Health and Nutrition. *Frontiers in Genetics*, 9(212): https://doi.org/10.3389/fgene.2018.00212.
17. Timmis, Kenneth, Ricardo Cavicchioli, José Luis Garcia, Balbina Nogales, Max Chavarría, Lisa Stein, Terry J. McGenity, Nicole Webster, Brajesh K. Singh, Jo Handelsman, Victor de Lorenzo, Carla Pruzzo, James Timmis, Juan Luis Ramos Martín, Willy Verstraete, Mike Jetten, Antoine Danchin, Wei Huang, Jack Gilbert, Rup Lal, Helena Santos, Sang Yup Lee, Angela Sessitsch, Paola Bonfante, Lone Gram, Raymond T. P. Lin, Eliora Ron, Z. Ceren Karahan, Jan Roelof van der Meer, Seza Artunkal, Dieter Jahn and Lucy Harper. (2019). The urgent need for microbiology literacy in society. Environmental Microbiology, 21(5), pp. 1513–1528.
18. van de Guchte, M, Blottière, H.M. and Doré, J. (2018). Humans as holobionts: implications for prevention and therapy. *Microbiome*, 6 (81), doi: https://doi.org/10.1186/s40168-018-0466-8.
19. McFall-Ngai, M. (2017). *Noticing Microbial worlds: The postmodern synthesis in biology*. In (eds). A. Tsing, H. Swanson, E. Gab and N. Bubandt: "Arts of Living on a Damaged Planet" Minneapolis: University of Minnesota Press, p.M65
20. Fischel, Stephanie, R. (2017) The Microbial State: Global Thriving and the Body Politic. Minneapolis: University of Minnesota Press.
21. Harmon, Catherine. (2009). Bugs Inside: What Happens When the Microbes That Keep Us Healthy Disappear? Scientific American, 16 December. [online]. Available at: https://www.scientificamerican.com/article/human-microbiome-change/. [Accessed 21 November 2020].
22. Kembel, Stephen W., Jones, Evan, Kline, Jeff, Northcutt, Dale, Stenson, Jason, Womack, Ann M., Bohannan, Brendan J.M., Brown, G.Z. and Green, Jessica (2012). Architectural design influences the diversity and structure of the built environment microbiome. *International Society of Microbial Ecology Journal*, 6, pp. 1469–1479.
23. Green, J. 2014. "Can bioinformed design promote healthy indoor ecosystems?" *Indoor Air*, 24, no.2, pp. 113–115.
24. Pike, Steve. (2008). Contaminant. *Architectural Design*, 78(6), pp. 24–29.
25. Beckett, R. (Not dated). Bioaugmented Design. [online]. Available at: http://www.richard-beckett.com. [Accessed 30 September 2019].
26. Kembel, Stephen W., Jones, Evan, Kline, Jeff, Northcutt, Dale, Stenson, Jason, Womack, Ann M., Bohannan, Brendan J.M., Brown, G.Z. and Green, Jessica (2012). Architectural design influences the diversity and structure of the built environment microbiome. *International Society of Microbial Ecology Journal*, 6, pp. 1469–1479.

27. Arundel, A.V., Sterling, E.M., Biggin, J.H. and Sterling T.D. (1986). Indirect health effects of relative humidity in indoor environments. *Environmental Health Perspec*tives, 65, pp. 351–361.
28. Kembel, Steve W., Meadow, James F., O'Connor, Timothy K., Mhuireach, Gwynne, Northcutt, Dale, Kline, Jeff, Moriyama, Maxwell, Brown, G.Z., Bohannan, Brendan J.M. and Green, Jessica L. (2014). Architectural Design drives the biogeography of indoor bacterial communities. *Public Library Of Science ONE*, 9(1), doi: e87093.
29. Stephens, Brent and Siegel, Jeffrey A. (2012). Penetration of ambient submicron particles into single-family residences and associations with building characteristics. *Indoor Air*, 22(6), pp. 501–513.
30. Prussin, Aaron J. and Marr, Linsey C. (2015). Sources of airborne microorganisms in the built environment. *Microbiome*, 3(78), doi: https://doi.org/10.1186/s40168-015-0144-z.
31. Brickner, Philip W., Vincent, Richard L., First, Melvin, Nardell, Edward, Murray, Megan, and Kauffman, Will (2003). The application of ultraviolet germicidal irradiation to control transmission of airborne disease: Bioterrorism countermeasure. *Public Health Reports*, 118(2), pp. 99–114.
32. Adams, Rachel I., Miletto, Marzia, Taylor, John W. and Bruns, Thomas D. (2003). Dispersal in microbes: Fungi in indoor air are dominated by outdoor air and show dispersal limitation at short distances. *The ISME Journal*, 7(7), pp. 1751–7362.
33. Rook, Graham A.W. (2013). Regulation of the immune system by biodiversity from the natural environment: An ecosystem service essential to health. *Proceedings of the National Academy of Sciences of the United States of America*, pp. 18360–18367.
34. von Mutius, Erika. (2016). The microbial environment and its influence on asthma prevention in early life. *Journal of Allergy and Clinical Immunology*, 137(3), pp. 680–699.
35. Wild, Christopher Paul. (2005). Complementing the Genome with an "Exposome": The Outstanding Challenge of Environmental Exposure Measurement in Molecular Epidemiology. *Cancer Epidemiology, Biomarkers and Prevention*, 14(8): 1847–1850.
36. Klevens, R.M., Edwards, J.R., Richards, C.L., Horan, T.C., Gaynes, R.P., Pollock, D.A. and Cardo, D.M. (2007). Estimating health care-associated infections and deaths in US hospitals, 2002. *Public Health Reports*, 122, pp. 160–166.
37. Turnbaugh, P.J., Ley, R., Hamady, M., Fraser-Liggett, C. Knight, R. and Gordon, J. 2007. The human microbiome project: exploring the microbial part of ourselves in a changing world. *Nature*, 449(7164), pp. 804–810.
38. Kembel, S.W., Jones, E., Kline, J., Northcutt, D., Stenson, J., Womack, A.M., Bohannan, B.J.M., Brown, G.Z. and Green, J. 2012. Architectural design influences the diversity and structure of the built environment microbiome. *International Society of Microbial Ecology Journal*, 6:1469–1479.
39. Blum, W.E.H., Zechmeister-Boltenstem, S. and Keiblinger, K.M. 2019. Does Soil Contribute to the Human Gut Microbiome? *Microorganisms*, 7, 287: doi:https://doi.org/10.3390/microorganisms7090287.
40. Potter, M.C. 1911. Electrical effects accompanying the decomposition of organic compounds. Proceedings of the Royal Society B, 571(84), pp.260–276.
41. Armstrong, R., Ferracina, S., Caldwell, G., Ieropoulos, I., Rimbu, G., Adamatzky, A., Phillips, N., De Lucrezia, D., Imhof, B., Hanczyc, M.M., Nogales, J. and Garcia, J. (2017). *Living Architecture (LIAR): Metabolically engineered building units*. In: Heisel, F; Hebel, D, (eds). "Cultivated Building Materials: Industrialized Natural Resources for Architecture and Construction". Berlin, Germany: Birkhauser, 170–177.
42. Ieropoulos, I., Pasternak, G. and Greenman, J. (2017). Urine disinfection and in situ pathogen killing using a Microbial Fuel Cell cascade system. PLOS ONE: https://doi.org/10.1371/journal.pone.0176475.
43. ALICE project website. 2021. Active Living Infrastructure: Controlled Environment: https://alice-interface.eu.
44. Latour, Bruno. (2012). Love your monsters: Why we must care for our technologies as we do our children. The *Breakthrough Institute*, Spring. [online]. Available at: https://thebreakthrough.org/journal/issue-2/love-your-monsters [Accessed 4 July 2020].

Rachel Armstrong is Professor of Experimental Architecture at the School of Architecture, Planning & Landscape, Newcastle University, UK. She was a Rising Waters II Fellow with the Robert Rauschenberg Foundation (April–May 2016), TWOTY futurist 2015, Fellow of the British Interplanetary Society, and a 2010 Senior TED Fellow. She is director and founder of the Experimental Architecture Group (EAG), whose work has been published widely as well as exhibited and performed at the Venice Art and Architecture Biennales, the Tallinn Architecture Biennale, the Trondheim Art Biennale, the Palais de Tokyo (Paris), the Institute of Advanced Architecture, Catalonia (IAAC), Aarhus Kuntshal, the University of the Underground (Amsterdam), the Gallatin School, New York University, Allenheads Contemporary Arts, and Culture Lab at Newcastle University. Rachel investigates a new approach to building materials called 'living architecture', which suggests it is possible for our buildings to share some of the properties of living systems. Collaboratively working across disciplines, she builds and develops prototypes that couple the computational properties of the natural world with matter at far from equilibrium. She calls the synthesis that occurs between these systems and their inhabitants "living" architecture. She is a coordinator for the €3.2 million Living Architecture project, which is an ongoing collaboration of experts. Rachel's work is widely published in both academic and popular press. Some of the titles include *Origamy*, *Soft Living Architecture: An Alternative View of Bio-informed Practice*, and *Liquid Life: On Non-linear Materiality.*

High Comfort – Low Impact: Integration of Thermal Mass in Pursuit of Designing Sustainable Buildings

Mona Azarbayjani and David Jacob Thaddeus

1 Introduction

Thermal mass (TM) is effective in improving building comfort in any location that experiences daily temperature fluctuations, both in the winter and in the summer. When properly combined with passive solar design, thermal mass can play an important role in significant reductions of energy utilization in active heating and cooling systems. Much like a sponge or a thermal battery, thermal mass diverts and stores heat away from space when it is hot outside and later releases it when it is cooler, thus ensuring thermal comfort of inhabitants. Through this process, large temperature swings are minimized thus guaranteeing occupant comfort.

The use of materials with thermal mass is most advantageous in locations where there is an appreciable difference in outdoor temperatures from day to night (or where nighttime temperatures are at least 10 degrees cooler than the thermostat set point) [1].

Thermal mass is the ability of a material to absorb, store, and later release heat energy. Extensive heat energy is required to change the temperature of high-density thermal mass materials such as water, concrete, bricks, and tiles. The appropriate use of thermal mass can make a big difference to comfort and sustainability in regard to heating and cooling consumption. When used appropriately, the thermal

M. Azarbayjani (✉)
University of North Carolina at Charlotte, Charlotte, NC, USA
e-mail: mazarbay@uncc.edu

D. J. Thaddeus
University of North Carolina at Charlotte, Charlotte, NC, USA

School of Architecture, College of Arts and Architecture, UNC Charlotte, Charlotte, NC, USA
e-mail: thaddeus@uncc.edu

A. Sayigh (ed.), *Achieving Building Comfort by Natural Means*, Innovative Renewable Energy, https://doi.org/10.1007/978-3-031-04714-5_3

mass has the capacity to moderate internal temperatures by averaging out diurnal (day-night) extremes. This increases comfort while reducing energy costs.

For thermal mass to be effective, it must be closely coordinated with suitable passive design strategies including orientation, shading of glazing, proper ventilation, and appropriate level of insulation. The use of thermal mass may not be appropriate if a location has poor air quality and high levels of noise pollution (characteristic of many city centers).

Natural ventilation integrated with thermal mass is a passive cooling process that can also be used to adjust the indoor environment to ensure indoor thermal comfort and maintain acceptable indoor air quality. To be effective, thermal mass must be integrated with sound passive design techniques. This means having adequate areas of glazing facing proper directions with an appropriate combination of shading, ventilation, insulation, and thermal mass.

When properly integrated into the building fabric, thermal modeling tools demonstrate that there are significant benefits to thermal mass in all climates. Furthermore, whole building savings range from 3 to 11% and the associated energy cost savings range from 2 to 9%. Energy codes in the United States recognize the contributions of thermal mass to passive strategies by reducing the amount of insulation required compared to construction in the same climate that has little or no thermal mass [2].

This chapter aims to advance the knowledge of integrated thermal mass leading to better human comfort and more sustainable building design. As a passive design strategy, thermal mass is an ancient practice that is currently experiencing a considerable resurgence.

From sacred to profane, this chapter will also describe a legacy of excellence that is recognized by old and new applications of thermal mass. This legacy will be affirmed by investigating case studies from around the world and also by exploring advances in nanotechnology that are applied to phase change materials (PCMs).

2 Looking Back

Since the dawn of civilization, man has built shelter from the environment with bounties gifted to his survival from the environment itself. In this chapter, we will survey what was born out of a quest for survival and an adaptation to human needs in a rainbow of climate conditions. Although the term thermal mass (TM) is relatively new, the concept and its application in buildings are not. Based on availability and climate appropriateness, ancient civilizations used dense and massive natural materials such as earth and stone for shelter. In addition to timber, these materials were prevalent in construction until the second industrial revolution introduced mass production processes in man-made materials such as brick, concrete, and steel. This popularized the use of machine-processed building materials rather than simple and handmade materials. This naturally ushered in modern architecture in steel and glass instead of traditional massive construction. With the vast amounts of glass in modern buildings, the use of thermal mass often becomes impractical.

Evolving from hunters and gatherers to herdsmen and from cave dwellings to nomadic tents, man sought to put down roots in more permanent and stable settlements instead of remaining in an ephemeral and temporal existence. This resulted in self-built small shelters constructed from readily available materials that could be molded by hand. With an abundance of clay, sand, water, and sun, adobe sun-dried bricks were the natural approach to constructing family shelters.

Adobe construction was popular in Mesopotamia, the Middle East, and North Africa. Among others, larger historic structures in adobe construction include the *Great Mosque of Djenne* in Mali and the *Arc e Bam* in Iran which dates back to 500 BC. In addition, evidence of ancient adobe structures may be found in Peru and other South American countries. Also, in the hot and arid American Southwest, the indigenous Pueblo people built magnificent structures in *Taos Pueblo* and other areas of New Mexico and Arizona. The Native American tribes created the most remarkable dwellings in the caves at Mesa Verde in Colorado, where the dwellings take advantage of the overhead rocks for a heat sink effect and the massive walls serve as thermal mass.

Cave dwellings in Mesa Verde National Park. CO. (Image credit: David Thaddeus)

Djenne Mosque. Mali. (Credit: Baron Reznik. Creative Commons. Flickr)

San Miguel Church 161–1628. Santa Fe. NM. (Image Credit: Shiny Things, CC BY 2.0)

3 Definition of Thermal Mass

Thermal mass is the property of a building material that facilitates the absorption, storage, and release of heat energy into an adjacent space. Thermal mass can be used year-round, although its greatest efficacy is demonstrated during the summer and winter months, as it serves as a storage of heat and cold, allowing the material to slowly heat or cool the space relative to seasonal need. The heat from the sun during daylight hours is conserved by the thermal mass then released at night to heat interior spaces in the winter months, thus maintaining a consistent air temperature. The exact process that occurs in the winter months is reversed in the summer months, where hot storage reverts to cool storage.

For passive solar heating design, the most common materials used are brick, concrete, masonry, tiles, rammed earth, or adobe. These materials are typically installed within interior walls or floors. Since water possesses a very high thermal capacity, it too can serve as an excellent thermal mass when placed in containers that are then used to construct walls. Additional materials and elements such as phase change materials (PCMs) or water, among others, may also be used as thermal massing. As their name suggests, PCMs store and release heat through the dynamic change of the material's state (i.e., solid to liquid or vice versa). Here, thermal storage capacity is dependent on the conductivity, specific heat, and density of the PCM. In the northern hemisphere, the most efficient orientation for a thermal mass in a wall is facing south, which provides maximum thermal performance throughout the year, regardless of the season.

The heating load itself can be the result of several factors including external loads such as direct solar radiation, or internal loads, such as occupancy, lighting, and electric equipment. Since internal loads cause the surrounding air to heat up, the primary benefit of using thermal mass is to delay the release of heat. Thermal mass absorbs heat without getting hot itself by acting as a conduit for the discharge of heat [2]. Additional advantages of thermal mass include reducing greenhouse gas emissions by avoiding peak electrical and air-conditioning demand.

Thermal mass materials act as heat sinks in warm periods and as heat sources during cold periods. High thermal mass materials reduce temperature fluctuations while maintaining comfortable interior temperatures and without consuming much energy [3]. In some climates, shading a thermal mass with overhanging eaves may be necessary to prevent it from overheating.

At this point, it is important to emphasize the difference between thermal mass and insulation. First and foremost, insulation *resists* heat and prevents its propagation into the interior of a space. Thermal mass, on the other hand, *conducts* heat by absorbing it and later releasing it. This fundamental difference demands that in order to function properly, thermal mass be placed *inside* the insulated thermal envelope and is never a replacement for insulation. Additionally, to serve its purpose, thermal mass must be left exposed to freely exchange heat with the surrounding interior air. If a concrete floor, for example, is to serve as thermal mass, it must not be covered with a nonconductive material such as carpet nor be covered by a raised floor. In contrast, stone, floor tiles, and other conductive materials assist in absorbing and later releasing heat [4].

In conformance with general sustainable practices, it is more beneficial if a thermal mass also serves another purpose such as supporting structural loads. Serving both purposes simultaneously save resources and expenses and are better for the environment.

4 An Essential Glossary of Thermal Mass Terms

The terms and definitions listed below are fundamental to a better understanding of the principles of thermodynamics that regulate the behavior of thermal mass.

Specific Heat Capacity Heat Capacity is the ability of a material to absorb and store heat. It is the energy required to raise and store the temperature of a material by one degree. The *specific* heat capacity is the ability to store heat energy per unit of volume for a certain material; in other words, it is the amount of energy needed to raise the temperature of 1 pound (1 kg) of the material by 1 degree Fahrenheit (1 degree Kelvin). As the temperature rises, so does the heat capacity of a material and thus it can store more heat. Heat capacity is measured in BTU/°F (J/°K).

Thermal Conductivity Is the measure of the efficacy of thermal mass to absorb and transfer stored heat. Conductivity is described as the amount of heat transferred in time through a material mass as a result of a temperature difference. It is the quantity of heat transmitted in a unit of time through a thickness of a certain material or the rate of heat transfer through a set thickness and due to a thermal differential. Thermal conductivity is measured in BTU/Hr.-FT-°F (W/m-°K).

Density It is the amount of mass or weight in a certain volume. In the context of thermal mass, density varies with temperature. Increasing the temperature of a substance makes it expand and thus decreases its density since heat increases its volume. Denser and heavier materials can store more heat (heat capacity).

Thermal Admittance It is the ability of a material to exchange heat with its surrounding under cyclical temperature variations (day/night fluctuations). Thermal admittance is measured in BTU/FT^2–°F–Hr. (W/m^2–°K).

Thermal Lag (Time Lag) It is the time delay for heat to be conducted through a material thickness. A material with high heat capacity and low conductivity will have a high thermal lag. Time lag is measured in hours.

Thermal Mass It is the ability of a material to absorb and dissipate heat based on temperature changes. The thermal mass of a material depends on its thickness and density, its thermal capacity and conductivity, and the texture and color of its surface [5].

To be considered suitable and efficient in the passive design of a building, thermal mass materials must **concurrently** maintain three properties, namely:

A. High specific heat capacity to maximize the amount of heat stored for the amount of mass
B. High density since heavier materials have more mass to store heat
C. An appropriate thermal conductivity that corresponds to the heating and cooling cycles of a building (thermal lag)

Good thermal mass materials will have a high product of (capacity x density x conductivity) [6].

Materials in which all of these properties exist simultaneously include water, earth materials (adobe, mud brick, rammed earth, stone, etc.), brick, and, of course, concrete. In contrast, wood has a high specific heat capacity but low thermal conductivity, whereas steel can store heat but has very high thermal conductivity and would lose that heat very rapidly.

In addition to material properties, it is appropriate to use thermal mass in certain occupancies and not in others. In institutional, commercial, and academic institutions, for example, the building use is predominantly in the daytime hours. This allows for night flushing when the building is unoccupied in preparation for another cycle of heat storage and release during the next day. In contrast, the use of thermal mass in hospitals may not be beneficial due to continuous occupancy and the difficult prospect of night ventilation to cool the thermal mass. It is essential that the heat storage and release cycle be "synchronized" with the building use.

Some materials that serve as good (and bad) thermal mass are discussed below.

In general, ideal thermal mass materials are dense and heavy.

Water Having the greatest volumetric heat capacity, water is one of the best thermal mass materials although incorporating it inside the insulated envelope in a discrete fashion may present a challenge. Any container of water must be made of a material that maximizes heat transfer into and out of the volume of water. There are great advantages to subjecting the water directly to solar exposure. With the high heat absorption rate of water and through convection currents and conduction, heat is transferred uniformly to the entire mass. It is significant to note that water has twice the heat storage capacity of concrete. Of course, leakage from containers of water is always a possible liability.

Concrete With twice the density of water, concrete has half the volumetric heat capacity.

Concrete is the second most-consumed substance in the world, second only to water [7].

Still, concrete is the foremost application of thermal mass for many reasons. It is ubiquitous and is among the best thermal mass materials since it is also capable of supporting structural loads while remaining cost-competitive. Unlike water, concrete has low heat conductance to the entire mass. Although substitutes for cement such as fly ash improve its eco-friendliness, the carbon emissions from the production of cement remain an eco-concern. Concrete mixes that substitute a portion of the cement with fly ash for applications such as autoclaved aerated concrete (AAC) has approximately a quarter of the volumetric heat capacity [1].

Rammed Earth, Adobe, and Masonry Earth materials represent the most ancient construction techniques that were developed as a response to hot arid climates. Rammed earth and adobe construction specifically are not appropriate for humid climates. In this type of construction, thick walls provide plenty of thermal mass and thermal lag to moderate interior temperature swings. Walls would have to be very thick if they are to support roof or floor loads. The earth materials are readily available, noncombustible, and have very low embodied energy. With similar thermal properties and advantages to rammed earth and adobe, masonry has much higher embodied energy.

Steel Although steel is approximately three times as dense as concrete and eight times as water, it is not efficient as thermal mass because its conductivity is too high which causes it to lose heat rapidly.

Wood Although wood has a high specific heat capacity, it has low thermal conductivity because of its moisture content. In general, wood products are slow to absorb heat. Mass timber, on the other hand, provides thermal mass with low embodied energy compared to concrete or masonry.

5 Phase Change Materials (PCM)

The development of emergent technologies is becoming more economical and is able to efficiently store considerable amounts of heat and cold within a limited volume and for an extended period of time. The advent of PCMs offers a lightweight, low mass substitute to traditionally heavy thermal mass materials. This makes the structure and foundation systems much lighter and thus more sustainable. This may offset the upfront cost of using this new technology. In addition, using PCMs such as paraffin encapsulated into gypsum wallboards presents lightweight construction with the ability to benefit from the advantages of thermal mass. PCMs can store three times the heat that can be stocked in water and nine times that in concrete. There is also research on the microencapsulation of PCMs in concrete mixes and photovoltaics. In changing phase from solid to liquid at constant temperature and pressure, paraffin is most efficient in storing *latent* heat. All other materials mentioned above store *sensible* heat.

The introduction of PCMs as an effective strategy for building energy conservation makes thermal storage play an increasingly important role. This is further augmented by the deployment of latent heat storage (LHS) technologies into building materials. The combination of LHS into PCMs is thus a very desirable solution for energy conservation due to its high storage density and low-temperature swing. Effective installation of PCMs within thermal mass components of buildings such as walls, ceilings, or floors greatly enhances thermal energy storage capacity. They can then either capture thermal energy through naturally occurring convection or by direct solar gain. The goal is to achieve maximum thermal comfort for building occupants by increasing the thermal storage capacity in the material, resulting in reduced indoor temperature swings and a more regulated, sustained, and desirable temperature for an extended period of time.

The demand for air conditioning has increased tremendously over the last few decades, thus large demands for electricity and limited fossil fuel reserves have led to an uptick in interest in efficient-energy solutions. Electric energy consumption varies drastically during the day and night hours based on the demand by commercial, residential, and industrial activities. A passive cooling strategy that harnesses thermal storage in phase change materials has proven to be an effective method for improving the thermal stability of rooms that are built within a lightweight envelope.

PCMs are environmentally friendly materials that include "benign salts or organic compounds which store and release latent heat energy by changing chemical bonds through a phase transformation" [8]. This process is antithetical to sensible heat storage materials, such as masonry or water, which alter a structure physically. At first, solid to liquid PCMs perform similar to typical latent heat storage materials; that is, their temperature increases while heat is absorbed. However, when PCMs achieve their melting point (or phase change temperature), they are able to absorb large amounts of heat without becoming hotter. Then, when the ambient temperature is reduced, the PCM is solidified, thus releasing the latent heat energy that was stored. In other words, PCMs are able to absorb and release latent heat energy while maintaining a relatively regulated, constant temperature.

PCMs are recognized as a possible option for the reduction of cooling and heating loads directly into the structure, thereby reducing the overall demand for energy. That said, incorporating PCMs directly into some materials may threaten their structural integrity. Accordingly, in order to avoid compromising the building integrity and structural performance, but rather to improve it, microencapsulated PCMs can be incorporated into the materials. By integrating microencapsulated PCM directly into the materials, the problem of leakage is resolved and a solid bond between the materials is achieved, which ensures enhanced structural material performance [9].

According to Madhumati et al., "Micro-encapsulation makes it possible to integrate PCMs into conventional building materials." Innovative phase change material products can be used in both new and existing buildings toward the improvement of their thermal qualities by adding to the amount of heat-storing mass [10].

The research demonstrates the enormous potential benefit of integrating PCMs into typical building materials in order to improve the regulation of temperature and to provide thermal mass to both new and existing buildings. This will allow lightweight buildings to have advanced thermal properties much like their higher-mass counterparts and further bring cost savings in the form of only the energy necessary to regulate comfortable indoor conditions [9].

As noted previously, thermal mass is an important factor in the passive solar design of buildings. It can best be described as a material's ability to resist temperature change as heat is added or removed. Building thermal mass into structures helps balance temperatures that would otherwise be fluctuating across diurnal cycles [10].

Frequently, lightweight materials with a considerably smaller thermal mass are used in modern structures. For this reason, the search has ramped up for a solution that will provide increased thermal mass without the undesired effect of subsequently increasing the structural loads. Accordingly, PCMs are proving to be an ideal choice to solve this problem due to their latent heat storage, because it offers high-capacity thermal storage in a narrower interval of temperature. Latent heat is that which is absorbed by a material or radiated during a phase change at constant pressure and temperature. PCMs are deployable in two particular areas of a structure. Firstly, they can be used in passive systems, for example, as energy storage in passive solar and heating or in cold storage for nighttime passive cooling [10]. Secondly, they can be used in areas dedicated to active systems, for example, heating and air conditioning where energy storage can be utilized, i.e., to "take advantage of off-peak electricity prices."

When comparing high- versus low-mass materials, low-mass materials have proven to be less capable of reaching daytime comfort conditions since they provide less time lag. Materials that are lightweight and low in density do not require very much heat to increase their temperature but they lose heat rapidly. Thus, they are called low thermal mass materials. According to one study, even where a building is continuously ventilated, thermal mass still plays an important role in reducing the maximum indoor temperature [11]. Subsequently, buildings with a high mass that are designed with an effective ventilation system can reach comfortable conditions even in a hot, humid environment. The natural cooling potential can be achieved by employing a thermal mass that is shaded during the daytime while exposed to a cooling breeze by night. In hot, humid regions with moderate diurnal ranges, the well-designed, well-shaded thermal mass has been shown to reduce night temperatures by 3–4 °C.

The two most common categories of PCMs are organic and inorganic. The majority of organic PCMs, like paraffin waxes, are noncorrosive and chemically stable. Furthermore, they demonstrate little to no supercooling properties which means that they do not have to be cooled to below their freezing point in order to begin crystallization. They are compatible with the majority of available building materials and they possess a high latent heat per unit volume and are recyclable. Inorganic compounds, like salt hydrates, demonstrate a much higher latent heat per unit volume, higher thermal conductivity, are nonflammable, and are lower in cost when compared with organic compounds. Inorganics may also be recycled after their life cycle is complete. One important drawback, however, is that they are corrosive toward most metals and are susceptible to decomposition and "supercooling," which can negatively impact their phase change properties.

PCMs serve the particular purpose of acting as a sort of bridge between the points where energy is needed and when it is available. One way of looking at PCMs is to see them as a thin version of mass. Where construction sites are difficult to build on, the use of phase change materials can be a successful alternative. PCM as thermal mass should only be added to those zones where overheating is rampant.

With regard to the implementation of PCM within buildings that operate at low temperatures near the desired ambient air temperature, we are only interested in solid-liquid and solid-solid PCMs. As noted, PCMs can be incorporated into walls, ceilings, and facade elements. Technological advancements and research have pushed forward studies in new areas like "integration of PCM with walls, gypsum wallboard, microencapsulation in concrete mix, thermal solar storage, and thermal control of photovoltaics" [10].

Research by Feldman and Hawes have contemplated methods of integration of PCM into the structure through "direct incorporation, immersion, and encapsulation" [12]. Richardson and Woods have concluded that construction materials built with PCMs can be much thinner and lighter than the same thermal mass without PCMs [13]. According to research by Castell et al., "incorporated macro-encapsulated PCMs in the typical Mediterranean brick construction system," measured savings of more than 15% of total energy consumption [14]. Building materials that incorporate PCMs can be integrated using the same construction methods and equipment used for typical building materials.

6 Thermal Lag

Thermal lag is described as the time required for heat to transfer from the outside surface of a building to the inside surface. The time lag of a material is influenced by thermal mass combinations such as their thickness and sequence, material layers, and thermophysical properties [15]. Thermal lag is further dependent on fluctuations of the ambient temperature throughout the day. It becomes much more difficult to regulate indoor temperature conditions under extreme thermal fluctuations.

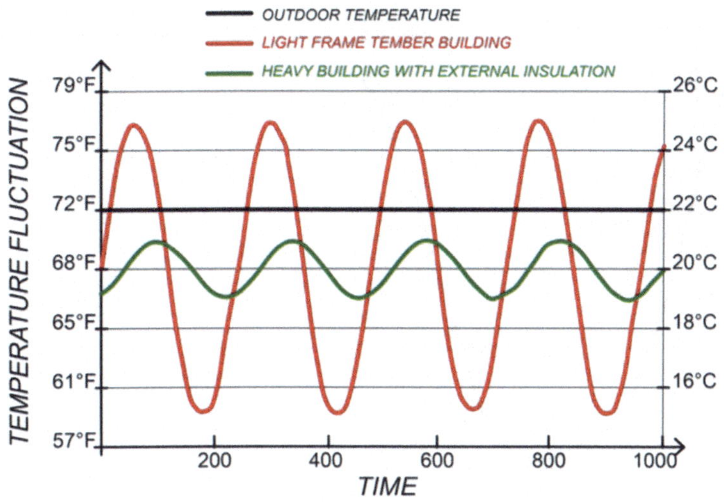

INFLUENCE OF THERMAL MASS

During warm weather conditions, thermal mass can absorb heat gained from sunlight. This will make the interior space more comfortable and greatly reduce the cooling demand and cost of air conditioning. Thermal mass is most beneficial in climates where there is a large fluctuation between the daytime and nighttime ambient temperatures. As a building cools the stored heat energy is then released during the night into the interior space thus reducing the heating demand. In some climates, the overnight temperature drop may be enhanced by radiant or evaporative cooling. In areas with high nighttime temperatures, thermal mass can still be utilized, and the building must then be ventilated at night with cooler night air to exhaust the stored heat energy [16].

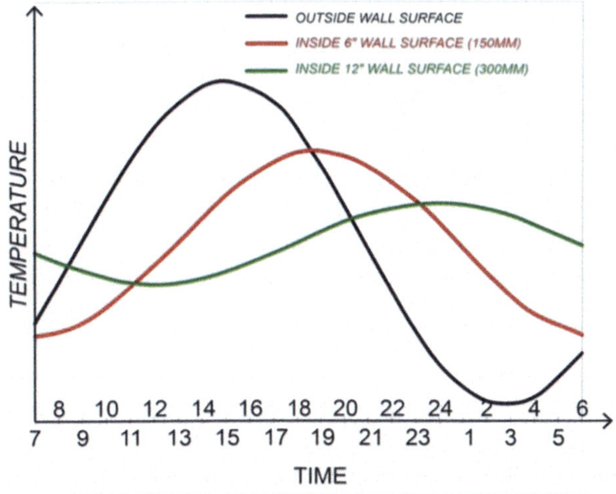

WALL SURFACE TEMPERATURE FLUCTUATION

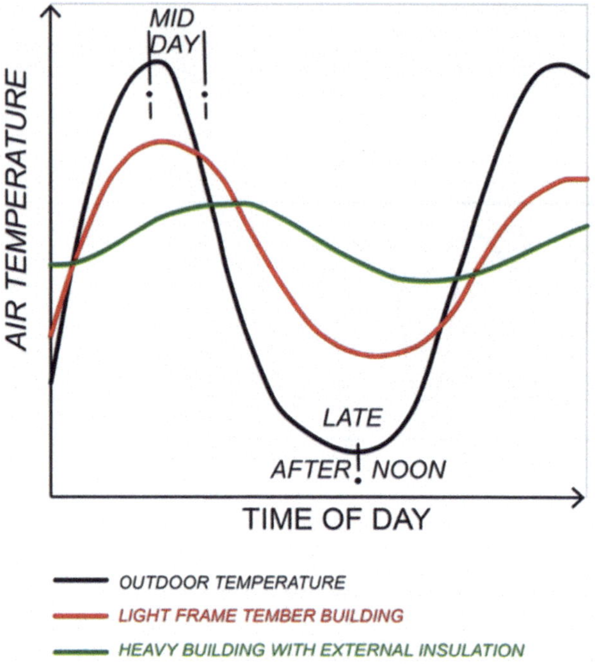

7 Where to Locate Thermal Mass

Thermal mass can be placed either internally or externally. External thermal masses typically consist of building envelopes such as the roof and exterior walls. "They often integrate heavy materials from the building's structure and have a significant thermal inertia" [17]. The material functions by essentially connecting the indoor and outdoor environments since it is exposed on both sides. The indoor thermal mass may include all of the indoor components that are not part of the building envelope. Examples of this type of thermal mass include furniture, partitions, and other accessories and finishes.

The location of the thermal mass is determined by, and dependent on, whether or not the building's energy consumption is dictated by heating or cooling demand. In order to determine the optimum positioning of the thermal mass, we must establish whether the greatest energy consumption is a result of winter heating or summer cooling.

Where heating the building is the primary concern, the thermal mass is located in areas that receive the most direct sunlight or radiant heat. If it is determined that both cooling and heating are to be equally prioritized, the thermal mass is localized in the interior of the building and preferably on the ground floor. Notably, the floor tends to be the most economical solution for locating a thermal mass. The ideal location for both summer and winter efficiency is to locate the mass toward the south of the structure for solar heat gain and in an orientation that allows it to catch any breeze for cooling.

Where the need for cooling is the primary concern, it is paramount that the thermal mass is protected from the sun through the use of shadings and insulation while simultaneously ensuring that cool breezes can pass unobstructed over the mass to remove the stored heat.

8 Climate Appropriateness

It is important to note that not all materials can serve as effective thermal mass. Furthermore, material selection will vary depending on the region and specific climate conditions as well as the particular location within the building.

In diurnal climates, there is a drastic temperature difference between day and night. Throughout the night, when the outside temperature drops considerably, the heat from the walls radiates out into the surrounding air helping to keep the interior spaces from getting too cold, then allowing the walls to absorb heat again the next day.

In humid climates where the temperature tends to remain more constant over a 24-h period, thermal mass, such as concrete, is able to successfully absorb low-angle winter sunlight while being insulated from heat loss. In such cases, additional methods of heating and cooling are needed to ensure an appropriate temperature range. For example, a house for a young couple designed to comply with the US Solar Decathlon requirements must be able to regulate thermal control within each room. Thus, by employing geopolymer concrete on the south-facing facade of the house, the inherent properties of dense materials are taken advantage of in order to provide well-regulated rooms that work with the environment.

Thermal mass is primarily used for storing heat in the daytime (during higher temperatures) to be released at night (during lower temperatures) resulting in a decrease in the effect of the high temperature by day on the structure and the subsequent temperature drop in the evening. Therefore, thermal mass as a passive strategy is most appropriate where outdoor climate conditions are warmer by day than the desired indoor temperature while considerably cooler in the evening than the desired indoor temperature.

According to research, a region's climate has the most impact on the efficiency of a building's thermal mass. For example, the temperate climate in California has shown a significant reduction in overall cooling load. Thus, the use of thermal mass could potentially reduce the cooling load by up to 82% in Los Angeles and up to 100% in San Francisco; however, many improvements in building performance have been noted in less temperate, more humid, or arid climates such as Hong Kong, Las Vegas, and New Zealand [18].

A region's climate influences thermal mass efficiency through different factors such as air temperature, solar radiation, wind speed, relative humidity, and temperature fluctuations. "Thermal mass benefit has the most impact when daily outdoor temperature variations are above and below the balance point of a building. Hence, cold climates benefit most in the summer season and hot climates in the winter season" [19].

8.1 Mild, Marine Climate

Use of walls with low diffusivity is beneficial for mild climates. The potential savings annually falls in the range of 22% for a standard wall. Furthermore, a wall with a lower diffusivity and a high conductivity can instead be exchanged for a wall with high diffusivity and low conductivity as a method of achieving a similar level of energy consumption. That is to say, there is a dual benefit to using walls with low diffusivity at lower conductivities. The greatest conservation of energy occurs during the summer. According to numerical simulations, the annual percentage reduction in sensible heat and cooling loads has been demonstrated to be as significant as 40% in more mild climates as a result of the thermal mass levels of the walls (here, floors and foundations are presumed to bear zero mass).

Importantly, high interior air and wall temperature variation is restricted by high mass buildings and thus are able to maintain a steadier overall thermal condition. The result is increased comfort, especially in the milder seasons (spring and fall), during drastic air-temperature swings (i.e., solar heat gain), and in areas with significant swings in temperature. In Southern European countries, traditional architecture has effectively demonstrated the aforementioned positive outcomes.

8.2 Hot, Dry Climate

According to research, the overall benefit of thermal mass falls in the range of 4.9% annually for a typical wall built in a hot, dry climate. Furthermore, a key aspect of reducing energy consumption is lowering conductivities. However, it is important to note that the impact of thermal mass benefit is reduced by lower conductivity. In this case, the greatest energy savings is observed in the wintertime.

Thermal mass serves as an effective passive strategy for indoor climate temperature control, particularly in hot, dry climates. Thermal mass can potentially improve extreme indoor variation. It may also serve as an efficient solution for protection against sudden weather events like those that are common in deserts and becoming more frequently observed in nearby regions. It is important that thermal mass be distributed evenly throughout a structure to ensure the facilitation of an indoor microclimate that is well balanced. On the more extreme side, solutions like excess thermal mass and Trombe walls ought only be considered with utmost skepticism, and if installed, they ought to be designed in such a way that would allow their isolation when needed.

In consideration of the aforementioned climate conditions, a pivotal method for controlling and improving a building's microclimate, and thus lowering its energy demand and consumption (which weighs heavily on infrastructure), is its capacity to store energy. Indoor maxima can be reduced by 35–45% of respective outdoor maxima by using high thermal capacity in shaded and insulated buildings when said buildings are unventilated [11]. Therefore, thermal mass arguably ensures the most beneficial and, oftentimes, the only solution when considering the potential for power outages caused by storms or extreme seasonal demand on the power grid. Similarly, when such outages caused by increased demand occur in the wintertime, thermal mass is able to take high indoor temperatures and store the heat [20]. As the global climate becomes increasingly unpredictable, the key ability of thermal mass to adapt is gaining importance and relevance, particularly where disparate hot and cold spells occur. Thus, it becomes considerably more difficult to provide and maintain conditioning demands in lightweight structures [21].

8.3 *Hot, Humid Climate*

Hot, humid climates demonstrate similar attributes to hot, dry climates; however, the overall impact of the thermal mass benefit is reduced. Here, the thermal mass benefit falls in the range of 3.1% annually for an average wall. When the thermal mass is added to all four sides of a building envelope, the indoor temperature profiles showed dramatic change. The addition of thermal mass managed to reduce the indoor maxima. Simultaneously, thermal mass increases indoor minima and the influence on maxima is greater than on minima. Secondly, thermal mass also influences and has a direct effect on time lag, where the heat storage properties of thermal mass have the ability to delay the occurrence of peak indoor temperature for several hours. Since the impact of thermal mass is two-way, its particular application is dependent on the circumstances in which it is installed [22].

According to the research of Y. V. Perez and I.G. Capeluto [23], simply placing the thermal mass of the roof on the inside and the insulation on the outside for a south-facing classroom produces a savings of 7 kW h/m2 in annual energy consumption as compared to insulation on the inside and thermal mass on the outside. Similar results were produced in varying orientations. The results of yet another study indicate that an optimization of thermal mass and window size coupled with the use of evening ventilation together reduces peak indoor air temperature in warm, humid climates and improves overall thermal comfort [24].

8.4 *Cold Climate*

The principal mechanism for decreasing energy consumption in a cold climate is the conductivity of an equivalent wall. When monitored annually, the benefit from the thermal mass is negligible in cold climates. The annual benefit falls in the range of

1.5% per year for a typical wall. As mentioned, the thermal mass benefit is entirely dependent on seasonal variation in temperature, where the greatest impact is observed in the summertime.

Results from additional research conclude that "while increased thermal mass does have advantages in all climates, such as a decrease in summer overheating, it is not an effective strategy for decreasing annual heat demand in typical residential buildings in Alaska." It should be noted here that increased thermal mass did indeed improve the level of comfort for occupants within the house as measured by predicted mean vote (PMV) and further triggered a decrease in summer temperature fluctuations. However, in Fairbanks, and Juneau, in Alaska, the variation in heat demand between a thick thermal mass and a thin wall was negligible, reported at less than 1% when compared to a 20% reduction in heat demand in Clayton, New Mexico. These findings corroborate those of previous studies which indicate that thermal mass is clearly not a worthwhile method of reducing heating demand annually in residential construction in cold climates. That said, further energy savings might be accomplished by the inclusion of passive solar design strategies. However, the aforementioned research only addresses the use of thermal mass in typical structures which do not employ passive solar design. For the typical home in a cold climate, the analysis demonstrates that increased thermal mass does not reduce heat demand annually by a significant margin [25].

9 Thermal Mass Impact on Comfort

Thermal mass is one of the passive solutions deployed for improving thermal comfort within buildings while simultaneously decreasing energy consumption. The studies show that buildings with larger amounts of thermal mass and thermal inertia help prevent the overheating of the buildings, which will improve overall thermal comfort [18]. In addition, reducing fluctuations in temperature throughout the day is another contributing factor to the thermal comfort in a building with high thermal inertia. Providing thermal comfort for a building with thermal mass requires a heating control system for allowing some degree of temperature swing. Therefore, a combination of thermal mass and a building management system could provide the best results, both economically and environmentally [26].

As discussed, thermal mass is a beneficial passive strategy for both summer and winter. During the summer, relieving heat stress by reducing the high temperature can also help reduce the energy consumption due to cooling. Similarly, in the wintertime, the peak heating demand decreases. In the summertime, thermal mass materials such as brick and concrete absorb the high air temperature that in turn decreases the cooling load in the daytime and regulates the night temperature by releasing the heat when the indoor temperature drops.

10 Rules of Thumb

Significant amounts of sunlight (i.e., heat gain) must be admitted into a space during the daytime and stored within that same space to be released during the cooler evening hours.

To best store a significant portion of the heat gained, construct walls, floors, and/or ceilings of masonry (concrete, brick, concrete block, adobe, etc.):

- A minimum of 10 cm (4 inches) in thickness.
- An exposed surface area to solar glazing area ratio of 3:1 to 9:1.

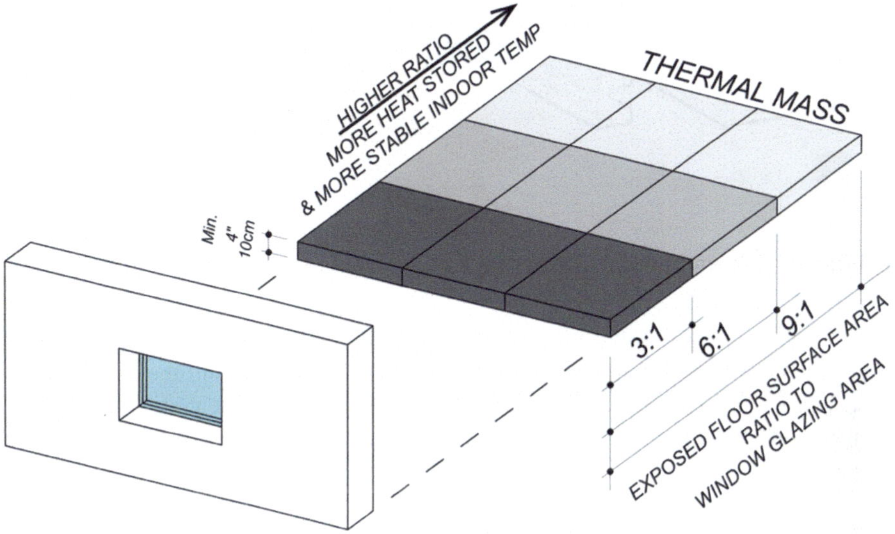

Floor Area to Glazing Area Ratio

The higher the ratio, the more heat stored and the more stable the indoor temperature becomes.

Ensure that ceilings and lightweight construction are kept a light color. Walls can be any color while floors must be a medium to dark color. For exterior masonry walls, locate the insulation on the exterior side of the wall so the masonry is exposed to the interior.

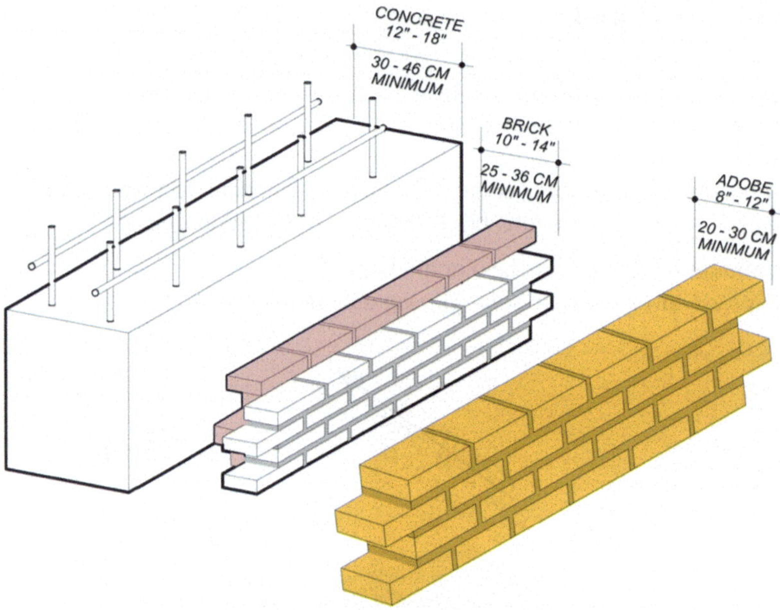

Minumum Recommended Thickness for Concrete and Masonry Thermal Mass

A Thermal storage wall is typically constructed in masonry and is located behind south-facing glazing (in the Northern Hemisphere). The wall and the south-facing glazing are sized as a percentage of the floor area that is to be heated. Two different climate conditions are illustrated below: cold (20 °F –30 °F / –6.7 °C to –1.1 °C) and temperate climates (35 °F –45 °F/2.2 °C –7.2 °C).

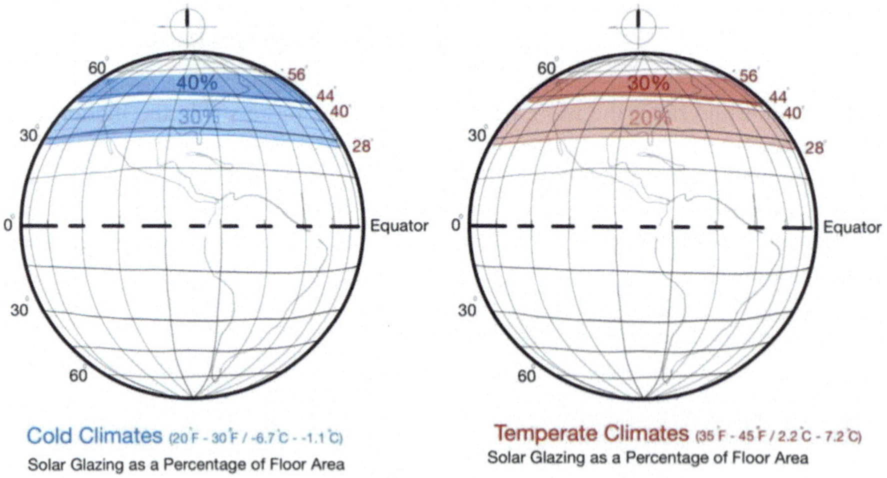

11 Looking Forward

In conclusion, the concept and use of thermal mass is a time-proven lesson that has evolved over centuries and that will continue to provide thermal comfort far into the future. In his dedication to the book *The Timeless Way of Building* [27], Christopher Alexander writes "To you mind of no mind in whom the timeless way was born." Thermal mass is one of those timeless ways of inhabiting the planet and living on it in harmony with the environment.

11.1 Case Studies

UrbanEden Solar Decathlon 2013- UNC-Charlotte. Charlotte, North Carolina.

UrbanEden is a net-zero, solar-powered prototype house designed for the climatic conditions of Charlotte, North Carolina in the USA, and consists of four integrated interior modules, each with a corresponding exterior component. In response to Charlotte's temperate climate and the lifestyles of southern living, UrbanEden increases the living space with a series of connected indoor and outdoor rooms, which create a versatile environment that can be adapted to multiple uses, allowing the home to be small but feel big. The vertical garden on the south side of the porch provides privacy and natural beauty. The house's canopy shelter at the entrance creates an inviting entry to the home while also providing a space for casual gathering.

Upon entering the house one finds a multipurpose living space, kitchen, and dining areas awash with natural light, with a lovely view to the exterior garden through the south-facing glass wall. Urban living defines the design, both interior and exterior. Emphasis is on adaptability and versatility, doing more with less. At 822 FT^2 (76.4 M^2), the interior of the home comfortably accommodates a single occupant or a couple, and by utilizing the outdoor space, the home doubles to 1644 FT^2 (153 M^2).

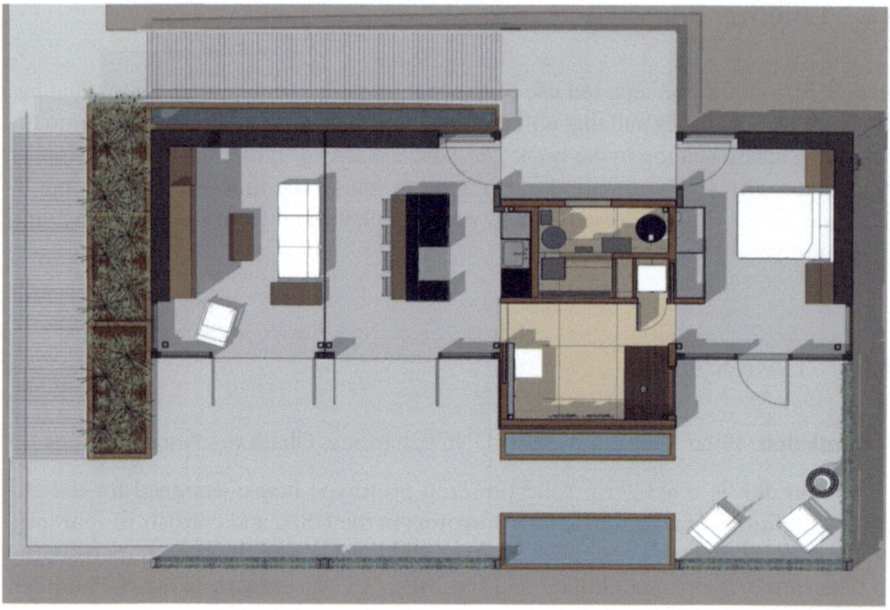

Photo Credit: Jason Flake / US Department of Energy Solar Decathlon

11.1.1 Modular

UrbanEden is designed to match the rhythms of urban living, maximizing customization and adaptability through a modular design. The house includes four basic modules—living room, kitchen, wet core (bathroom and mechanical room), and bedroom. Homeowners can upgrade each of these modules or add additional modules linearly. Each of these modules has their independent function with the possibility of rearrangement, mass production, and endless opportunities to reconfigure the house. As shown in Fig. 1, the modules are all lined up with the exception of one, which sets back and creates a welcoming entrance to the house.

The modularity of the house allows the owner(s) to make several decisions that cater to their needs and thus leading to an endless number of potential design configurations. The owner starts by choosing the number of bedroom and living modules, then chooses in which order to arrange the volumes in a linear scheme, and finally chooses which modules to shift forward or backward. Arrangement possibilities include four, five, and six modules. Adding an additional bedroom or living module makes the home large enough for families or for those who desire additional space. In order to sift out the designs that meet the objectives of the UrbanEden house, three essential rules can be employed:

1. The kitchen module and the living room module should be placed next to each other and not be separated by any walls to ensure an open floor plan.
2. The kitchen module should be placed next to the core module where mechanical and plumbing functions are housed for ease of connection.
3. If there are two-bedroom modules in a certain design, they should be placed on opposite sides of the house so that hallways do not cut through them.

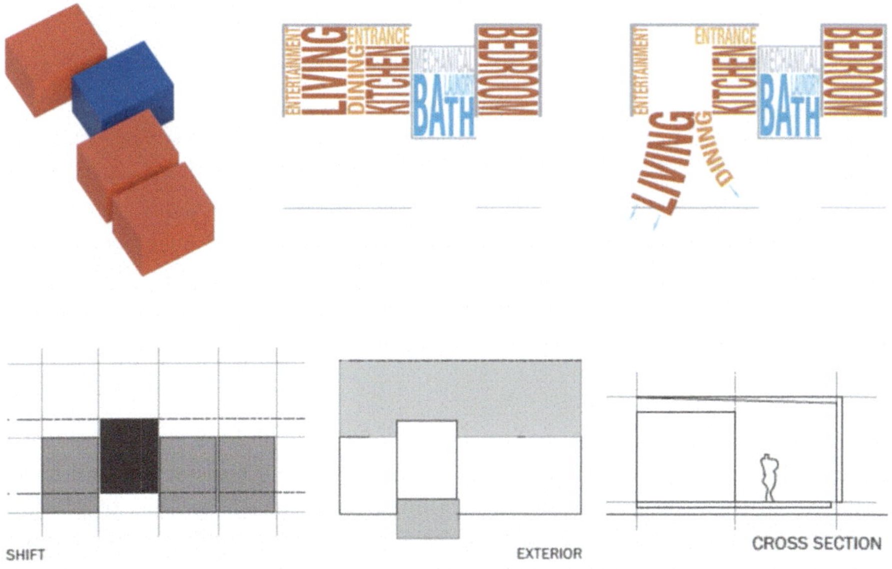

Fig. 1 Solar Decathlon House Diagrams

Photo Credit: Jason Flake / US Department of Energy Solar Decathlon

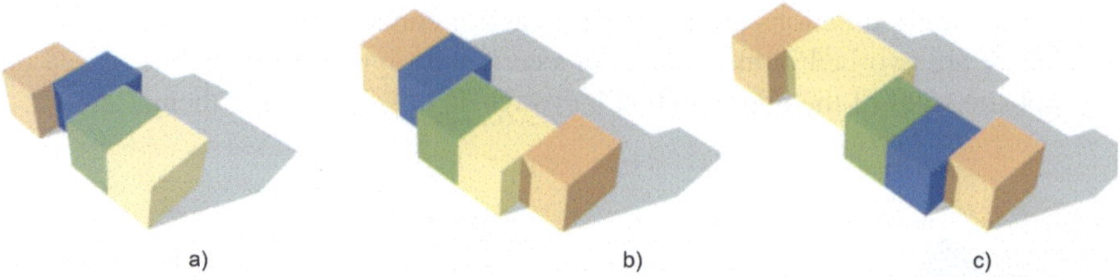

Fig. 2 Examples of house plans with more modules. (**a**) 4-module arrangement—1 bedroom, 1 living room, 1 kitchen, 1 core. (**b**) 5-module arrangement—2 bedrooms, 1 living room, 1 kitchen, 1 core. (**c**) 6-module arrangement—2 bedrooms, 2 living rooms, 1 kitchen, 1 core

Even when adhering to these rules, the number of possible arrangements is enormous. The calculations for the module numbers below do not include mirrored or flipped plan scenarios (4 modules = 46 arrangements, 5 modules = 112 arrangements) (Fig. 2).

11.1.2 Wall System

UrbanEden's envelope was designed to minimize the heat transfer through well-insulated, air-tight construction, while maximizing heat storage capacity by placing considerable mass in the living space. UrbanEden's precast concrete walls consist of 6-inch styrofoam insulation sandwiched between two layers (or wythes) of geopolymer concrete. Unlike post-and-beam or metal/wood stud construction systems, there are no columns in the wall to interrupt the insulation.

Traditional Portland cement concrete is responsible for 5–8% of the collective worldwide carbon footprint since manufacturing a ton of cement takes about 6.5 billion BTUs of energy. By completely eliminating Portland cement from its composition, the geopolymer cement concrete used in UrbanEden becomes a much greener product, reducing both energy consumption and carbon dioxide emissions in its production while recycling a waste product (fly ash) to beneficial use. The geopolymer concrete completely replaces Portland cement with an advanced fly ash mix that not only yields a stronger material but also makes safe use of a waste product of coal combustion. This process lowers the carbon footprint of concrete by up to 90%. Geopolymer cement concrete is made by reacting aluminate and silicate bearing materials with a caustic activator. Commonly found waste materials such as coal combustion fly ash or slag from iron and metal production are used in the manufacture of geopolymer cement.

Photo Credit: Jason Flake / US Department of Energy Solar Decathlon

Another passive strategy incorporated into UrbanEden's design is thermal mass that has moved to the walls in the form of fully insulated, precast panels instead of just being on the floor in order to increase the surface area. By markedly increasing the area and related volume of the thermal mass in the passive solar context, we were able to implement a hybrid passive/active hydronic cooling system that, unlike conventional hydronics, uses only pump energy to accomplish temperature changes. UrbanEden's high-mass geopolymer cement concrete walls provide a container in which to store large quantities of thermal energy. In the summer ambient heat soaks

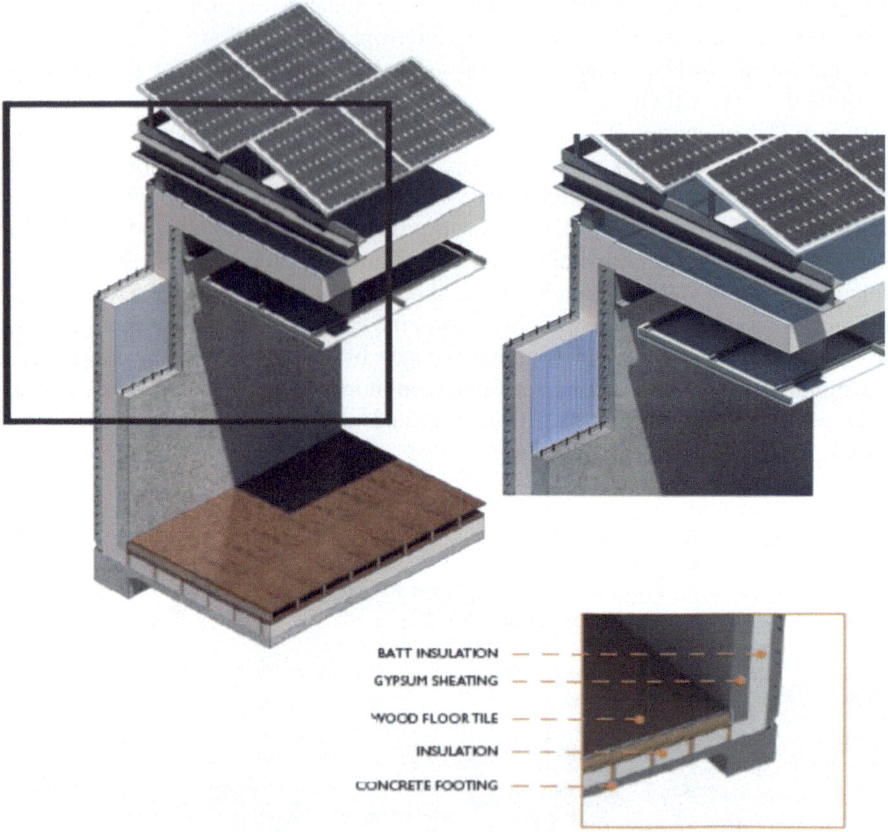

Fig. 3 Wall section

into the thick walls throughout the day primarily by means of radiation as shown in Fig. 3. At the end of each day, the stored thermal energy can be released from the walls by an innovative hybrid system. Hydronic or capillary tube mats embedded within concrete walls provide a vessel for the circulation of a water-based working fluid throughout the building's envelope, which flushes out the heat accrued during daylight hours. Once captured within the water, the thermal energy extracted from the walls must be dissipated. Heat dissipation is made possible by a network of custom-fabricated, rooftop-mounted copper-fin tube heat exchangers, designed to serve as radiation emitters. After the sun sets each evening, the water in the wall tubes is pumped onto the roof of the house and circulated through the heat-exchanger network. Passing through the serpentine heat-exchanger network, the warm water releases its stored thermal energy into the cool night sky. In fact, the exchange of radiant heat to the night sky is so efficient, the temperature of the fluid on the discharge side of the heat-exchanger network could potentially drop below that of ambient air (Fig. 4).

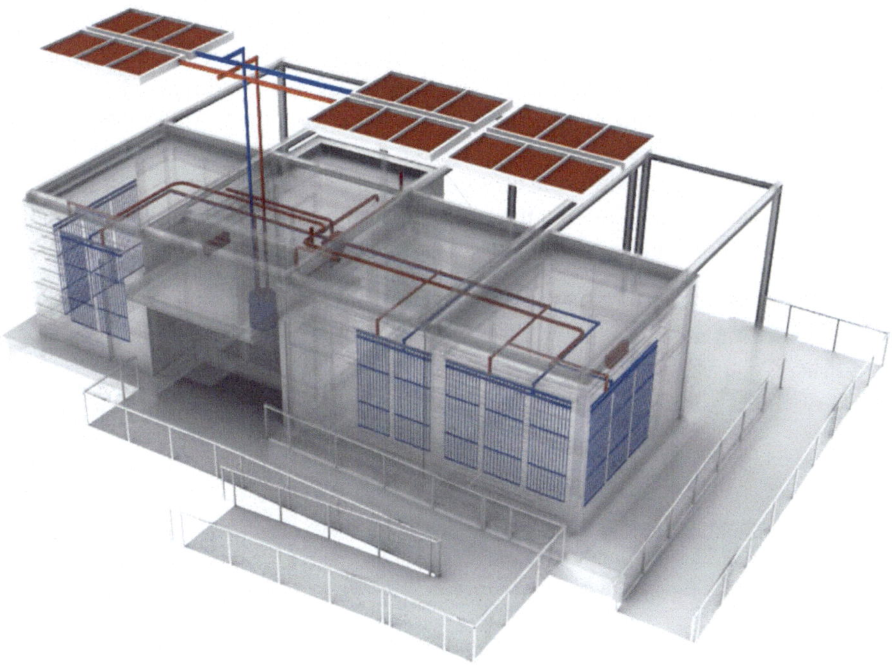

Fig. 4 System integration

11.1.3 Dynamic Thermal Property Calculator

Design Builder has a time lag built into it. The graphs shown below are the findings of the potential thermal time lags of the 4″ (10.2 cm), 5″ (12.7 cm), and 6″ (15.25 cm) interior geopolymer walls. Each test was over a 24-h prior and an internal surface resistance of 0.13 m^2 K/W, which is a default for ISO 6946. For the calculations, we used an exposed dense concrete with the following constants:

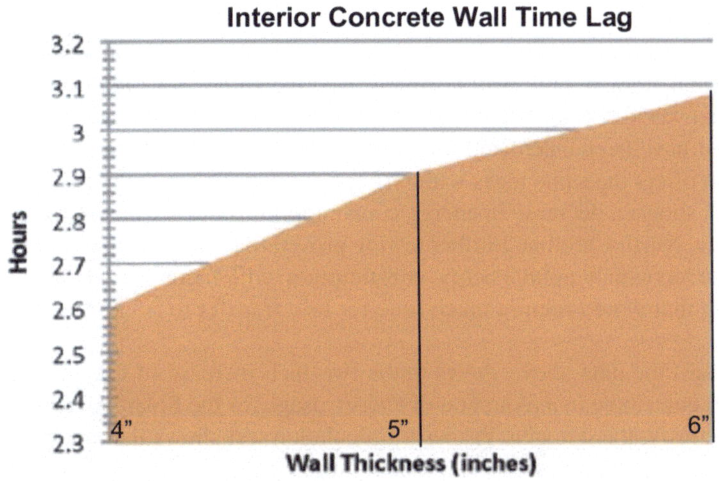

The results of the analysis shows that a 4″ thick concrete wall will have a time lag of 2.6 h, a 5″ thick concrete wall will have a time lag of 2.9 h, and a 6″ thick concrete wall will have a 3.08-h time lag. There was a 10.4% increase in time lag between the 4″ and 5″ thick walls while there was a 5.9% increase from the 5″ wall to the 6″ wall. There was a 15.6% increase in time lag between the 4″ and 6″ concrete walls.

The following graphs show the heat flow over a 24-h period for each of the three wall thicknesses. Each shows the internal temperature of the room in relation to the internal and external surface heat flow over the 24-h time period.

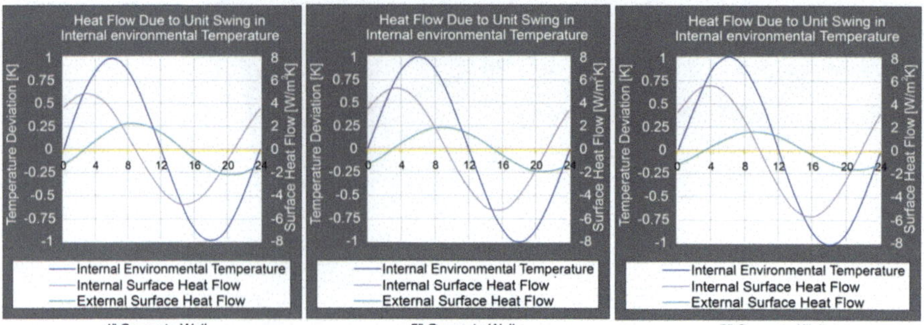

4" Concrete Wall 5" Concrete Wall 6" Concrete Wall

11.1.4 Analysis

The thicker wall required less energy usage by the heat generation and cooling loads.

12 Energy Consumption

Our focus in this study was to see if by increasing the interior concrete wall thickness, energy consumption would be reduced, we then ran calculations to see how the time lag was affected by the thickness change and if it was significant enough to affect the energy consumption. While the study proved that insulated wall systems work well as thermal storage, built wall testing proved the simulations not to be as accurate as the real world resting, requiring extra cooling energy to be used.

The results that have been calculated for the hot humid climate have consistent information. Each offers the same ideas with respective information.

Our simulation shows a decrease in energy consumption from four inches to six inches in Charlotte, North Carolina. Further testing proved that there is a maximum depth of concrete at which point energy consumption will increase due to an increased time lag that would require much time for heat transfer to occur over the 24-h period.

For our purposes, the data above shows that a two-inch increase of the interior concrete wall will contribute to a reduction of energy usage for the Solar Decathlon House. Simultaneously, the dynamic Property Calculation test shows that the two-inch increase in concrete thickness is a significant jump for the time lag. Combined this data shows that a 6″ thick interior concrete would be more beneficial than a four- or five-inch interior concrete wall due to an increased time lag that would require too much time for heat transfer to occur over the 24-h period.

13 Lycee Schorge Secondary School/Kéré Architecture

Photo Credit: Iwan Baan

- Location: Ouagadougou, Burkina Faso
- Heating degree days: 0
- Cooling degree days: 3242
- Solar radiation: 5372 w/m^2 (average daily)
- Annual precipitation: 1000 mm
- Building type: secondary school
- Area: 11,660 sqm (built area)
- Client: Stern Stewart Institute & Friends
- Design team: Kéré Architecture
- Completion: 2016

In Africa, designers and architects are taking the lead in an effort to rethink the use of concrete in construction, with particular consideration given to its ecological footprint and structural inequity. Additionally, research on materials that are locally sourced, readily available, easily constructed and maintained, and more effectively ventilated, cooled resulting in better energy efficiency [28]. Certain uses for these materials are founded on common traditions as demonstrated in the design of Lycée Schorge Secondary School designed by Francis Kéré in Burkina Faso [28]. The Lycée Schorge Secondary School, built in Burkina Faso's capital and most populated city, Ouagadougou, highlights an architectural precedent for locally sourced materials in an inventive and modern way while also serving as a center for education [29]. Along with the strategic use of locally sourced materials, other efficient methods like the use of cooling towers and innovative ceiling structures render this building effective for quick response to problems and other potential challenges while keeping costs low.

Nine modules form the plan and overall geometry of the school. The modules support a system of classrooms and administration zones; one of which houses a dental clinic that provides dental health services to students. The architect selected a local stone called laterite stone for the module walls. This stone is easily cut and formed into bricks after being quarried. The stones harden when given time above ground and proper exposure to the atmosphere [29]. "It has a high thermal mass, plus a beautiful color and a porous, textured surface," says the architect.

Serving as a secondary facade for the school is another system of wooden elements beyond the modules' walls. This facade is built from locally sourced and fast-growing wood that provides the structure with shaded, semi-open spaces between the classroom walls and the courtyard. From the outside, the secondary facade is perceived to be transparent in nature; however, in reality, it consists of a system of wooden screens [29]. These screens serve the dual purpose of securing protection for the classrooms against wind and dust while simultaneously providing spaces for students to engage in learning activities while waiting for their classes [29]. A unique screen is deployed to cover the outside of the school. It is made from a fast-growing tree that is typically used for scaffolding or firewood. As previously mentioned, there are several reasons that warrant its use based on sustainable practices. "I wanted to give this material a more powerful and sustainable purpose," says

Kéré when discussing the enclosure, which is used to frame the walkways and patios [28].

Contrarily, beyond reasons of sustainability for selecting locally sourced materials to build the school, economical considerations were also considered. The need for expensive transportation can be avoided by having access to several local materials. This method has proven less harmful to the environment locally and may also provide an obvious solution for furniture making. Furniture inside the classrooms is constructed from locally sourced hardwoods and other leftover materials like steel scraps [29].

A courtyard in the center of the school plays an important role in both educational and sustainable objectives. It serves as a common gathering space for the school. However, the various environmental impacts of this project can also be enumerated.

According to studies, the average size of a courtyard in a particular region is determined by latitude [30]. Here the courtyard is thin enough to allow for a shaded area during the heat of summer days, yet wide enough to ensure sufficient penetration of sunlight during the wintertime. Furthermore, the courtyard can function as a zone of security, privacy, and comfort. A pleasant environment and beautiful setting are ensured by landscaping the courtyard with trees, flowers, and shrubs [30]. The students of the Lycée Schorge Secondary School are exposed regularly to these serene features and may well serve to enhance learning outcomes.

The project also managed to make good use of daylight and maximize its potential. This on top of the passive heating and cooling elements that have been successfully incorporated into the earlier design phases.

As previously discussed, the secondary facade built from locally sourced wood has several gaps in it to allow sunlight to reach through to the walkways. However, the interior spaces of the classrooms are treated differently with masonry facades punctuated by tall, unglazed windows covered with colorful metallic shutters that ensure copious amounts of light and air to pass through the classrooms [28].

The architect takes advantage of a unique system of classroom ceilings to maximize natural light with the ability to penetrate into the interior areas. The architect deployed one of his "signature design strategies" when it came to designing the corrugated steel roof [26]. Circulation of natural air is aided by a clever system of trusses creating a gap between the walls of the modules and the ceilings. With the use of trusses made from rebar, the architect elevated the roof above the building envelope, which allows hot air to escape more easily.

The materials used for the ceilings are constructed from a type of perforated plaster. With the selection of this material, indirect sunlight is diffused by the ceilings of the module while radiation is blocked [29]. The architect further takes advantage of the light-reflecting properties of white-colored materials which make it an effective reflective material for interior spaces, thus allowing the diffusion of light.

Above the roof, 8 ft. (2.5 M) high concrete wind towers were designed to capture fresh air and send it downward, thus improving the indoor climate [26]. A wind-powered, renewable energy source is deployed to ensure successful ventilation via wind catchers. Wind towers serve as a potent solution for augmenting the temperature in a living space by using the thermal energy of wind, coupled with natural convection along with naturally occurring temperature variations.

These undulating ceilings are built from gypsum fiberboard, cement mortar, and straw. Lastly, cream-colored surfaces with long slits enable classrooms to breathe and release heat while allowing diffused daylight.

The general consensus is that the school has had a positive impact. "I cried when I saw it," said Pertle, who noted that buildings of lesser quality would not have been as successful. "The landmark shows our students that we want them to work hard, dream big, and achieve great goals" [30]. As a whole, the project takes advantage of local resources to offset costs of construction and maintenance. Although constructed on a low budget, it manages to serve as an effective monument, which demonstrates to students how creative design and innovation make big goals that much more achievable.

14 Turtle Sanctuary at Kalba Mangrove Reserve, Sharjah, United Arab Emirates (UAE)

Photo Credit: Marc Goodwin

Photo Credit: Marc Goodwin.

Photo Credit: Marc Goodwin

- Location: Kalba Mangrove Reserve, Sharjah, United Arab Emirates (UAE)
- Heating degree days: 137
- Cooling degree days: 3507
- Solar radiation: 2285 kWh/m2 (Annual Average)
- Annual precipitation: 106.9 mm
- Building type: aquarium/research center
- Area: n/a
- Client: Sharjah Environment and Protected Areas Authority (EPAA)
- Design team: Hopkins Architects
- Completion: 2021

Khor Kalba Turtle and Wildlife Sanctuary is located on a nature reserve in the Persian Gulf (one of the most fragile and diverse such reserves) that serves as a center for the rehabilitation of turtles, care of endangered birds, and collaboration with local governments [32]. At a certain point in the building construction, a method of prefabrication is deployed to reduce the overall effect on the environment and existing landscape. Passive design strategies such as sealed shells as thermal mass are able to cope in the warm climate of the region as well as the challenges brought on by close proximity to water.

Photo Credit: Marc Goodwin.

Photo Credit: Marc Goodwin

An important aspect of the project's initiative is that Sharjah's Environmental Protected Authority (EPAA) will help raise environmental awareness and provide education and visitor facilities for such programs. The important work done by the EPAA will be enhanced by a research and monitoring facility dedicated to the protected Kalba Reserve [33].

Photo Credit: Marc Goodwin

As visitors approach the Reserve, they are greeted by an inspired, semi-enclosed ribbed pod which acts as a space for orientations. It boasts glass window openings oriented toward key vistas. A color palette of light, beachy tones is used to soften the interior which is further lit by skylight oculi [32].

Seven interconnected pods and tensile structures allow guests to explore a visitor center offering panoramic views on the mangrove forest and mountains in the distance. This method minimizes the negative impact of construction on the site by using prefabricated parts. The center boasts exhibition areas, amenities for visitors, offices for staff, veterinary facilities, classrooms, gift shops, a cafe, and aquariums. "A carefully set out nature trail encourages visitors to explore the reserve's rich biodiversity of indigenous mangrove forests and mudflats and the species it supports including turtles, stingrays, gazelles, and the rare Arabian Collared Kingfisher" [32].

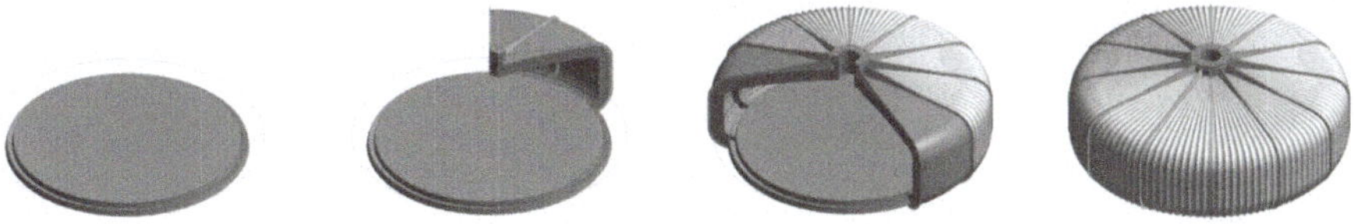

Image Credit: Marc Goodwin

Photo Credit: Marc Goodwin

"The geometry of the pods is inspired by urchin exoskeletons and purposefully echoes those of the Buhais Geology Museum, with which the Sanctuary is paired" [32]. Due to the tidal location and to protect the pods from the tide, simple concrete foundations of robust, elevated discs are used—their design helps to keep the pods from disturbing the adjacent terrain.

> The modular buildings invite the landscape into the spaces, using framed panoramic views out and rays of natural light from above. The pods are clad with segments of white scalloped precast concrete referencing the shells found on the local shoreline and creating subtle variations of light and texture. An array of steel ribs accentuates the sculptural cantilevered forms and completes this robust cladding system, itself designed to withstand the site's unforgiving coastal conditions [33].

Photo Credit: Marc Goodwin

Photo Credit: Marc Goodwin

Passive design methodologies were applied during the construction phase to ensure interior spaces are well protected from the intense desert heat and to reduce overall levels of energy consumption [32]. Precast concrete shells, ribs, and in-site foundation discs across the pod floor provide a thorough thermal massing method.

An insulating membrane and a waterproof membrane are combined within the cladding cavity.

In summary, the primary catalyst behind building a project like this is in consideration of environmental concerns and to strive toward the protection of the ecosystem of turtles in their natural environment. The method of construction of the shelter also served to achieve this primary goal. The construction process itself was inclusive of sustainable methodologies. Prefabricated elements were selected in response to the particular nature of the project. The construction methods deployed here are environmentally friendly and able to help block radiation and are thus more sustainable than traditional construction methods.

15 Las Vegas Spring Preserve: Rammed Earth as Thermal Mass

A Leadership in Energy and Environmental Design (LEED) Platinum building in the middle of the Las Vegas desert; no, not a mirage; an impossible feat for sure though. The complex includes a visitor center, a learning center, and a broad range of gardens. For more than 15,000 years, springs broke through the desert floor to create grassy meadows, an oasis on the Old Spanish Trail. This 180-acre site is the birthplace of Las Vegas (Spanish for meadows) and is protected by the National Register of Historic Places since 1978 [40]. The Las Vegas Springs Preserve has won many awards since its opening including the Best Green Project in 2007 and continues to lead by example.

There are many ways in which sustainability was incorporated into this project. Through the use of rammed earth as thermal mass, coupled with natural ventilation, the façade allows for proper solar intake in the winter and keeps interior space cool in the summer. Sustainability was first and foremost in the initial stages of design and until the very end. The many strategies employed coalesce to produce much more than the sum of the individual approaches. The focus in this study will mostly be on the benefits of deploying a thermal mass strategy in the harsh environment of the Mojave Desert in Nevada.

The Desert Living Center (DLC) is an action-based public outreach and applied research facility that serves as a catalyst for individual and community change from being "In the desert" to being "of the desert." It comprises a complex of five LEED Platinum-rated buildings. Exhibits illustrate the benefits of recycling, conservation, and alternative energy sources through an interactive environment of galleries, design labs, and training centers where the architecture and construction techniques of the buildings also double up as a learning platform. The interpretive educational storylines are carried throughout the facility with items ranging from the 43 hands-on interactive environmental exhibits, 20 static sustainable design principles exhibit integrated into the buildings, two temporary galleries, four classrooms, a dialogue center, library/research center, design lab, and a technical training studio [41].

This project has all the sustainability bells and whistles including the use of such materials as straw bales, soda cans, recycled glass, shredded blue jeans for insulation, and much more. Rammed earth coupled with cooling towers feature prominently in the façade as they secure a cool daytime interior environment.

The Desert Living Center at the Las Vegas Springs Preserve

Designed by Lucchesi Galati Architects of Las Vegas, the 46,000 FT^2 (4274 m^2) Desert Living Center (DLC) is an environmental education campus that champions and teaches sustainable building practices, celebrates recycled materials, and technologies specific to life in the harsh desert environment [42]. With engineering consultant LERA (Leslie E. Robertson Associates, NYC), several of the green strategies and materials were put to the test when the complex opened in 2007. This includes one of the first uses of self-consolidating concrete (SCC) which minimizes any damage to the formwork and is sustainable since the concrete does not need any vibration [44].

Even though technology changes very rapidly, the DLC continues to be a beacon for climate-sensitive design for its application of time-proven methods of surviving in the desert and on the planet in general. For example, the orientation of the DLC optimizes solar daylighting and heat. Passive technologies for the Desert Living Center are incorporated throughout the facility but are not the primary focus of the design. In living up to sustainability ideals, the roof of the DLC was built with recycled railroad trusses. Furthermore, the building does not use any active mechanical systems since it relies entirely on passive strategies to maintain the thermal comfort of visitors.

Shaded parking lot with tracking photovoltaics

The Las Vegas Springs Preserve project is a showcase for high-tech implementation of low-tech principles. One such example is in the parking lot that displays high-tech (for its time at least) tracking photovoltaic cells to generate power while simultaneously shading the cars below it with the aid of a light-reflecting shade canopy.

Next to the rammed earth envelope, the predominant feature of the Las Vegas Springs Preserve is the conservation and collection of precious and scarce rainwater to augment the irrigation of gardens in the Preserve meadow [42]. This is most evident in the butterfly roofscapes and water collection cisterns that are prevalent in the architecture. Also, garden exhibits are used to further Mojave Desert education, as are the constructed wetlands for the treatment of all gray and black water for the entire Preserve site. The treated water is reused in the DLC toilets and gardens which is yet another example of how the building function is on display for education and awareness for visitors and Las Vegas alike.

Butterfly roof and collection cistern

Starting from the conceptual design itself, the DLC is intended to fit with the local environment, beginning with solar orientation and continuing down to earth by utilizing it as a thermal insulator for the structures by integrating the buildings into the land and subsequently above ground, where the mass and thickness of walls assist in protecting from heat gain and loss. The thermal mass of rammed earth and cast-in-place concrete is most suitable in the dry arid desert climate.

Additionally, through the use of straw bale construction and shredded blue jeans, the buildings on the campus are able to achieve very high insulation values, which protects the indoor environment from the intense Las Vegas heat. The long roof overhangs protect the buildings from summer heat gain while allowing the low winter sun to warm the interior spaces [42].

Cooling towers and courtyard designs are intrinsic to reducing heating and cooling loads. In addition, a radiant heating system of hot solar water heats slabs, which in turn are cooled by the rammed earth. The technologies that are used serve as a complement to the basic sustainable design principles. A responsive façade allows windows to open automatically for natural ventilation and cooling of the rammed earth thermal mass [42].

Along with automatic windows, cooling towers aid in the night flushing of the rammed earth thermal mass

The estimated annual energy savings for the DLC alone approximate about 279,591 kWh avoiding 258 tons of carbon emissions annually. Water savings are estimated at 380,099 gallons per year and about 7748 tons of waste has been diverted from landfills [42].

Using recycled content and sourcing materials locally, construction materials are manufactured nearby and are not on the red List. Furthermore, any non-repurposed wood products used in the manufacturing process come from certified sustainable forests and rapidly renewable crops. Visitors will be able to see firsthand examples of sustainable design in the Preserve, including straw bale walls throughout all five Desert Living Center buildings and carpeted walkways made of recycled soda bottles [44]. Materials from the old trestles of the old railroad crossing Salt Lake were used in creating the support trusses for the Desert Living Center [44].

By using the earth as a thermal insulator, buildings can be protected from solar radiation by being embedded in the ground. As one of the main materials in the project, straw rice also provides noise reduction in addition to heat insulation [43].

The roof framing structure is repurposed from old railroad ties

Another initiative that was development in the Las Vegas Valley is a region-specific carbon-emissions mapping tool [39] that accurately baselines residential energy use per street and provides a comprehensive picture of local hotspots. The tool utilizes valley-wide databases of energy consumption, taking a single dwelling as its smallest unit of measurement. Thus, a bottom-up inventory of resultant carbon emissions was created. Using web-enabled geographic information systems (GIS) functionality, the tool further lends itself to mass outreach programs where consumers are able to compare their individual carbon footprint to neighborhood, city, state, and national averages. This continues to instigate a behavioral change among consumers by providing them with specific information on the effects of their individual energy use on the environment. The mapping tool is designed to delineate local, state, and national averages of electricity usage and generated carbon dioxide emissions [42].

Automatic windows and ventilation arrows "teach" the time-proven strategy of night ventilation of cooling thermal mass

There are sustainable elements in different areas and parts of the building that demonstrate that sustainability was incorporated into design plans from the very beginning, which is another way to say sustainability was prioritized. In response to climate challenges, the Las Vegas Springs Preserve has introduced several new sustainable strategies.

The Las Vegas region has two places that one can refer to as "places" with regard to a sociological perspective, which is intangible and involves a sense of belonging. Las Vegas is not just the mythical tourist heaven but speaks to the reality of the working class, families, and persons living outside of the Las Vegas strip. As the Preserve continues to gain foothold and popularity as a nationally recognized symbol of sustainable design, it continues to act as a bridging ground for this disconnect [42].

"The Springs Preserve was developed to create a sense of place, a sense of identity for Las Vegas," said director Francis N. Beland. The Visitor Center, where guests begin their tour of this unique facility, and the Desert Learning Center both achieve that goal. The first of its kind, the Preserve is designed to set new standards in "green building" [44]. Although the Preserve is from 2007, it serves as an inspiration for projects and technologies that followed, including the prefabricated rammed earth panels of the Morocco Pavilion at the EXPO 2020 in Dubai, UAE.

The Las Vegas Springs Preserve has earned its stellar reputation and its LEED Platinum status. It is of the earth and is just as timeless.

Gabion walls assembled with local stones provide screening of work areas. (Credit for Las Vegas Springs Preserve Photos: David Thaddeus)

16 The British Pavilion at Seville: World EXPO 1992

Photo Credit: Jo Reid & John Peck. By permission of Nicholas Grimshaw and Partners

The British Pavilion at the 1992 World Expo in Seville, Spain, by architect Nicholas Grimshaw and Partners (NGP) was a tour de force of sustainable building design and construction. The sustainability movement was still in its infancy and the design for this pavilion was well ahead of its time. Designed specifically for the hot and arid climate of Seville, the design showcased environmental strategies and technologies that were and continue to be an inspiration to many architects and designers around the world.

The "92 Expo was intended to commemorate the 500th anniversary of Christopher Columbus" discovery of the New World. It was held on the Island of Cartuja on the Guadalquivir River at Seville. This was the location from which Columbus set sail to discover the Americas [51].

The official theme of the last Expo of the twentieth Century was "Discovery." The unofficial theme, though, was "Man's troubled relationship with climate and ecology of the planet" and is the one that more accurately reflected global concerns at the end of the twentieth Century [47]. Not unlike the present, it was the troubling realization that the planet's energy resources are finite and are rapidly being depleted after the second industrial revolution.

Grimshaw's approach to the design of the British Pavilion in his commitment to creating a comfortable environment for an estimated 12,000 visitors/day without wasting precious energy and creating unnecessary pollution. In the spirit of the age and the high-tech movement, Grimshaw strongly believed that modernist architecture is rooted in expressing function and technology instead of mere form. "The British Pavilion is unashamedly Modern and a product of industrial technology." In the end, it is "an expression of function and the character of the building is derived from climate-moderating technology" [47].

The challenge for Grimshaw was to create a cool environment for visitors with minimal energy consumption. Seville is the hottest and driest city in continental Europe where daytime temperatures could reach as high as 115 °F (46 °C) and as low as 40 °F (5 °C) on the same night [48]. As a result, The British Pavilion had to respond to Seville's typically hot climate with huge diurnal changes in temperature. "Its simple rectilinear façades were clad with a range of materials to moderate diurnal change and extreme temperatures" [49].

Grimshaw was inspired by the local vernacular building traditions of Andalucía, ones that specifically addressed the climate of Seville. The dryness of the air causes large swings in temperature and low humidity allows major heat gain. Windows were small and well-shaded. In addition, any opportunity for evaporative cooling by incorporating fountains and courtyards is common practice in the region. Heavy thermal mass in the form of stone was prevalent to passively mitigate the afternoon and evening scorching heat. Any good thermal mass has high heat capacity and will act as a thermal flywheel to moderate internal air temperatures despite rising heat outside.

Having the highest heat capacity and ideal thermal lag, the architect selected water as the ultimate thermal mass for the pavilion. "Grimshaw's competition entry proposed that the exhibits should be interpretations of a unifying theme: WATER!" [47]. His approach was to take full advantage of the cooling properties of water. The pavilion incorporates running water, evaporative cooling, and water as thermal mass. The pavilion is one of the "earliest structures to use water in a wide variety of ways for passive cooling" [50].

The entire envelope of the pavilion is conceived as a passive climate moderator since energy conservation and the recyclability of building components were forefront in Grimshaw's mind. As with most high-tech architects, Grimshaw conceived of a building that would be prefabricated with minimal on-site assembly and without the need for the on-location skilled craft. For this reason, the building was conceived as a kit-of-parts with pinned joints that would facilitate its design for disassembly (DfD). The prefabrication of the pavilion had to be tailored to accommodate the climate and context of Seville. With all the structure, details, and technological devices expressed on the envelope, "the architect was sharing all the information about the building installations and structure with the visitor" [46].

To enter the pavilion, visitors had to walk over a moat and through the "water wall" on the east façade that provided a cool and refreshing mist (this was way before Diller and Scofidio's Blur project!). The water in the half-in and half-out pool cools visitors and reduces the heat on the east-facing curtain wall and is recirculated using submerged pumps that are powered by the photovoltaic cells on the roof. Using the *heat* of the sun to *cool* the air inside the pavilion is brilliant. "… but in an exhibition pavilion such as this, the poetic idea of a cooling device directly powered by the heat of the sun is as important as the actual measurable performance." Air conditioning was limited to pods that were suspended from the structure to house the exhibits [47]. "Patterns forming in the falling water gave the glass a dynamic, translucent quality which enhanced views in both directions. In engineering terms, the water also limited the maximum glass temperature to around 24 °C (75 °F so that it remained below skin temperature, thus contributing to visitor comfort" [52].

The Water Wall on the East façade with overhanging fabric sails that provide shading to the water wall and roof. (Photo credit : Richard Bryant. By permission of Nicholas Grimshaw and Partners)

The most inspiring innovation on this project graces the west façade where the loading docks were located. Here water is used in a most unconventional manner and represents the main reason we elected to include this project as a case study in this chapter. On this façade, Grimshaw deployed a "shield" [49] of 24 water-filled metal shipping containers filled with water to a height of 49.2′ (15 M) [48]. The shipping containers were painted white to reflect the afternoon sun and minimize the heat gain and were fitted with a butyl rubber lining for waterproofing. The thermal mass of approximately 1.2 million BTU/°F kept the indoor environment cool without the use of any air-conditioning in the main volume. According to engineers at the consultant firm Arup & Associates, "... the inside surface stayed at a constant 86 °F (30 °C) while the exposed face experienced a 77 °F (25 °C) temperature swing between day and night extremes" [48]. This allowed the heat from the west wall to radiate into the cool night sky. As in Samuel Taylor Coleridge's The Rime of the Ancient Mariner, "Water, Water, everywhere, nor any drop to drink." Not quite the case here as seven of the shipping containers stored potable water.

In contrast to the heavy west façade, the north and south façades were made of elegant and ultra-lightweight PVC-coated polyester fabric sails that provided filtered soft light. The roof was also covered with photovoltaics and fabric that provided shading and reduced the heat gain on the roof. In addition, the roof housed photovoltaic cells that also provided shading to the roof while generating power to run the pool pumps. The cells were elevated off the plane of the roof to allow air to flow freely and cool both the cells and the roof. Another climate-specific feature of the roof was the bold decision to omit roof drains based on annual rainfall rates for the region. Up to 3.2″ (80 mm) of rainwater was allowed to accumulate and would provide additional evaporative cooling of the roof surface [47].

The British Pavilion in Seville of 1992 remains an inspiration to many architects and designers as it served as a bell-weather for sustainable design and implementing passive strategies that abound today. "Ultimately, 11 illustrates that there is no inherent conflict between high technology and 'green' issues and in any future use this fundamental statement should not be lost" [52]. The reuse of the building was successful as the pavilion was disassembled and shipped back to Britain for use as a museum. The Asian Sky TV corporation purchased it from the British Department of Trade and Industry (DTI) for £750,000 but the costs grew exponentially with shipping and storage costs. The pavilion sat with each component labeled in containers for a long time. Mr. Sharad Patel of the proposed Asian Sky and owner of the building said: "Mr. Grimshaw visited the site twice. He is excited about it. At least we were able to show him that it was not going to turn into a curry house" [53].

The ultimate fate of the building remains unknown but does not take away from the genius and the elegance of the work.

References

1. "Thermal mass – Wikipedia." [Online]. Available: https://en.wikipedia.org/wiki/Thermal_mass. [Accessed: 21-Jul-2021].
2. Medgar Marceau, "Understanding Thermal Mass I Architect Magazine," architect magazine, 2009. [Online]. Available: https://www.architectmagazine.com/technology/understanding-thermal-mass_o. [Accessed: 21-Jul-2021].
3. P. Shafigh, I. Asadi, and N. B. Mahyuddin, "Concrete as a thermal mass material for building applications – A review," J. Build. Eng., vol. 19, pp. 14–25, Sep. 2018, doi: https://doi.org/10.1016/J.JOBE.2018.04.021.
4. N. Lechner, "Heating, Cooling, Lighting," Wiley, vol. 1, no. 4, p. 53, 2015.
5. "Thermal mass in buildings - Designing Buildings Wiki." [Online]. Available: https://www.designingbuildings.co.uk/wiki/Thermal_mass_in_buildings. [Accessed: 21-Jul-2021].
6. J. D. Balcomb, W. Wray, R. W. Jones, E. D. Best, and C. Peck, "PASSIVE SOLAR HEATING ANALYSIS ADDITIONAL SLR CORRELATIONS," no. November 1987.
7. ["Eco-Friendly Alternatives To Traditional Concrete I Specify Concrete." [Online]. Available: https://www.specifyconcrete.org/blog/eco-friendly-alternatives-to-traditional-concrete. [Accessed: 21-Jul-2021].
8. [P. Sheppard and P. Sheppard, "Phase Change Mate r i a l s – O ve r v i e w Actions Towards Sustainable Outcomes," no. August, pp. 1–4, 2006.
9. A. Hassan, M. S. Laghari, and Y. Rashid, "Micro-Encapsulated Phase Change Materials: A Review of Encapsulation, Safety and Thermal Characteristics," Sustain. 2016, Vol. 8, Page 1046, vol. 8, no. 10, p. 1046, Oct. 2016, doi: https://doi.org/10.3390/SU8101046.
10. A. Madhumathi and M. C. Sundarraja, "Energy Efficiency in Buildings in Hot Humid Climatic Regions Using Phase Change Materials as Thermal Mass in Building Envelope:," https://doi.org/10.1260/0958-305X.25.8.1405, vol. 25, no. 8, pp. 1405–1421, Dec. 2014, doi: https://doi.org/10.1260/0958-305X.25.8.1405.
11. B. Givoni, "Building design principles for hot humid regions," Renew. Energy, vol. 5, no. 5–8, pp. 908–916, Aug. 1994, doi: https://doi.org/10.1016/0960-1481(94)90111-2.
12. D. W. Hawes and D. Feldman, "Absorption of phase change materials in concrete," Sol. Energy Mater. Sol. Cells, vol. 27, no. 2, pp. 91–101, Jul. 1992, doi: https://doi.org/10.1016/0927-0248(92)90112-3.
13. M. J. Richardson and A. W. Woods, "An analysis of phase change material as thermal mass," Proc. R. Soc. A Math. Phys. Eng. Sci., vol. 464, no. 2092, pp. 1029–1056, 2008, doi: https://doi.org/10.1098/rspa.2007.0312.
14. A. Castell, I. Martorell, M. Medrano, G. Pérez, and L. F. Cabeza, "Experimental study of using PCM in brick constructive solutions for passive cooling," Energy Build., vol. 42, no. 4, pp. 534–540, Apr. 2010, doi: https://doi.org/10.1016/J.ENBUILD.2009.10.022.
15. H. Asan, "Investigation of wall's optimum insulation position from maximum time lag and minimum decrement factor point of view," Energy Build., vol. 32, no. 2, pp. 197–203, Jul. 2000, doi: https://doi.org/10.1016/S0378-7788(00)00044-X.
16. "Thermal mass – Energy Education." [Online]. Available: https://energyeducation.ca/encyclopedia/Thermal_mass#cite_note-4. [Accessed: 21-Jul-2021].
17. H. Johra and P. Heiselberg, "Influence of internal thermal mass on the indoor thermal dynamics and integration of phase change materials in furniture for building energy storage: A review," Renew. Sustain. Energy Rev., vol. 69, pp. 19–32, Mar. 2017, doi: https://doi.org/10.1016/J.RSER.2016.11.145.

18. T. Kuczyński and A. Staszczuk, "Experimental study of the influence of thermal mass on thermal comfort and cooling energy demand in residential buildings," Energy, vol. 195, p. 116984, Mar. 2020, doi: https://doi.org/10.1016/J.ENERGY.2020.116984.
19. R. Ghattas, F.-J. Ulm, and A. Ledwith, "Mapping Thermal Mass Benefit," 2013.
20. I. Meir, "Thermal comfort – Thermal mass: Housing in hot dry climates." .
21. A. Meir and S. C. Roaf, "Thermal Comfort – Thermal Mass: Housing in Hot Dry Climates," Architecture, vol. In Proc. 9, no. Santa Cruz, CA, pp. 1050–1055, 2002.
22. V. Cheng, E. Ng, and B. Givoni, "Effect of envelope colour and thermal mass on indoor temperatures in hot humid climate," Sol. Energy, vol. 78, no. 4, pp. 528–534, Apr. 2005, doi: https://doi.org/10.1016/J.SOLENER.2004.05.005.
23. Y. V. Perez and I. G. Capeluto, "Climatic considerations in school building design in the hot–humid climate for reducing energy consumption," Appl. Energy, vol. 86, no. 3, pp. 340–348, Mar. 2009, doi: https://doi.org/10.1016/J.APENERGY.2008.05.007.
24. S. Amos-Abanyie, F. O. Akuffo, and V. Kutin-Sanwu, "Effects of thermal mass, window size, and night-time ventilation on peak indoor air temperature in the warm-humid climate of ghana," Sci. World J., vol. 2013, 2013,doi: https://doi.org/10.1155/2013/621095.
25. V. Stevens, M. Kotol, B. Grunau, and C. Craven, "The Effect of Thermal Mass on Annual Heat Load and Thermal Comfort in Cold Climate Construction," J. Cold Reg. Eng., vol. 30, no. 1, p. 04015002, Feb. 2015, doi: https://doi.org/10.1061/(ASCE)CR.1943-5495.0000092.
26. J. Karlsson, "Possibilities of using thermal mass in buildings to save energy, cut power consumption peaks and increase the thermal comfort."
27. Alexander, Christopher. The Timeless Way of Building. Oxford University Press, 1979.
28. https://www.berghahnjournals.com/view/journals/cja/38/2/cja380209.xml
29. https://www.archdaily.com/885677/lycee-schorge-secondary-school-kere-architecture
30. Shokouhian, M., F. Soflaee, and F. Nikkhah. "Environmental effect of courtyard in sustainable architecture of Iran (Cold regions)." 2nd PALENC Conference and 28th AIVC Conference on Building Low Energy Cooling and Advanced Ventilation Technologies in the 21st Century. 2007
31. Zarandi, Mahnaz Mahmoudi. "Analysis on Iranian wind catcher and its effect on natural ventilation as a solution towards sustainable architecture (Case Study: Yazd)." Eng Technol 54 (2009): 574-579
32. https://www.archdaily.com/961350/turtle-sanctuary-at-kalba-mangrove-reserve-Hopkins-architects
33. Hopkins Architects_Khor Kalba Turtle and Wildlife Sanctuary Press Release_April 2021.
34. Las Vegas Review Journal. "Las Vegas History Outlined." July 20th, 1989.
35. Moehring, E.P. Resort City in the Sun Belt: Las Vegas 1930-1970. Reno/ Las Vegas: University of Nevada Press. 1989. pp 2.
36. Gottdeiner, M., Collins C., and Dickens, David. Las Vegas: The Social Production of an All-American City. Blackwell Publishers. 1999.
37. The Las Vegas Springs Preserve. Principal Architects: Lucchesi Galati Architects, Inc. Lucchesi Galati Architects . [Accessed: 27-July-2021].
38. Wadhwa, A. "Baselining Carbon Dioxide Emissions of the Las Vegas Metropolitan Area Residential Building Sector". Master's Thesis. University of Nevada Las Vegas. 2007.
39. The Las Vegas Springs Preserve. Las Vegas Springs Preserve. [Accessed: 27-July-2021].
40. Las Vegas Springs. Wikipedia. LVS History. [Accessed: 27-July-2021].

41. Navis, I., et al. "Influencing community mindsets: the role of outreach programs in achieving urban sustainability." Tourism as a Tool for Development. WIT Transactions on Ecology and the Environment, Vol 139, © 2010 WIT Press
42. "Desert Living Center Exhibits Arid Green". [Online]. Available: Solaripedia [Accessed: 27-July-2021].
43. Cornelius, B., PE. "Exposed Structural Systems Enhance Sustainability.
44. Las Vegas Springs, Las Vegas, NV". Structure Magazine. April 2008.
45. Chadwick, Michael. "The hottest technology under the sun (Expo'92)." IEE Review 38.7. (1992): 251-253
46. Krotowski, Tomasz, PhD. Architecture Of Expo Pavilions – Beauty Or Rationality? Technical University of Lodz. Poland
47. Davies, Colin, and Jo Reid. *British Pavilion – SEVILLE Exposition 1992: Nicholas Grimshaw and Partners*. Phaidon, 1993.
48. Austrian. All Vienna. BRITISH PAVILION, 1988-1992. [Online]. Available: British Pavilion, 1988-1992. [Accessed: 30-July-2021].
49. British Pavilion EXPO 1992. Seville. Spain [Online]. Available: NGP website: British Pavilion at Seville. [Accessed: 30-July-2021].
50. Brownell, Blaine: "Cooling with Water". *Material Strategies: Innovative Applications in Architecture* (New York: Princeton Architectural Press, 2012)
51. Wikipedia. Seville Expo '92. [Online]. Available: Seville Expo '92. [Accessed: 30-July-2021]. "Seville Expo '92." *Wikipedia*, Wikimedia Foundation, [Online]. Available: Seville Expo '92. [Accessed: 30-July-2021].
52. Gardner, Ian, and David Hadden. "UK Pavilion, Expo '92, Seville. Autumn 1992." *Arup Journal*, vol. 27, no. 3, 1992, doi:10.5040/9781350991538.
53. Niesewand, Nonie. "Open the Glass Box." The Independent, Independent Digital News and Media, 26 June 1997. [Online]. Available: The Independent. [Accessed: 01-August-2021].

Mona Azarbayjani is Associate Professor of Architecture (May 2016 onwards) and Graduate Program Director (2020–present) at the University of North Carolina at Charlotte, USA. She obtained her bachelor's degree in 2003, master's degree in 2006 and PhD in 2010, all from the University of Illinois at Urbana-Champaign, where she was a graduate research assistant and a teaching assistant from 2006 to 2010.

She has also worked as a design architect at Naghshe Jahan Pars Engineering Consultant Company from 2003 to 2006. Prof Azarbayjani has many publications in recognized journals, in addition to many written chapters in books and technical reports about buildings. The following are her synergistic activities: research faculty associate, Energy Production and Infrastructure Center (EPIC), EPIC), University of North Carolina at Charlotte (UNCC) 2012–present; LEED (Leadership in Energy and Environmental Design) Associate Professor, 2009–present; reviewer—conferences and journals—Building Technology Educators Society (BTES), Architecture Research Centers Consortium (ARCC) conferences, the *Journal of Sustainability*, *Perkins+Will Research Journal*, and *Energy and Buildings*; member of the American Society of Heating, Refrigerating and Air-Conditioning Engineers (ASHRAE), member of the Society of Building Science Educators (SBSE) 2007–present, member of the International Building Performance Simulation Association (IBPSA) 2007–present.

David Jacob Thaddeus Professor Thaddeus holds a Bachelor of Engineering and a Master of Architecture and has taught architectural design and structures courses at the University of Houston and the University of North Carolina at Charlotte. In addition, he has taught seminars on structures for the Architect Registration Exam (ARE) across the United States, Puerto Rico, Canada, and the UK. His research interest are centered on sustainable construction materials and systems with a focus on mass timber.

Enhancing the Microclimate Toward Outdoor Thermal Comfort in Urban Isles of the Mediterranean Region

Despina Serghides, Stella Dimitriou, Ioanna Kyprianou, and Costas Papanicolas

1 Introduction

The trend of increasing urbanization has induced anthropogenic climate change and contributed to increased energy consumption patterns in cities, accompanied by high CO_2 emissions and the phenomenon of urban heat island [8, 10]. Urban heat islands are caused due to the heat imbalance of urban areas, where heat is created and absorbed by the surroundings but is not equally lost, leaving a positive balance of heat [16]. This is especially pronounced in urban areas with typical Mediterranean climates and long dry summers [19]. Heatwaves and extremely high temperatures in poorly developed urban environments are imposing a risk on human health as well as leading to poor thermal energy performances for urban isles [14]. It is estimated that for every 1 °C increase in temperature, electricity demand in cities increases by 2–4% and therefore with continuously increasing temperatures, the energy consumption and the associated carbon emissions increase exponentially in a short period of time [1]. With cities being the major emitters of CO_2, accounting up to 70% of the global carbon emissions [15], action needs to be taken to improve urban sustainability. Therefore, urban centers should be assessed and measures need to be taken in order to limit the heat island effect, increase outdoor thermal comfort, and decrease energy consumption and CO_2 emissions, in accordance to the European Union's (EU) regulations. In an empirical study using climatic data collected in the duration of almost three decades (1983–2010), support was provided for the existence and development of the urban heat island effect in Nicosia, Cyprus [20]. Although Cyprus holds a small share of energy use and impact on climate change on a global scale, its urban environment influences the temperature profile of the

D. Serghides (✉) · S. Dimitriou · I. Kyprianou · C. Papanicolas
Energy, Environment and Water Research Center (EEWRC), The Cyprus Institute, Nicosia, Cyprus
e-mail: d.serghides@cyi.ac.cy

A. Sayigh (ed.), *Achieving Building Comfort by Natural Means*, Innovative Renewable Energy, https://doi.org/10.1007/978-3-031-04714-5_4

island as a whole, which in turn affects the greater Eastern Mediterranean region due to the island's geographical position. Urban centers stand for areas par excellence in the pursuit of enhanced urban energy performance and pedestrian thermal comfort. Their role is especially prominent in Mediterranean countries due to the extreme heat stress of summertime in this region.

Thermal comfort at its basis is the condition of the mind, which expresses satisfaction with the thermal environment [3]. However, it can be considered in three ways. Thermal comfort can be defined from a psychological standpoint, at which the mind is satisfied with the environmental thermal conditions; from a physiological perspective, where the human body reacts to the environmental thermal stimuli in biological ways through skin receptors and finally from the energy outlook, where thermal comfort is relative to the heat flows between the human body and the external environment [18]. Outdoor thermal comfort is harder to define and analyze than indoor thermal comfort, due to various additional affecting variables (e.g., solar radiation, wind speed, and urban geometry).

When it comes to urban geometry, many variables could alter the outdoor thermal environment and therefore influence thermal comfort. Landscaping of urban areas, grouping of individual buildings into blocks, as well as specific features such as shading mechanisms, surface materials and building arrangement (affecting solar radiation, wind orientation, as well as the heating balance of the buildings themselves) are a few of the most significant features that should be considered when urban areas are developed [6, 11, 15, 21]. Improper design and not accounting for the abovementioned factors often contribute to thermal discomfort at the pedestrian level [2]. On the other hand, carefully considering the building and insulation materials could go a long way in improving public health and the impact of urban centers on the environment [12].

Middel et al. [9] demonstrated that shading in the hot-arid urban area of Tempe, Arizona, could lead to heat stress mitigation at the pedestrian level. This is possible by means of either natural or artificial shading. For example, a photovoltaic canopy is proposed to be used for shading and at the same time to generate energy and contribute to the improvement of the energy performance of urban isles. The importance of shading and wind speed was highlighted as well by Rodríguez Algeciras and Matzarakis [7] in their study on the impact of these parameters on the outdoor thermal comfort conditions in Barcelona, Spain.

Regarding outdoor materiality and vegetation, Turini et al. [22] investigated different renovation measures for the improvement of outdoor thermal comfort conditions in two case study areas in Madrid, Spain. These included substitution of the existing asphalt with cool materials or white asphalt and increasing green bands with shrubs and trees. Through simulations, they concluded that the green area increase using tall trees is the best solution in terms of reducing heat stress during summer. Moreover, in a study of Perini and Magliocco [13], ground areas with trees, shrubs, and grass were shown to have a higher positive impact on thermal comfort at the pedestrian level, relative to green roofs. However, the role of green roofs should not be disregarded, whether they are employed as vegetative areas or to increase the albedo of cities, since they can offer significantly decreased ambient temperatures in urban settings [17]. In fact, green roofs were found to be more effective in decreasing the cooling load of buildings, which is significant within the context of energy management systems in urban isles [13]. Still, in addition to the development of urban isles in order to be sustainable and energy efficient, the design of urban areas should take into account the human behavior. Particularly, urban centers should be designed with a certain degree of social responsibility in mind, in order to prompt the corresponding human behavior that is suitable for such locations [5]. For instance, the decreasing physical activity and increasing obesity of urban dwellers have been linked to poor construction and esthetics [23].

It is therefore essential that the initial design, as well as later developments and refurbishment of urban areas, takes into account the various implications regarding the energy performance of urban isles, while considering the human behavior and activities that are intended to take place at the specific locations.

This study examines the current situation in terms of outdoor thermal discomfort for a region impacted by increasingly frequent extreme heat phenomena. Through simulations of proposed interventions on the surrounding built environment in a case study of an urban isle in Cyprus, results indicate optimal and nonoptimal adaptations of the measures, as they present variations on both seasonal and spatial aspects. The following sections describe the methodology employed as well as the case study. The results of the simulations for the measures introduced as interventions to enhance the microclimate and thereof improve the pedestrian thermal comfort are analyzed, and recommendations are made accordingly.

2 Methodology

The present study aims to determine the current outdoor thermal conditions and provide targeted measures to be taken, on a specific urban isle under study in the coastal city of Limassol, in Cyprus. It aspires to improve the overall experience of pedestrians, mitigate heat stress, and at the same time improve the energy performance of the surrounding buildings.

Based on the Smart Urban Isle (SUI) project criteria, a typical urban isle from the historic, cultural, and commercial center of Limassol was selected as a case study. Among the selection criteria of the isle was the number of buildings, which was to include 10–100 buildings, of which at least one would be public. In addition to the size and the geometry of the site, the decision for its selection was based on the coexistence of older stone buildings with newer concrete ones as well as on its capacity to accommodate multiple uses, including leisure, retail, office, and services.

The 3D model of the Urban Isle was created based on architectural drawings provided by the Cyprus University of Technology (CUT), who is the owner and tenant of the largest building on the site. The thermal properties of the building's envelope its energy performance and its impact on the outdoor environment and comfort was investigated in depth in other studies. The height of the remaining buildings, the type of vegetation, and the outdoor materials were recorded based on on-site observation. The weather data information was collected from a weather station installed by the Cyprus Institute (CyI) SUI project research team on the roof of the CUT building.

The simulations were performed using the ENVI-met V4.1.2 software, a software used to simulate urban environments and to assess the effects of green architecture on the outdoor thermal comfort.[1] The simulation was run for a time interval of 6 h, from 10 AM to 4 PM, for typical summer and winter days. The hours selected for the simulations are the most compromising in terms of outdoor thermal comfort during summer and were used for both the summer and the winter simulations, in order to allow comparisons of the seasonal outputs. Some of the important assumptions made in the ENVI-met simulation process were that the terrain was flat, the buildings' geometry was adjusted to the cubic grid of 1 m, which is the maximum resolution available, and the wind profile had a constant value.

The results from the simulations were processed with ENVI-met application of Biomet for a model of a 35-year-old male, of 80 kg, walking at a speed of 1.2 m/s. The clothing was set to 1 clo for winter and 0.6 clo for summer, in order to obtain the predicted mean vote (PMV). The PMV is a stationary value, which means that a person is assumed to be exposed long enough to a constant climate situation until all energy exchange processes of the human body have become stationary. This is, of course, only true if this person stands exposed to the same climate conditions for up to 20 min in some cases [4]. The outcome was visualized with ENVI-met application of Leonardo, which produces layered data maps. Through the graphic output, PMV was correlated with other parameters, such as the mean radiant temperature, the wind speed, the air temperature, etc., in order to assess their impact on thermal comfort and propose alternatives toward improving it.

From the outdoor thermal comfort analysis, conclusions were drawn regarding the impact of the built environment on it and how it can be enhanced thus mitigating the outdoor discomfort in the city center of Limassol, for the sunlight working hours, both in summer and winter.

Subsequently, an intervention scenario is proposed with measures aiming at minimizing outdoor thermal discomfort, and a revised, improved model was created. The effectiveness of the scenario in improving thermal comfort conditions was assessed through comparison of the PMV outcome, between the revised model and the current state model, for the most compromising hour interval to be outdoors, both for summer and winter.

3 Case Study Overview

The case study isle is located in the built-up area of the coastal city of Limassol, adjacent to the main commercial street and the headquarters of the Cyprus University of Technology.

[1] ENVI-MET official website: https://www.envi-met.com/

There are various open spaces nearby, such as squares and parking lots. The buildings comprising the isle are a mixture of one- and two-story-old stone buildings and newer concrete structures of more than three floors. There are mainly retail stores and education buildings in the area, along with cafeterias and wine and cocktail bars, thus propelling the number of visitors.

From the drawings and the survey questionnaire conducted, the geometrical characteristics and the occupancy motif (users and visitors/day) of the area were extracted (Table 1). Of special interest are the two short-length covered galleries.

The materiality of the environment, buildings, and road and pavement surfaces, together with the type and height of the trees were recorded during on-site observations. The existing materials of the horizontal surfaces are asphalt on the road, granite cube stones on the pavement, and gray concrete on the ground of the open space in the center of the isle. For the unsealed soil, the default material is used. The buildings were mainly identified and modeled as concrete/brickwork structures (common practice in Cyprus), the trees of lower height as deciduous, and those of higher stature as coniferous. The graphical representation of the urban isle is depicted in Fig. 1.

4 Results

From the results, it is indicated that during winter, from 10 am to 4 pm, the most compromising hour to be outside (according to the Biomet results) is between 3 pm and 4 pm (Fig. 2). Nonetheless, the PMV does not fall below −2 at any part of the understudy area. Overall, the PMV ranges from −1.8 to +0.2 and is directly positively affected by the mean radiant temperature.

Table 1 Urban Isle case study characteristics

Area characteristics		
Urban location:	City center	
Area (including pavement):	4267	m^2
Built area (ground floor):	2374	m^2
Green area:	22	m^2
Green area/area:	0.5	%
Ground space/area (GSI):	55.6	%
Total floor space/area (FSI):	126.5	%
OSR (spaciousness) {OSR = (1–GSI)/FSI}	35.1	%
Occupancy		
Nr. of occupants (employees):	≈130	
Nr. of visitors/day (customers, students):	≈970–1240	

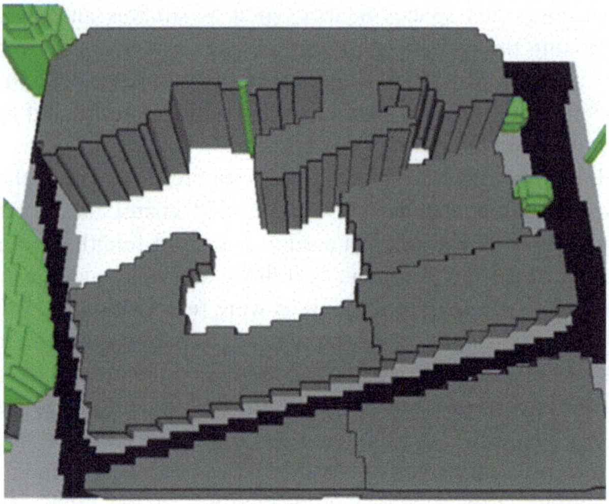

Fig. 1 ENVI-met model

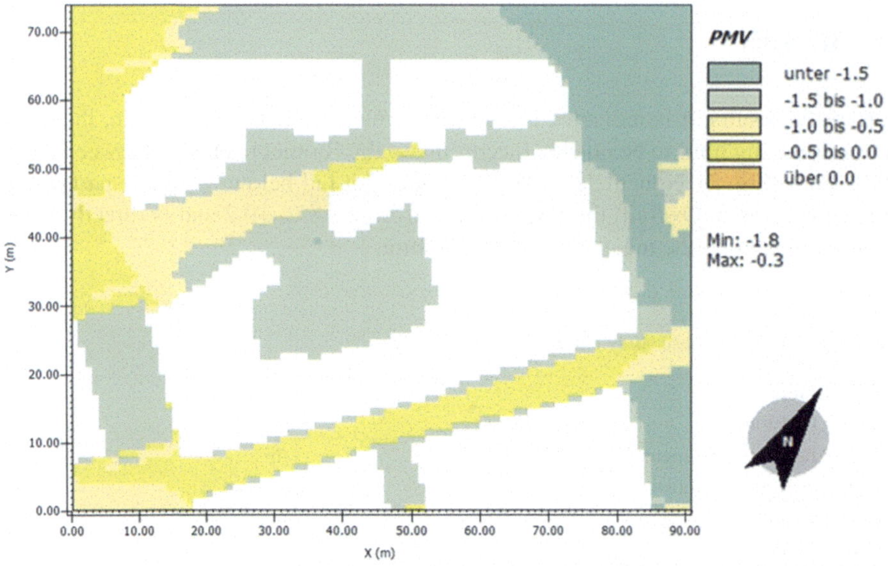

Fig. 2 PMV: typical winter day at 3 pm

During wintertime, it is expected to experience worse thermal conditions prior to 10 am and after 4 pm, since there is reduced sunlight during early morning and late afternoon; however, this study focuses on the sunlight hours, when the highest crowds of people in the area, take place.

The wind speed also affects the PMV value in a negative way, yet not significantly, due its low velocity. Relative humidity and temperature, barely oscillate through the simulation model and do not therefore contribute to the changes observed in the PMV of the area at a specific hour frame. Overall, pedestrians can move around the isle relatively comfortably during a typical winter day from 10 am to 4 pm, with the roads on the southwest to northeast axis having the best conditions.

In contrast, during a typical summer day, at the same time range, the PMV does not fall below +2, ranging from +2.1 to +5.7, being over +4 (classified as extremely hot) during most of the time and in most of the space. The most compromising time to be outside is between 1 and 2 pm (Fig. 3).

In summer, mean radiant temperature is once more the main impact factor for outdoor comfort, with the least uncomfortable spaces being the covered galleries. Nonetheless, even when walking through a covered gallery, the PMV exceeds the accepted value range.

Aiming at minimizing the thermal discomfort conditions during summer, without adversely affecting the winter comfort, a revised model was developed, in which specific design measures were examined in an integrated approach, focusing on the combination of the materiality of the horizontal surfaces and the prospects for shading provision. The two measures concern the replacement of the road and pavement materials with surfacings of higher albedo, increasing the value by up to 0.6 and adding shading via horizontal opaque awnings around the isle under study and the neighboring ones at a height of 4 m. The width of the shading varies from 1 m to 2 m, depending on the geometry of the pavement.

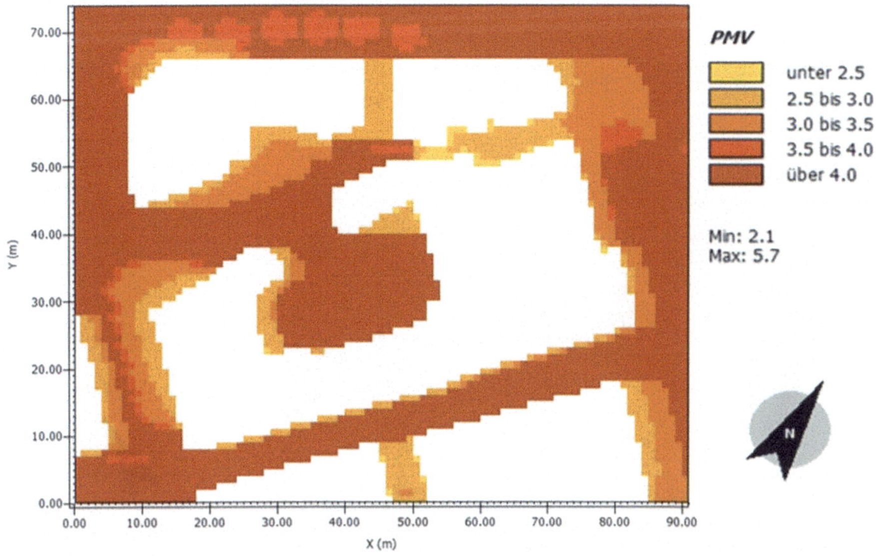

Fig. 3 PMV: typical summer day at 2 pm

The simulations were repeated for the same hour interval for winter and summer, using the same variables for the person's age, sex, clothing, and activity. The PMV during winter in the revised model ranges from −0.4 to −1.8, retaining almost the same range as before. By comparing this with the existing state during winter, the difference observed in the PMV ranges from slight improvement (0.2) to deterioration, up to −1.10, but only at a few limited spots under the shade from the canopy (Fig. 4). The resulting thermal discomfort could be mitigated by installing adaptable awnings, which can be designed to be removed during winter in order to allow the solar interception.

The summer results of the revised model give a PMV range of 2.4–5.2, increasing the lowest value by 0.3 and reducing the highest prior value by 0.5.

When the revised model is compared with the existing state, significant reduction of the PMV is shown under the awnings, up to 2 points. In addition, less prominent but significant reductions in PMV are present in the parts of the road that were assigned previously as asphalt (Fig. 5). Despite all this, the PMV is still over 2 in the entirety of the isle, a value which is considered to be very hot, leading to overheating of the pedestrians and extreme dissatisfaction. Although there is overall notable improvement during summertime, the measures proposed are not sufficient to reach outdoor thermal comfort during summer. This is mostly due to the very high temperatures prevailing during the summer, which might exceed the 40 degrees centigrade, especially during the critical interval of a hot day selected for the purpose of the present study.

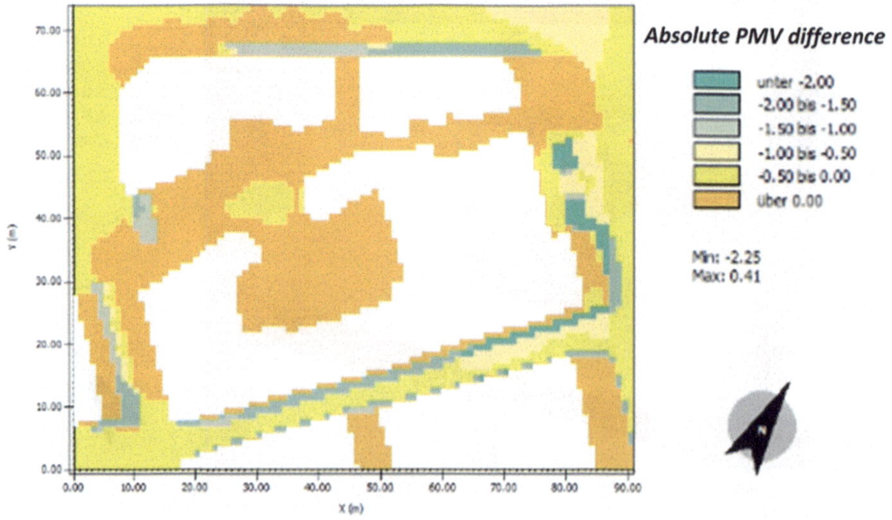

Fig. 4 Absolute difference in PMV in summer at 3 pm: comparison before and after the measures

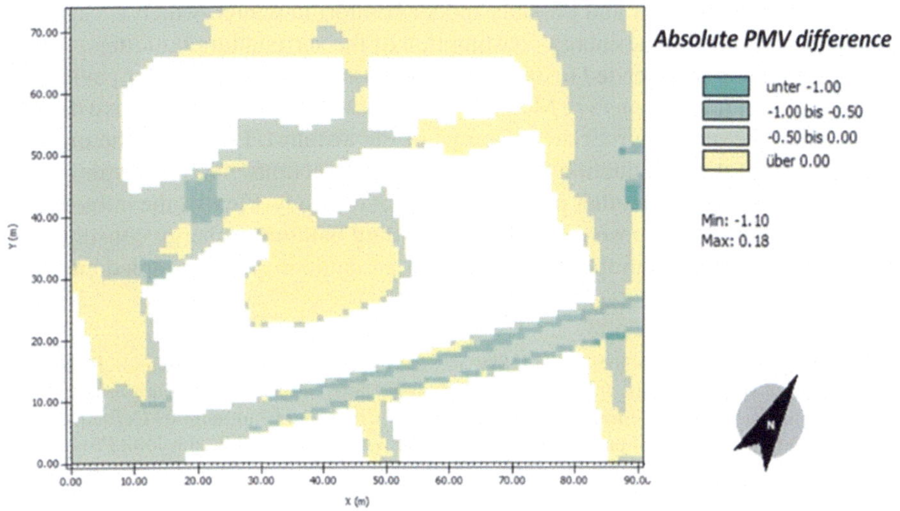

Fig. 5 Absolute difference in PMV in winter at 2 pm: comparison before and after the measures

5 Conclusions

In Cyprus, urban thermal discomfort during summertime is pronounced and has to be mitigated, so as to create a more pleasant outdoor environment. The predicted mean vote (PMV) for outdoor comfort was used as an index of comfort evaluation, along with environmental parameters such as air temperature, radiant temperature, relative humidity, and wind. Simulations were performed with ENVI-met V4.1.2, on a selected urban isle, for summer and winter. The results confirm high or extreme discomfort during summer, which discourages people to visit the city centers especially during midday, with PMV over 2.1 in the whole area and over 4 in most of the space. On the contrary, during the sunlight hours of a typical winter day, pedestrians feel comfortable to slightly cold.

After the modification of the simulation model with the introduction of awnings around the isle and the replacement of the entire road and pavement materials with ones of increased albedo value, the positive impact on the summer PMV values was evident, as targeted, reaching local improvements up to 2.25 points. The results for winter are not as prominent, with the PMV only slightly affected and mostly in a negative way.

The combination of the awnings with the increased albedo is effective in mitigating thermal discomfort during summer. However, it was observed that the albedo increase has negative impact on comfort, if the replaced material has less than 0.4 difference in albedo value, since the decrease in air temperature is counteracted by the increase in mean radiant temperature. Also, it is critical for the awnings to be retractable, so as to allow solar access in winter, since shading affects in a noticeably negative way the outdoor and possibly indoor comfort in winter, which is expected to impact additionally the energy consumption of the surrounding buildings.

Future studies are intended to determine the optimal albedo values for paving the roads in coastal urban centers in Mediterranean, as well as the most effective dimensioning and the most appropriate materials of the awnings. Further to the outdoor materiality and the introduction of small-scale constructions, such as awnings, the impact of vegetation is another prospective aspect for study. Finally, the influence of the proposed outdoor measures on the surrounding buildings' energy consumption will be examined through the development of an outdoor-indoor-coupled environmental simulation.

Acknowledgments The authors would like to thank the EU and the Research Promotion Foundation for funding, in the frame of the collaborative international Consortium Smart Urban Isle (SUI) financed under the ERA-NET Cofund Smart Cities and Communities Cofunded Call. This ERA-NET is an integral part of the 2014 Joint Activities developed by the Joint Programme Initiative Urban Europe (JPI Urban Europe).

References

1. Akbari, H., M. Pomerantz, and H. Taha. 2001. "Cool surfaces and shade trees to reduce energy use and improve air quality in urban areas." Solar energy 70(3): 295–310. https://doi.org/10.1016/S0038-092X(00)00089-X
2. Ali-Toudert, F., and H. Mayer. 2007. "Effects of asymmetry, galleries, overhanging façades and vegetation on thermal comfort in urban street canyons." Solar Energy 81(6): 742–754. https://doi.org/10.1016/j.solener.2006.10.007.
3. ASHRAE. 2004. Standard 55-2004 – Thermal environmental conditions for human occupancy. 1791 Tullie Circle NE, Atlanta, GA 30329.
4. ENVI-met website. "PMV and PPD." Accessed 2 March 2017. http://www.envi-met.com/documents/onlinehelpv3/hs550.htm
5. Chen, L., and E. Ng. 2012. "Outdoor thermal comfort and outdoor activities: A review of research in the past decade." Cities 29: 118–125. https://doi.org/10.1016/j.cities.2011.08.006.
6. Jamei, E., P. Rrajagopalan, S. Seyedmahmoudian, and Y. Jamei. 2016. "Review on the impact of urban geometry and pedestrian level greening on outdoor thermal comfort." Renewable and Sustainable Energy Reviews 54: 1002–1017. https://doi.org/10.1016/j.rser.2015.10.104
7. Rodríguez Algeciras, J. A., and A. Matzarakis. 2015. "Quantification of thermal bioclimate for the management of urban design in Mediterranean climate of Barcelona, Spain." International Journal of Biometeorology 60(8): 1261–1270. https://doi.org/10.1007/s00484-015-1121-8.
8. Karl, T. R., and K. E. Trenberth. 2003. "Modern global climate change." Science 302(5651): 1719–1723. https://doi.org/10.1126/science.1090228.
9. Middel, A., N. J. Selover, B. Hagen, and N. Chhetri. 2016. "Impact of shade on outdoor thermal comfort – a seasonal field study in Tempe, Arizona." International journal of biometeorology 60(12): 1849–1861. https://doi.org/10.1007/s00484-016-1172-5.
10. Neophytou, M., P. Fokaides, M. Sandberg, E. Batchvarova, H. J. S. Fernando, J. Lelieveld and G. Zittis. 2014. "Toward designing strategies for urban heat island mitigation based on multiscale flow considerations." Paper presented at the 94th American Meteorological Society Annual Meeting, February 2–6.
11. Papadopoulos, A. 2001. "The influence of street canyons on the cooling loads of buildings and the performance of air conditioning systems." Energy and Buildings 33(6): 601–607. https://doi.org/10.1016/S0378-7788(00)00123-7.

12. Papadopoulos, A. 2005. "State of the art in thermal insulation materials and aims for future developments." Energy and Buildings 37(1): 77–86. https://doi.org/10.1016/j.enbuild.2004.05.006
13. Perini, K., and A. Magliocco. 2014. "Effects of vegetation, urban density, building height, and atmospheric conditions on local temperatures and thermal comfort." Urban Forestry & Urban Greening 13(3): 495–506. https://doi.org/10.1016/j.ufug.2014.03.003
14. Pyrgou, A., V. L. Castaldo, A. L. Pisello, F. Cotana, and M. Santamouris. 2016. "On the effect of summer heatwaves and urban overheating on building thermal-energy performance in central Italy." Sustainable Cities and Society 28: 187–200. https://doi.org/10.1016/j.scs.2016.09.012
15. Sanaieian, H., M. Tenpierik, K. van den Linden, F. Mehdizadeh, and S. M. M. Shemrani. 2014. "Review of the impact of urban block form on thermal performance, solar access and ventilation." Renewable and Sustainable Energy Reviews 38: 551–560. https://doi.org/10.1016/j.rser.2014.06.007
16. Santamouris, M. 2013. "Using cool pavements as a mitigation strategy to fight urban heat island – A review of the actual developments." Renewable and Sustainable Energy Reviews 26: 224–240. https://doi.org/10.1016/j.rser.2013.05.047.
17. Santamouris, M. 2014. "Cooling the cities – A review of reflective and green roof mitigation technologies to fight heat island and improve comfort in urban environments." Solar Energy 103, pp. 682–703. https://doi.org/10.1016/j.solener.2012.07.003.
18. Schellen, L., M. G. L. C. Loomans, B. R. M. Kingma, M. H. de Wit, A. J. H. Frijns, and W. D. van Marken Lichtenbelt. 2013. "The use of a thermophysiological model in the built environment to predict thermal sensation: Coupling with the indoor environment and thermal sensation." Building and Environment 59: 10–22. https://doi.org/10.1016/j.buildenv.2012.07.010
19. Theophilou, Marios. 2015. "Θεωρητική και πειραματική μελέτη της τοπικής κλιματικής μεταβολής στην Κύπρο και επιπτώσεις στο αστικό περιβάλλον (in Greek), PhD diss., Cyprus University of Technology.
20. Theophilou, M., and D. Serghides. 2014. "Heat island effect for Nicosia, Cyprus." Advances in Building Energy Research 8(1): 63–73. https://doi.org/10.1016/j.enbuild.2015.08.034
21. Theophilou, M., and D. Serghides. 2015. "Estimating the characteristics of the Urban Heat Island Effect in Nicosia, Cyprus, using multilayer urban and rural climatic data and analysis." Energy and Buildings 108: 137–144. https://doi.org/10.1016/j.enbuild.2015.08.034
22. Tumini, I., E. Higueras García, and S. Baereswyl Rada. 2016. "Urban microclimate and thermal comfort modelling: strategies for urban renovation." International Journal of Sustainable Building Technology and Urban Development 7(1): 22–37. https://doi.org/10.1080/2093761X.2016.1152204.
23. Vanos, J. K., J. S. Warland, T. J. Gillespie, and N. A. Kenny. 2010. "Review of the physiology of human thermal comfort while exercising in urban landscapes and implications for bioclimatic design." International journal of biometeorology 54(4): 319–334. https://doi.org/10.1007/s00484-010-0301-9.

Despina Serghides is an adjunct professor at the Cyprus Institute (CyI), where she is heading the Sustainable Built Environment Unit. She is also the project leader of several European projects and carries out research in architecture, energy and the urban environment. Up until recently, she was a professor in the Department of Science Technology and the Environment, Cyprus University of Technology (CUT). Her research focuses in the field of bioclimatic architecture, sustainable cities, zero-energy buildings, and energy-efficient building design for sustainability and comfort for indoor and outdoor environments. In the early '90s, the results of her research for 'Zero Energy for the Cyprus House' hit the headlines of the Cyprus press, and were reported internationally. The architectural design and concepts of her research have been presented and discussed at international congresses, forums, and conferences, and also published in international scientific journals and proceedings.

She obtained her doctorate and master's degrees in Architecture, Energy & Environment from the Architectural Association School of Architecture (AA), London, UK, from where she also obtained her Diploma in Architecture and completed her postgraduate studies in Planning at the Planning Department.

She is the president of the International Solar Energy Society of Cyprus (ISES), member of the Executive Committee, and member of the Board of Directors of ISES-Europe, of which she was the president. She acted as a consultant for the Cyprus government and parliamentary committees and has been the national scientific representative of the National Scientific and Technical Co-operations with Greece and China. She gives lectures at national and international universities, academies, and institutes. She is editor and reviewer of scientific journals and international conferences' proceedings. She has a lot of scientific publications and was honoured with a lot of scholarships, awards, and offices. She chairs and participates in organizing scientific committees of international and national conferences, seminars, and congresses, in which she is invited as plenary or/and keynote speaker to present her scientific research.

Stella Dimitriou is an affiliated researcher with the Cyprus Institute (CyI), where she held the position of external expert and research assistant on several European research programmes. Previously she held a position as a research fellow at the Cyprus University of Technology (CUT) and as an architect at Yiannos Anastasiou Architects & Designers studio in Limassol. She graduated with honours from the National Technical University of Athens in 2012 and holds a master's degree in Architecture, Energy and Environment from UPC (Universitat Politecnica de Catalunya). In 2012, she took part in the fuel poverty research in Cyprus as a member of the CUT investigating team.

Her research interests include sustainability in the built environment with a focus on energy efficiency, energy poverty, and bioclimatic design.

Ioanna Kyprianou joined the Energy, Environment and Water Research Center (EEWRC) of the Cyprus Institute as a PhD student in February 2018. She completed her MSc in Energy Science at the University of Utrecht in 2015 and her BSc in Biology at the University of Sheffield in 2012. Her main interests involve sustainable energy use in the urban environment, the phenomenon of energy poverty, urban vulnerability, and sustainable built environment solutions.

Prof. Costas N. Papanicolas is the president of the Cyprus Institute and the CEO of the Cyprus Research and Educational Foundation, professor emeritus at the University of Athens Greece, and adjunct professor at the University of Illinois at Urbana Champaign, USA. He is advisor to the President of the Republic of Cyprus on climate change and Presidential Special Envoy for Climate Change.

He holds BSc and PhD degrees in Physics from the Massachusetts Institute of Technology. He has served as a researcher at CEA (Saclay, France) and as a full professor at the University of Illinois (U-C) and at the University of Athens (Greece). He directed the Institute of Accelerating Systems and Applications of Greece of the National Technical University of Athens and the University of Athens and the Physics Laboratory of the University of Athens. He chaired the Accreditation Council for Tertiary Education of the Republic of Cyprus (1998–2006) and the Committee that established the Departments of Physical Sciences at the University of Cyprus. He served on the National Advisory Board on Science and Technology of the Hellenic Republic and on the Research and Innovation Council of the Republic of Cyprus.

He has researched and published extensively in the area of Nuclear, Particle and Medical physics, in Energy Policy and Solar Energy, and in Educational Policy.

He is a fellow of the American Physical Society, member of the European Academy of Sciences (Academia Europea), member of the Silk Road Academy of Sciences (China), and founding member of the Cyprus Academy of Sciences, Letters and the Arts.

He has been awarded the "Medal of Excellence for Service to the Republic" by the Republic of Cyprus for his contributions to research and education to the Republic of Cyprus and ordained as "Commendatore dell'Ordine della Stella d'Italia" by the President of Italy.

Energy Management and Savings in the Domestic Situation: A Case Study

Tony Book

Description of a house that was recently refurbished to a high specification of insulation and windows is given.

In addition, a 5.4 kWp PV system was installed on the southeast facing roof, an underfloor heating system and various controls which can be monitored by smart phone apps.

The hot water comes from an immersion heater (3 kW) primarily from the roof through a smart switching system. This detects when the house electricity load is exporting and instead diverts this energy to the hot water cylinder (210 liter).

The underfloor system which is zoned in every room is heated by gas. There is also a 7 kw log burner in the main living area, which due to the efficiency of all the above is hardly ever used. There is also a 10 kW battery storage system (3200 mAh).

The paper demonstrates several points.

The baseline of about 7000 kWh per annum is reduced by about 80% with the use of the PV running in tandem with the battery system. The dwelling is virtually self-sufficient most of the summer months.

The efficiency of energy diversion for hot water, battery charging, and running an electric vehicle is demonstrated by reducing the export of PV to less than 10%.

The financials of the above are discussed together with discussion of using smart variable tariffs now becoming increasingly available (Fig. 1).

The house was recently refurbished to a high specification of insulation and K glass windows.

Additional installations were a 5.4 kWp PV array on the southeast facing roof, underfloor heating, and various system controls which can be monitored by smart phone apps.

T. Book (✉)
Hove, UK

A. Sayigh (ed.), *Achieving Building Comfort by Natural Means*, Innovative Renewable Energy, https://doi.org/10.1007/978-3-031-04714-5_5

Fig. 1 5.4 kWp PV system

Fig. 2 Distribution boards and control

The hot water comes from an immersion heater (3.2 kW) powered primarily from the roof through a smart switching system. This detects when the PV-generated electricity is exporting and instead diverts this energy to the hot water cylinder (210 liter).

The underfloor system, which is zoned in every room, is heated by gas. There is also a 7 kw log burner in the main living area, which due to the efficiency of all the above, is hardly ever lit. When it has been used, the living area, which is quite large (7 m × 6 m), overheats to about 25 C even if the outside temperature is around 0 C (Fig. 2).

New electrical distribution boards were installed, future proofing for installation of battery storage.

All high power requirements such as the oven and hob are on one set of circuits, and all the lower power systems, TV, fridge, lighting, phone hubs, computers, etc. are on the other ring circuits (Figs. 3 and 4).

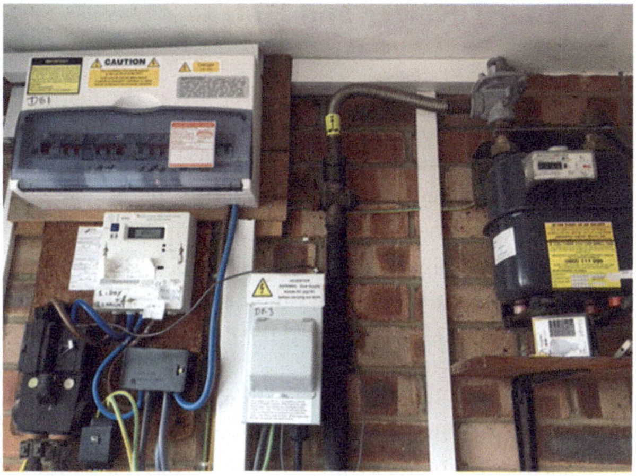

Fig. 3 Upper distribution board and 100 amp supply

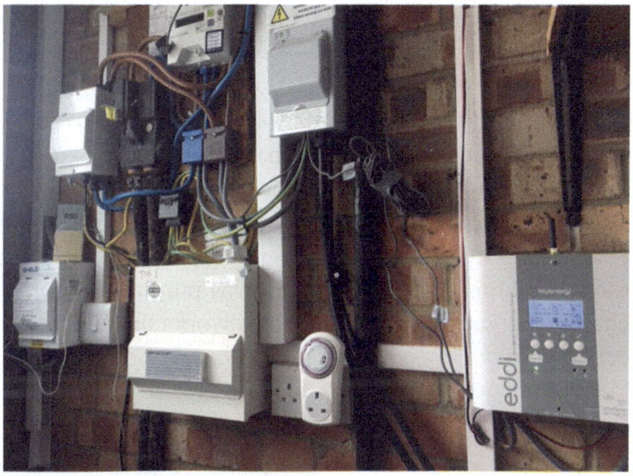

Fig. 4 Lower distribution board (EDDI (DWH diverter) (lower right)

1 Electricity Meters

There are two electricity meters: One is the smart meter for the power readings recorded half hourly and the other is the generation meter to record the harvesting of energy from the PV system. The former is for billing from the utility company, and the latter used to collect the feed in tariff from the energy produced from the PV generation.

2 Power Diverter and Hub

The white and gray box on the lower left is the "smart" controller which diverts the exportable energy. The system is connected via a hub which balances the excess load between the hot water tank and the car charger (Figs. 5 and 6).

This readout on the simple screen of the power diverter shows that about 5 kW have diverted to the hot water tank by 2 pm on a mid-March day saving £1 of energy and average saving over a year is about £400.

The data can also be seen on an app (see later).

This example shows, toward the end of a day in May, 7.75 kWh has been diverted into the 210 liter tank.

Hybrid car being charged is shown in Fig. 7. About 75% of the energy over the year came from the PV panels, the remainder from the grid, usually off peak.

The car had a battery range of 30 miles in the winter months and up to 50–55 miles in the summer months. Additionally there was a 35 liter petrol tank which provided energy to a small generator which powered the electric motor extending total range by another 300 miles.

The car was driven about 10,000 miles per year (Fig. 8).

The hybrid was exchanged for an EV which averages 5 miles (8 km) per kW in the summer and 3 miles (4.8 km) per kW in the winter.

The 50 kW battery has a range of about 225 miles in the summer.

At the time of publication, the car has traveled about 5000 miles and consumed 1306 kW, 589 kW from the grid at 5 p/kWh, 546 kW from the PV system on the roof. In addition 150 KW @25p was purchased on journeys (the balance was free charging at various locations). This works out at about 1.3 pence per mile or 1 euro-cent per km.

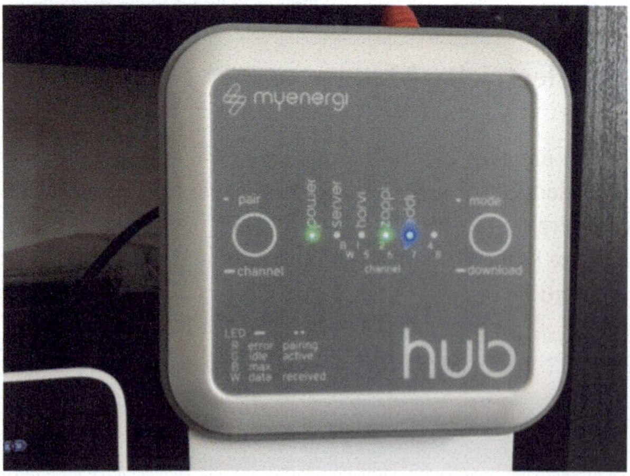

Fig. 5 Blue light shows tank up to temperature (60 C)

Fig. 6 Diverter screen

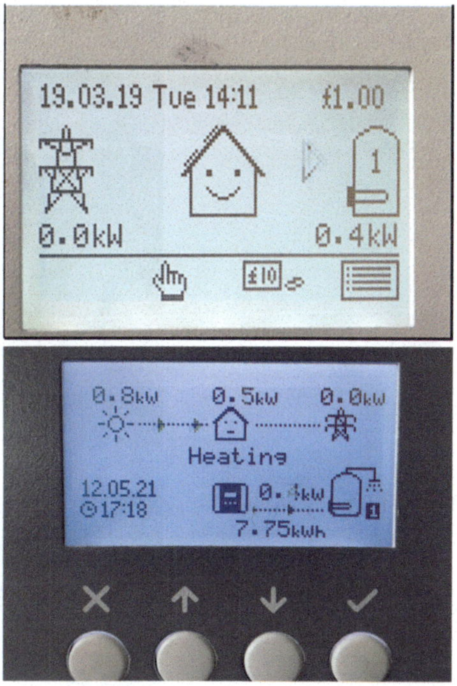

Although there is a price premium on EV purchase when compared to ICE cars (internal combustion engines), EVs have very low maintenance expenses and running costs. The financial break-even is about 30,000 miles when compared with an equivalent ICE car (Fig. 9).

The hot water tank and underfloor heating manifold on the left.

The hot water tank is heated by PV-generated energy about 300 days a year. There are coils in the tank which can be heated by the gas boiler but this system has been used on less than a dozen occasions in 4 years (Fig. 10).

Each room has its own control and is a controllable zone by itself.

In terms of energy management, not a lot of thought is given to water usage other than heating it.

It pays to be mindful of how water volume is used. There are two people in this household, who mainly shower. The average volume of water usage is about 240 liters per day in the winter months and 400 liters per day during the summer period. This extra amount in the summer is due to watering the garden, including irrigation for growing vegetables.

Fig. 7 A picture of electric plug for car charger

Fig. 8 EV being charged

3 Water Usage

In order to ameliorate the water use in summer (4 × 250 liter) water butts filled via down pipes from the roof have been added to reduce the water volume purchased (Figs. 11 and 12).

4 Further Improvements After 3 Years

There is incorporation of a charge inverter and battery pack to "time shift" electricity consumption throughout the year.

There is purchase of an all-electric car (EV).

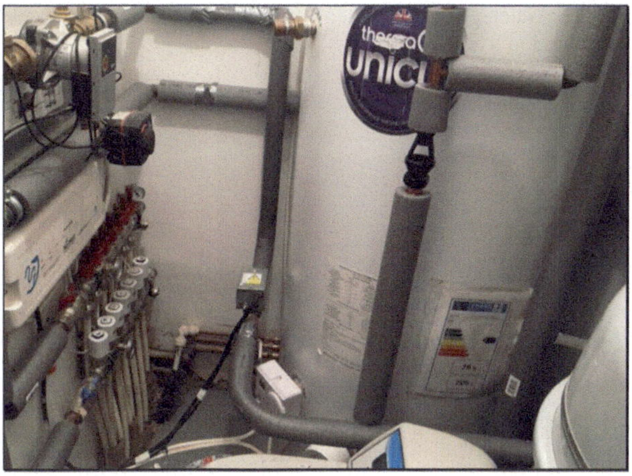

Fig. 9 Plant room

Fig. 10 Underfloor heating system

There is acquisition of a smart car charger which "fuels" the car only when there is surplus on the house. This works in tandem with the smart hot water diverter and is very efficient.

There is the opportunity to obtain split plunge pricing of electricity from a utility company and pay for power by the half-hour and even get paid for using it in the off-peak times. This is dependent on the carbon intensity which is difficult to predict from week to week. By definition the carbon intensity tends to be higher in the summer as the grid sources more "clean" energy from PV, wind, and other renewables (Figs. 13, 14, and 15).

The storage system comprises 16 × Calb 2000 mAh lithium-ion phosphate batteries controlled by a set of 123Smartbms microprocessors. These cards calibrate

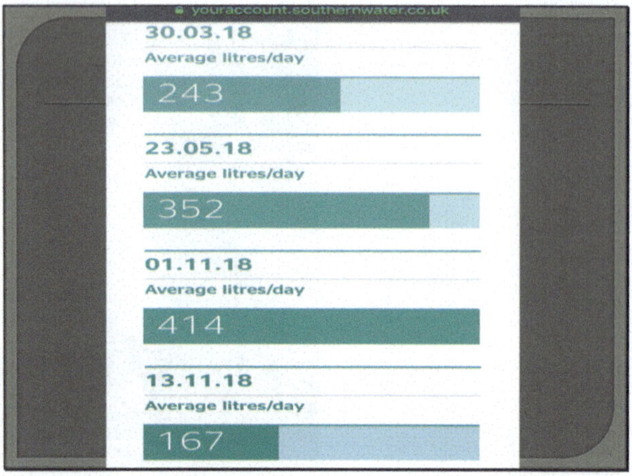

Fig. 11 Water consumption readings

Fig. 12 Three of the four 250 liter water butts

the batteries to ensure they neither are overcharged nor over discharged. The parameters are set at 90% and 20%, respectively. In addition the BMS prevents the batteries from overvoltage and over/under temperature. The charge-inverter was sized to match the PV array (5.4 kWp) with charge and discharge rates of 4 kW from the grid or the PV system.

In the winter months, the charge-inverter is scheduled to come on during off-peak times at 2:30 am. To charge from 20% to 90%, it generally takes about 2.5 h and uses about 10 kW and costs 50p (5 p/kWh on the Octopus GO-Faster Tariff). At 5:00 am the hot water is boosted for 90 min. This uses 5 kW (15p). The charge-inverter is "held" until 0630 before being switched back on. This stops the batteries supplying the hot water and using up some of the battery storage. It also allows the BMS time for absorption and to float the cell units, conditioning them, and increasing their life. The EV can be charged during this time if it is required, but that depends on driving plans and current car range levels.

Fig. 13 Two inverters: On the left, the PV inverter; on the right, a charge inverter which controls the 10 kW battery system

Fig. 14 The CALB 16 × 2000 mAh battery set

During the day in the winter, the property runs on off-peak electricity stored overnight and released during the day. Without any added PV energy, on a typical winter's day, the batteries will last until early evening and may or may not cover the whole of the evening meal cooking.

Gradually as the PV energy increases in late January into February, the battery supply goes through to at least 0230, with a combination of off-peak grid and solar energy.

As the spring comes round, the battery state of charge (SOC) is reduced to 60–80% allowing the increasing solar radiation to make up the difference during the day.

Similarly, the grid-sourced water heating is cut back too. By late March the batteries and solar generation are capable of meeting all the system charging, hot water

Fig. 15 The BMS smart card ensures the batteries run between 20% and 90% charge and do not overvoltage or overheat

heating, and house requirements of cooking, washing, drying, and the EV, around the clock.

This continues until late September when the grid starts to be used more frequently.

5 Examples of Spring and Summer Energy Usage

Note: The baseline when away on holiday is 2 kWh. This is the amount required just to keep the tank hot at 60 C from one day to the next (Fig. 16).

The hot water was boosted for an hour around 6 am as there had been no sunshine the previous day.

The car was charged during the morning and more water was heated around the middle of the day and early afternoon.

Only 13 kW was imported (purchased). The evening consumption shows a small increase around supper time and the rest for lighting and evening TV (Fig. 17).

A small car boost first thing followed by a lot of water heating on a sunny day.

Only 4 kW imported all day and 7.5 kW actually used (Fig. 18).

The car was charged overnight with off-peak electricity and there was washing machine usage/cooking in the morning (Fig. 19).

A good summer's day produces over 30 kW (the annual amount is 5500 kWh).

Once again plenty of hot water was produced. There is 84% green contribution on this day which is typical for the summer months (twice that gained in the winter), and additional saving was made due to energy management using the diverter (Figs. 20 and 21).

The running costs are very attractive.

The hybrid car was filled with petrol about every 5–6 weeks. Most journeys were local except for occasional 400 mile round trips.

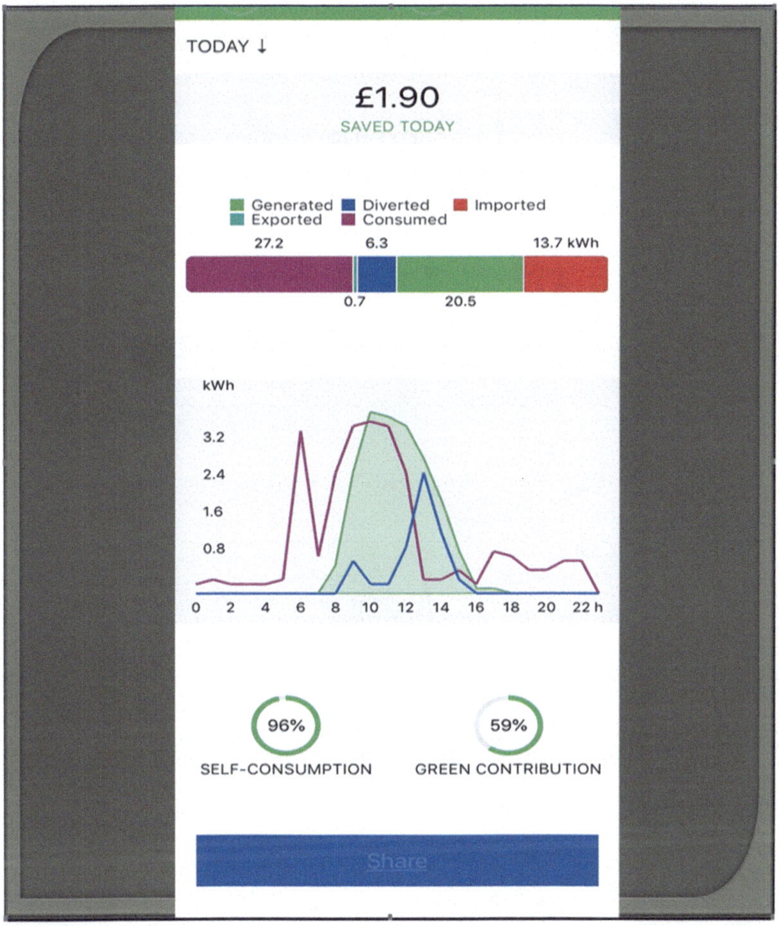

Fig. 16 App screen of performance: 27 kWh consumed, of which 20 kWh generated

During the recent Corona Virus lockdown, from March 2020 to July 2020, the car was used for short journeys only. About 900 miles were covered in up to 40-mile journeys using electricity only. No petrol was purchased for 5 months. Indeed you have to be mindful of the gasoline deteriorating in that time period.

It is estimated that about a third of the PV generated off the roof is used to charge the car. This produces a very economical result… (see above Fig. 22).

This is self-explanatory.

About a quarter of the PV output heats the hot water and the "roof" contributes significant savings to the utility bills and running the car…

This is self-explanatory.

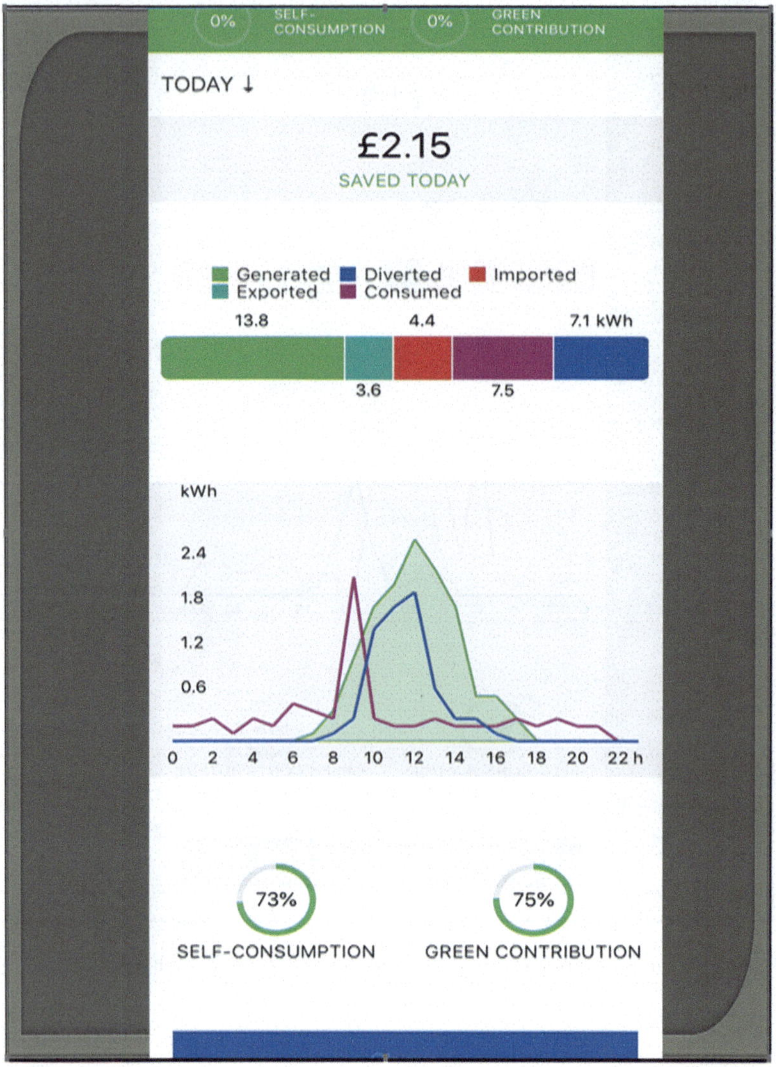

Fig. 17 Typical spring day

About a quarter of the PV output heats the hot water and the "roof" contributes significant savings to the utility bills and running the car… (see above Fig. 23)

On a slightly cloudy day, the PV system is generating 2.6 kW, the batteries are floating, and the hub is balancing the car and hot water… (see above Fig. 24)

A winter's day (off peak 00.30–04.30): Eddi = DHW, Zappi = car, home = battery charger. At 00.30 am, the batteries get charged at 02.30, the car receives 5.9 kW, and the hot water about 5 kW. The PV heats the water a further 1 kW around midday

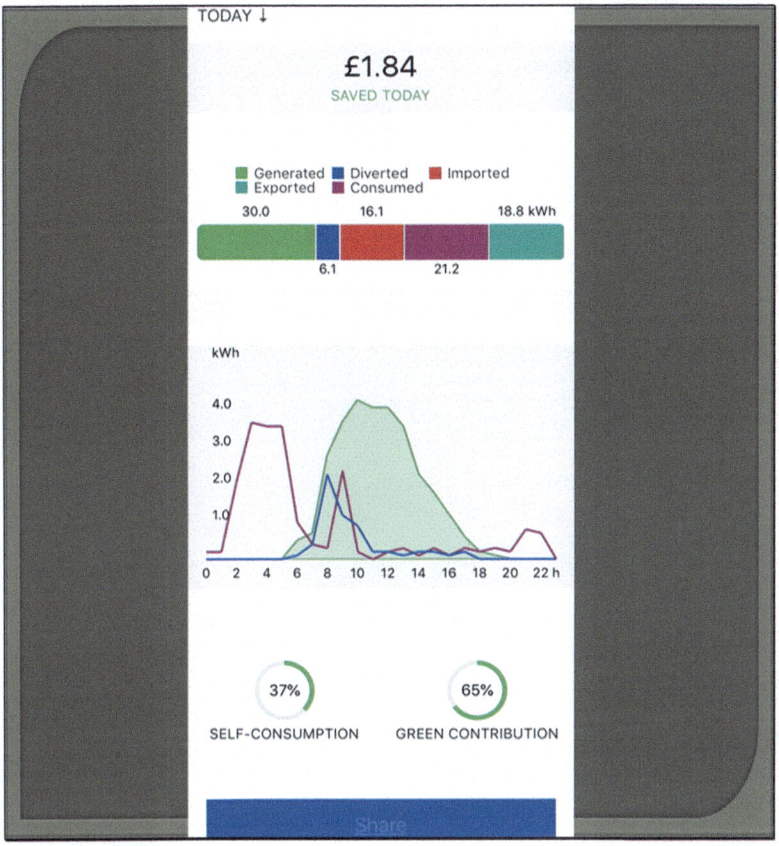

Fig. 18 A summer's day

from the winters sun, which produces 5.3 kWh over the day. It is a heavy day of use; overall 27 kWh was imported.

Off peak was changed to 02.30–0630 after 6 months to ensure the DHW was at its hottest at late as possible… (see above Fig. 25).

From 6 am the PV starts generating and the batteries (home) get charged up first.

The car gets a good 15.6 kW which is the equivalent of about 62 miles (100 Km) and the hot water takes nearly 6 kW. In all 34.5 kW was generated by the PV and 2.2 kw imported. Only 0.9 kW was exported, so that 95% of the PV-generated energy was used up. Once the sun goes down, the batteries cover evening cooking, TV, lighting, and all consumption through the night.

In the winter the batteries don't get much charge from the winter sunshine and need topping up overnight to get through the day. On the night tariff (Octopus Go)

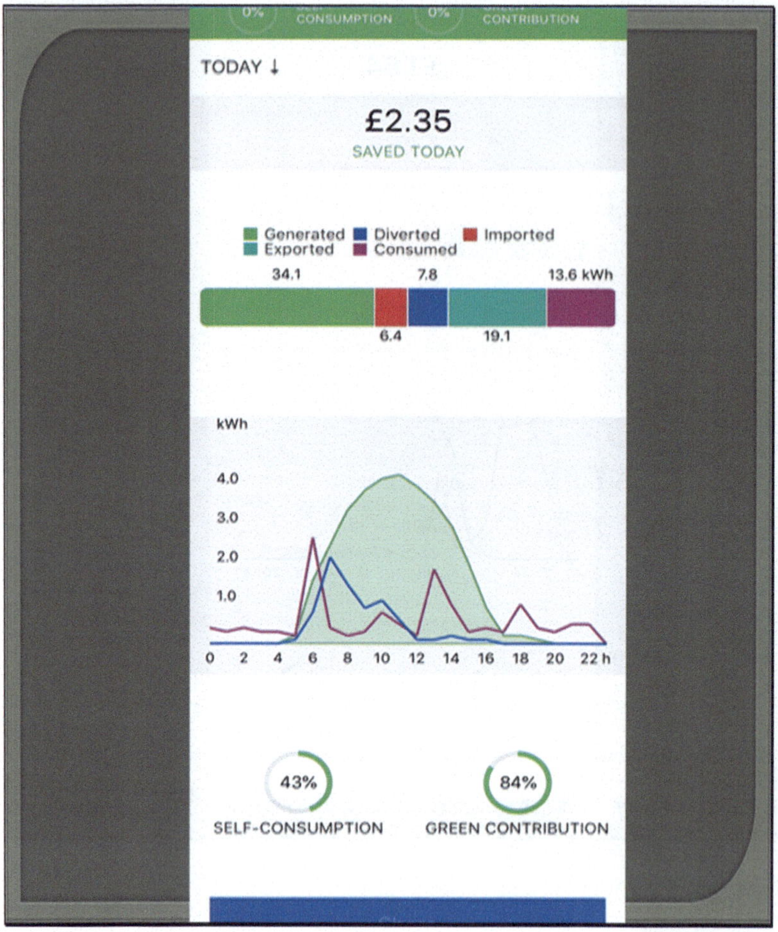

Fig. 19 More activity

at 5p per unit, it only costs about 50p, which is a third of the daytime tariff. The savings are considerable.

The household consumes 7500 kWh per year excluding space heating (gas).

About 5400 kWh comes from the PV system. About half the year the property is self-sufficient, defined as importing of energy of 1 Kwh or less, per day. It takes 1 kW per day to run the 16 battery control cards. About 11 kWh per day is the winter average consumption, most of that from the batteries. There is a lot of debate about the economics of the PV and or batteries. It depends a lot on life style choice. These days, and it is improving all the time, PV will pay back in 5–7 years and batteries in 7–9 years.

Buying this technology is equivalent to buying energy forward which stays at today's price while the actual tariffs rise with inflation and governments energy policies not only in this country but throughout the world.

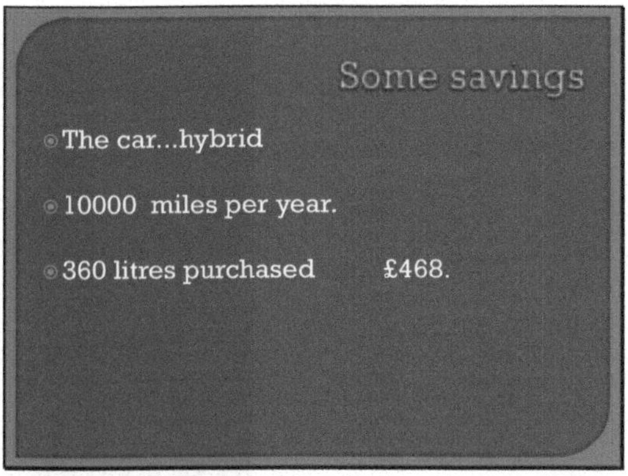

Fig. 20 Car financials

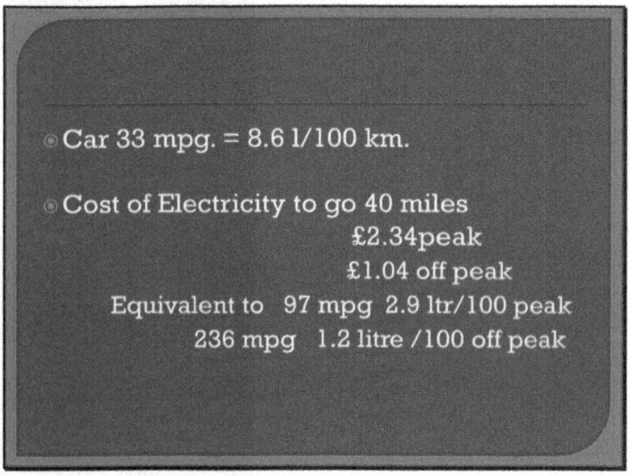

Fig. 21 Hybrid car costs

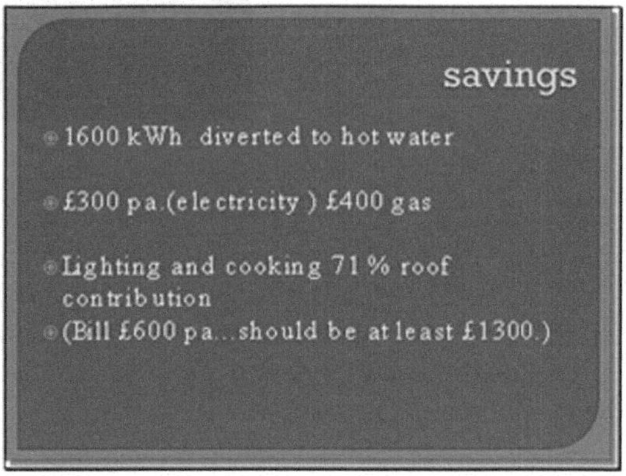

Fig. 22 Financial Saving from Electric Car during the year

It is important to choose the most economical electricity tariffs, and this is best researched from the price comparison websites. The best prices can be found from products that use smart meters where supplies are priced by the half hour and can be monitored by an IDH (in-home display).

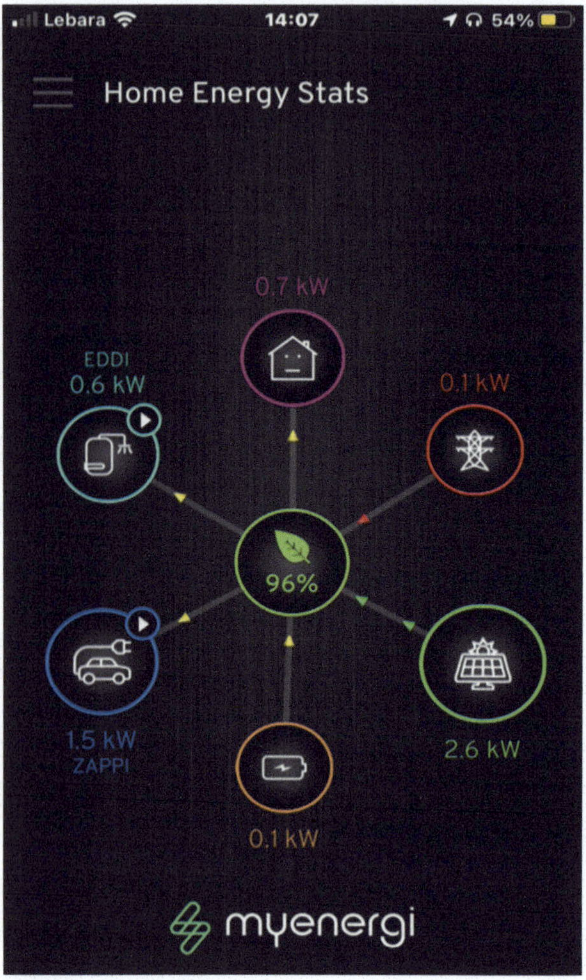

Fig. 23 Home Energy Stats

For space heating which is still using a gas boiler (at least for another decade), shopping around for the most economic tariff is important. Gas at 10000–20000 kWhs are being purchased per year so even small variations in tariff can make a difference to the budget also taking into consideration the daily standing charge.

The monthly consumption of the household is 600 kWh/month throughout the year.

Apart from gas space heating, all cooking, washing, drying, TV, etc. are electric.

All lighting is by LED.

The consumption of grid electricity in the winter months:

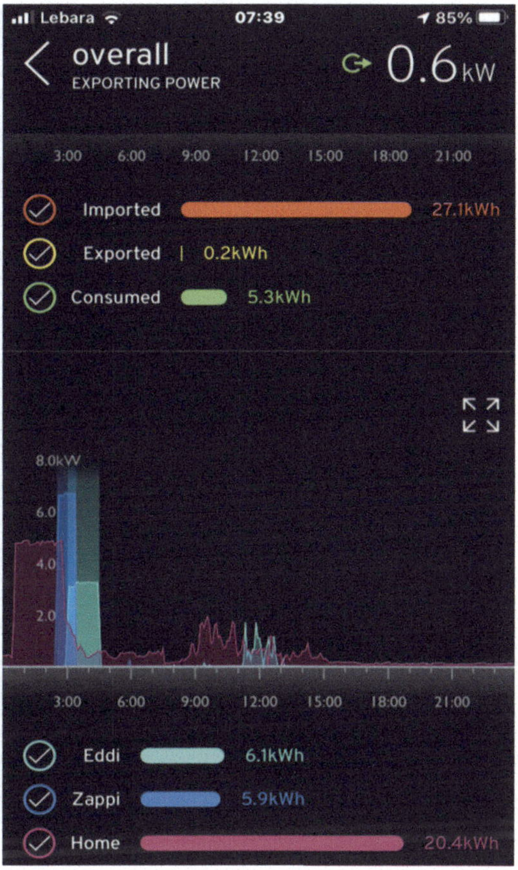

Fig. 24 Overall EXPORTING POWER Winter Readings (Actual Meter Reading).

Per month.

About 200 kWh comes from the PV system, the rest from the batteries.

The time shift of supply from off-peak to peak times has changed the ratio of grid consumption from 60:40 to 10:90 with commensurate savings.

Summer months:

Per month.

700 kWh is supplied by the PV. On average the only 35 kW is imported and monthly bill is £12–£14 of which £7.50 is the standing charge.

Below are sample reports from a smartphone app analyzing the smart meter data.

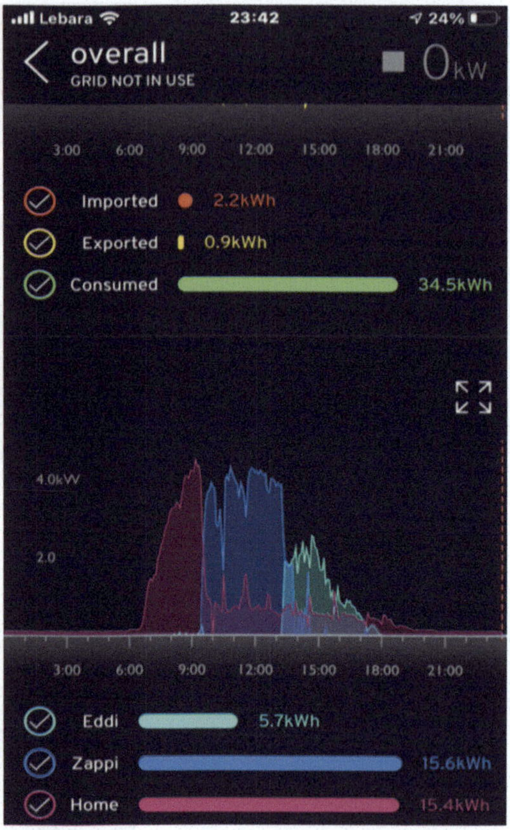

Fig. 25 Overall EXPORTING POWER Summer Readings (Actual Meter Reading)

6 Typical Winter Consumption

6.1 10th Jan 2021–9th Feb 2021 (31 Days)

Your total electricity cost is £47.77 inc 5% VAT.
The unit consumption cost is £40.02 inc VAT.
The standing charge cost is £7.75 inc VAT.
On Go, you would have paid an estimated £47.77 inc 5% VAT – est. 6.81p/kWh.
Your electricity usage is STABLE during this period.
Your electricity cost is STABLE during this period.

You have a total consumption of 587.5 kWh.
Your average unit rate is 6.81p/kWh.
Your average factual unit rate is 8.13p/kWh (includes your standing charge).
Your average standing charge is 25.00p/day.
Your national carbon footprint is 98.68 $kgCO_2$, or 170 gCO_2/kWh.
Your regional carbon footprint is 88.55 $kgCO_2$, or 151 gCO_2/kWh.

> Both the national carbon footprint (across Great Britain) as well as your regional footprint (for your distribution area) is calculated. An average value between 0–59 gCO_2/kWh is very low, 60–159 gCO_2/kWh is low, 160–259 gCO_2/kWh is average, 260–359 gCO_2/kWh is high, and 360 gCO_2/kWh or more is very high. You can improve your carbon footprint by avoiding peak hours, reducing your base load, and planning consumption when carbon intensity on the grid is the lowest.

Your estimated base load is 52 W, totaling 39.1 kWh.

> This is an estimate of the amount of electricity used when you are not actively using anything. This includes devices that are kept running all the time such as routers and fridges, smart lights, standby power for plugged in appliances, etc. An average home has a base load of 105–120 W. Lowering your base load can make a big impact on your electricity cost and can be as simple as turning a device off at the socket when not in.

7 Typical Spring Consumption

7.1 14th Apr 2021–14th May 2021 (31 Days)

Your total electricity cost is £15.53 inc 5% VAT.
The unit consumption cost is £8.42 inc VAT.
The standing charge cost is £7.11 inc VAT.

On Go, you would have paid an estimated £14.87 inc 5% VAT – est. 9.92p/kWh avg. import unit rate in p/kWh avg. import in kWh (one standard deviation error bar).
Your electricity usage is STABLE during this period.
Your electricity cost is STABLE during this period.
You have a total consumption of 71.8 kWh.
Your average unit rate is 11.72 p/kWh.
Your average factual unit rate is 21.62 p/kWh (includes your standing charge).
Your average standing charge is 22.94 p/day.
Your national carbon footprint is 14.83 $kgCO_2$, or 204 gCO_2/kWh.*
Your regional carbon footprint is 10.86 $kgCO_2$, or 163 gCO_2/kWh.

* Some of the CO2 data was not available. Instead, estimates were calculated based on the available data.

Both the national carbon footprint (across Great Britain) as well as your regional footprint (for your distribution area) is calculated. An average value between 0–59gCO_2/kWh is very low, 60–159gCO_2/kWh is low, 160–259 gCO_2/kWh is average, 260–359 gCO_2/kWh is high, and 360 gCO_2/kWh or more is very high. You can improve your carbon footprint by avoiding peak hours, reducing your base load, and planning consumption when carbon intensity on the grid is the lowest.

Your estimated base load is 6 W, totaling 4.5 kWh.

This is an estimate of the amount of electricity used when you are not actively using anything. This includes devices that are kept running all the time such as routers and fridges, smart lights, standby power for plugged in appliances, etc. An average home has a base load of 105–120 W. Lowering your base load can make a big impact on your electricity cost and can be as simple as turning a device off at the socket when not in use.

8 Conclusions

The system has been operating for 4 years and has been constantly monitored, using smart phone apps. It is proving to be very efficient in making significant savings of both energy and running costs. The return on investment is about 20%, and the residence uses about half the normal fossil fuel energy compared to a similar sized house without such a system.

The system is saving about 6 tons of CO_2 per year. The gas boiler uses about 3 tons of CO_2 per year. If it was realistic to equip half the homes in the UK similarly over the next 10–15 years, the savings (of CO_2) would be over 30 million tons per year.

Tony Book, JP, BSc, F Inst M, M.IDM. Born on Tyne side where he went to Newcastle RGS, Tony was awarded a Knowlson Trust Scholarship whilst at the University of Bristol.

From university he went into marketing, spending eight years with Lever Brothers in brand management and market research before joining the American Express as Marketing Director for Europe, Middle East, and Africa. He was later promoted to Director of Consumer Services. He was the winner of the British Computer Society Award for Applications 1983, UK Renewable Energy Award for installations 2008, and South-east Business Man of the Year 2012.

He went on to run a renewable energy business for 25 years, becoming an advocate for climate change and presenting at conferences across the world. He is also author of numerous articles on renewable energy applications for industrial and domestic use.

Adaptive and Sustainability for Visual and Perceptive Comfort: Neuroscience and Architecture

Carla Balocco

1 Introduction

The connection between light, lighting, and neurosciences should not make people forget that they concern three different but interrelated domains: light, as the physical quantity, i.e., a small portion of the electromagnetic spectrum, between 380 nm and 800 nm, that can evoke a neural response from photoreceptors in the human retina and brain; lighting as the design and application of electromagnetic radiation by means of the visual and thermal effects control for architectural and social benefits; neuroscience as the quest for understanding physiological and behavioral responses to the electromagnetic radiation and/or different indoor/outdoor luminous climate. Tackling a lighting project is always a complex operation, because with light (natural and/or artificial) it is possible not only to respond to functional needs, connected to the user safety and visual performance, but also to create or recover atmospheres and suggestions, as well as bright, pleasant, and comfortable environments. If the lighting project concerns historical buildings, with a high cultural, artistic, and symbolic value, the lighting designer must combine the technical-scientific skills with a solid historical-humanistic culture, as well as a strong artistic sensitivity to manage the difficult plant system installation, contemporaries in an architectural context that did not foresee them [1–3]. Many literature researches have demonstrated that technologies belonging to cognitive neuroscience can be used for lighting design [4–7]. In particular, the new neuroesthetics wave, which refers to cognitive neurosciences applications and to the well-known embodied cognition, is a basic instrument for all the research contexts concerning perception, aesthetics, teaching, art, education, and culture spread communication [4–7]. On the other hand, physics, optics, information theory, and neuroscience applications can

---

C. Balocco (✉)
Department of Architecture DiDA, University of Florence, Florence, Italy
e-mail: carla.balocco@unifi.it

A. Sayigh (ed.), *Achieving Building Comfort by Natural Means*, Innovative Renewable Energy, https://doi.org/10.1007/978-3-031-04714-5_6

provide an important support to theoretical and practical lighting research [8–10]. It is well known that the interaction between observer and object is a complex system of top-down and bottom-up cognitive processes. The top-down processes are guided by knowledge, empirical, tactile, visual and previous kinesthetic experiences, semantic/cognitive representations, and syntactic representations of the subject. The bottom-up processes are guided by the optical, photometric, and colorimetric properties of the object but above all by light, luminance, spectral distribution, and color that invest it in the "perceptive neutrality" of the subject. Health benefits, learning and cognitive improvement due to quality of light and lighting, and optimal combination between natural and artificial light, occurring at the right time of day, can help to regulate our circadian clock, mood, alertness, and activity level. Starting from practical investigations, analyses, and comparisons of crucial literature evidences, an integrated methodological approach was defined for a new lighting design based on adaptivity, resilience, and dynamic brain plasticity, rhythm in the light in space and time for quantity and quality of vision and perception [1, 2]. This method allows the analysis of architectural models and their semantic content referred to sociocultural factors but mainly to embodied cognition and perception, i.e., the feedback signals from the regions that involve the visual cortical areas and the cerebral sensomotory and premotory circuits in the presence of visual affordance of displayed stimulus. The proposed method uses an experimental and numerical simulation integrated approach. Therefore, it is based on photometric, colorimetric, and radiometric measurements and psychophysical tests of ergonomics of vision but also eye tracker and visual mapping techniques, visible light communication (VLC) application combined with COB and DALI control regulation systems, connected to numerical transient simulations [11, 12].

Research findings showed that all the visual explorations, the greater number of fixations, vision and perception quality, mental elaboration, interpretation and representation of cognitive processes, and perception of different building architectural areas, objects and/or artworks, and people movements and paths inside different environments happen because of their greatest informative content [12, 13]. The information theory was applied to assess the adaptivity and sustainability of lighting design: it is based on messages' transmission (set of signals) from a source (light), through a communication channel (object and/or artwork, visual exploration path) to a receiver (subject/observer). This involves the visual stimuli transmission and the perception as information transmission which also requires a source (light), a transmission channel (the visual system), and a receiver (the brain that processes the information received). For this reason, the application of our method can be useful to understand how much, where, and when a different illumination of the signifier (object/artwork) can alter the meaning and therefore the interpretation and perception of the one who uses it and observes it. In other words, the choice of the illuminant and light color is connected to the relationship between the number of favorable cases to the occurrence of the perceptual event and the number of possible cases or the probability of making common the mental representation of the perceptual-cognitive experience. By means of practical experiences and tests, it was possible to evaluate the quality of light in different environments, which people surely appreciated because it is a part of that common sense of the communication experience. The latter constitutes the information transmitted that can be measured and assessed by means of the knowledge of all the visual and perceptive data concerning visual path and visual performance changes [1, 2, 4, 11–14]. The method applications show how the concepts of sustainability and adaptivity derive from the sensation of beauty, physiologically mapped in the hippocampus. The sensation and awareness (consciousness) of beauty that light can trigger as an activator of brain and visual-perceptive, cognitive-interpretative functions, allow to outline important and new connectomic maps that are able of activating the well-being and development of a prospective optimism that intensifies and strengthens the resilience differential.

2 Methods and Materials

2.1 Knowledge Base

Nowadays, lighting engineering for both research and design is a very wide and complex discipline. It requires interdisciplinary approaches and it is constantly expanding and evolving. Lighting design should be understood as a project with light and of light: this fact is a sign of success but also of acquired complexity. Any lighting project should be set on a methodological approach for quality and sustainable, adaptive lighting design that can be applied to any historical building and newly designed, guaranteeing the optimal use and control of natural light, preventive conservation, visual comfort, and visual ergonomics, well-lighting. A deep knowledge of the luminous climate due to natural and artificial light existing in the building and its environment is a necessary starting condition. However, especially in the case of historical buildings or cultural heritage buildings, museums, exhibition spaces, etc., it is also necessary to know their history, transformations of functional organization, and use that they have undergone over time. It is well known that we do not see light, but we experience luminous, virtual, synergistic, dynamic, inferential spaces. It is necessary to start from the study of light, on the possibilities of natural/artificial light control, but also of the connected physical, optical, photometric, colorimetric phenomena that give rise to vision and perception and that trigger the "eye shot" (i.e., the understanding of an object, artwork, space/environment, sense of beauty, imagination, mental elaboration, "common sense"). This concept means that psychophysical attitude, through which people imagine stories, starts from the "simple vision" of a color, a particular scene, and any object. Light investigation turns out that "normality" (i.e., light and lighting that everyone likes) is often surprising. Color perception is a complex phenomenon that refers to physiological and functional mechanisms through which the brain perceives color and has a psychophysical dimension. The neurosciences and deep sciences applied to vision show how the two types of photoreceptors, rods, and cones, respectively, used for vision at low luminance levels (night vision) and "blind" to color and the seconds for vision at high luminance levels (daylight) are the first neural structures involved in the color perception. The color of a light radiation or an object is defined by the relative activity of three types of cones, each sensitive to short, medium, and long wavelengths. No cellular structure can mark that a certain color has been perceived. A photoreceptor creates an electrical code that depends on how much energy (number of photons) it has managed to capture from a specific radiation. For this reason, the cells of a single class of photoreceptors would confuse the wavelength of the light radiation with its intensity (completely different size), making it unable to "see" the different colors ("univariance problem"). This question is solved by comparing the neural activity of three different classes of photoreceptor cones. But since this comparison is the true equivalent of the perceived color, color has not a physical dimension: this means that it does not belong to the external world but to a psychophysical dimension, "created" by the brain. As a consequence, visual and color information is transmitted by specific neural mechanisms to the occipital (backside) part of the head, where the first cortical area used for visual analysis is located. This area, called V1 (Vision1), contains a full spatial representation of the external world (a spatial map) but also neural structures with the task of analyzing color at a higher level of complexity than what is done by peripheral visual structures. From the primary visual cortex, information on color follows a so-called ventral path to reach brain structures located in the temporal lobe (the one near the ears) where color analyses reach a higher level of complexity, e.g., it has been shown that due to the activity of the brain area V4 (Vision4), we are able to see always the same colors even when the lighting characteristics change. This phenomenon is called color constancy [6–9]. Brain damage on this brain area produces serious perceptual problems and sometimes total color blindness, i.e., the achromatopsia phenomenon. Retinal images, vision, and perception coexist with the self-image or body scheme, i.e., the understanding, the awareness that the mind has about the body, as well as its encumbrance in space, perceptions, and visions, in a continuous and reciprocal semantic, physical, perceptive, visuospatial, and topological contamination [10]. Light should be understood as a physical element that makes it possible to view and receive the information associated with the scene or the illuminated objects, representing an essential vehicle of information and knowledge. Vision can be considered an interpretation and transformation process of the external world that begins in the retina, with the excitation of millions of light-sensitive receptors. It has its final stages in the cerebral cortex and in the behavioral responses that guide our interaction with the outside world [4, 5]. It is difficult to distinguish between what we actually perceive and what is inference, interpretation, and integration with the long-term memories of our brain. The contact with external objects, although phenomenally immediate, is actually mediated by a series of transformations, which make up the psychophysical chain, consisting of distal stimulus (physical bodies, delimited by surfaces and immersed in a medium); proximal stimulus (two-dimensional image, generated by the projection on the retina of light reflected from the surface of objects, color of things, and surfaces); activation pattern of retinal receptors; decoding/coding, organization, and interpretation processes; and precepts, thoughts, and actions. Through vision, we create an image of the outside world starting from the luminous information collected and processed. The relationship between light and matter has always been twofold and rather controversial: if, on the one hand, it proves indispensable to be able to see the surrounding world and, above all, fill it with colors, on the other hand, it reveals its nature capable of irreversibly, by means of modifications, with more or less long times, of the chemical-physical properties of a substance (e.g., photodegradation phenomena).

The process of transduction of analogic signals into digital ones includes the encoding of the visual signals into neural signals, their transmission along the nerve pathways, and their parallel processing in the different visual stations, with the selection of that part of the information that is considered most important and attenuation or rejection of the others considered redundant.

This dynamic constitutes what the functional neuroimaging, the basic tool of cognitive neuroscience, calls the activation process of a network of brain areas (i.e., feedback, synchrony, and plasticity of the brain). These brain areas, activated and producing the connectome, when a certain task is performed, play a key role for understanding behavior, emotions, cognitive functions, and neuronal substrate, e.g., an activation of the occipital lobes is typical for those tasks in which there is a visual stimulation, which receives cortical afferents from the retina, so that it is the first substrate for visual perception. There is always interaction between logical cognitive processes with hot/cold ones (emotional, affective, interpretative functions) which lead, by means of the mind operation of cleaning, pruning, and construction to the three understanding directions of the information flows: i.e., from inside to outside, from outside to inside, and from inside to inside. These processes are triggered by activators, such as light, and synchronously modify the connectome. The connectome is the informational neuro-electric current (electrical-wiring) that determines the dynamic traces of memories, transforming the mental structures into molecular structures [4, 5].

If we read these processes by means of information theory, we can see that these three mechanisms are the link between source, signal, transmission channel, and receiver. They are in turn very complex systems because they are linked to an interactive and constructive social dynamic with a continuous transformation (visual communication). The information theory connected to the ergonomics of the multi-perceptive learning and optical physics can be used for the assessment of the correct light sources in terms of spectral emission and color light temperature but also vision and perception quality. If interaction between observer and object is a complex system of top-down and bottom-up cognitive processes, then sustainable lighting design should be conceived in its ability to carry information: i.e., light is an information channel that carries signals with informational and cultural content. This fact means that lighting design must be based on dynamic light control to reorganize spaces and environments, and create new pathways, by means of information and new perceptions carried by light. Any lighting solution, with its quantity and quality rhythm over time, can describe a newly designed or historical space and its different use, particularities and preciousness, and cultural and social value by means of that prospective optimism that only a plastic and resilient lighting design can trace in the neural networks. Nowadays, lighting design can "write" by tracing on the retina, in the memory networks, the enormous transformations we are experiencing, causing them to become cognitive retinal images and guiding us to the rediscovery of the peculiarity of spaces and places. Effective lighting control and regulation systems, filtering and transforming natural light (e.g., different kinds of plants, films of advanced nanostructured materials, absorbent opaline fabric materials, matte textiles, and optical fiber textiles) through digital technologies application, provide a visual and perceptive setting where reading and writing, museum itineraries, shopping, working, and any visual task becomes a new proximity as a new visual and perceptive physical distance and therefore a new creative emotional experience.

3 Light Measurement and Simulation Approaches

3.1 Experimental Measurements

If we consider the triple function of light radiation (i.e., visual, circadian, cognitive), it is necessary to change the traditional lighting design, using interdisciplinary methodological approaches both from the point of view of measurements and lighting simulation. The knowledge and assessment of the existing luminous climate of any environment require specific direct measurements carried out by means of lux meters, luminance meters, and colorimeters on a number of points for certain dimensions of grids suggested by the standards [15, 16] (Fig. 1). This is possible only when the type of environment has been identified for architectural, historical, and functional features; different types of materials present, for photometric, optical, radiometric, and colorimetric characteristics; different accesses of solar radiation; and existing artificial light sources. Experimental measurements should be done repetitively and sequentially and for defined time steps. The greatest difficulty arises in conditions of mixing between natural and artificial light. In this case, measurements must be made for different external climatic conditions (sky conditions and solar radiation for thermal and luminous aspects) and with continuity over time, with invasiveness and cost reduction, but ensuring sequentiality, repetition, and reliability. The compliance with these conditions suggests the possibility of using

Fig. 1 Some basic instruments for experimental measurements: from left to right, photoradiometer, luxmeter, luminance meter, colorimeter, digital camera with V_λ filter, hardware platform Arduino, fundamentals of eye tracker technique

advanced techniques such as that one we developed by a recent research project carried out at our Department in cooperation with the National Institute of Optics-CNR in Florence, titled "Simultaneous illuminance measurements on large surfaces." A rapid illuminance measurement technique on large surfaces was developed by means of the use of a digital camera with V-lambda filter, coupled to a data processing system on a PC and to a sample panel, which exploits the appropriate coupling of the characteristics of the sensor, optics, and reference panel. A digital camera and a calibrated luxmeter and/or luminance meter are the instruments involved (Fig. 1). Generally, the luminance measurements require the use of dedicated photometric instruments, such as luminance meters and video photometers. The former allows luminance measurements for small solid angles, requiring, for surveys on large surfaces, measurement repetitions for numerous points with a consequent increase in times and costs and problems about the detection simultaneity in the presence of light fields that vary over time (e.g., natural lighting measures). Video photometers exploit CCD detectors to obtain measurements of large surfaces allowing to reduce survey times and guaranteeing the measurement simultaneity for all the points of the field of view. Using perfectly diffusing surfaces (Lambertian) positioned at the site to be analyzed as targets, both instruments allow to obtain illuminance measurements, once the reflectance of the surface is known and used as target. The video photometer and the calibrated targets are replaced with a digital camera flanked by a calibrated luxmeter. The digital camera is used to perform luminance measurements. The illuminance value obtained is used for fixing the proportionality coefficient between the value of the single pixel and that one of the illuminance of the corresponding portion of the diffusing surface (each pixel is reached by the radiation coming from a small area). This technique allows carrying out measurements at the same time as the presence of visitors, making the authorization process easier.

An important detection phase concerns all the measurements dedicated for identifying the chromatic aspects of light on surfaces from any direct or indirect light source, colorimetric measurements for determining the color both in transmission and reflection. The color characterization of a material/object is not a simple operation. Usually, a certain type of organic material is combined with other means, such as wood or metal, with the function of support and base, and undergoes degradation effects, including indirect and irreversible, due to their chemical and thermal-mechanical alteration. For example, if the pigments are organic in nature (e.g., oxidized copper resins, copper resins, and organic metals), they are very sensitive to environmental conditions due to light and thermohygrometric variations. The use of colorimeters for incident light to measure the chromatic aspects of light on surfaces, starting from any light source, direct or indirect, allows the evaluation of illuminance, tristimulus values, X,Z,CIEx,y and color coordinates, and correlated color temperature CCT. It is also possible to observe that the CCT detected on any surface depends on the color temperatures of the lamps installed and to a lesser extent on the chromaticity of the indirect lighting produced by the different surfaces of the environment. In these cases, it can be more useful to use the spectrophotometer. It is an instrument that allows the measurements of the spectral energy distribution of a

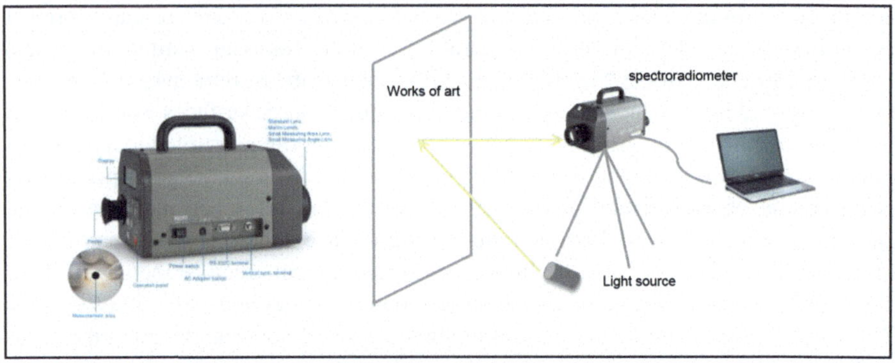

Fig. 2 Experimental setup for identifying the chromatic aspects of light on surfaces from any source, direct or indirect, colorimetric measurements for determining the color both in transmission and reflection

light source for identifying its radiometric, photometric, and colorimetric properties (Fig. 2). In particular, for the measurements of the emission spectra of the studied sources, the Spectralon, a material whose reflectance is greater than 99%, calibrated with the spectrophotometer, was used as the reference surface/point to carry out the measurements. First of all, it was necessary to choose a repeatable and possibly internationally coded method of measurement. For the measurement of any diffusing object, the light source was put at 45° with respect to the object, and measurements from its front (configuration 45/0), without the specular light component, were carried out in the visual field of the instrument, which is the part of the radiation specularly reflected from a particularly shiny object. Figure 2 shows the experimental setup.

The instrument, equipped with a camera lens, "looks" at a certain area on the studied surface and analyzes the diffuse light component. Diffuse light component depends on the light with which the object is illuminated and on the reflective characteristics of the specific pigment of the surface/object (e.g., blue, red, and gold).

If the illuminated surface is not a painting or an object, but a white diffusing surface (i.e., the white reference tile, Spectralon Labsphere), then the reflected light is the same as the one emitted by the light source: it is possible to trace the spectral characteristics of the single light source. When it is used with its own white reference tile, it can also measure the spectral distribution, luminance, chromaticity, and color temperature of different light sources (Fig. 3) [17]. In Fig. 3, the experimental error is not represented: it is smaller than the size of the experimental point in the corresponding graph. The LEDs characterization is based on the color of the light emitted, the emission spectrum and relative white point (i.e., the black body curve), and the correlated color temperature (Fig. 3). Compared to other sources (discharge or induction lamps) for LEDs, the ANSI/IES TM-30-18 [17] suggests metrics based on two indices, the fidelity (Rf) and the gamut (Rg) indices. Rf, similar to Ra, is an accurate measurement of the average color fidelity with range values from 0 to 100 in comparison with a reference illuminant.

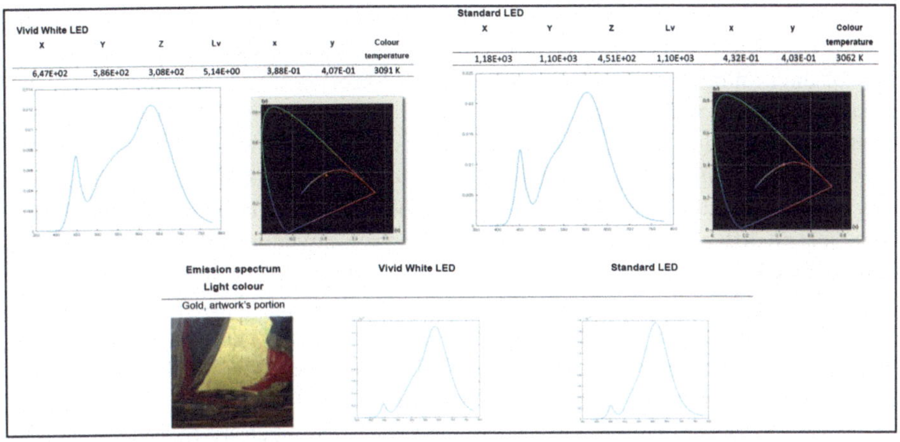

Fig. 3 Results of color characterization (emission spectrum) of the light of two LEDs and color of a portion of the pictorial surface

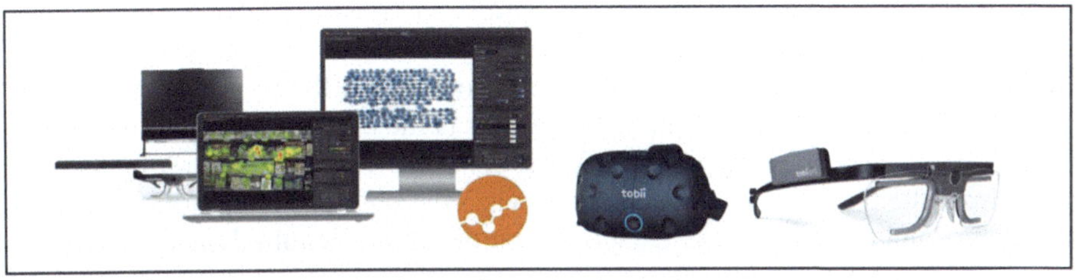

Fig. 4 The tools and components needed for the eye-tracking technique application

Rf index does not allow color quality measurements related to perception, in addition to fidelity, or other effects related to color memory, or it is a measure of human color preference or naturalness perception. Therefore, maximizing Rf does not necessarily correspond to the increase of desirability, utility, or any other perceptual attribute of naturalness. Rg index is an index of the gamut: an increase (greater than 100) or a decrease (less than 100) in the average saturation level of the source with respect to the reference illuminant. Rg index provides information about the relative range of colors that can be produced (with reflection) by a white light source. An Rg score close to 100 indicates that the light source produces colors with saturation levels similar to those of an incandescent bulb (i.e., 2700 K color temperature) or daylight (i.e., 5600 K–6500 K color temperature).

For LEDs with good color quality, Rg can typically range between 80 and 120, with higher scores representing higher overall levels of saturation. From the lighting point of view, visual perception happens if the observed object, light source, and the observing subject are interacting. Visual perception measurement can be obtained with the use of eye-tracking techniques (Fig. 4) [18].

First of all, a sample of visually impaired subjects without anomalies in the perception of color, of different sex and age, must be defined. This set depends on the type of environment studied (museum, office, library, etc.). Each sample subject must be instructed with respect to the task to be performed: e.g., "During the experiment it is required to concentrate on the works shown, trying to exclude, as far as possible, the rest and express one's preference for the work itself, as the light scenes vary choices;" each participant after wearing the Glasses, which are calibrated each time on each different participant, can access the room where lighting is only on the studied area. Each subject is positioned at a certain distance in relation to the dimensions of the surface, artwork, and object (e.g., usually 1 m, 1,5 m, 2 m).

The observed object is lighted with different sources previously calibrated for color temperature, emission spectrum, and white point or black body curve. Generally, a relatively short time is used for preventing the adaptation phenomenon (10–15 s) during which all the eye movements of the subject are recorded with the eye tracker. This measurement is repeated for each subject in relation to each different light source. Everything must be repeated for each sample person. By measuring the fixations and visits duration for each participant, it is possible to deduce the visual preference in terms of vision, perception, and reading, identifying the most effective source in attracting the observer's gaze and attention. The Tobii Pro Glasses 2 system connected to the Analyzer software is the device used for data recording (Fig. 4). When the aim of the project is the evaluation of the light influence on perception, the filter used is I-VT (Fixation). This filter, separating the saccades from the fixations, can also be used when the observer is in motion (i.e., for static and dynamic vision). For each light source, all the acquired data are mapped in order to identify the exploratory pattern of the subjects. The Heat Maps, obtained for each participant and light source, are a function of the fixations duration and different Gaze Plot images.

The Gaze Plots are images that provide information about the greatest number of fixations per surface/object observed and their duration. These investigations are developed after the identification of the areas of interest (AOI) on the surface, object, pear, etc. The AOIs must be the same for each snapshot of different light sources. This allows the evaluation of the differences between fixations, visits, and their duration within each AOI. The visit is the time interval between the beginning of the first eye movement (fixation) within an AOI and the end of the last eye movement within the same AOI. In particular, during a fixation, defined as the ability to maintain the image of an immobile object on the fovea, the four systems of eye movement control (saccadic, tracking, vestibular-oculomotor, and convergence) participate simultaneously.

For example, in the reading process, each saccade is followed by a fixation pause, lasting 200–250 milliseconds, which represents the phase in which the letters that make up the words are decoded and interpreted. Postprocessing results allow to obtain the path of the eyes both in the dynamic and static observation phase and in relation to the type of light source.

Therefore, it is possible to obtain the different Gaze Plots in relation to the light source used. Data processing allows one to obtain graphs of the number of fixations in the considered AOIs, graphs of the number of visits to the same AOIs, the average value of the duration of each visit in each AOI, the average value of the duration of the fixations in each AOI, the relationship between the visits detected with respect to the analysis time, the relationship between the fixations detected with respect to the analysis time, and the relationship between the highest (total) number of fixations with the lowest number of visits [18].

A particular measurement system developed by the LUCE5 company (engineers Federico Bartoli and Matteo Castelli), called "Gaze," allows to detect the user behavior within an environment (i.e., position, tracking, gaze, age, gender, emotional state, Fig. 5). It is a very interesting system for the acquisition of information on people in a space and their postprocessing. It allows any user analysis, which can be consulted by the manager/owner of the space (i.e., the most used path, the average age of visitors, which object arouses the most interest, and how many visitors in certain time slots). Gaze was integrated with control systems on the turnout and maximum number of people in an environment but also to regulation systems for dynamic lighting control.

3.2 *Lighting Simulation*

The real lighting simulation tools are not those that produce renderings without connection with the physics of light, considering only direct lighting and neglecting the inter-reflections between surfaces. This type of instrument uses the first-order algorithms that evaluate only direct illumination and the first reflection of light on surfaces. They are known as local lighting models. The tools based on models that take into account the multiple reflections between the surfaces, rendering the secondary effects for perception, are known as global illumination models. They can simulate the real conditions, or potential effects of a project, using two calculation algorithms, radiosity and raytracing. Raytracing is a global computation technique that incorporates, within the same computational scheme, the aspects of interaction between light and objects using different algorithms for local reflection models. Raytracing allows to trace the path of light rays specularly reflected, transmitted,

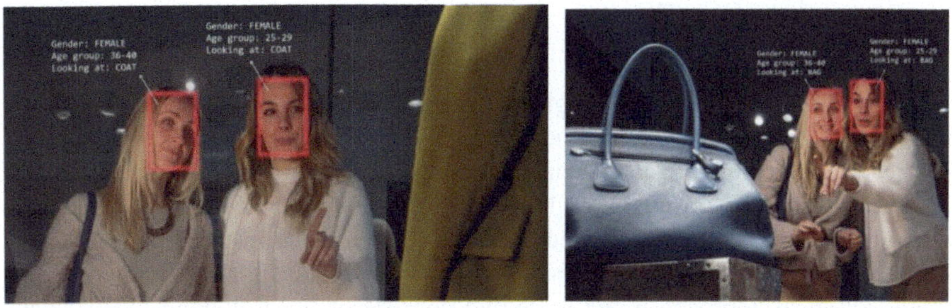

Fig. 5 Two photos of the real operation conditions of the Gaze system

and refracted within any environment. One ray is drawn from a fixed point of view, through each pixel in the direction of the scene that must be simulated. This is the backward ray tracing technique, which considers only photons reaching the observer's eye. A series of rays is drawn back from the position of the observer's eye, until it intersects a surface. The light rays are traced from this point of view to the light source. Furthermore, multiple reflections and transparency are always considered tracing back rays within the same algorithm. Radiosity simulates diffuse interreflection phenomena by calculating the lighting characteristics of discrete points in the environment, whose surfaces have been divided into meshes combining smaller polygons with homogeneous photometric properties (e.g., reflection coefficient). All surfaces are considered Lambertian type, i.e., homogeneous isotropic diffusers. Therefore, it is not possible to simulate the specular reflection characterized by a preponderant directionality. Each element receives the light energy and returns a quantity, until all the reflected energy has not been totally absorbed. After the distribution of illuminance and luminance has been calculated, the analysis can be developed in real time, because it is independent from the observation point. Whichever simulation tool is used, model validation is required by comparing simulated data with those measured. Two numerical simulation models can be implemented: the first obtained using experimental data, the second based on the knowledge of the physical laws expressed by numerical governing equations and defining the correct boundary conditions. In the first case, the numerical model is calibrated by experimental data, i.e., the latter will be directly used in the simulation to reduce the difference between the real measured and calculated data; in the second case, the numerical model is validated, and comparison between measured data and calculated/simulated ones is carried out by means of the assessment of standard deviation, percentage error, and Chi-Square test. Any lighting simulation must use precise boundary conditions: definition and possible simplification of the architectural three-dimensional solid model, built-up urban context or surrounding green, sky model used, typical and precautionary day and its hourly data (i.e., maximum and/or minimum values of solar radiation and external illuminance; number of hours of max and/or min illuminance and luminance), and reference color model or color space model.

3.3 Lighting Design

In this section, some lighting designs, developed by means of the integrated multidisciplinary approach linking the adaptivity and sustainability concepts to the neuroscience for architecture, visual, and perception comfort are shown and discussed. All the projects have a common thread. Regardless of the setup and protocol of experimental measurements carried out, type of measuring instruments used, specific simulation performed, and specific software used for lighting simulation, the methodological approach is the same and consists of the following phases: data collection on historical references, limits and constraints due to current standards

[19–21], external ambient climate, architectural-functional aspects, and different uses of the ambient; identification of different volumes and paths; technical data on the existing lighting system and evaluation of the possible recovery and reuse of the existing electrical, suspending, and control systems; existing luminous climate assessment by experimental measurement on natural and artificial light; 3D building/environment solid architectural modeling; lighting simulation with color model integration; luminous climate analysis at the existing state and comparison with current standards; new lighting proposal by means of adaptivity, sustainability, and reversibility concepts; and simulations of the new lighting solutions and comparisons. In particular, all the issues concerning the "history" of the building and its context, all the materials, plants, and artificial lighting systems, regulation, and control system, when come to newly designed environment, refer to concepts and project choices that can change together with lighting. The current standards refer to different requirements, depending on the intended use, visual tasks, and any limits due to the constraints of conservation and preventive protection but always with reference to health, visual comfort, and ergonomics of vision, as well as well-lighting and energy saving [19–22]. The examples shown in the present section belong to recent researches extensively analyzed and discussed. Fundamentals of all the proposed lighting solutions, differently used and applied depending on the context, type of building and its use, are the following: to conceive light as a channel of information and transmission of contents and meanings (signs/signals); to define an informational lighting project because of quality inside quantity and relational lighting because it is characterized by communication and transmission of intersubjective information that constitute the "common sense;" to use LEDs for effective, dynamic, flexible, and adaptive light together with high-quality control/regulation systems and visual mapping by space syntax techniques, taking into account the use of environments, different people activities and paths, visual comfort, and ergonomics of vision; to obtain well-lighting and human-centric lighting solutions but also well-being and energy sustainability; to refer to sustainability of a lighting design understood as its ability to assure energy-saving and carry information (i.e., light is an information channel that carries signals with informational and social/cultural content); to obtain dynamic light control and light design for space reorganization and new paths creation, user/visitor information, communication, and new perceptions; to apply visible light communication on lighting design, oriented to the critical and dramatic period we are experiencing; to use light and lighting for a prospective optimism along new paths of proximity and physical but creative distances; to propose solutions of new light rhythmic over time for quantity and quality, inside workspaces, museums, and exhibition spaces of newly designed and/or historic/listed buildings; to provide luminous environment/architecture or architectures of light, tracking new cognitive retinal images in the memory networks; to propose control/regulation systems for an optimal combination between natural and artificial light, effective integration with natural elements, which filter and transform it, such as plants and/or digital technologies; and to create a visual and perceptive frame, where every human experience and activity can become a new visual, perceptive, kinesthetic, proximal, and immersive-emotional experience, and rhythmic by light over time and space.

3.4 Light for a Florentine Historical Monastery Used as University Library: Diffusing Sail Systems for Solar Radiation and Daylighting Control

This is a lighting design focusing on lighting quality, adequate lighting conditions for visual tasks, vision ergonomics, and well-being and guarantying the preventive conservation and protection of heritage books [2]. An optimal toolset for lighting with the aim of sustainability was proposed decomposing the whole space into different illumination volumes, according to different occupant activities, visual tasks, and uses (Fig. 6). Three zones were identified inside the historical monastery used as library, taking into account the corresponding illuminance values and photosensitivity classes of different materials: reading room (reading books on open shelves, online consultation, reception), apse room (display of paper material, manuscripts, etc. with important historical value), and media room (activities related to software applications use). Experimental measurements of illuminance and luminance at the existing state, for different external climatic conditions, were used for lighting models calibration.

Three lighting scenarios were identified and simulated: museum-philological-historical, functional-library, and exhibition. In particular, for the exhibition scenario, a passive infrared system (PIR), able to detect any very limited people movement, was used for light dimming and control. The solar radiation control was obtained by means of the integration of matte surfaces, which exploit the splaying of the windows, with custom LED sources, designing and checking their specific beam and aiming angles, position, and the connected efficient control system (Table 1). The existing transversal electrified tracks were used for the suspension and fixing of two sail systems made of polyester fabric with braided and thermo-fixed PVC-coated thread, which has light diffusion (opaline surface) and sound-absorbing properties. The installation of the sail systems together with the covering of all the splayed arched windows with a matte surface provided an optimal control of natural and artificial lighting (Fig. 7 first and second block; Fig. 8).

Fig. 6 Photo: Three views of the University Library housed in the Seminary Chapel

Table 1 The new lighting system: characteristic of the LED used

Name	Typology	Control System	Photometric curve	Luminous Flux/Power Ratio(lm/W)	Ra	Rf	Rg	Colour temperature (K)	N. installed lamps	Height from Floor
Optec Wallwasher	LED	DALI	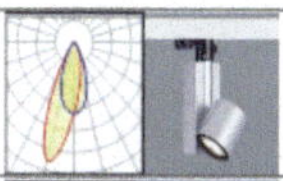	511/ 8.6	92	90	100	3000	8	5.51
Optec Washer	LED	DALI	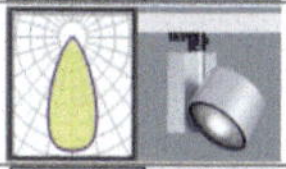	4685 / 36	92	90	99	4000	4	6.51
Parscan Zoom	LED	DALI	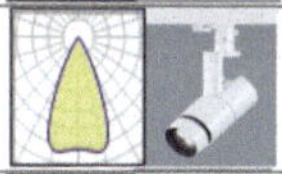	630 / 6	92	90	100	3000	14	5.51
Parscan Zoom	LED	DALI	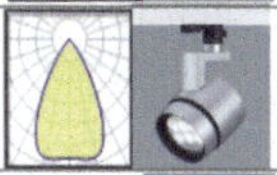	1161/ 14	92	90	99	4000	16	6.51
Lucy	LED	Dimmer	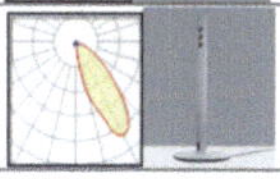	221 / 13	92	90	100	3000	18	1.50

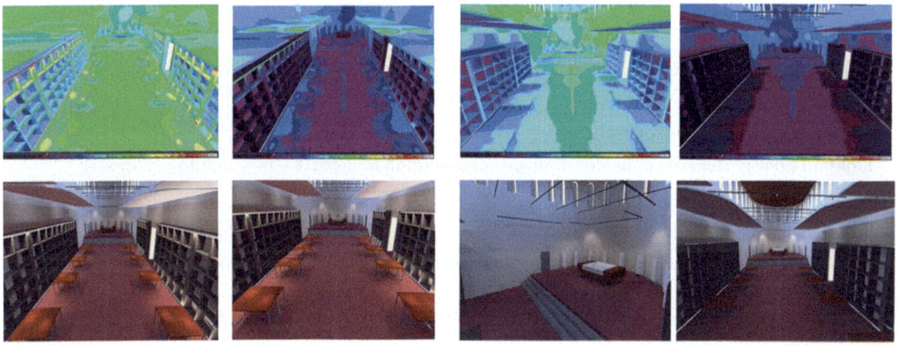

Fig. 7 The first four figures on the left. Rendering of new lighting design (functional-library scenario) for cautionary conditions with mixing between natural and artificial light and all the LED luminaries 100% working. At 10 am: top left false colors of illuminance (lx) and top right of illuminance (cd/m^2); bottom left at 10 am and bottom right at 2 pm
The second four figures on the right. Renders in false colors of the illuminance values in the afternoon 2 pm (top left; lx) and the luminance values (top right; cd/m^2) – exhibition scenario design; Exhibition scenario rendering: view from the apse (bottom left) and view from the entrance (bottom right) 6 pm

The proposed reversible solutions maximize the use of natural light with a correct and effective control for its mingling with the artificial one (Fig. 9), allowing to obtain not only visual comfort, reduction of energy consumption, and costs (especially due to maintenance) but also compliance with the limit values envisaged for different visual tasks, vision ergonomics, and quality of perception (Table 2). Energy performance and efficiency of the lighting proposal (and for all the lighting scenarios) were evaluated by means of the lighting energy numerical indicator

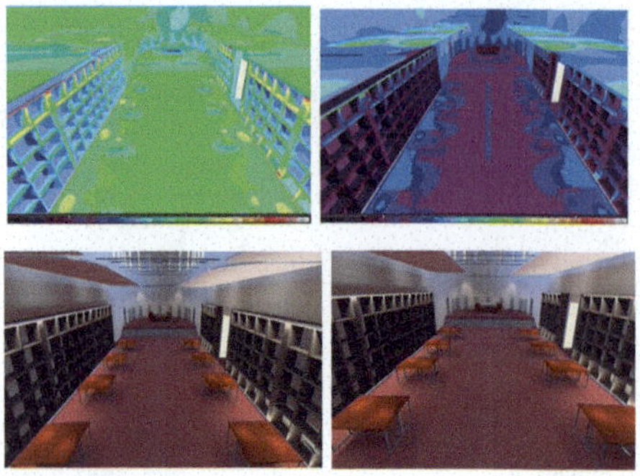

Fig 8 Rendering new lighting design (functional-library scenario), mixing between natural and artificial light, all the LED luminaries 100% working. 10 am: top left illuminance (lx); top right luminance (cd/m^2); bottom left at 10 am and bottom right at 2 pm

Table 2 Basic illumination and modelling index values – new lighting proposal

	Parameters	100% working	Museum-Philological Historical	Functional Library	Exhibition
Apse	Esc (1.2m) lx	17.5	42.6	8.03	51.6
	Esc (1.6m) lx	17.2	45.7	8.25	58.3
	Ev (1.2m) lx	14.4	46.8	5.19	62.4
	Ev (1.6m) lx	13.5	48.7	6.09	67.3
Reading Room	Esc (1.2m) lx	58.1	22.2	54.2	17.5
	Esc (1.6m) lx	54.2	22.1	51.9	18.0
	Ev (1.2m) lx	45.0	20.1	44.8	14.2
	Ev (1.6m) lx	42.7	21.4	42.5	14.1
Modelling Index (M=Esc/Ev)	Apse (1.2 m)	1.21	0.91	1.55	0.83
	Apse (1.6 m)	1.27	0.94	1.35	0.87
	Reading room (1.2 m)	1.29	1.11	1.20	1.23
	Reading room (1.6 m)	1.27	1.04	1.22	1.27

Lighting Energy Numeric Indicator, Energy consumption and costs – new lighting proposal

	Existing State	Museum-Philological-Historical	Functional Library	Exhibition
Consumption (kWh/year)	10300-15100	4650-7000	4200-6300	4200-6300
Costs (€/year)	3093-4528	1400-2100	1260-1890	1260-1890
LENI (kWh/year/m^2)	22-23	11-17	10-15	10-15

Fig. 9 New lighting proposal: Illuminance (left; lx) and luminance (right; cd/m^2) at 10 am – only natural light (artificial light off) for the functional library scenario design

(LENI [22]) for total energy consumption. Table 3 shows results comparison with the existing state: the appreciable, strong reduction of LENI, annual energy consumption, and costs can be noted.

Table 3 Lighting Energy Numeric Indicator, energy consumption and costs – new lighting proposal

	Existing State	Museum-Philological-History	Functional Library	Exhibition
Consumption (kWh/year)	10300-15100	4650-7000	4200-6300	4200-6300
Costs (€/year)	3093-4528	1400-2100	1260-1890	1260-1890
LENI (kWh/year/m^2)	22-23	11-17	10-15	10-15

Table 4 Technical and photometric data of the new proposed artificial lighting system

Commercial name	Lamp type	Control System	Luminous flux [lm]	Power [W]	Ra	Rf	Rg	Correlated colour temperature [K]	Number installed lamps	Height from floor [m]
Optec Wallwasher	LED	DALI	511	8,6	92	90	100	3000	8	5,51
Optec Washer	LED	DALI	4685	36	92	90	99	4000	4	6,51
Parscan Zoom	LED	DALI	630	6	92	90	100	3000	14	5,51
Parscan Zoom	LED	DALI	1161	14	92	90	99	4000	8	6,51
Lucy	LED	DIMMER	221	13	92	90	100	3000	18	1,50

3.5 Combining Natural Light Control with Optic Fiber Textiles

Natural light control combined with artificial light regulation by means of optic fiber textile integration acted as fundamentals of the proposed sustainable and adaptive lighting design [13]. This is a deepening of previous studies applied to a historical monastery used as university library. The integrated use of natural light with LED systems and optic fibers by means of a command structure made with supervision and home automation systems based on Konnex, the first open building automation standard, allows lighting solutions for quality and environmental and energy sustainability. Solar radiation and daylight were assessed by the calculation of the hourly variation of the incidence angle on the splayed arched windows using the method suggested [23]. For simulations, the transparence index results were used and suggested "asymmetrical" artificial light sources for the two sail systems of optic fiber textile. Due to the asymmetry in the intensity distribution of natural light, two different control and regulation groups for the dimming LEDs, installed in the optic fiber textiles, were designed. At the entrance zone and in the center of the church (i.e., in the nave zones), emergency lights were connected to the electrified tracks. Taking into account the natural light availability, different dimming systems were used for LED sources grouping luminaires in different control groups. This choice provides important energy savings guarantying light quality, in compliance with standards [19–22] (Table 4). Results showed that the proposed lighting assures preventive protection for people and book heritage "health and well-being" but also visual quality and ergonomics. The efficiency and efficacy of the new lighting were evaluated for two scenarios, in the precautionary condition of mix between the highest levels of natural and artificial light (i.e., artificial lighting system operating at 100% without dimming and control and at 10 a.m.) and referring to 1,2 m height from the floor for people sitting and 1,6 m height for people standing as required for the average cylindrical illuminance (Esc) maintained evaluation (Table 5). The modeling index M that describes the balance between diffuse and directed light, i.e., the ratio of semi-cylindrical illuminance to vertical illuminance at a defined point was calculated M index value for the whole library/reading/apse zone, is within the limits which is in the range of 0,8–1,3 (Table 5). M index results compared with luminance and illuminance distribution, the local contrast index, and unified glare index prove light optimal mix, good distribution and uniformity, and visual ergonomics but also goodness of light modeling.

Our findings highlighted that sustainable, reversible, and adaptive relighting solutions, oriented to the integration of human-centric lighting with preventive conservation of cultural heritage, can be achieved by the optimal mix between natural and artificial light combining efficient control and dimming systems with innovative application of LED technologies and smart-advanced textiles (Figs. 10 and 11).

Table 5 Illumination and modelling index value – new lighting design

	Parameters	100% working	Museum - Philological Historical	Functional Library
Apse	Esc (1,2m) lx	59,0	58,9	46,6
	Esc (1,6m) lx	63,4	63,3	52,7
	Ev (1,2m) lx	70,7	70,7	57,9
	Ev (1,6m) lx	73,2	73,2	63,6
Reading Room	Esc (1,2m) lx	125	123	137
	Esc (1,6m) lx	116	125	124
	Ev (1,2m) lx	160	155	168
	Ev (1,6m) lx	144	153	153
Modelling Index (M=Esc/Ev)	Apse (1,2 m)	0,83	0,83	0,8
	Apse (1,6 m)	0,87	0,86	0,83
	Reading room (1,2 m)	0,80	0,80	0,81
	Reading room (1,6 m)	0,80	0,82	0,81

Lighting Energy Numeric Indicator, Energy consumption and costs – new lighting design

	Existing State	100%	Museum-Philological-History	Functional Library
Consumption (kWh/year)	10300-15100	3800-5700	2500-3600	3400-5150
Costs (€/year)	3093-4528	1139-1709	800-1200	1025-1538
LENI (kWh/year/m^2)	22-23	9-14	6-8	8-12

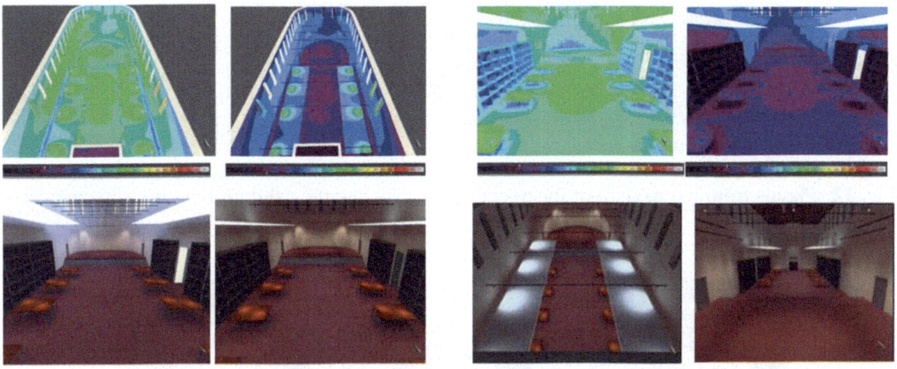

Fig. 10 First four images (left): new lighting design for the functional-library scenario at cautionary conditions with mixing between natural and artificial light and all the LEDs 100% working at 10 am: (top left) false colors render of illuminance values (lx); (top right) false colors render of luminance values (cd/m^2); (bottom left) rendering at 10 am and (bottom right) rendering at 6 pm (with natural light)
Second four images (right): new lighting design; two images at the top: Illuminance values (left; lx) and luminance values (right; cd/m^2) at 10 am – only natural light (artificial light off) for the functional library scenario a view below the two sails system with optic fiber textile; two images at the bottom: from left to right rendering of functional-library scenario all the LEDs at 100% working at 6 pm (without natural light)

Fig. 11 Rendering views of the lighting proposal: from left to right rendering of new lighting design (museum-philological-historical scenario) obtained at cautionary conditions with mixing between natural and artificial light and all the LEDs 100% working, at 10 am

3.6 Science of Anatomy and Light: The Example of the Anatomical Waxes of the Specola Museum in Florence

The new light for anatomical waxes of the Specola Museum (Fig. 12) is based on the dynamic integration between quantity and quality [24]. The lighting solution is flexible and adaptable, with low energy consumption intended for works sensitive to light and in compliance with the limits required by the legislation for conservation and preventive protection. The environment of an operating theater was reproduced, i.e., a seventeenth century anatomy laboratory, with these aims: rediscovery of the pedagogical purpose of waxes; dynamic, flexible, easily maintainable, and removable but also involving display of works of art that are very sensitive to the action of

Fig. 12 Photos from left to right of the Specola Museum and the exhibition halls of the anatomical waxes

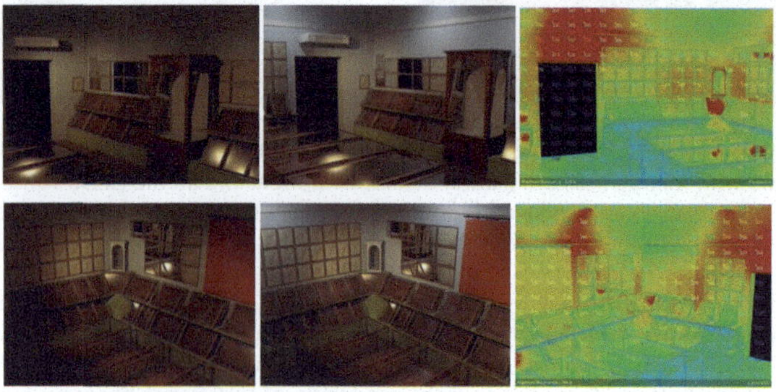

Fig. 13 New lighting. At the top: (left) rendering of the accent light, (center) rendering of the basic light, (right) illuminance levels distribution. At the bottom, different views: (left) rendering of the accent light, (center) rendering of the basic light, (right) illuminance levels distribution

microclimatic and light factors; and safety, quality, and ergonomics of vision. Experimental measurements of illuminance and luminance levels at the existing state but also of the color of light and different materials, in particular the anatomical waxes, were carried out.

The first scenario was designed by means of the installation of eight Parscan LED pointers (spotlights on electrified tracks, 12 W, 1590 lm, 4000 K) dimmable, with wide flood light distribution, and two Parscan LED washer mounted on electrified tracks, dimmable, with oval flood light distribution (12 W, 1590 lm, 4000 K) direct the light respectively on the perimeter display cases and on the central horizontal ones.

Transient simulations were implemented on the 3D model of the ambient and calibrated by means of experimental data. The assessment of the existing luminous climate highlighted the possibility of design two new lighting scenarios (Figs. 13 and 14).

To allow visitors to see the individual anatomical details and the representations hanging on the walls, a second light scenario is proposed, regulated by a presence detector sensor (Figs. 13 and 14). The project is therefore oriented toward the modern concept of interactive museum, according to which the visitor, the container (the exhibition space), and the content (the works of art) of the museum play different but all interacting roles. Here the visitor is the director and can change the lighting scenario: at his entrance six lights (two for each long wall and one for both shorter

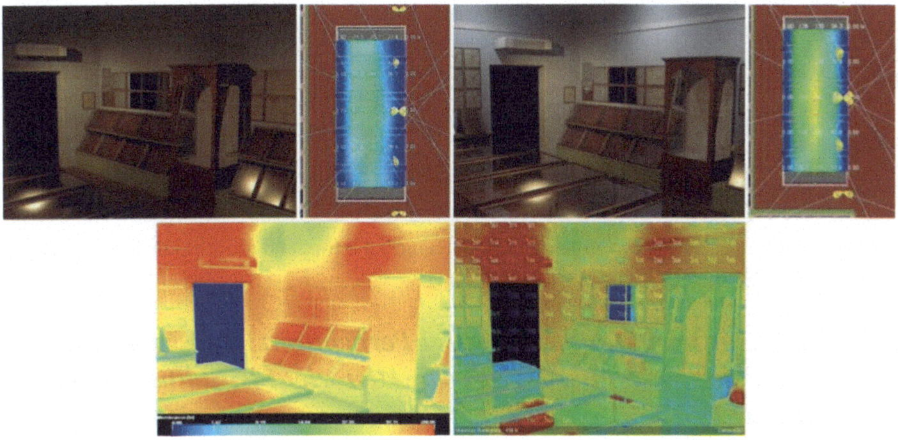

Fig. 14 New lighting. At the top: (left) rendering of the accent light, with illuminance levels distribution on the display cases, (right) rendering of the basic light, with illuminance levels distribution on the display cases. At the bottom, the same 3D views: (left) illuminance levels due to the accent light, (right) illuminance levels due to basic light

Fig. 15 Rendering lighting simulation of the lighting scenes connected to the visit (entrance-exit) of a visitor

walls) LED wall-washers Quadra recessed luminaires in the ceiling equipped with lenses and dimmable high the walls and activate the spectacular cascades of light. The LENI value for the new design is 7,53 kWh/m^2 year, for the existing state 32,30 kWh/m^2 year.

The visit to the museum is not identified as an observation of an almost serial succession of works but as a real perceptive, tactile, kinesthetic, emotional, and historical-cultural evocative experience guided by light capable of arousing interest, the imagination, and perhaps even the enjoyment of visitors (Fig. 15).

Fig. 16 Photos: (left) a view of Villa La Quiete, (center) the frescoed room studied, (right) detail of the grotesque frescoes on the ceiling

3.7 Light, Information, and Visual Perception: An Experience Inside the Florentine Medicean Historical Building "Villa La Quiete"

This is a crucial example of physics, optics, information theory, and neuroscience applications for providing an important support to applied research for lighting our Cultural Heritage, the Medicean historical building Villa La Quiete in Florence [1, 25] (Fig. 16). Quantitative measurements of the relationship between observer and artworks were performed, with eye-tracking technique application. Information theory connected to the ergonomics of multiperceptive learning and optical physics was used for the assessment of the correct light sources in terms of spectral emission and color light temperature. Results showed that different colors of light change the observer's perception and, from the information theory point of view, the communication and interpretation process of the connected signals. It was also possible to measure the informative content of the interaction of light with works of art by assessing the perceptive data of the visual path.

Three lighting sources were used for optical measurements: a LED vivid warm plus (VWP), a LED vivid white (VWh), and a LED standard (STD). Their main component is a COB (chip on board) light source with a combination of specific LEDs with specific phosphor mix and with nominal light beam opening up to 120°. Their secondary optics are provided by a combined diffusing/shielding and diffractive/shielding system that works with a front closure in extraclear tempered glass, up to 60° nominal opening of the light beam. Nominal color temperatures of the three LED sources are 3200 K for VWP, 3000 K for STD, and 4100 K for VWh LED.

Results obtained for the three LED sources lighting the white reference tile, and their spectral characteristics are provided in Fig. 17. From light experience and perceptive experimental measurements visit number, visit duration, gaze number (number of times that the eye stops on different areas), fixation number (number of micromovements of fovea occurring during visual path), time from the first gaze, and time from the first were obtained. Comparing visual patterns of the visitors sample and Gaze data collected, the definition of the areas of interest (AOI, targeted by fixation count and visualized by the HeatMap) that is the fresco zones with the highest number of visits was obtained. In particular, 54% of observers preferred VWh. The ratio between higher fixations number and less visits number provided

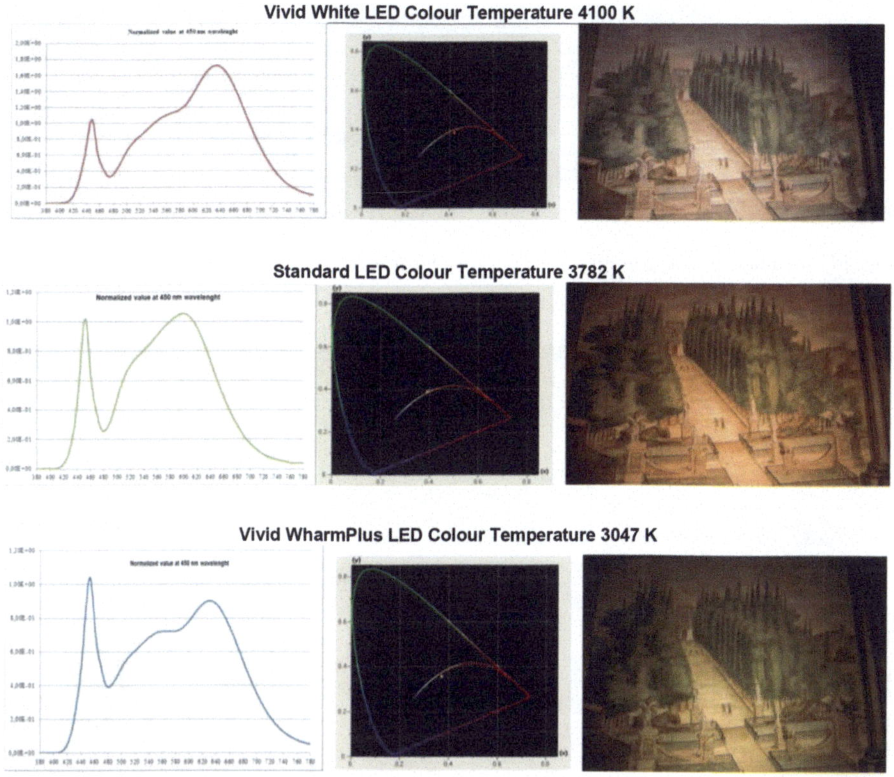

Fig. 17 Results for each LED: at the top, the first three images, center second three images, bottom third three images, respectively, for VWh, ST, VWP, emission spectrum, white point compared to black body, and light effect on a portion of the fresco

Fig. 18 Eye tracker measurement result: the AOI (left); HeatMaps (center); Gaze Plots (right) for a fresco studied

indication about the information content, because fixations tend to be concentrated on a certain AOI that appears to be more interesting.

The greatest variations in time, and fixation numbers of each area, depend on the type of light source: higher concentration, processing of visual stimulus, visual scanning patterns with many fixations oriented to an AOI before moving to another, perception and then higher transmission of information content are obtained with

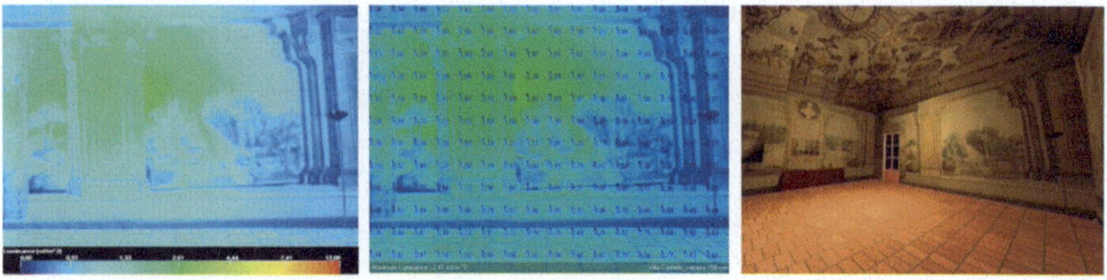

Fig. 19 Luminance (left) and illuminance (center) levels at the existing state. A rendering (right) of the room studied

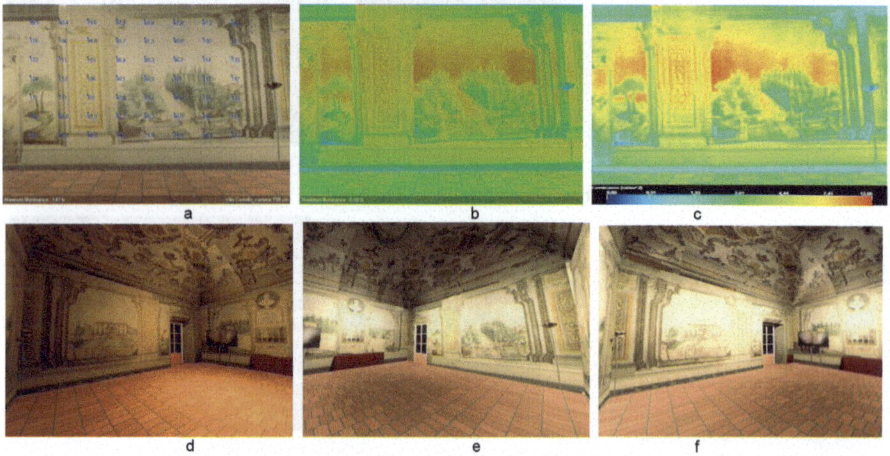

Fig. 20 Simulation results. Lighting proposal: (**a**) render with light meters; illuminance (**b**) and luminance (**c**) levels for accent lighting; (**d**) rendering at the existing state; (**e**) rendering for accent lighting; (**f**) rendering for basic lighting

colder light, i.e., the VWh. With the vivid white LED, the best and consistent results were obtained: greater concentration, greater processing of visual stimuli, increased perception, and visual scanning pattern with numerous fixations oriented to a specific AOI before moving to another. Lighting simulations, using a 3D model of the rooms studied, were calibrated by means of these experimental results. Figures 18, 19, and 20 show the simulation results.

One of the main findings of this study concerned the knowledge on how, when, and how much light color affects perception [25]. In particular, if the AOIs were common to most visitors, the order and observation priority (visual pathway) was different, i.e., lighting design can be linked to the "common sense" that light, as a visual-perceptive experience, can return in the transmission-communication of information content.

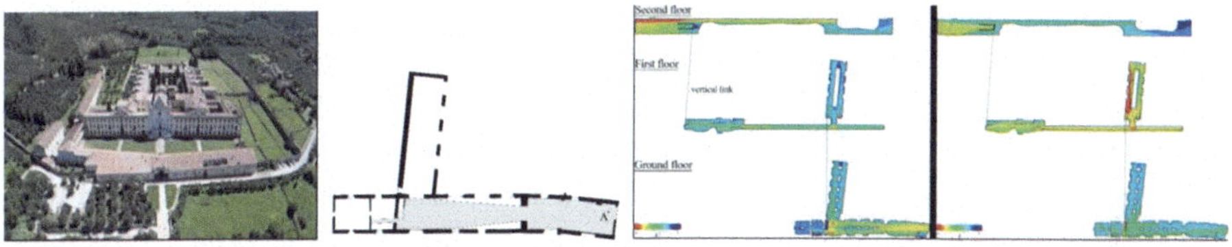

Fig. 21 Carthusian Monastery housing History Museum of Pisa University; isovist for point A at the museum entrance; visibility graph for the galleries with connectivity index trend (left) and integration index trend (right)

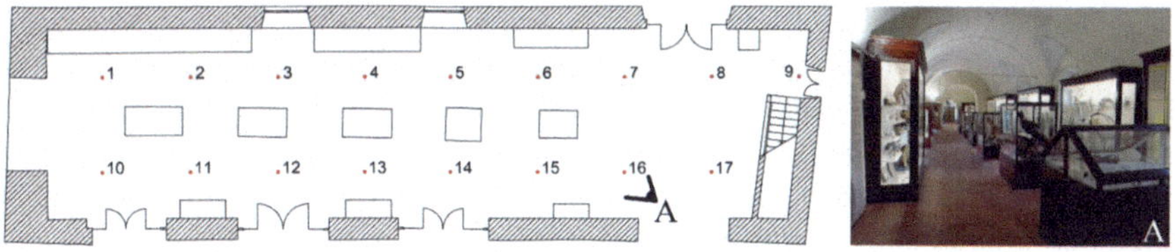

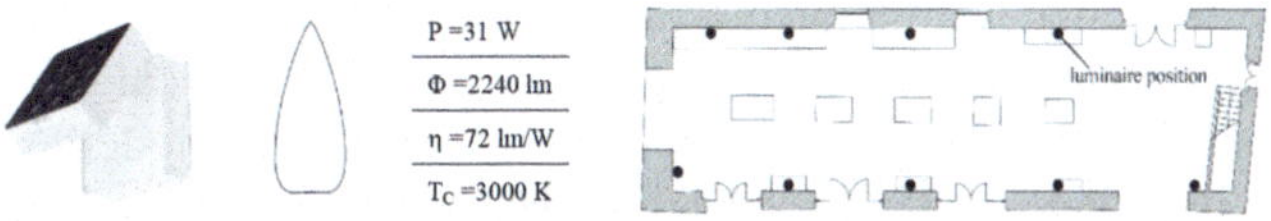

Points ID	E_{sc}	E_v	Points ID	E_{sc}	E_v
1	62.8	54.5	10	54.2	63.9
2	55.1	38.9	11	51.8	52.9
3	52.4	46.2	12	68.7	66.6
4	48.8	27.4	13	52.3	30.5
5	66.7	33.4	14	40.3	31.5
6	35.2	23.1	15	51.5	41.3
7	28.2	18.7	16	20.9	15.5
8	15.5	15.7	17	13.1	13.6
9	11.5	11.9			

Fig. 22 At the top: (left) map with measurement points in the exhibition space, (right) a photo of one of the meaningful environments. At the bottom: (left) a photo of the light source and its technical data; (center) map of the luminaries and artworks positions; (right) experimental data measured at each points

3.8 *Light and Visual Mapping for People Visual Behavior and Movement: Space Syntax Modeling for Cultural Heritage*

This is an operative example of the use of space syntax technique for improving indoor lighting and understand where and how enhance visitors experience, perception, and visual well-being and also guarantee preventive conservation, health, and comfort protection [14]. The gallery of the Natural Museum of the University of Pisa was the case study [14] (Fig. 21). A grid on the analyzed spatial layout was superimposed and two synthetic parameters were calculated, i.e., the connectivity and the integration index.

The connectivity index counts how many points of the grid are visible from an analyzed point considering the walls restrictions. It derives from the concept of isovist, which is a polygon drawn on a building layout around a reference point (Fig. 21). The integration index concerns the accessibility of a given point within the analyzed layout and describes the mean depth of one point with respect to all the other points.

The depth of a single point is defined as the topological (not geometrical) distance separating a pair of points, and it is measured as the number of points that divides them along the shortest path between the two points. The integration index quantifies how close a given point is from all the other points (Fig. 22).

The greater the integration index value of a point in a spatial layout is, the more accessible and integrated is in the system; vice versa the lower the value of the integration index is, the more isolated and segregated the point results. Space syntax analysis can be automated by means of dedicated software. All the plans were used as input data for the software, in a vector format (e.g., dxf files). All the boundaries, given by walls and museum display cases, were used to reproduce the viewer's eye limits (at the height of 1,50 m; Fig. 22).

The outputs, i.e., color maps (visual maps) and numerical tables provided comparison between the two synthetic parameters connected to different environments. Lighting simulations were carried out on a 3D model of the gallery and luminaires arrangements. Simulation model was validated by the experimental measurements carried out on luminance and illuminance levels at the existing state.

The percentage difference between the modeling indices calculated from the measured values and those calculated/simulated is lower than 14%. In Fig. 23 a false color rendering of the vertical illuminance mapping and semi-cylindrical illuminance mapping for the gallery are shown at the height of 1,50 m. The modeling index can be used for light modeling effect evaluation. The modeling index was chosen to be connected to the spatial properties of the museum environments. It describes the balance between diffuse and direct light, i.e., the ratio between semi-cylindrical illuminance and vertical illuminance at one defined point. In particular, the grid points for semi-cylindrical and vertical illuminance coincide showing the quality of lighting design. For uniform arrangement of luminaires and roof lights, the value between 0,8 and 1,3 is an indicator of good modeling (Fig. 23).

Additional benefits of daylighting with respect to artificial lighting can compensate for its effect on modeling values. As a consequence, modeling values from daylight can be extended from the range previously indicated. The modeling effect improves as the modeling index decreases: for modeling index values higher than 1,3, shadows appear very soft and the 3D objects contour are not clearly distinguishable from the background; for values lower than 0,8, the perceived images are with a high contrast and shadows are very harsh. A correlation between the modeling

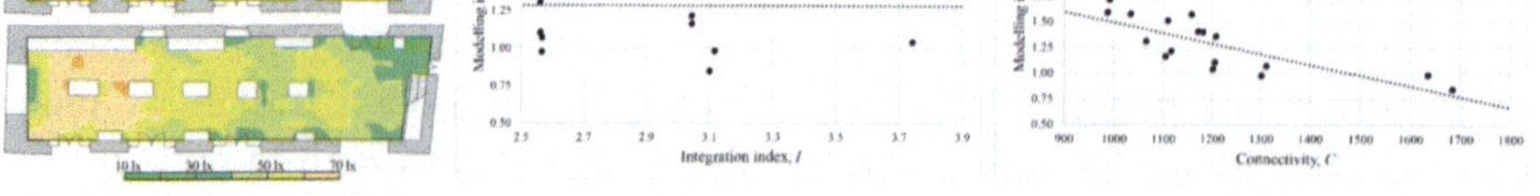

Fig. 23 False colors rendering of the vertical illuminance mapping (top) and semi-cylindrical illuminance mapping (bottom); correlation between the integration and modeling index (center); correlation between the connectivity and modeling index (right)

Fig. 24 Two planimetric views (left): the first, most followed path (continuous lines) and main deviations (dashed lines); the second, VGA graph of the I index. Example of three eye-tracking results: (left) the AOI; (center GazePlot (circles are fixation points; circle numbers are gaze sequence; different colors are different observers; circle size is fixation duration); (right) heat maps, from green to red, from higher to lower interest (concentration)

Fig. 25 Photo of shop window at west (left) and south (center); illuminance level of the new lighting proposal (right)

index and the space syntax connectivity index was carried out (Fig. 23). It was stronger than that one obtained for the integration index. This fact means that in museum spaces human behavior is more influenced by the visual connections and the visibility interactions (meaning the property, expressed by the connectivity index of a point to be visible to all the other points in space) than by the shortest paths linking different points (property expressed by the integration index; Fig. 23). Integration with eye-tracking techniques allowed to know the visual-spatial and temporal sampling, cognitive processes analysis, data collection on attention, visibility, mental processing, and understanding. Experimental setup and measurements protocol were those explained in Sect. 3.1.

The exploratory pattern calculation, by means of the measurements assessment and postprocessing obtained from a sample of more than 31 participants, with respect to each AOI, and the duration and number of fixations and visits obtained show that their greatest number and their maximum duration concerned the faces of the main subject of each artwork (Fig. 24). The results assessment, combining visual mapping with eye tracker data (Fig. 24), highlighted that when light is not designed to see well (i.e., for well-lighting and well-being), the energy linked to the vision is dispersed on boundaries (e.g., areas of little significance of the artwork and/or the surrounding museum space); but when it is designed for well-lighting, it can be a fundamental guide for the project of setting up, exhibition, and museography, based on quality light that everyone likes (the historian, museum exhibition curator, conservator, and any visitor).

3.9 Natural Lighting Design for the Enhancement of Historical Spaces, Sustainability, and Health in the Proximity: The Scervino Space in Florence

A sustainable and well-lighting solution was proposed in a space of historical and architectural value for a precious and exclusive tailoring art, like that of Ermanno Scervino [11] (Fig. 25). The new lighting, based on the optimal mix of natural and artificial light, was obtained with technical control/regulation solutions (PIR systems integrated with COB/DALI), with bioclimatic solutions (i.e., tropical plants that filter and transform it), minimally invasive, energetically sustainable and easily maintainable, guaranteeing the protection of particular fabrics and materials. It is also capable of guiding the purchase along paths of new visual experiences, but in proximity, because the dramatic situation we are experiencing. The proposed solution kept the existing luminaires, modifying their arrangement and pointing angles, but introduced regulation and dimming systems for the luminous flux control, correlated color temperature and light color of LEDs linked to the new daylight sensors, photoresist twilight switches, with on/off and dimmer actuators, and people presence sensors. The LED lighting fixtures were reorganized into custom groups to obtain a base lighting, 30% reduced in terms of intensity, in comparison with existing condition, and a selective accent lighting, respectively for the whole space, the exhibition spaces, and different routes, was designed (Fig. 25). The natural light and photodegradation risk control were achieved by inserting groups of different tropical plants in the window area (Fig. 26). These plant groups, from the first to the

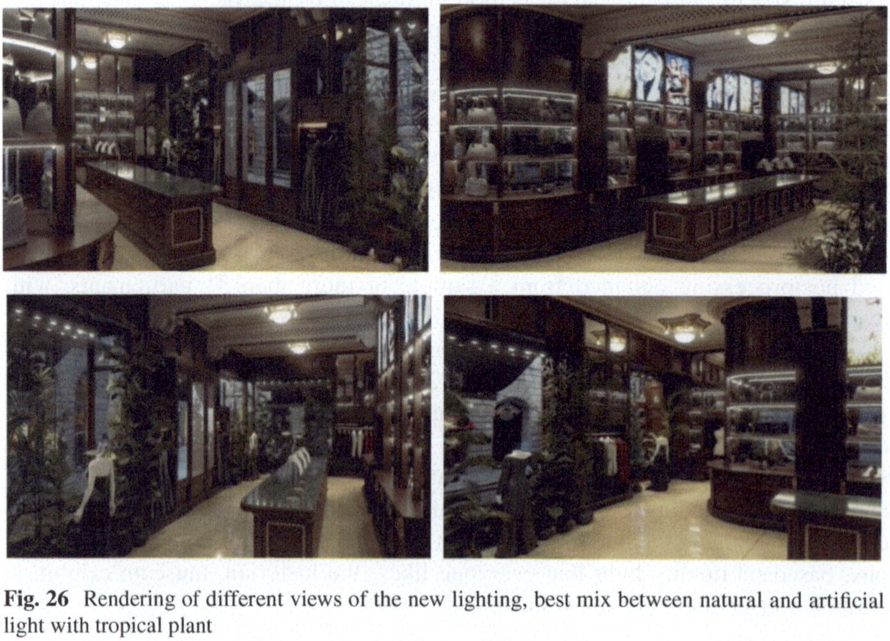

Fig. 26 Rendering of different views of the new lighting, best mix between natural and artificial light with tropical plant

third, include *Areca* palm, bamboo, and *Strelitzia nicolai*, with 4 m height; *Philodendron*, *Ficus*, and banana with 2–3 m height; and *Alocasia*, *Strelitzia reginae*, and the braided *Ficus* (weeping fig tree) with 1,5–2 m height. This is a minimally invasive solution with low maintenance costs but also suited to stimulate creativity, a state of well-being (visual and perceptive), and contribute to air purification. There is no doubt that plants cannot be a mere replacement of controlled mechanical ventilation systems, but they can really be a useful addition to support environmental quality and health requirements, since this kind of environment is characterized by a reduced number of people, whose access would be entirely planned.

Tropical plants were subdivided according to height and leaf area index and arranged in successive layers, for controlling, screening, and absorbing solar radiation. Referring to recent research, an average value of the absorption coefficient of 80–85% and reflection coefficient of 20–15% for all three groups of plants was deduced. Taking into account the precious and photosensitive materials of building, items of clothing, and objects displayed inside (e.g., silk, linen, wool, special yarns, damask, and lace), basic degradation parameters were calculated and compared with the photosensitivity classes defined in Blue-Wool ISO standards and Italian MiBAC standard [21], the maximum levels of annual dose of light is always within the limits imposed by the current standards (Table 6).

Although the illuminance levels near doors and windows are appreciably higher than in the surrounding zones, the corresponding values show a reduction of more than 50% compared to the existing state, due to the introduction of the three layer-groups of tropical plants. In particular, likewise, the direct illuminance value on the south wall, which is the most important area of the shop windows, is reduced by 33,83%; the corresponding value on the west wall, which has just as many shop windows, is reduced by 35,43% (Figs.25 and 26). Significant results were therefore achieved through the use of tropical plants that allow solar radiation control integrated with the limitation of UV penetration. Considering the opening hours of the boutique throughout the year, the LENI value obtained for the lighting solution was 42,92 kW/m^2 year, compared to 89,11 kW/m^2 year of the existing state; this fact shows an important energy consumption reduction of 51,84%.

Table 6 Maximum value of the annual dose of light (LO)

	LO shop-windows		LO Shelves
wall	**Summer natural&artificial light**	**Winter natural&artificial light**	**Only artificial light**
South	64642.5 lx/h year	45330 lx/h year	-
West	80167.5 lx/h year	61755 lx/h year	-
East	-	-	3915 lx/h year
North	-	-	4320 lx/h year
North-East	-	-	8235 lx/h year

3.10 Structural Lighting: VLC Application for Well-Lighting Mixing Natural and Artificial Lighting with Information and Communication

A structural lighting solution, i.e., quality of vision and perception by means of the transmission of information and therefore communication, designed for a standard office, belonging to a recent research, is shown and discussed [12]. The environment studied is a typical office with a lighting system with linear compact fluorescent discharge lamps. The proposed lighting solution is based on the integration of LEDs, which are essentially electronic components, with VLC techniques. The VLCs allow important applications of well-lighting and sustainable lighting: With the same energy, it lights up and communicates. In a VLC device, the light radiation emitted by an LED is modulated at frequencies invisible to the human eye but identifiable by a photodetector. The demodulated signal transfers information with a band potentially a few orders of magnitude higher than that one allowed by Wi-Fi devices, without additional electromagnetic pollution, since the signal carrier is the light radiation. This technology can be used for localization systems, with high precision, especially in indoor environments where radiofrequency (RF) systems encounter some critical issues. This study started from the VLC application to the existing lighting systems, i.e., linear compact fluorescent discharge lamps. The innovative solution proposed is an integrated system with six desk lamps to communicate by means of visible light. The LEDs whose characteristics are shown in Fig. 27 are table lamps modified with VLC, manufactured, and tested in the LENS European Laboratory for nonlinear spectroscopy in Florence [12]. This choice guarantees average illuminance values in the office and on the work surfaces in compliance with those suggested by standards [15–17], system reversibility, easy handling and replacement, and structured light with VLC (Fig. 27). Photometric and radiometric characteristics were measured by a spectroradiometer/photometer in modulation and nonmodulation conditions: the test results are shown in Fig. 27. The connection-combination of the LED lamps with the VLC system required a previous modulation check of the digital signal performed in the laboratory. Standard experimental measurements were carried out, with digital lux meters and luminance meters, on grids located on significant levels to define the luminous climate present

Photometric parameters of the LED built-in Spotlight measured with the spectrofotometer

Parameters	VLC Off	VLC On
Power	2 W	2 W
Luminous flux	170 lumen	170 lumen
Emission spectrum	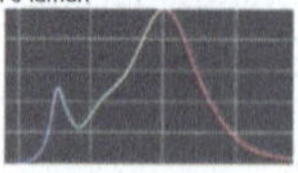	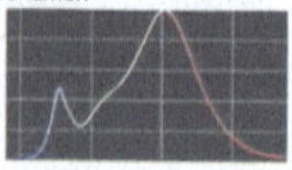
Colour temperature	2912 K	2984 K
Life	30 000 hours	≤ 30 000 hours

Fig. 27 Two photos of connection between LED and VLC system realized at LENS (left); LED built-in spot light measured data (right)

due to the mix of natural light and artificial light, luminous climate obtained with VLC integration on the artificial lighting systems present, and luminous climate obtained with VLC integration on new LED systems. Natural lighting is always guaranteed and controlled with a low-emission glass for all the conditions studied. A 3D model of the office with all the architectural, dimensional, and functional/distributive characteristics of the spaces and furnishings and optical, photometric, radiometric, and colorimetric characteristics of all surfaces was used for the lighting simulations of the two VLC integration solutions. Measurements allowed the construction and calibration of all the simulation models.

The comparison between two solutions, the first with VLC integration on linear fluorescent systems and the second with VLC integration on LED systems, shows a more uniform distribution of illuminance values and lower UGR on the worktops and floors for the second solution (Fig. 28).

The structured light solution (LED with VLC) provides quality of vision on all workstations, even for critical visual tasks, visual comfort, high contrast index, and ergonomics of vision even for long times (Fig. 28). The LENI value calculated at

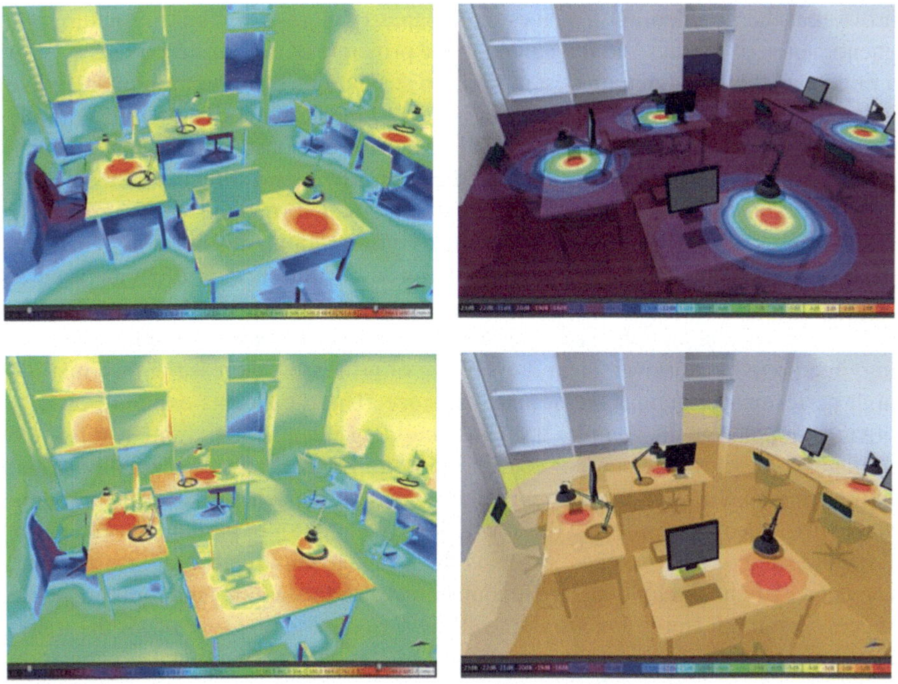

Fig. 28 VLC connection to the existing lighting system: on top left quality of light from the lighting point of view (lux), top right, dispersion of the received signal in dB compared to the maximum value, from the telecommunications point of view (dB); new lighting system VLC-LED integration: bottom left quality of light from the lighting point of view (lux), bottom right, dispersion of the received signal in dB compared to the maximum value, from the telecommunications point of view (dB)

precautionary conditions (maximum power of the sources and without regulation, hours of the day of maximum value of solar radiation, and natural light intensity) for the VLC-LED solution is 20,04 kWh/m^2 year. This value is higher but absolutely comparable with that one calculated for VLC-fluorescent which is 19,04 kWh/m^2 year. If basic VLC stations were also introduced instead of traditional Wi-Fi systems, energy savings would be greater and would guarantee a comprehensive integration of lighting and communication.

4 Conclusions

Any project is never definitively concluded. It could not be thought nor interpreted as definitively concluded, because it would not be a "human-centric lighting." On the other hand, *neuroscience* combined with *deep learning* studies shows, by means of complex connectomic maps, how the plasticity and differential resilience of the brain are constantly changing despite some functions and characteristics remaining constant over time (homeostasis). Neuroscience applications for lighting design prove that the production of images and shapes (patterns, rhythms) in nature is a phenomenon of physics and it is linked to the principle of the constructive law: "For a finite-size flow system to persist in time (to survive) its configuration must evolve in such a way that it provides an easier access to the currents that flow through it" (i.e., the constructal theory by Bejan [26]). This law demonstrates that any physical phenomenon and system have its own shape and temporal direction. Sustainable lighting and well-lighting allow to read architectural forms and building-lighting designs, by means of constructal theory, as information transmission. This fact means that they maximize the access of energy and matter flows and/or can produce characteristics to prevent them. The temporal evolution of these configurations, both for biological and physical systems, can be described through light and lighting, interpreted as survival, natural evolution of processes, increase in information, high performance, compactness, and resilience. These concepts explain how any organization/design of complex systems, building-plant system, and natural and artificial lighting system are susceptible to continuous changes, adaptations, and improvements, the more they are experienced and introspectively seen (i.e., thought and experienced by people and those which created them). In particular, optogenetics, connected to neuroscience, explains the fact that sensations and introspection, activated by light and lighting, are fundamental to understanding and thinking about spaces and architectures that are above all *biological structure* can follow the traces and the epigenetic footprints outlined, e.g., by the natural functional organization and form of any biological system (with energy and environmental effects). The lighting solutions provided and discussed show how the concepts of sustainability and adaptivity could derive from the sensation of beauty, physiologically mapped in the hippocampus. The concept of beauty (in architecture but also in nature) should be understood as harmony, right balance, and transmission of signals with a high information content, which can provide visual and perception quality and comfort, but above all that well-lighting, strictly connected to well-being, that constitute that common sense-sensation, to which everyone believes and from which starts to experiment and design new ones.

Acknowledgments Avoiding the name dropping, with the risk of forgetting someone, the author thanks very much all the collaborators and colleagues, as well as companies and research institutes that have contributed to the research and projects development and that are mentioned in literature references.

References

1. Balocco, C., Farini, A., Baldanzi, E., Volante, G. "Light, Information and Perception inside Historical Buildings. A case study", IOP Conf. Ser. Mater. Sci. Eng. 364, 2018, 012007.
2. Balocco, C., Volante, G. "A Method for Sustainable Lighting, Preventive Conservation, Energy Design and Technology. Lighting a Historical Church Converted into a University Library", Sustainability Int.J., 11, 2019, 3145-3163. doi:https://doi.org/10.3390/su11113145
3. Ciampi, G., Rosato, A., Scorpio, M., Sibilio, S. "Retrofitting Solutions for Energy Saving in a Historical Building Lighting System", Energy Proced. 78, 2015, 2669–2674.
4. Coburn, A., Vartanian, O., Chatterjee, A. "Buildings, Beauty, and the Brain: A Neuroscience of Architectural Experience" Journal of Cognitive Neuroscience, May 11, 2017, 1–11.
5. De Paiva, A. "Neuroscience for Architecture: How Building Design Can Influence Behaviors and Performance", Journal of Civil Engineering and Architecture 12 (2018) 132-138 . doi: https://doi.org/10.17265/1934-7359/2018.02.007
6. Fuentes, S.M., Pullen, S.J., Kaufman, T.K., Selby, A.J., Wolfe, D.J. "Are short (blue) wavelengths necessary for light treatment of seasonal affective disorder?", Chronobiology Int., 33 (9), 2016, 1267-1279. https://doi.org/10.1080/07420528.2016.1207660
7. Hendrick, C., Martyniuk, O., Spencer, T.J., Flynn, J.E. "Procedures for Investigating the Effect of Light on Impression: Simulation of a Real Space by Slides", Environment and Behavior, 9(4), 1977, 491–510. https://doi.org/10.1177/001391657794003
8. Seulki, H., Daehee, L. "The effects of treatment room lighting color on time perception and emotion. Int.J. Phys. Ther. Sci. 29, 2017, 1247–1249. https://creativecommons.org/licenses/by-nc-nd/4.0/
9. Suk, H.J., Kim, G.M. "The influence of chromacity of LED lighting on time perception", Int. J. Korean Saf Emotion Sens, 13, 2010, 69–78.
10. Houser, K.W., Esposito, T. "Human-Centric Lighting: Foundational Considerations and a Five-Step Design Process", Front Neurol.,12, 2021, 630553. doi: https://doi.org/10.3389/fneur.2021.630553
11. Balocco, C., Carbone, S. "Intensity and quality of light for the enhancement of historic spaces, sustainability and health in proximity. The Scervino space in Florence" LUCE, 335, 2021,56-61.
12. Balocco, C., Cataliotti, F.S., Mucchi, L., Caputo, S., Scacchi, A. "Lighting Design and Visible Light Communication. An example of structured light and vision quality", LUCE, 324, 2018, 106-111.
13. Balocco, C., Cecchi, M. "Adaptive reuse, refurbishment and conservative rehabilitation of Cultural Heritage by means of Quality and Energy Sustainable Lighting", Int. Journal of Physics: IOP Conf. Ser.: Mater. Sci. Eng. 949, 2020, 012046.
14. Balocco, C., Leccese, F., Volante, G., Salvadori, G. "Modelling Sustainable Lighting with Eye tracker and Spatial Syntax techniques", Int. Journal of Physics: IOP Conf. Ser.: Mater. Sci. Eng., 2020.

15. UNI EN-12464-1. Light and Lighting–Lighting of Work Places–Part 1: Indoor Work Places; EU: Brussels, Belgium, 2011.
16. I.E.S. Lighting Handbook: The Standard Lighting Guide; Illuminating Engineering Society of North America (Classic reprint): New York, NY, USA, 2018.
17. ANSI/IES TM-30-18. IES Method for Evaluating Light Source Color Rendition; ANSI/IES: Washington, DC, USA, 2018.
18. Volante, G. "Light, Information and Perception inside Historical Building. A case study: Villa la Quiete alle Montalve". Degree Thesis at the University of Florence, Civil Engineering Faculty, 2018. Tutors: Prof. Carla Balocco, Prof Mario Cozzi.
19. CEN TC 346-UNI EN 16883. Conservation of Cultural Heritage-Guidelines for Improving the Energy Performance of Historic Buildings; EU: Brussels, Belgium, 2017.
20. CEN/TS 16163. Conservation of Cultural Heritage-Guidelines and Procedures for Choosing Appropriate Lighting for Indoor Exhibitions; EU: Brussels, Belgium, 2014.
21. Ministry of Cultural Heritage and Activities in Italian language D.M. 10/05/2001 Guidance Document on Technical-Scientific Criteria and Museum Operating and Development Standards. (D. Lgs. n.112/98 Art. 150 Comma 6, MiBAC: Roma, Italy, 2001.
22. UNI-EN 15193-1. Energy Performance of Buildings-Energy Requirements for Lighting-Part 1: Specifications, Module M9; EU: Brussels, Belgium, 2017.
23. Duffie, J.A., Beckman, W.A. Solar Engineering of Thermal Processes. John Wiley & Sons Inc; four edition, USA, 2012
24. Balocco, C., Nistri, V., Pratesi G. Luce per la Scienza dell'Anatomia: le cere anatomiche del Museo della Specola, OPD Restauro, 27, 2015 (Book of the Opificio Pietre Dure Florence, Italy. in Italian)
25. Balocco, C., Baldanzi, E., Farini, A., Mercatelli, L. Luce, colore e percezione: studio illuminotecnico. In: Donatella Pegazzano, Capolavori a Villa La Quiete. Botticelli e Ridolfo del Ghirlandaio in mostra. 1-21, University Press Firenze, 2016.
26. Bejan, A., Lorente, S. Design with Constructal Theory. Wiley Ed. 2008

Carla Balocco is Associate Professor of Thermodynamics and Heat and Mass Transfer in the Department of Architecture, DiDA, Architecture Faculty, University of Florence (Firenze), Italy.

Main research interest: building physics and plant systems; historical buildings and cultural heritage; natural and artificial lighting; solar radiation control; energy sustainability; indoor air quality; people health and well-being. *Career*: PhD, Energy Engineering, Engineering Department, University of Florence (Italy), from 1991 to 1995. Assistant Professor and Lecturer in Thermodynamics and Heat and Mass Transfer at the University of Florence (Italy), from 1991; postdoctoral researcher at the University of Florence (Italy) from 1995 to 1997, postdoctoral research programme at the Technical University of Denmark (DTU), from September 1997; postdoctoral researcher at the Chemical Department, University of Siena (Italy), from 1998 to 1999; postdoctoral researcher in the 'Florence University Research Programme – 1999', studying on Thermodynamics Indicators for Energy-Environmental Sustainability, from 2000; postdoctoral research programme for Applied Thermodynamics – Heating and Air Conditioning at the University of Florence (Italy), from 2001; researcher and Senior Lecturer in Thermodynamics and Heat and Mass Transfer, ING/IND11 at Department of Industrial Engineering of Florence (DIEF), University of Florence, from 2002. Member of Centro Ricerca Energie Alternative e Rinnovabili (CREAR). From 2009, member of the Doctoral Committee in 'Energy and Innovative Industrial and Environmental Technologies', University of Florence. 2016–2017, member of the technical and scientific group for the Project concerning the Museums System of the University of Florence. 2018, referent and scientific member of the University Strategic Project (University of Florence).

From 2017, member of the technical and scientific group for the Project concerning the Basic Guidelines for Lighting Design in Italian Prisons 'HCL' Human Centric Lighting for the Ministero di Giustizia with the cooperation of ENEA (Agenzia nazionale per le nuove tecnologie, l'energia e lo sviluppo economico sostenibile). From 2016, member of the Technical Board UNI/GL 02 e NC CIE Italia DIV 3. From 2014, editorial board member of several international journals. 2014–2018, president of the Tuscany Section of the Italian Association of Lighting (AIDI). From 2014, member of the Scientific Committee of the International Congress on Heritage and Energy Efficiency. *Awards and recognition*: received several awards since 2012. *Research references*: has more than 80 scientific publications and papers in refereed journals on heat transfer, applied thermodynamics, thermal measurements, sustainable lighting, and solar radiation control. *Professional member*: of the Associazione Italiana Condizionamento dell'Aria Riscaldamento e Refrigerazione (AICARR), Unione Italiana Termofluidodinamica (UIT), Associazione Italiana Gestione Energia (AIGE), Centro Ricerca Energie Alternative e Rinnovabili (CREAR), Associazione Italiana di Illuminazione (AIDI), Associazione Termotecnica Italiana (ATI), Comissione Internazionale sulla Illuminazione (CIE), Società Italiana di Igiene Medicina Preventiva e Salute pubblica (SITI), and Fisica Tecnica Italiana (FTI).

Keeping Cool Under the Hot Arewa Sun: Natural Cooling Systems in Traditional Hausa Architecture of Nigeria

Amina Batagarawa and Huraira Umar Baba

1 Introduction

Traditional Hausa residential architecture can be loosely classified into urban and rural dwellings [14]. The difference in style between the urban and rural landscapes is not established as one finds similar styles within both landscapes. The urban style is mostly copied from international trends and as a result of availability of affordable building materials and fire safety codes in Nigeria.

This chapter discusses cooling in traditional Hausa residential architecture. Hausa land, also known as Arewa, is situated in the north of Nigeria. The region covers a belt from latitude 8° north to latitude 12° north of the equator.

The structure of the chapter is based on building components, environmental and cultural factors. The information used in this chapter is from a combination of sources; building materials and construction data from a literature review and the cooling strategies employed by the *magina*, the traditional professional builders in Hausa land, are described from information collected in a series of interviews.

The traditional construction process in the north of Nigeria is a cultural process of apprenticeship training in the art of building by the more seasoned experts to young boys coming of age. The building profession is seen as a family craft learnt through generations. Out of the seven professional builders interviewed, six are related. They are the chief or head builder of the Zaria emirate *masarauta*, his deputy, the *Wazirin magina*, and four heads of the ruling houses.

A. Batagarawa (✉)
Baze University, Abuja, Nigeria

H. U. Baba
De Montfort University, Leicester, UK

A. Sayigh (ed.), *Achieving Building Comfort by Natural Means*, Innovative Renewable Energy, https://doi.org/10.1007/978-3-031-04714-5_7

1.1 *Enviro-cultural Factors*

A combination of environment and social factors determine the different architectural design solutions that emerged in the precolonial traditional setting. The north of Nigeria, commonly known as *Arewa*, covers a belt from latitude 8° north to latitude 12° north of the equator. This makes the region a mostly hot and dry climate. The religion or philosophical learning of each community, their social order together with culture, local builder's expertise, industrialization, and even colonization influence the forms, grouping of buildings, as well as the process of their production (2001) [1, 4, 7].

1.1.1 Environment and Climate

The climate in the northern part of Nigeria falls within the savannah area and is characterized by very little rainfall and extreme temperature difference between day and night. Bright sunshine and hot, dry air dominate the daytime, whereas nights are cold. Trade winds originate from the northeast and southwest direction (see Fig. 1), known locally as *gabas* and *yamma* trade wind directions, respectively.

Windows are placed in the north and south generally to harness the trade winds. They are also placed high in the walls for air circulation through stack effect in a style called *ffuska-fuska*. High windows are also utilized for privacy, as discussed further in the next section.

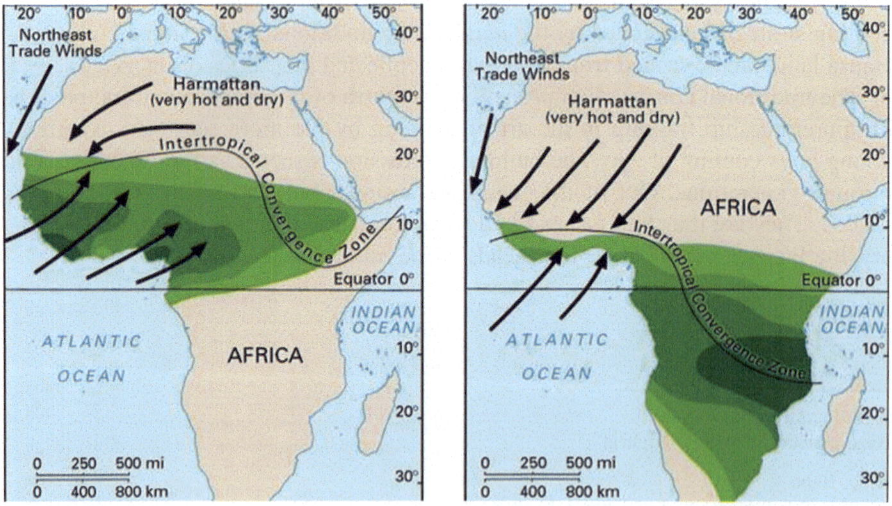

Fig. 1 Trade winds for Nigeria (Encyclopædia Britannica, Inc., 2012). (Source: Invalid source specified)

1.1.2 Socioculture

The rural landscape remains dominated by nucleated villages, possibly within the remains of old settlement walls or isolated and dispersed in more recent settlements [14]. In terms of hierarchy, three settlement types are identified as village *qauye*, town *gari*, and city *birni*. Culturally, there exists a concept of the triple cores in compound family houses which is replicated in the old concept of town or cities surrounded by a compound wall and accessed through the entrance gate.

Each family compound is made of organically round huts made of mud with thatch, usually protected by an entrance porch of thatch on forked sticks. Sun-dried mud strengthened with cut straw and thatch grass are the basic construction material for huts. Various types of sticks and mountain bamboo are used in constructing the thatch roof, often crowned with a plaited straw cap.

There are marked differences with the more urban compound structure, chief of which is that compounds are vanishing. This is due to a combination of factors such as price and availability of land, family structure morphing from a polygamous unit living in the same compound to individual units as one moves from rural to urban landscapes. In addition, the infrastructure required to operate such complex and multitiered compounds are inadequate and costly.

Another major factor is the influence of Islam, which leads to the use of curvilinear conical and mud dome roof structures as well as the ideology of seclusion and privacy for women [1]. The Hausa courtyard is encircled by numerous rooms that facilitate the expansion to accommodate more inhabitants such as wives and children [10]. The Hausa tradition of seclusion for women and children is a substantial factor in room orientation and placement of openings. Primary research shows that window openings are placed above eye level to seclude room occupants and soak up internal noise and conversations such that outsiders do not have a view inside a room and are not privy to conversations in a room.

Doorways are placed very low of a height of a little over a meter to prevent direct view into rooms as shown in. The absence of a pre-room or anteroom is a factor that leaves rooms naked and calls for such measures. A concept the expert and traditional head of builders, the "Sarkin-magina," is termed as "sirrin daki," – directly translated as "room privacy." The air from the doorway exits through the higher placed windows, further circulating the air, thereby improving thermal comfort. High placement of windows is for noise reduction or absorption, as well as for security as it prevents a view into rooms and as much for ventilation and air circulation (Fig. 2).

An apparent willingness to embrace change and new technology brought through formal or informal channels of technology transfer from a more developed to a less developed region, in this instance technology transfer through colonialism is embedded in some religious inclination. This was highlighted during the interviews with the professional builders who revealed that they do not reject change because God described himself as "zamani" a term that alludes to time, change, or modernity. This may be with reference to the Islamic Prophetic hadith (Huraira) that says

Fig. 2 Low level doorway

"Allah, the Exalted and Glorious, said: The son of Adam displeases Me by abusing Dahr (time), whereas I am Dahr--I alternate the night and the day."

1.2 *Traditional Hausa Construction*

The predominant materials used in African traditional building include earth, clay, sticks, corn stalk, thatch, water for mixing, stone, reeds, and grass [14]. These materials are used to make building components like the roofs, walls, foundations, floors, and openings like doors and windows.

1.2.1 Roofing

The timber beam used in traditional Hausa construction is commonly called *azara* beam. The common types of mud roofs are flat, usually supported on *azara* beams that sit on mud walls. Wider flat roofs are achieved by introducing corbels of mud at the top of the wall. Similar to mud walls, s*hafe* is a fine soil screeding that is laid on the roof, followed by horse manure *kashin doki*, which is used as waterproofing because it gives a shine and gloss that is waterproof.

Another roof type is the domed roofs called *tuluwa*, which combines a number of the elements found in flat roofs. The simplest *tuluwa* covers a square interior and is supported by *daurin guga*, the crown of the walls at the four corners of the room, which is covered with triangles of diagonally laid *azaras* (*tauyi*). On top of these were laid beams, also called *tauyi*, running parallel to the four. The apex of the crossed beams are plastered and covered with four triangular layers of azara, set parallel to the walls of the room. A dark mortar of earth and straw is used while working on the ribs supporting the domes of buildings.

The waterproofing of the flat roofs is done with stiff zana mats whereas that of the dome differs with the use of much more flexible asabari mats were used, made chiefly of strong grass (tsaure).

Other types of roofs are described [10, 11] *and the roofing system is called* rufin lema *which means umbrella cover. If it has eight columns, it is called* rufin lema mai qafa takwas, *which means umbrella cover with eight legs and shows a roofing type called* dami, *one with six columns and ten columns called* dami mai qafa shidda *and* goma, *respectively (Saad, Climate and Human Settlements – Integrating Climate into Urban Planning and Building Design in Africa,* [11]. Another *roofing system is called* rufin dami babba (Figs. 3, 4, 5 and 6).

Due to the thermophysical properties of thatch and earth, traditional roofs protect the building user against extreme temperatures [1]. However, the introduction of new building materials such as cement and corrugated iron sheet from Europe quickly gained wide acceptance, hence the abandonment of traditional thatch roofs which are highly inflammable. In addition, circular buildings are difficult to roof with rectangular iron sheets leading to the transformation from circular forms, which promote better ventilation through air movement to more rectangular forms.

1.2.2 Walling

The method of walling commonly employed is the wattle-and-daub earth technology locally called *tubali* (see Fig. 7). These bricks are made from excavated earth *kasa* thoroughly mixed with grass and thatch. The mixture is molded and left in the sun to dry. *Tubali* walls are vertical element of stacked blocks of either about 100 × 250 mm or 100 × 150 mm. The mud walls are constructed with a depth of about 5 to 6 pieces of *tubali* at the bottom which gradually tapers to a depth of two *tubali* at the top, meaning that the bottom about 600 mm which tapers to about

Fig. 3 Damin bakin rijiya mai qafa shidda, with six columns

Fig. 4 Damin bakin rijiyamai qafa goma, with ten colums

Fig. 5 Zaria mosque columns with mai bakin rijiya (well opening) roofing

200 mm in depth. This approach is employed to ensure strength and stability and, according to the professional builders interviewed, prevent dampness from traveling from the exterior into the interior (Fig. 8).

Tubali is laid in courses and then covered with a specially prepared mud mortar for joining and bonding unit until the wall reaches the required height. The mortar is processed from earth of the highest quality. It forms a shell around each brick, creating a three-dimensional structural network forming a composite element with the brick. It is common practice to cover the wet earth *birji* with horse manure and be left for about 3 days with more water poured daily. After being trampled upon, another layer of manure is spread over the composite, watered, and further trampled on. This process is repeated continuously about four times over three weeks.

Another type of mud wall construction used to construct round, square, and sometimes rectangular huts is a monolithic element and not tiered like the *tubali*. It is also made with mud sourced from selected pits that are sometimes far away from the building site. Wet and broken down lope grass or rice leaves are added to the mud to make a sticky paste and used like adobe.

Fig. 6 Rufin dani babba

Fig. 7 View from extension of Friday mosque, Masallacin Jumma'a showing more modern architecture Plater of Paris style of ceiling under a conventional aluminium sheets that are supported by wooden trusses

A special wall plaster, *chafe,* made from black earth to which a glutinous fluid, *gabaruwa,* obtained from acacia tree is added, while the plaster is still wet on the wall. The plaster may also be mixed with a fluid *makuba* from fruit pods of locust bean trees. *Gabaruwa* and *Makuba* add a waterproofing property to the plaster, similar to horse manure as discussed in the previous section. The performance of the plaster may be improved by the addition of *gashim juna,* which consists of goat hair mixed with grease scraped off previously soaked skins (Fig. 9).

Powdered wild root *dafara* is soaked to produce gum that binds the plaster. Two coats of *gabaruwa* and *makuba* allow the plaster to last for years without maintenance.

A type of temporary wall in nomadic homes or used as a fence is made of cornstalk or reed called *zana*. The wall is constructed by digging holes along the perimeter of the area and burying the pole in the holes with the thicker ends downward.

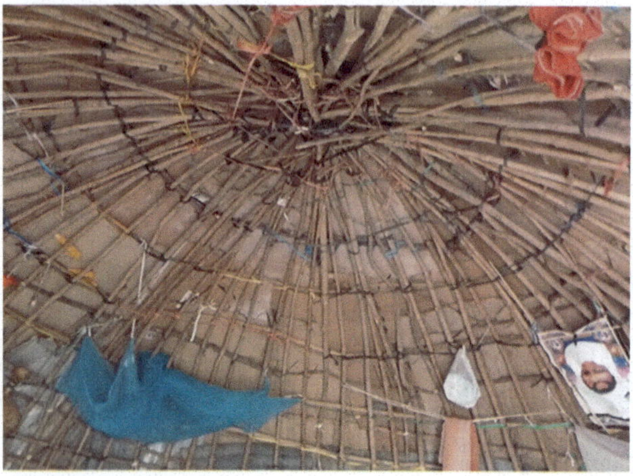

Fig. 8 Tubali wall construction

Fig. 9 Temporary zana construction

Reed mats are then tied to the poles for support by ropes made from grass. The walls are further strengthened with horizontal members, also made of corn stalk.

1.2.3 Openings

The openings found in traditional Hausa architecture are usually small door spaces with little or no window openings, in order to eliminate the hot, dry and dusty air ambient in the region [1]. The small doors also minimize bright daylight, dust, flies, and nighttime cold air (see Fig. 10). However, research shows that culture and religion play a major part in the configurations of the openings. Segregation between males and females and privacy dictated that openings are small and placed at a level not aligned to the average human's line of sight (Fig. 11).

Fig. 10 Sarkin Magina demonstrating entrance to a doorway to his court

Fig. 11 Openings on interior walls at Masallacin Juma'a Zaria

Doors

Doorways are of two kinds: outer doorways enclosed by wood or metal [2] and hung on pivots or hinges and inner doorless doorways that are occasionally screened with grass-plaited curtains. Outer doorways are generally rectangular, with a horizontal lintel made of a wooden beam supported at each end.

A typical doorless entrance consists of two parts as shown in Fig. 12. The lower part is rectangular and covered by a semicircular arch. When the floor of the room is not raised above ground level, the lower part of the doorway is barred with a *dangarama* to prevent the entry of rainwater. An overhang, *gemu* protects the doorway from the rain by projecting some distance from the wall above the opening as well as acting as a decorative element in the façade by emphasizing the entrance to the building.

Fig. 12 Head Builder's 'Sarkin magina' court entrance showing two-part doorway (right) and series of doorway through the house at Sarkin magina's quarters (left). (Source: Authors)

Windows

Windows, *Taga,* are slim and narrow between 4 and 6 inches wide. They are simple openings usually set in the uppermost part of walls on the side of the building less affected by driving rain. The location and small size of the windows help in curtailing the incursion of dust and flies. Most often the shape of windows is rectangular, although in more elaborate buildings they are often topped with an arch, *kandame*, or with a triangle. They can be hinged to open outward and angled in order to direct breezes into the building (Figs. 13 and 14).

2 Conclusion

Cultural and religious factors seem to play a substantive role in building construction decisions, more so than bioclimatic factors. This moves away from earlier studies that suggest that knowledge and technology were diffused through associations and integrations brought about by conflicts and as the case with traditional Hausa architecture, colonialism, or both.

The supremacy of culture over even climate as a factor for building decisions and eventual house structures is verified in all the interviews conducted, where emphasis on decisions of openings and details such as thickness of walls suggests that cultural considerations and religious positioning take precedence over climatic factors with the exception of building materials.

Fig. 13 A typical 'taga' or high-level openings

Fig. 14 High level openings or 'taga' depicted at Zauren Sarkin Magina (Head Builders' Court)

According to the professional builders interviewed, the main coolant is the earth material itself. The builder's choice of using earth as a building material is for its thermal qualities and abundance.

Apart from the use of earth, cooling in traditional Hausa architecture is achieved naturally by placing multiple rows of walling units, placing windows to harness the south and northerly trade winds, and limiting the size of openings to reduce flow of hot, dusty air into the cooled interiors.

A comparison with conventional buildings shows that walling materials are changing from the cheaper and more abundant earth to commercialized concrete blocks. The practice of using thatch is also disappearing due to improve fire safety and pest control. Roofs are more commonly made of metal roofing that is supported on roof trusses and an air gap. Flat roofs are more commonly made of concrete. The biggest factor affecting thermal comfort may be changes in openings.

Openings are made of doors and windows. Doors are more commonly made of wood in rectangular frames. On the other hand, windows are made of glass and placed with little recourse to sun-path of trade winds. Cooling of interiors has become mainly through mechanical means.

Further work is recommended through empirical comparison of the thermal performance between traditional and conventional Hausa architecture. The results will add to knowledge required for decision-making by designers and builders and in formulating building codes and policy as well as contribute to the body of knowledge required to bridge the gap between vernacular architecture of the Hausa and frontier innovative building practices.

References

1. Agboola, O. P., & Zango, M. S. (2014). Development of Traditional Architecture in NIgeria: A case Study of Hausa house form. *International Journal of African Society Cultures and Traditions, 1*(1), International Journal of African Society Cultures and Traditions. Retrieved from https://d1wqtxts1xzle7.cloudfront.net/35152825/Development-of-Traditional-Architecture-in-Nigeria-A-Case-Study-of-Hausa-House-Form-with-cover-page-v2.pdf?Expires=1624884445&Signature=g3AKlkWyk4d60V-D5sAHWueC8n5qRhGAgP~r6MLzuOvnt-rjkItgJrZ9puXvMb9jzSDKQGbm
2. Batagarawa, A., & Tukur, R. B. (2019). Hausa Traditional Architecture. In A. Sayigh, *Sustainable Vernacular Architecture: How the Past Can Enrich the Future* (pp. 206–228). Springer. Retrieved from https://books.google.com.ng/books?hl=en&lr=&id=va6PDwAAQBAJ&oi=fnd&pg=PA206&ots=Wzms1HaS3n&sig=5_37uxRo0XlW2yfcsYDn31_BLzw&redir_esc=y#v=onepage&q&f=false
3. Eneh, A. E., & Ati, O. F. (2010). The Influence of Rainfall on Hausa Traditional Architecture. *Research Journal of Applied Sciences, Engineering and Technology, 2*(8), 695–702. Retrieved from https://maxwellsci.com/print/rjaset/v2-695-702.pdf
4. Fatiregun, A. (1999). *A brief history of traditional African architecture (with 50 illustrations).* Ilesa: Fatiregun.
5. Gupta, S. K., & Nathani, N. (2016). Thermal Performance of Wall & Roof – Efficiency of Building Materials. *International Journal of Science and Research (IJSR), 5*(5), 1354–1357. Retrieved from https://www.ijsr.net/archive/v5i5/NOV163707.pdf
6. Huraira, A. (n.d.). *Sahih Muslim* (Vol. 40:2). Retrieved 2021, from https://sunnah.com/muslim:2246b
7. Opoko, A. (2001). Low energy features of traditional buildings in the Hot-Dry Climatic. *Nigerian Institute of Architects Journal, 2*(8–12), 29–300.
8. Peel, C. (1958). Thermal conditions in traditional mud dwellings in northern Nigeria. *Journal of Tropical Medicine and Hygiene*, 189–203.
9. Rapoport, A. (1969). *House Form and Culture: Foundations of Cultural Geography Series.* New Jersey: Englewood Cliffs, Prentice Hall. Retrieved 07 21, 2021

10. Saad, T. (1991a) Folk Culture and Architecture in Northeastern Nigeria. *Mitteilungen zur Kulturkunde*, 253–289. Retrieved from http://www.jstor.org/stable/40341634.
11. Saad, T. (1991b). *Climate and Human Settlements – Integrating Climate into Urban Planning and Building Design in Africa.* United Nations Environment Programme. Retrieved 2021, from http://www.nzdl.org/cgi-bin/library.cgi?e=d-00000-00%2D%2D-off-0unescoen%2D%2D00-0%2D%2D%2D%2D0-10-0%2D%2D-0%2D%2D-0direct-10%2D%2D-4%2D%2D%2D%2D%2D%2D-0-0l%2D%2D11-en-50%2D%2D-20-about%2D%2D-00-0-1-00-0-0-11-1-0utfZz-8-10&cl=CL3.123&d=HASH01b752746b83885cf0c460de.8.3&x=1
12. Schwerdtfeger, F. W. (1984). Thermal conditions in traditional urban houses in northern Nigeria. *Habitat International VIII, 3*(4), 43–76.
13. Thatching Advisory Services, UK. (2021, 07 11). *U-Value & Condensation Reports*. Retrieved from Thatching Advisory Services: http://thatchingadvisoryservices.co.uk/fire-barriers-systems/new-building-extensions/u-value-condensation-reports/
14. Umar, G. K., Yusuf, D. A., & Abdullah, A. A. (2019). The practice of Hausa traditional architecture: Towards conservation and restoration of spatial morphology and techniques. *Scientific African, 5*, e00142. https://doi.org/10.1016/j.sciaf.2019.e00142.
15. Wigton Physics. (2018, 06 25). *U Value of a Thatch Roof*. Retrieved from 0.29 and 0.23: http://wigtonphysics.blogspot.com/2018/06/u-value-of-thatched-roof.html

Amina Batagarawa holds a BSc in Architecture from Ahmadu Bello University, Zaria, Nigeria; an MSc in Environmental Design of Buildings from Cardiff University, Wales; and a PhD in Architecture from Newcastle University, England.

Amina is a lecturer in the Department of Architecture, Ahmadu Bello University, Zaria, Nigeria, at undergraduate and postgraduate levels. Her teaching and research are focused on sustainable energy use in the built environment. She is the author of various articles, including the following:

Benefit of conducting energy calculations in the built environment of Nigeria, at the 4th West Africa Built Environment Research (WABER) Conference, 24–26 July 2012, Abuja, Nigeria, 389–397;

Disaggregating primary electricity consumption for office buildings in Nigeria, at the 12th Conference of International Building Performance Simulation Association, Sydney, 14–16 November 2011; and

A comparison of lightweight and heavyweight construction incorporating Phase Change Materials for Office buildings in a composite hot climate, at the 6th Annual Conference of the West African Built Environment Research Institute, Ghana, which won Winner of Best Industry-Related Paper.

She is a global advisory board member of the Green Campus Institute (www.greenthecampus.org); a trustee of the Board of Sustainable Energy Practitioners Association of Nigeria, where she heads the Research Support Services Department of the association (www.sepan.org.ng); and a member of the Energy Institute, UK.

The Role of Shading, Natural Ventilation, Daylighting, and Comfort in Enhancing Indoor Environmental Quality and Liveability in the Age of COVID-19

Mohsen Aboulnaga and Elsharkawy Maryam

1 Introduction

The impact of Coronavirus (COVID-19) on buildings, urban areas, and cities proved the significance of healthy spaces in all building types [1, 2]. However, post-pandemic-transformed spaces are mostly related to the users and new functional space requirements. In office spaces, teleworking has occupied a major role in changing the working conditions to suit the pandemic period. Most office spaces were fully or partially abandoned, and workers relied on teleworking conditions [3]. In housing units, biophilic design approach and the use of balconies have been considered [4]. While, in public spaces, reforming social distance rings in parks and widening pedestrian paths with emphasis on larger cycling lanes have been suggested and introduced in many countries [5]. Indeed, fewer studies discussed the indoor environmental quality (IEQ) basic principles and recommended strategies to adapt for both existing and new building construction during the pandemic period [6]. The IEQ also describes the healthy indoor environmental conditions that cover the air quality, the daylighting and access, the lighting and thermal comfort of users, as well as the view and noise level [7] as shown in Fig. 1. Many studies examined building spaces and users' satisfaction in terms of IEQ to highlight the significance of all aforementioned factors in providing healthy indoor spaces [7–9].

Indoor environmental quality (IEQ) is essential to assure the provision of healthy spaces in buildings and cities amid COVID-19 pandemic, specifically in highly populated buildings such as educational and public buildings, but the absence of clear standards that measure the effectiveness of indoor conditions in providing

M. Aboulnaga (✉)
Department of Architecture, Faculty of Engineering, Cairo University, Giza, Egypt
e-mail: maboulnaga@eng.cu.edu.eg

E. Maryam
AET Program, Faculty of Engineering, Cairo University, Giza, Egypt

A. Sayigh (ed.), *Achieving Building Comfort by Natural Means*, Innovative Renewable Energy, https://doi.org/10.1007/978-3-031-04714-5_8

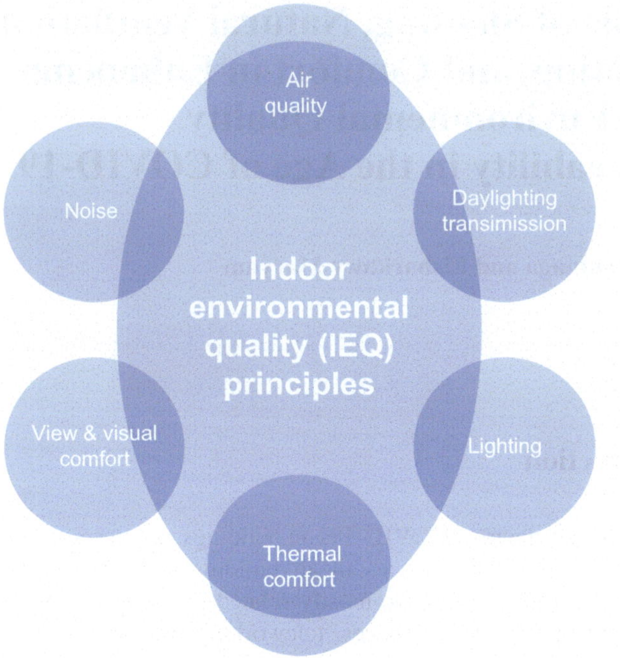

Fig. 1 Principles of indoor environmental quality (IEQ). (Image source: Developed by authors)

healthy spaces resulted in the need for benchmarks for related IEQ [10]. Many strategies and techniques are incorporated within buildings to enhance their IEQ. Such strategies encompass shading on building facades, achieving natural ventilation through inner spaces, adequate daylighting provision, and thermal comfort of occupants. Different factors affect the IEQ inside the building spaces [11, 12]. Thus, a thorough review of previous cases is essential to understand and learn the good lessons and highlight the most effective techniques that could be reimplemented in future cases with special requirements to suit the age of COVID-19. Even with the lifting of lockdown and confinement measures in the Netherlands and the United Kingdom (June and July 2021), an additional need to ensure healthy spaces and buildings in cities is posed [13, 14].

2 Objectives

The work aims at highlighting the significance of IEQ and assessing different global case studies that provide adequate and healthy spaces for users in the age of COVID-19. The research work enriches the knowledge base and explores the different factors involved in achieving liveability for occupants inside building space, encouraging the sustainable experience in buildings, and learning lessons from previous cases.

3 Methodology

A comparative case study analysis approach is provided through assessment and review of various global experiences with integrated different design approaches and strategies to improve the building indoor environmental quality. Analysis of existing global experiences and their related design impact on liveability reveals the significance of the applied strategies.

4 Benefits of Indoor Environmental Quality (IEQ)

High indoor environmental quality has countless benefits for users and has additional enhanced values for the building asset as well. Users' benefits are not limited to the positive psychological effects, stress reduction, and health improvements but also on the productivity and social status [11]. However, there are different factors that participate in improving IEQ in building spaces, and they are explained briefly below [12]. Figure 2 presents these benefits in brief.

4.1 Shading Strategies and Building's Facades

Building facades is considered the main barrier between indoor and outdoor environments and the effect of shading strategies in improving mainly the indoor daylighting conditions, while participating in improving thermal comfort and ventilation as well. Indoor daylight condition is critical, and achieving both adequacy of daylight and visual comfort levels inside building spaces has been studied to optimize the effect of the shading strategies utilized in buildings [15]. This is made possible through various techniques and technologies utilized and fitted specifically for each building location, orientation, and function as well [16]. Shading strategies come in various forms, and they can further be separated into static and dynamic shading strategies [17].

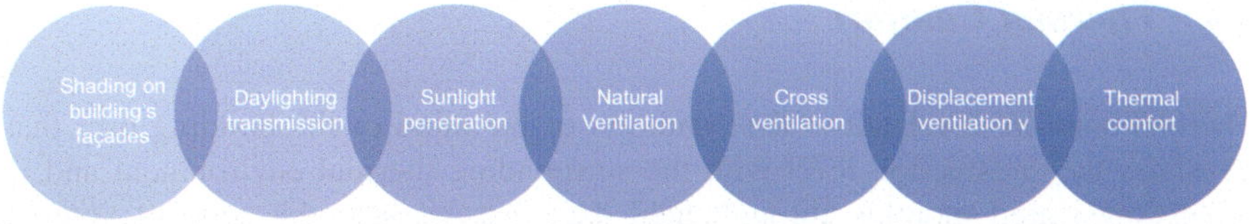

Fig. 2 Benefits of indoor environmental quality (IEQ). (Image source: Developed by authors)

4.2 Natural Ventilation, Cross Ventilation, and Displacement Ventilation

The ventilation term is originally derived from the Greek word vent, which means opening or window [18]. Natural ventilation is considered in building spaces with potential good airflow circulation, such as in hot regions where cool air at night is allowed to diffuse through open windows into building spaces, and this helps ventilate the warm air outside spaces [19]. The main benefit of good natural ventilation is the provision of fresh air, and subsequently, it would reduce the operational costs due to less usage of electricity for mechanical ventilation in building spaces [20]. However, this would be effective when air temperature is below the body temperature (33 °C to 36.5 °C). Nevertheless, Centers for Disease Control and Prevention (CDC) recommends a layered approach and exploiting multiple strategies to reduce the spread of Coronavirus (COVID-19), including improving ventilation in buildings and reducing the risk of exposure [21]. Tools to enhance ventilation and improve IEQ are listed in [21].

In addition, cross ventilation happens when fresh air enters the space through two opposite openings. Cross ventilation is becoming even far essential to provide fresh air at a certain volume per person (1 cubic meter per person) [22]. In the time of COVID-19 pandemic, this volume is recommended to increase for health reasons. Moreover, displacement ventilation is essentially needed in large spaces [23].

4.3 Daylighting and Sunlight

Daylight and sunlight are often neglected in building design, because in most cases the positive effect of them is intangible. However, the psychological effect of daylight availability on building occupants has been studied in all building types and has proven to be of great significance on the occupants' well-being [24], but nevertheless, the negative effect of solar heat gain through translucent elements has limited the incorporation of daylight into building spaces. The challenge in design to achieve optimum daylight and sunlight availability inside building spaces has been tackled by various researchers and supported by technologies that found their way to arrive at optimum solutions [25].

4.4 Thermal Comfort

Thermal comfort is related to the condition of the mind of the building user that translates the satisfaction level with the surrounding thermal environment and is usually measured monthly through field data that assesses the comfort vote of participants [26] and through simulation software that depends on the thermal comfort rates such as predicted mean value (PMV) where the highest value of 1.15 means extremely discomfort with high thermal heat exposure [27]. Optimum thermal comfort levels in buildings could be achieved by passive means such as improving the external thermal envelope efficiency. Indeed, thermal comfort enhances the users' thermal satisfaction and improves quality of life for occupants. Also, the capability of the building design and envelope efficiency to maintain acceptable thermal comfort levels inside building space without relying on cooling/heating systems is considered one of the main sources of reducing building operational energy costs and harmful carbon emissions [28].

5 Benchmark for IEQ in the Time of COVID-19

During COVID-19 pandemic, the benchmark for IEQ has been clearly stated in standards and rating systems [29] that included acceptable levels of indoor finishing materials with low harmful emissions, thermal and visual comfort levels, CO_2 monitoring levels, effect daylight, and external views [29]. In the age of COVID-19, the benchmark for IEQ has changed to adapt to the new measures enforced by governments worldwide, to meet the challenging requirements, and to minimize the disease transmission by users into spaces mainly closed spaces [30]. In most cases, the connection to nature, natural ventilation, sufficient daylighting, adequate shared common spaces with social and physical distance, and external view is prioritized to ensure livability of building spaces in the age of COVID-19 [31].

This section depicts selected contemporary and traditional buildings that include the main features of IEQ in order to draw the lessons learned from such buildings. The section also highlights how IEQ achieves liveability in the age of COVID-19 and presents a comparative analysis of assessed buildings (case studies).

6 Global Examples of Best Practices

Global examples of significant building design and construction have exempted huge transition in terms of innovation and creativity to ensure high IEQ inside spaces. There are multiple techniques and strategies applied that serve the needs of each case to adapt to the surrounding conditions. Figure 3 illustrates the examples selected for studies that address shading strategies.

Fig. 3 Examples of examined buildings depicting shading Strategies in the USA, Denmark, Spain, Sweden, Australia, France, UAE, and the Czech Republic. (Source: Developed by authors)

6.1 Shading Strategies and Building Facades

Building facades are considered the most vital building element to alter the indoor building conditions and contribute to the IEQ. Different cases have utilized different shading strategies on building facades to enhance the building performance. These techniques and their related performance are reviewed.

6.1.1 Aqua Tower Chicago Condos, USA

In the USA, the Aqua Tower Chicago condos have a unique design [32]. Semi-outdoor areas are created randomly to provide self-shading structures on the building facade that protect the indoor from direct daylight through the large glazing areas [33] as presented in Fig. 4a, b.

6.1.2 Syddansk University, Kolding Campus in Copenhagen, Denmark

The facade incorporates perforated metal shades that automatically move in response to sunlight to adjust daylight exposure in the interior space (Fig. 5). Figure 6 also shows the details of the polycarbonated louvers for shading. These colorful shades are unique in design and help the building to integrate mixed mode

(a) General view of the Aqua Tower (b) Shading louvers in the façade of the Aqua Tower

Fig. 4 The Aqua Tower Chicago condos, USA (**a**) General view of the Aqua Tower. (Image credit and source: George Showman, https://commons.m.wikimedia.org/wiki/File:Aqua_Tower_Chicago.jpg) (**b**) Shading louvers in the facade of the Aqua Tower. (Image credit and source: Paperclips0701, https://commons.m.wikimedia.org/wiki/File:AquaTower2009_02_08_image2.jpg#)

(a) General view of the University building from both sides (b) Vertical triangle-shaped shading louvers in the main façade to provide daylight and control glare

Fig. 5 Syddansk University, Kolding Campus in Copenhagen, Denmark (**a**) General view of the university building from both sides. (Image credit and source: S. Juhl, https://commons.m.wikimedia.org/wiki/File:Syddansk_universitet.Campus_Kolding.Denmark.2014_(37).JPG) (**b**) Vertical triangle-shaped shading louvers in the main facade to provide daylight and control glare. (Image credit and source: S. Juhl, https://commons.m.wikimedia.org/wiki/File:Syddansk_universitet.Campus_Kolding.Denmark.2014_(35).JPG)

lighting into the interior. The shades are composed of 1600 adjustable triangular shutters that respond to daylight and are fully closed when they lie flat to the external facade by means of sensors and a small motor; thus, the kinetic mechanism of the shutters could be achieved (Fig. 6a–d). With 30% shutter opening angle, optimum daylight efficiency is provided [34]. The building encounters the provision of natural ventilation through the middle atrium which enhances the thermal comfort of occupants with low energy [35].

(**a**) Close up view of the coloured louvres from outside (**b**) Dynamic louvers seen from inside the building

(**c**) A view of the louvers shading the main façade (**d**) Detailed view of the dynamic coloured louvers

Fig. 6 Vertical louvers of the dynamic facade, Syddansk University, Kolding Campus, Copenhagen, Denmark (**a**) Close up view of the colored louvers from outside. (Image credit and source: S. Juhl, https://commons.m.wikimedia.org/wiki/File:Syddansk_universitet.Campus_Kolding.Denmark.2014_(46).JPG) (**b**) Dynamic louvers seen from inside the building. (Image credit and source: S. Juhl, https://commons.m.wikimedia.org/wiki/File:Syddansk_universitet.Campus_Kolding.Denmark.2014_(50).JPG) (**c**) A view of the louvers shading the main facade. (Image credit and source: S. Juhl, https://commons.m.wikimedia.org/wiki/File:Syddansk_universitet.Campus_Kolding.Denmark.2014_(40).JPG) (**d**) Detailed view of the dynamic colored louvers. (Image credit and source: S. Juhl, https://commons.m.wikimedia.org/wiki/File:Syddansk_universitet.Campus_Kolding.Denmark.2014_(44).JPG)

6.1.3 ICTA-ICP UAB Research Centre, Barcelona

The ICTA-ICP Research Centre at UAB Barcelona (Universitat Autònomous de Barcelona) is a low-cost exterior bioclimatic skin. It operates to regulate the internal air temperature, and it enables sun heat to warm the building whenever required in summer, creating a greenhouse effect in the roof (Fig. 7). Inside the building, four patios are utilized to enhance daylighting and ventilation in addition to several plantations that regulate the indoor humidity [36, 37]. The ICTA building has smart shading technologies to provide excellent IEQ. The smart facades and main entrance of the ICTA building are shown in Figs. 8 and 9, including:

- Double-skin facades
- Polycarbonate smart glass louvers
- Wood acts as an insulation for the facades' windows
- Louvers connected to computerized system to operate them in response to climate

Fig. 7 ICTA-ICP Research Centre building at UAB, Barcelona, Spain. (Image credit and source: Pietro Tonini and Mohsen Aboulnaga, 31 January 2022)

(a) The façade of the ICTA building with shading elements and smart technologies

(b) General view of the façade with the kinetic shutters closed in winter

(c) Polycarbonate smart louvers to provide shading

(d) Polycarbonate pyramids on the roof for daylight

Fig. 8 Smart shading technologies exploited in the facades of the ICTA-ICP, UAB Barcelona (**a**) The facade of the ICTA building with shading elements and smart technologies (**b**) General view of the facade with the kinetic shutters closed in winter (**c**) Polycarbonate smart louvers to provide shading (**d**) Polycarbonate pyramids on the roof for daylight. (Images' credit and source: Pietro Tonini and Mohsen Aboulnaga, 31 January 2022)

Fig. 9 Main entrance of the ICTA-ICP Research Centre at UAB. (Image credit and source: Pietro Tonini and Mohsen Aboulnaga, 31 January 2022)

The smart envelope of the ICTA-ICP Research Centre at UAB Barcelona (Universitat Autonomous de Barcelona) includes many features such as the installed greenhouse smart system that opens and closes automatically where the solar gain and ventilation are regulated. It is also possible to control and raise the interior temperature naturally and guarantee a base of comfort in the circulation spaces as well as in the in-between spaces. In addition to the shading strategy in the ICTA-ICP UAB Research Centre in Barcelona, the building provides many benefits in energy efficiency, energy saving, and comfort as well as generating clean food through urban farming in the upper floor. These benefits are:

- Optimize both comfort and energy consumption through controlling and raising the indoor temperature naturally and guarantee a base of comfort by circulating the air into spaces
- Minimize the use of nonrenewable energy sources
- React and adapt constantly, opening and closing itself
- Activate and deactivate and manage itself to use all the natural possibilities within the outside/inside environment
- Utilize a greenhouse smart system that relies on mechanisms that open and close the system automatically; thus, the solar gain and ventilation are regulated and daylighting is enhanced
- Provide real comfort perception with less artificial than usual [36, 37]

6.1.4 Kuggen Building at Chalmers Lindholmen, Sweden

The Kuggen building at Chalmers Lindholmen in Sweden building is colorfully located within an office district as illustrated in Fig. 10. The rounded building does not look similar from each side, the upper floors project outward to shade the lower floors, and the projection increases on the southern side than on the northern side. Also, a rotated upper screen element shades the top floors following the sun path to

Fig. 10 Kuggen building at Chalmers Lindholmen in Sweden. (Image credit and source: Mats Kristoffersen, https://commons.m.wikimedia.org/wiki/File:Kuggen.JPG#)

reduce the amount of daylight entering the building spaces and to ensure visual comfort for all building occupants with improved thermal comfort impacts [38].

6.1.5 Surry Hills Library and Community Centre, Sydney, Australia

In Australia, the Surry Hills Library and Community Centre in Sydney is an innovative environmentally friendly building. It incorporates an automated louver system, which is located on the southern facade as illustrated in Fig. 11a. The vertical timber louver system automatically functions to track sun rays, and the daylight, while ensuring maximum benefit of daylighting to the occupants inside the building [39, 40]. As presented in Fig. 11, the computerized smart sun-travel system is integrated in the main facade to regulate:

- Natural ventilation
- Sun-shade louvers to adjust air movement, daylight, and heat

6.1.6 The Ministry of Culture in Paris, France

The old building of the Ministry of Culture in Paris, France, was modified with adding a contemporary metallic screen that illustrates an abstracted renaissance painting as shown in Fig. 12. The golden screen only shines in the daylight and seems to disappear in absence of daylight. It also provides shade and shadows on the facades. In addition, the metallic screen reflects an architectural value and helps regulate sunlight and daylight conditions into the interior spaces [41].

(a) General view of Surry Hills Library and Community Centre in Sydney, Australia

(b) The automated vertical timber louver system to adjust to solar heat, track the daylight ensuring maximum benefit of natural light

Fig. 11 Surry Hills Library and Community Centre in Sydney, Australia. (**a**) General view of Surry Hills Library and Community Centre in Sydney, Australia. (Image credit and source: Elekhh, https://commons.m.wikimedia.org/wiki/File:Surry_Hills_Library_2010.jpg) (**b**) The automated vertical timber louver system to adjust to solar heat, track the daylight ensuring maximum benefit of natural light. (Image credit and source: Nick-D https://commons.m.wikimedia.org/wiki/File:Surry_Hills_Library_and_Community_Centre_July_2017.jpg)

6.1.7 Al Bahr Towers, UAE

The adaptive parametric facade designed to operate in response to sun exposure level during the year. The screen lies on an independent frame that holds the triangular elements programmed to respond to sun movement to reduce solar heat gain and discomfort glare levels shown in Fig. 13 [42]. The system is very accurate in

(a) A metallic screen on the façade to provide shading on windows

(b) Golden screen shines in the daylight and such shine disappears in absence of daylight.

Fig. 12 The Ministry of Culture building in Paris, France (**a**) A metallic screen on the facade to provide shading on windows (**b**) Golden screen shines in the daylight and such shine disappears in absence of daylight. (Images' credit and source: Mohsen Aboulnaga)

(a) View of the kinetic façades of the twin towers

(b) Details of façades' kinetic elements linked with sensors to provide real-time data for computerised systems to regulate indoor temperature and daylight

Fig. 13 Al Bahr smart twin towers in Abu Dhabi, UAE (**a**) View of the kinetic facades of the twin towers. (Image credit and source: Mohsen Aboulnaga) (**b**) Details of facades' kinetic elements linked with sensors to provide real-time data for computerized systems to regulate indoor temperature and daylight. (Image source: https://www.sciencedirect.com/science/article/pii/S1110016818302230)

responding to the building's shading and daylight requirements. It also attains maximum daylight adequacy with minimum glare levels while reducing solar heat gain through the large glazed facade area. During the night, all the shading elements are closed to keep a clear view to the outside of the building [43]. The double-skin and dynamic facades, which are inspired by traditional mashrabiya, encompass the following smart features:

- Dynamic shading system
- The external cladding also provide shading
- Smart dynamic lenses reduce the building's solar gain by 50%
- Facades are linked with sensors to provide real time data for computerized systems that adapt to changing weather conditions daily

6.1.8 Hotel Building in Prague, Czech Republic

This building in Prague, Czech Republic, was renovated and serves as a hotel in the capital. The facade is integrated with shading titled screens to protect the rooms' glazing windows from the western sun and provide adequate daylighting and prevent glare in the summer months, as shown in Fig. 14. On the western facade, louvers are fixed at an angle of 15 degrees to protect the widows' glazing for solar radiation and heat by providing enough shading. Each widow has a large horizontal louver to:

- Control solar radiation
- The use of bright white color facade to assist in reflecting the solar radiation

(a) General view of louvers on the main façade **(b)** Shading louvers to control sunlight and daylight

Fig. 14 A hotel building in Prague, Czech Republic (**a**) General view of louvers on the main facade (**b**) Shading louvers to control sunlight and daylight. (Images' credit and source: Mohsen Aboulnaga)

- Provide shading
- Reduce heat gain through the glass and walls near the widow
- Protect from summer heat
- Control daylight and glare
- Contribute to indoor thermal comfort

6.2 Thermal Comfort

Thermal comfort is humans' perception of thermal well-being related to four environmental factors which are: (a) air temperature surrounding the body; (b) air speed; (c) mean radiant temperature (MRT), which is related to surfaces' temperature in the space, normally six surfaces (depends on the configuration of the indoor space); and (d) relative humidity.

Thermal comfort is also affected by other personal factors like metabolic rate activities/work rate) and clothing insulation (layers of cloth covering the body). Other factors are body surface area (slim or obese), skin color, age, and location in the space (near window or external wall or far from the surfaces) [44].

Thermal comfort can be assessed based on a checklist according to British and European international standards relevant to working in thermal environment [45]. According to Fanger, tools to measure thermal comfort in buildings are listed in [46]. However, the predicted mean vote (PMV) is a 7-point scale of thermal sensation in a function of measured physical parameters [47]. Thermal comfort can be achieved by adapting the outdoor environment of the building, i.e., coupling indoor comfort with that of outdoor. This is achieved by balancing interdependent factors to achieve low-energy thermal comfort for occupants [48].

The following section presents examples showing how thermal comfort is tackled, namely, the Beehive building in Sydney Australia; the Beehive Parliament building in New Zealand; the Kendeda Building in Georgia, USA; Luxor Temple and Habu Temple in Luxor, Egypt, as illustrated in Fig. 15.

6.2.1 The Beehive in Sydney, Australia

Thermal comfort has been tackled in the Beehive building through the form of the building and its envelope design. A wooden screen enfolds the exterior of the building in a dynamic way, which filters sunlight and heat into the semi-outdoor spaces. This solution enables indirect shading that creates a buffer environment for the outdoor air to cool down before entering the building inner spaces, thus participating in improved thermal comfort [48].

Fig. 15 Examples of examined buildings depicting shading Strategies in New Zealand, the USA, and Egypt. (Source: Developed by authors)

Fig. 16 The Wellington Beehive Parliament Building's conical shape with shading screen in New Zealand. (Image credit and source: Michael Coghlan, https://commons.m.wikimedia.org/wiki/File:The_Beehive_(26564318913).jpg#)

6.2.2 The Beehive Parliament Building in Wellington, New Zealand

The Wellington Beehive Parliament Building concedes thermal comfort in the same way through the conical style design of the Beehive building. The building envelope design which is made of wooden screen wraps the exterior of the building in a dynamic manner which filters sunlight and sun heat into the semi-outdoor spaces; this enables indirect shading that also creates a buffer environment for the outdoor air to cool down before entering building inner spaces, thus participating in improved thermal comfort levels [49] as shown in Fig. 16.

6.2.3 Kendeda Building in Atlanta – Georgia, USA

The Kendeda Building – an education-type building – has a broader thermal comfort range that operates by considering air velocity, air temperature, humidity levels, clothing, and activities (Fig. 17). The building is able to operate at higher thermal comfort level while maintaining the comfort levels for occupants, thus reducing energy needed for the mechanical systems (Fig. 17a). The envelope of this building is continuously insulated walls and slabs, with triple windowpane that participated in insulating the whole envelope to outdoor conditions [50]. Also, shade and shadow are provided by the exterior PV canopy. In addition, the exterior blinds on the western facade provide solar control to the interior as illustrated in Fig. 17b, c.

(a) The large roof with integrated Solar PV panles to provide shades on the building's facades

(b) The solar PV canopy providing shading on the building's western façade

(c) The dynamic roof acting as a chimney to cool the interiors via stack effect and the 330 kW (DC) roof comprising of 917 solar PV panles

Fig. 17 Kendeda Sustainable Building at the Georgia Institute of Technology in Atlanta – Georgia, USA (**a**) The large roof with integrated solar PV panels to provide shades on the building's facades (**b**) The solar PV canopy providing shading on the building's western facade (**c**) The dynamic roof acting as a chimney to cool the interiors via stack effect and the 330 kW (DC) roof comprising of 917 solar PV panels. (Images' credit and source: (**a**) J. Jonah Jackalope, https://commons.m.wikimedia.org/wiki/File:Kendeda_Building_1.jpg# (**b**) KBISD, https: https://commons.m.wikimedia.org/wiki/File:The_Kendeda_Building.jpg# (**c**) KNISD, https://commons.m.wikimedia.org/wiki/File:Kendeda_Building_Solar_Canopy.jpg.

6.2.4 Habu Temple in Luxor, Egypt

Shading and ventilation were depicted in early Egyptian civilization particularly in Habu Temple in Luxor (Fig. 18). The ancient temple uses thick stone walls with high thermal capacity and time lag to minimize heat gain through external walls. Figure 18a. shows the Habu Temple and its walls. Also, Habu Temple provides comfort by the recessed gateways that were built to funnel air through to create natural cooling as well as provide shade and shadow in summer days as shown in Fig. 18b.

6.2.5 The Luxor Temple in Luxor, Egypt

The Luxor Temple in Luxor, Egypt, is another example of the Egyptian civilization, where the Lotus columns were designed to provide shade and shadow in summer days as shown in Fig. 19. Moreover, the use of thick stones is characterized by thermal capacity to reduce heat gain and offer comfort.

6.3 *Natural and Cross Ventilation*

Ventilation is one of the most building life essential strategies, and usually it is considered natural or induced wind-driven procedure. Cross ventilation happens due to pressure difference between two opposite sides of a building, allowing air to move or cross from one side to the other creating a favorable air movement inside the building space that enhances the IEQ [51]. In addition, natural and cross ventilations are important strategies in buildings especially in the age of COVID-19

(a) High gate and thick stones of the walls of Habu Temple to provide comfort in summer heat

(b) Gateways providing shadows and comfort naturally

Fig. 18 Habu Temple in Luxor, Egypt (**a**) High gate and thick stones of the walls of Habu Temple to provide comfort in the summer heat (**b**) Gateways providing shadows and comfort naturally. (Images' credit and source: Mohsen Aboulnaga)

Fig. 19 Luxor Temple in Luxor, Egypt (**a**) The Lotus columns were designed to provide shade and shadow, and comfort in summer days (**b**) Entrance of Ramesses II Temple (**c**) The Lotus columns at night. (Images' credit and source: Mohsen Aboulnaga)

Fig. 20 Examples of examined buildings depicting shading Strategies in Russia and the USA. (Source: Developed by authors)

pandemic. The following three buildings' cases depended on the natural ventilation to enhance the indoor conditions for occupants (Fig. 20).

6.3.1 Russia Tower, Russia

The Russia Tower is deemed the largest naturally ventilated tower in the world [52]. The tall building's core as the lung by maximizing filtration and circulation of air through the wall system (Fig. 21). The wall system houses hydroponic plants that regulate the temperature and air quality throughout the interior spaces. In addition, the main void fosters stack effect for air to flow and to turn the wind turbines at the top of tall building. Moreover, such open space increases the daylight adequacy into the interior. All these features participate in reducing the operational energy use by 65% [52].

(a) The façades of the tower

(b) The tower at night showing the shaft and turbines

Fig. 21 Russia Tower, Russia (**a**) The facades of the tower. (Image credit and source: Milkomède, https://commons.m.wikimedia.org/wiki/File:Russia_Tower.png) (**b**) The tower at night showing the shaft and turbines. (Image credit and source: https://en.m.wikipedia.org/wiki/Russia_Tower#/media/File%3ARussiaTower-render.jpg)

6.3.2 Apple Park and Jobs' Theatre, California, USA

The Apple Park and Jobs' Theatre building designed by Sir Norman Foster depends mainly on natural ventilation for cooling and switches to mechanical ventilation only during the hottest summer days. The smooth perimeter helps circulate the air, and the external shades reduce the heat gain and ensure high daylight efficiency for the interior as presented in Fig. 22. The building complex energy production from PV is greater than it can consume [53]. The smart and real example exhibits how the users can be inside and outside at the same time. In addition, the use of smart glass with the distinctive 155-foot metallic carbon fiber ("flying saucer lid") contributes to achieving comfort.

6.3.3 NASA Sustainable Development Base, CA, USA

The NASA Sustainable Development base in Moffett Field, California, USA, is an iconic and innovative design by Architect William McDonough + Partners. The LEED platinum iconic building depends on the structural shape to create a wind tunnel for air circulation along with maximum daylight efficiency through efficient shaded glazing envelope, as shown in Fig. 23. The shape of the two buildings allow

(a) The Cylinder form and cantilever roof provide shades, cooling, and ventilation

(b) The innovative design of smart glass with the distinctive155-foot metallic carbon fibre 'Flying saucer lid' provides comfort inside the theatre

Fig. 22 Apple Pavilion Park and Steve Jobs' Theater in Silicon Valley, California, USA (**a**) The cylinder form and cantilever roof provide shades, cooling, and ventilation. (Image credit and source: Justin Ormont, https://en.m.wikipedia.org/wiki/File:Steve_Jobs_Theater_-_external.jpg) (**b**) The innovative design of smart glass with the distinctive155-foot metallic carbon fiber "flying saucer lid" provides comfort inside the theatre. (Image credit and source: Justin Ormont, https://commons.m.wikimedia.org/wiki/File:Steve_Jobs_Theater_-_Auditorium.jpg)

outside air to channel between these curve lines (Fig. 23a, b). The building can maximize both daylight and ventilation, while depending on mechanical heating systems to maintain thermal comfort for occupants throughout the year. The super-insulated exterior walls with high glazing performance increase the thermal performance of the building envelope, and smart lighting technology can balance the interior lighting with daylight. Solar PV panels installed on roofs to general electrical energy as seen in Fig. 23b [54]. The building's features can be summarized as follows:

- This building engulfs intelligent technology, inspired by the air safety program of the agency which supplies, most notably, air flight controls; such technology has been used to control different zones of the building to provide comfort
- Provides real-time data on the flows through the structure to control radiation by shades, which allow daylighting and ventilation and cooling
- Incorporates renewable energy
- Utilizes recyclable or recycled materials in the building's elements
- Exploiting other devices, sensors, and technologies in order to optimize energy use

(a) Bird's-eye view of the twin buildings with a space in between to channel air and solar PV panles on roofs to generate electrical energy

(b) Double-skin façade with external shading louvers connected to sensors for real-time data to control solar heat, provide indoor comfort, and solar panles on roofs

Fig. 23 NASA sustainable development base surrounded by green areas in California, USA (**a**) Bird's-eye view of the twin buildings with a space in between to channel air and solar PV panels on roofs to generate electrical energy. (Image credit and source: William McDonough + Partners, https://www.nasa.gov/sites/default/files/styles/full_width_feature/public/thumbnails/image/scene1-aerial.jpg) (**b**) Double-skin facade with external shading louvers connected to sensors for real-time data to control solar heat, to provide indoor comfort, and to control solar panels on roofs. (Images' credit and source: NASA, https://www.nasa.gov/feature/ames/space-technology-and-conservation-the-efforts-of-reducing-nasa-ames-water-use)

6.3.4 Morocco's Buildings, City of Casablanca, Morocco

The Urban Agency of Casablanca, city of Casablanca, Morocco, relied on the use of white color facades to reflect about 95% of the solar radiation impinging on them and may utilize recessed ground floor to provide shadow in summer and create cooler air for ventilation and recessed windows to provide shadow and reduce heat gain, as presented in Fig. 24a.

The Bank of Morocco's building in the city of Casablanca, Morocco, also uses bright color facades to reflect about 90–95% of the solar radiation impinging on them, hence reducing heat gain through external walls and cooling load to provide comfort, as shown in Fig. 24b.

In addition, a residential building in Casablanca, Morocco, is able to lower direct heat gain, exploiting awnings/canopies to provide shades/shadow and comfort for people sitting at the café, and uses recessed windows to provide shadow to reduce heat gain and cooling in summer, as illustrated in Fig. 24c.

6.4 Daylight and Sunlight

Daylight is a significant factor in achieving IEQ, and it is usually coupled with natural ventilation or other factors that enhance the building experience for occupants. Daylight is the clear light coming inside a northern window with no direct sunlight, while sunlight is the direct sun rays coming inside a space with a warmer, yellower quality.

Examples assessed for daylighting are six buildings in South Korea, China, Germany, Egypt, and Kingdom of Saudi Arabia, as depicted in Fig. 25.

(a) White façade of Urban Agency of Casablanca **(b)** Bank of Morocco's building with white façade and recessed windows, Casablanca **(c)** Residential building, city of Casablanca, Morocco

Fig. 24 Bright color material of buildings in Casablanca, Morocco (**a**) White facade of Urban Agency of Casablanca (**b**) Bank of Morocco's building with white facade and recessed windows, Casablanca (**c**) Residential building, city of Casablanca, Morocco. (Images' credit and source: Mohsen Aboulnaga)

1. Theme Pavilion, Yeosu, South Korea
2. City Hall building, Seoul, South Korea
3. Sanbaopeng LKKER Jingdezhen Ceramic Design Centre, China
4. Darmstadt University of Technology, Germany
4. The New Library of Alexanderia, Egypt
6. King Fahd Cultural Centre and Library in Riyadh, KSA

Fig. 25 Examples of examined buildings depicting shading Strategies in South Korea, China, Germany, Egypt, and KSA. (Source: Developed by authors)

(a) The Theme Pavilion mimicking the wave's movement and a large fish built on an island

(b) A Kinetic façade with hinged louvers closing and opening to control heat and provide daylighting

Fig. 26 The Theme Pavilion, Expo 2012 Yeosu, South Korea (**a**) The Theme Pavilion mimicking the wave's movement and a large fish built on an island (**b**) A Kinetic façade with hinged louvers closing and opening to control heat and provide daylighting. (Image credit and source: (**a**) Kang Giho – Korea.net/Korean Culture and Information Service, https://commons.m.wikimedia.org/wiki/File:KOCIS_Korea_Yeosu_Expo_Sketch_Kang04_(7589658734).jpg (**b**) Hyolee2, https://commons.m.wikimedia.org/wiki/File:Expo2012_Theme_pavilion.JPG#)

6.4.1 The Theme Pavilion in Expo 2012 Yeosu, South Korea

The building skin is composed of glass fiber-reinforced polymers (GFRP) that can be formed into different animated patterns. The movable hinged fish-like skin incorporates a daylight-driven mechanism that responds to the occupant's needs in terms of daylight adequacy [55, 56]. Figure 26 presents the iconic building's facades with kinetic-hinged louvers.

6.4.2 The City Hall Building in Seoul, South Korea

In this building, the main facade is built with high-reflective Low-E glazing to reduce heat gain through the building's skin, but it is inclined from the top creating a self-canopy to provide shadow as well as using landscaping to shade the lower part of the façade as shown in Fig. 27. Also, the facade has operable windows for natural ventilation.

6.4.3 Sanbaopeng LKKER Jingdezhen Ceramic Design Centre, China

The Sanbaopeng LKKER Jingdezhen Ceramic Design Centre in China incorporates a screen-like pattern made of copper tiles that controls the daylight into building spaces as shown in Fig. 28. The circular envelope provides external view and daylighting, with the addition of an interior courtyard that maintains daylight adequacy for building occupants [57].

6.4.4 Darmstadt University of Technology, Germany

The renovated building southern façade is composed of folded anodized aluminum sheets that regulate daylight into the interior spaces [58] as presented in Fig. 29. The building's facade has enabled the occupants to achieve higher thermal comfort levels with adequate provision of daylight. Such form educes head gain through the building envelope as well as the roof is inclined to reduce the impact of radiation. The courtyard also assists in cooling and provides comfort indoors.

(a) High reflective Low-E glazing façade to reduce heat gain and inclined from the top creating a self-canopy to provide shadow

(b) Glazed façade to offer daylight and operable windows to provide ventilation

Fig. 27 The city hall building in Seoul, South Korea (**a**) High-reflective Low-E glazing facade to reduce heat gain and inclined from the top creating a self-canopy to provide shadow (**b**) Glazed facade to offer daylight and operable windows to provide ventilation. (Images credit and source: Mohsen Aboulnaga)

(a) The circular façade with a copper tiles screen controlling daylight into building spaces

(b) Building's interior showing the dazzling controlled daylight into the indoor spaces

Fig. 28 Sanbaopeng LKKER Jingdezhen Ceramic Design Centre in China (**a**) The circular facade with a copper tiles screen controlling daylight into building spaces (**b**) Building's interior showing the dazzling controlled daylight into the indoor spaces. (Images credit and source: Idea 2587595: Sanbaopeng LKKER Jingdezhen Ceramic Design Center by Office Mass in Jingdezhen, China, https://www.architizer.com)

(a) Haus C10 building's with 3-D light metal folding elements as sunshades for widows' balancing light and sun protection

(b) The insulated façade with operable widows for natural ventilation and Brise Soleil for sun protection

(c) Close up view of the 3-D light metal folding elements to allow for daylighting

Fig. 29 The H-da C10 high-rise building at Darmstadt University of Technology, Germany (**a**) Haus C10 building's with 3D light metal folding elements as sunshades for widows' balancing light and sun protection (**b**) The insulated facade with operable widows for natural ventilation and brise soleil for sun protection. (Image credit and source: Guenni88, https://commons.m.wikimedia.org/wiki/File:H-da_C10_south.jpg) (**c**) Close up view of the 3D light metal folding elements to allow for daylighting. (Images' credit and source: (**a**, **c**) LSDSL, https://commons.m.wikimedia.org/wiki/Category:H_da_High-rise#/media/File%3AC10_South.jpg)

Another building in Darmstadt University of Technology incorporates integrated PV [59] as presented in Fig. 30 to generate energy as well and regulate heat and provide daylight.

6.4.5 New Library of Alexandria, Egypt

The New Library of Alexandria (Bibliotheca Alexandrina) in Alexandria, Egypt, exhibits an innovative model for daylighting provision coupled with sun protection and indoor thermal comfort (Fig. 31). The library hall is well-lit utilizing daylight from triangular-shaped skylight openings that can direct daylight deep into the interior space of the reading areas and book racks in the library building, thus maximizing the daylight benefits into the space (Fig. 31c–f). Moreover, the roof is inclined toward the north to cool it with sea breath. It is worth mentioning that the external curved granite wall provides shading and cooling on the inclined roof surface of the main reading hall of the library [60] as shown in Fig. 31b.

6.4.6 King Fahd's Library and Culture Centre in Riyadh, KSA

The iconic facade of King Fahd's Library and Culture Centre in Riyadh is made up of an additional layer to the glazed skin, composed of marine echoes that protect the building from direct solar sunlight as presented in Fig. 32. The envelope is a double-skin facade, where the outer skin is fixed to the structure via steel cables with a traditional pattern-like skin similar to latticework as shown in Fig. 32a–c.

(a) Intelligent facades with integrated photovoltaics on automated wood louvers

(b) Automated louvers with solar PV to generate energy, regulate heat and daylight

Fig. 30 Sustainable and smart building in Darmstadt University of Technology in Darmstadt, Germany (**a**) Intelligent facades with integrated photovoltaics on automated wood louvers (**b**) Automated louvers with solar PV to generate energy, regulate heat and daylight. (Image credit and source: Jeff Kubina, https://commons.m.wikimedia.org/wiki/File:Technische _Univerit%C3 % A4t_Darmstadt__Solar_Decathlon_2007.jpg#)

(a) External view of the inclined roof with integrated triangular shaped skylight for daylighting

(b) The external curved granite wall provides shading on the inclined roof surface

(c) Skylights providing daylight without glare to the readers in the main reading area inside the library

(d) Triangular shaped skylight for daylighting from inside the library main hall

(e) Book stacks with enough daylighting and no glare

(f) Inclined ceiling with skylights for daylighting

Fig. 31 Exterior and Interior of the New Library of Alexandria, Egypt (**a**) External view of the inclined roof with integrated triangular-shaped skylight for daylighting (**b**) The external curved granite wall provides shading on the inclined roof surface (**c**) Skylights providing daylight without glare to the readers in the main reading area inside the library (**d**) Triangular-shaped skylight for daylighting from inside the library main hall (**e**) Book stacks with enough daylighting and no glare (**f**) Inclined ceiling with skylights for daylighting. (Images' credit and source: Mohsen Aboulnaga)

The outer skin which is made of Teflon provides shades and responds to the dynamic daylight and change in pattern to regulate daylight into spaces [61]. In addition, a three-dimensional tensile-stressed steel cables structure is developed on the library envelope to act as sunshades. The facade combined a ventilation and cooling system consisting of layered ventilation and floor cooling.

(a) Side view of the library's double-skin lightweight textile façades

(b) 3-D tensile-stressed steel cables structure on the envelop as sunshades

(c)Teflon façade to provide ventilation and cooling

Fig. 32 Exterior and interior of King Fahd National Library and Cultural Centre in Riyadh, KSA (**a**) Side view of the library's double-skin lightweight textile facades. (Image credit and source: Yasser Bakhsh, https://commons.m.wikimedia.org/wiki/File:King_Fahad_National_Library1.jpg) (**b**) 3D tensile-stressed steel cables structure on the envelop as sunshades (**c**)Teflon facade to provide ventilation and cooling. (Image credit and source (**b**) and (**c**) Mrcosch, https://commons.m.wikimedia.org/wiki/File:King_Fahad_National_Library.jpg)

7 Comparative Analysis of Assessed Buildings: Case Studies in Egypt

A comparative analysis between four local case studies is carried out to identify the main features in these buildings which enhance the overall IEQ in these buildings (Table 1); these case studies are (a) the Grand Egyptian Museum in Giza; (b) Credit Agricole Egypt Headquarters in New Cairo; (c) New Al-Gorna village by late Hassan Fathy in Luxor; and (d) Dar Al-Handasah Headquarters in Smart Village in Giza – Greater Cairo, Egypt.

7.1 Results of the Comparative Analysis

The results of the comparative analysis of the four assessed local case studies of buildings in Greater Cairo metropolis (Giza and Cairo) in Egypt revealed that only one building; the New Egyptian Grand Museum fully achieved the four features of IEQ (shading, thermal comfort, natural ventilation, and daylighting). The third building (Hassan Fathy example of Al Gorna in Luxor) achieved three features of IEQ, but not daylighting due to large areas of walls and very small openings. The second case study (Credit Agricole Egypt Headquarters in Cairo) achieved only two features of IEQ (shading and thermal comfort), but not natural ventilation and daylighting due to airtight building (no operable windows) and small openings. The fourth case study (Dar Al-Handasah Headquarters in Smart Village, 6th October City, Giza) accomplished only one feature of IEQ, which is shading. Moreover, none of the case studies can achieve livability in the time of COVID-19 except the New Egyptian Grand Museum due to the fact it is built in the age of the Coronavirus (COVID-19).

Table 1 Assessment of the four local case studies in Egypt

Status	Name	Features				Can achieve livability in the age of COVID-19?
		Shading	Thermal comfort	Natural ventilation	Daylighting	
 Image credit and source: Djehouty, https://en.m.wikimedia.org/wiki/Grand_Egyptian_Museum#/media/File:Grand_Egyptian_Museum_2019-11-17.jpg	The Grand Egyptian Museum in Giza	✓	✓	✓	✓	Yes
 Image source: https://www.egypttoday.com/Article/3/73654/Credit-Agricole-records-profits-of-LE-1-3B-during-1st	Credit Agricole Egypt-Headquarters in Cairo	✓	✓	x	x	No

Status	Name	Features				Can achieve livability in the age of COVID-19?
		Shading	Thermal comfort	Natural ventilation	Daylighting	
 Image source: https://doi.org/10.1007/978-3-030-87794-1_3 [62]	Hassan Fathy (Al-Gourna) in Luxor	✓	✓	✓	x	No
 Image source: https://www.glassdoor. com/Photos/Dar-Group-Office-Photos-E147602.htm	Dar Al-Handasah Headquarters – Smart Village, Giza	✓	x	x	x	No

Source: Developed by authors

8 Conclusion

The role of shading, natural ventilation, daylighting, and comfort in enhancing indoor environmental quality IEQ and liveability in the age of COVID-19 has been assessed. The principles of Indoor Environmental Quality – IEQ were presented and reviewed. Also, the benefits of IEQ were reviewed and highlighted, and benchmark for IEQ in the time of COVID-19 has been appraised. Twenty-six buildings were presented and assessed in terms of shading strategies, ventilation, and daylighting and sunlight as well as thermal comfort. The review of these successful case studies was considered unique with varied sustainability measures and high environmental performance. It, therefore, is essential to reassess these buildings and ensure their liveability features and their adequacy in the time of pandemic COVID-19. Moreover, the four examined local examples were analyzed to draw a comparison of the IEQ feature and explore the capability of these buildings in achieving livability in the time of COVID-19. It is deduced that only one building is able to accomplish that which is the New Egyptian Grand Museum in Giza. Finally, IEQ is vitally essential to be achieved in buildings by utilizing multiple strategies, especially in large buildings with high occupancy rates to ensure healthy and livable spaces, and to reduce the spread of the Coronavirus (COVID-19).

Acknowledgment The authors sincerely express their heartfelt appreciation to Professor Ali Sayigh for his benevolent invitation to contribute to this important book, especially during the time of the Coronavirus (COVID-19) crisis. The authors would also like to convey their gratitude to Mr. Pietro Tonini, who is a PhD researcher and works at the ICTA-ICP Research Centre at UAB Barcelona (Universitat Autonomous de Barcelona), for his effort and assistance in capturing the images of this iconic and smart building in Barcelona. Finally, we thank Springer Nature for publishing this book.

References

1. Megahed, N. A., & Ghoneim, E. M. (2020). Antivirus-built environment: Lessons learned from Covid-19 pandemic. *Sustainable cities and society,* 61:102350.
2. Cheval, S., Mihai Adamescu, C., Georgiadis, T., Herrnegger, M., Piticar, A., & Legates, D. R. (2020). Observed and Potential Impacts of the COVID-19 Pandemic on the Environment. *International journal of environmental research and public health*, 17(11), 4140
3. Belzunegui-Eraso, A., & Erro-Garces, A. (2020). Tele-working in the context of the Covid-19 crisis. *Sustainability*, 12(9), 3662. https://doi.org/10.3390/su12093662
4. Eltarabily, S., & Elghezanwy, D. (2020). Post-pandemic cities-the impact of COVID-19 on cities and urban design. *Architecture Research*, 10(3), pp. 75–84, https://doi.org/10.5923/j.arch.20201003.02
5. Salama, A. M. (2020). Coronavirus questions that will not go away: interrogating urban and socio-spatial implications of COVID-19 measures. Emerald Open Research, 2.
6. D'alessandro, D., Gola, M., Appolloni, L., Dettori, M., Fara, G. M., Rebecchi, A., & Capolongo, S. (2020). COVID-19 and living space challenge. Well-being and public health recommendations for a healthy, safe, and sustainable housing. *Acta Bio Medica: Atenei Parmensis*, 91(9-S), 61.

7. Larsen, T. S., Rohde, L., Jønsson, K. T., Rasmussen, B., Jensen, R. L., Knudsen, H. N. & Bekö, G. (2020). IEQ-Compass–A tool for holistic evaluation of potential indoor environmental quality. *Building and Environment,* 172, 106707.
8. Bortolini, R., & Forcada, N. (2021). Association between Building Characteristics and Indoor Environmental Quality through Post-Occupancy Evaluation. *Energies,* 14(6), 1659.
9. Sant'Anna, D. O., Dos Santos, P. H., Vianna, N. S., & Romero, M. A. (2018). Indoor environmental quality perception and users' satisfaction of conventional and green buildings in Brazil. *Sustainable Cities and Society*, 43, 95–110.
10. Nundy, S., Ghosh, A., Mesloub, A., Albaqawy, G. A., & Alnaim, M. M. (2021). Impact of COVID-19 pandemic on socio-economic, energy-environment and transport sector globally and sustainable development goal (SDG). Journal of Cleaner Production, 127705.
11. Kim, J., & De Dear, R. (2012). Impact of different building ventilation modes on occupant expectations of the main IEQ factors. *Building and Environment*, 57, 184–193.
12. Parkinson, T., Parkinson, A., & de Dear, R. (2019). Continuous IEQ monitoring system: Context and development. *Building and Environment, 149*, 15–25.
13. Sadler, K. (2021). *Safely increasing capacity in our buildings from 19 July – Inside HMCTS. Insidehmcts.blog.gov.uk. Retrieved 25 July 2021, from* https://insidehmcts.blog.gov.uk/2021/07/15/safely-increasing-capacity-in-our-buildings-from-19-july/.
14. Elgot, J., & Sample, I. (2021). *Boris Johnson to revoke Covid rules on 19 July*. The Guardian. Retrieved 25 July 2021, from https://www.theguardian.com/world/2021/jul/05/boris-johnson-says-most-covid-rules-likely-to-end-in-england-on-19-july.
15. Liu, S., Kwok, Y. T., Lau, K. K. L., Chan, P. W., & Ng, E. (2019). Investigating the energy saving potential of applying shading panels on opaque façades: A case study for residential buildings in Hong Kong. Energy and Buildings, 193, 78–91.
16. Bellia, L., Marino, C., Minichiello, F., & Pedace, A. (2014). An overview on solar shading systems for buildings. Energy Procedia, 62, 309–317.
17. De Luca, F., Voll, H., & Thalfeldt, M. (2018). Comparison of static and dynamic shading systems for office building energy consumption and cooling load assessment. *Management of Environmental Quality: An International Journal.*
18. Atkinson J, Chartier Y, Pessoa-Silva CL, et al., editors. Natural Ventilation for Infection Control in Health-Care Settings. Geneva: World Health Organization; 2009. 2, Concept and types of ventilation. Retrieved 25 July 2021, from https://www.ncbi.nlm.nih.gov/books/NBK143277/#!po=1.56250
19. Sakiyama, N. R. M., Carlo, J. C., Frick, J., & Garrecht, H. (2020). Perspectives of naturally ventilated buildings: A review. *Renewable and Sustainable Energy Reviews*, 130, 109933.
20. Aviv, D., Chen, K. W., Teitelbaum, E., Sheppard, D., Pantelic, J., Rysanek, A., & Meggers, F. (2021). A fresh (air) look at ventilation for COVID-19: Estimating the global energy savings potential of coupling natural ventilation with novel radiant cooling strategies. *Applied Energy*, 292, 116848.
21. *Community, Work, and School*. Centers for Disease Control and Prevention. (2021). Retrieved 25 July 2021, from https://www.cdc.gov/coronavirus/2019ncov/community/ventilation.html
22. Khan, N., Su, Y., & Riffat, S. B. (2008). A review on wind driven ventilation techniques. *Energy and buildings*, *40*(8), 1586–1604.
23. Bhagat, R. K., Wykes, M. D., Dalziel, S. B., & Linden, P. F. (2020). Effects of ventilation on the indoor spread of COVID-19. *Journal of Fluid Mechanics*, *903*.
24. Elkhatieb, M. A. (2017). *A Performance-Driven Design Model of Territorial Adaptive Building Skin (TABS) for Daylighting Performance Optimisation in Office Buildings in Egypt*. The University of Liverpool (United Kingdom).
25. Halawa, E., Ghaffarianhoseini, A., Ghaffarianhoseini, A., Trombley, J., Hassan, N., Baig, M & Ismail, A. (2018). A review on energy conscious designs of building façades in hot and humid climates: Lessons for (and from) Kuala Lumpur and Darwin. *Renewable and Sustainable Energy Reviews*, (82) 2147–2161.

26. Wang, Z., de Dear, R., Luo, M., Lin, B., He, Y., Ghahramani, A., & Zhu, Y. (2018). Individual difference in thermal comfort: A literature review. *Building and Environment*, (138) 181–193.
27. Nicol, J. F., & Humphreys, M. A. (2002). Adaptive thermal comfort and sustainable thermal standards for buildings. *Energy and buildings*, 34(6), 563–572.
28. Hurwitz, Z. L., Dubief, Y., & Almassalkhi, M. (2020). Economic efficiency and carbon emissions in multi-energy systems with flexible buildings. *International Journal of Electrical Power & Energy Systems*, 123, 106114.
29. Altomonte, S., Schiavon, S., Kent, M. G., & Brager, G. (2019). Indoor environmental quality and occupant satisfaction in green-certified buildings. *Building Research & Information*, 47(3), 255–274.
30. Awada, M., Becerik-Gerber, B., Hoque, S., O'Neill, Z., Pedrielli, G., Wen, J., & Wu, T. (2021). Ten questions concerning occupant health in buildings during normal operations and extreme events including the COVID-19 pandemic. *Building and Environment*, 188, 107480.
31. Peters, T., & Halleran, A. (2020). How our homes impact our health: using a COVID-19 informed approach to examine urban apartment housing. *Archnet-IJAR: International Journal of Architectural Research*. (15)1, 10–27.
32. Parker, D., & Wood, A. (2013). Aqua Tower, Chicago, USA. In *The Tall Buildings Reference Book* (pp. 380–385). Routledge.
33. Gang, J. (2016). Three points of the residential high-rise: designing for social connectivity. *International Journal of High-Rise Buildings*, 5(2), 117–125.
34. Fakourian, F., & Asefi, M. (2019). Environmentally responsive kinetic façade for educational buildings. *Journal of Green Building*, 14(1), p. 165–186.
35. Abdelwahed, M. A. & Halim, H. A. (2021). Kinetic Facades: How Nature of Components Affects the Applications of Different Transformation Strategies. *Engineering Research Journal*, 170, 11–25.
36. Laboy, M. M. (2019). Temporal reciprocities of building and site: Structural patterns for resilient future-use structures. In *Structures and Architecture: Bridging the Gap & Crossing Borders* (pp. 1031–1040). CRC Press.
37. Abdullah, Y. S., & Al-Alwan, H. A. (2019). Smart material systems and adaptiveness in architecture. *Ain Shams Engineering Journal*, 10(3), 623–638.
38. Endonio, G. S., & Bringas, E. N. (2015). Positive Impact on Wellbeing and Energy Related Behaviors on Office Buildings: The Challenge Lab 2015 (Master's thesis).
39. Wahab, I., Azman, A. N., Abd Aziz, H., Hamberi, M. J. M., & Ismail, L. H. (2020). Adjustable Folding Aluminium Louvres Application on 'Hop (B): Nob Marina Quay'. *Recent Trends in Civil Engineering & Built Environment*, 1(1), 88–94.
40. Gosztonyi, S. (2018). The role of geometry for adaptability: Comparison of shading systems and biological role models. *Journal of Facade Design and Engineering*, *6*(3), 163–174.
41. *The Silver Net on the French Ministry of Culture | Aussie in France*. Aussie in France I from the Tropics to the City of Light. (2021). Retrieved 25 July 2021, from http://www.aussieinfrance.com/2011/12/1029/.
42. Attia, S. (2018). Evaluation of adaptive facades: The case study of Al Bahr Towers in the UAE. *QScience*
43. Connect, 2017 (2). *Special Issue on Shaping Qatar's Sustainable Built Environment – Part I)*, 6
44. *HSE – Thermal comfort: The six basic factors*. Hse.gov.uk. (2021). Retrieved 25 July 2021, from https://www.hse.gov.uk/temperature/thermal/factors.htm.
45. Hse.gov.uk. (2021). Retrieved 25 July 2021, from https://www.hse.gov.uk/temperature/assets/docs/british-european-int-standards.pdf.
46. Van Hoof, J. (2008). Forty years of Fanger's model of thermal comfort: comfort for all? *Indoor air*, 18(3), 182–201.
47. Piasecki, M., Fedorzcah-Cisak, M., Furtack, M., & Biskupski, J. (2019). Experimental Confirmation of the Reliability of Fanger's Thermal Comfort Model—Case Study of a Near-Zero Energy Building (NZEB) Office Building. *Sustainability*, 11(2461), p. 6. https://doi.org/10.3390/su11092461

48. Hellwig, R. T., Teli, D., Schweiker, M., Choi, J. H., Lee, M. J., Mora, R., & Al-Atrash, F. (2019). A framework for adopting adaptive thermal comfort principles in design and operation of buildings. *Energy and Buildings*, 205, 109476.
49. Romano, R. (2021). Kinetic Adaptive Façades. A Systematic Review of Technological and Adaptive Features. *Bioclimatic Approaches in Urban and Building Design*, 499–519.
50. Choi, Y. O., & Ahn, Y. H. (2012). Implementation and Evaluation of Green Building Features: Case Study of Trees Atlanta Kendeda Center. *KIEAE Journal*, 12(2), 59–64.
51. Aldawoud, A. (2017). Windows design for maximum cross-ventilation in buildings. *Advances in building energy research*, 11(1), 67–86.
52. Avdonina, L., Alekhin, V., & Galiev, D. (2018, November). Innovative solutions implemented in design of iset tower. In *IOP Conference Series: Materials Science and Engineering*, 451(1) 012049. IOP Publishing.
53. Mundo-Hernández, J., Valerdi-Nochebuena, C., & Sosa-Oliver, J. (2015). Post-occupancy evaluation study: occupant's perception vs. Occupancy survey. *European Scientific Journal*, 11(2), 110–122
54. O'Donnell, J. T. (2013). Transforming BIM to BEM: Generation of building geometry for the NASA Ames sustainability base BIM.
55. Khezri, M., & Rasmussen, K. (2021). Shading Module With Buckling as Driver for Shape Morphing. *Available at: SSRN 3881757.*
56. Lee, S. (2021). A Study on the Trends for Expression in Korean Contemporary Architectural Facade Design: Focusing on Large Buildings in the City Center. *Buildings*, *1*(7), 274.
57. *Solar Decathlon – CUBITY*. CUBITY. (2021). Retrieved 25 July 2021, from https://cubity.de/en/solar-decathlon/.
58. Tang, Q. (2020). Endless Cycle: Perspective on Mythological Time and Space in Yueling. In *Ritual Civilization and Mythological Coding* (pp. 231–291). Springer, Singapore.
59. Kozminska, U., & Rynska, E. D. (2018). Existing Buildings: How to Meet an nZEB Standard–The Architect's Perspective. In *Design Solutions for nZEB Retrofit Buildings* (pp. 228–247). IGI Global.
60. Aboulnaga, M. M. (2018). Sustainability Measures of Public Buildings in Seaside Cities: The New Library of Alexandria (New Bibliotheca Alexandrina), Egypt. In *Seaside Building Design: Principles and Practice*, pp. 65–111. Springer – Cham.
61. Baik, A. H. A., Yaagoubi, R., & Boehm, J. (2015) Integration of Jeddah historical BIM and 3D GIS for documentation and restoration of historical monument, International Society for Photogrammetry and Remote Sensing (ISPRS).
62. Aboulnaga M. M., Badran M. F., Barakat M. M. (2021) Informal Urban Settlements and Slums' Upgrading: Global Case Studies. In: Resilience of Informal Areas in Megacities – Magnitude, Challenges and Policies, Springer, Cham. https://doi.org/10.1007/978-3-030-87794-1_3

Mohsen Aboulnaga has more than 35 years of vast experience in higher education, high-level government positions, senior management, and consultancy in sustainable cities, climate change adaptation, and sustainable development as well as sustainable energy (energy efficiency and renewable energy). He is a qualified architect who graduated from Cairo University with BSc Arch (1979) and MSc Housing Economic (1985). He also holds a PhD degree from the University of Leeds, UK (1991), https://www.leeds.ac.uk/. Prof Dr Aboulnaga is Professor of Sustainable Built Environments at Faculty of Engineering, Cairo University, Egypt, https://scholar.cu.edu.eg/mohsen_aboulnaga/. He is a senior expert on sustainable urban development, including green and smart cities and green building. He was involved in developing many concept notes for European Union (EU)–funded projects and the Union for Mediterranean (UfM). Dr Aboulnaga's areas of competency encompass sustainable development attainment, strategy planning and policy development, low-carbon society, and sustainable cities as well as climate change mitigation and urban farming in cities. Other areas of expertise include executive capacity building for local governments and coaching/training on strategic planning, leadership, strategic planning and Sustainable Development Goals (SDGs), and innovation. Prof Aboulnaga has many milestones jobs and publications, and he is a WREN (World Renewable Energy Network) Pioneer, permanent member, and main advisor. He has had more than 100 published articles in recognized journals and more than 50 books he was involved in writing. The most recent publication is a journal paper in Sustainability by MDPI; it is entitled: "Sustainability Assessment of the National Museum of Egyptian Civilization (NMEC): Environmental, Social, Economic, and Cultural Analysis" https://www.mdpi.com/2071-1050/14/20/13080.

Elsharkawy Maryam is a Ph.D. holder from Cairo University, Faculty of Engineering, Department of Architecture; with the thesis entitled Enhancing Daylight Utilization in Adaptive Reuse of Heritage Buildings Using Building Information Modeling. Dr. Maryam is also working as a part-time lecturer and researcher at the Faculty of Engineering, Cairo University, and Nile University. She also worked at the School of Science and Engineering, American University in Cairo (AUC). She acquired a Master of Science degree in Environmental Design from Cairo University in 2015, with the thesis title 'Design Guide for Educational Building Retrofitted Facades to Achieve Low Energy Performance in Egypt'. She also graduated with a Bachelor's degree (BSc) in Architecture and Engineering Technology (AET) from Cairo University in 2012. During this period she attended two international outreach summer programs that are related to her study. Eng. Maryam worked as an architect for ECG Engineering Consultants Group S.A. between 2012 and 2016. She received her LEED certificate from the US Green Building Council (US-GBC) in 2014. This certificate allows her to expand her professional knowledge and experience further in the field. Her experience in building performance simulation and enhancement has enabled her to carry out our many research projects and professional works. She has attended many professional workshops, notably the Technical Building Construction workshop in Berlin for the application of building technologies in 2015 and similar workshops in London between 2016 and 2019. Moreover, she contributed, with Prof. Dr. Mohsen Aboulnaga, to the World Renewable Energy Congress (WREC) 2020 in Lisbon, Portugal from 26-29 July 2021 through a paper entitled: Integrating Building Information Modeling Tool for Energy and Daylighting Enhancement in Buildings Reuse and Attaining Sustainability, in addition to another paper presented at Sustainable Built Heritage 21 (SBE21) conference in Bolzano, Italy from 14-16 April 2021 entitled: Sustainability assessment of restored historic buildings: a case study of Baron Empain Palace in Heliopolis, Cairo, Egypt. Dr. Maryam has also published three book chapters with Prof Mohsen Aboulnaga by Springer and another book chapter by IOPscience Publishing. More collaboration on different research projects is available at: https://www.researchgate.net/profile/Maryam-Elsharkawy.

Diachronic Analysis of Daylight Design and Management Techniques in Mediterranean Constructions

Katerina Tsikaloudaki

1 Introduction

The adoption of techniques found in historic buildings or vernacular architecture has been proposed by numerous researchers as the key to enhancing energy efficiency and upgrading the environmental performance of contemporary buildings [1]. Undoubtedly, vernacular architecture has its routes on the temporal development of building construction, which is based on the perpetual need of people to control their environment and create safe and improved indoor conditions. The relationship between humans and buildings can be characterized as dynamic and mutual [2], given that the people's identity, ways of life, social and religious functions, etc. are closely tied to the built environment.

In parallel, the sequential changes in the buildings' evolvement provide an insight into the achieved level of technology of each era. Such connections are very evident in the past constructions, especially when no means of automatic control existed and people had to rely on their intelligence and acuity to make the best use of the available technological means and materials; through trial and error, they ultimately managed to configure the best possible indoor conditions.

In that sense, daylighting techniques were born and evolved with architecture; as for many centuries, daylight was the main light source together with torches and oil lamps [3]. Inside the settlements, daylight was derived from the luminance of the sky vault seen through doors or windows [4]. The techniques, though simple at first, were able to guarantee the required light levels in order to perform indoor daily activities. Gradually, these techniques and design strategies became more eloquent,

K. Tsikaloudaki (✉)
Laboratory of Building Construction and building Physics, Department of Civil Engineering, Aristotle University of Thessaloniki, Thessaloniki, Greece
e-mail: katgt@civil.auth.gr

A. Sayigh (ed.), *Achieving Building Comfort by Natural Means*, Innovative Renewable Energy, https://doi.org/10.1007/978-3-031-04714-5_9

and in certain building types, they were used in order to express symbolisms and special meanings.

In the bibliography, not much is said about this evolution of building morphology in relation to daylight. There are many analyses that attempt to study the indoor visual environment in historic buildings as part of a conservation project (e.g., in [5–9]) but only a few attempts to interrelate the architectural and morphological characteristics of historical constructions with daylighting, especially as regards the distant past.

Within this context, this study attempts to explore the diachronic relationship between the building envelope configuration and daylighting. The analysis identifies the daylight strategies that are found in characteristic buildings of several historic periods. It is focused in the geographical area of Greece, as this part of Europe was inhabited since the early ages and hosted many civilizations till today.

Beyond the architectural elements that determined the indoor visual environment, a "holistic" approach is employed, in order to understand better the objectives and the intentions of the builders/architects, following the approach of the ancient Greeks: for them, science, philosophy, art, politics, and ethics were interwoven and combined into one worldview, which is often depicted in the building design.

More specifically, science was integrated in the buildings in the form of technological achievements; philosophy offered the theoretical basis for special applications used in order to accomplish particular impressions; art was an inextricable element in buildings of particular importance; politics often dictated the inner arrangement of rooms in domestic constructions, while ethics represented the social factors that determined the morphology of the private and public buildings [10].

The interrelationships between these features of everyday life and the constructions that housed them may differ in each historic period; however, they exist both in the remaining buildings of the past and in the written works of history, philosophy, geography, or literature.

2 Approach and Influences

The diachronic analysis of daylight design and management techniques found in Mediterranean constructions concerns four major periods of Greek history:

- The prehistoric period, represented by the Minoan civilization (2000 BC–1400 BC)
- The Classical (500BC-336BC) and the Hellenistic period (336 BC–146 BC)
- The era of the Byzantine Empire (330 AD–1453 AD)
- The post-Byzantine/Ottoman period, which ultimately shaped the concepts of the vernacular architecture

Within these timelines, two main building types are examined: the residential and the public ones. It is essential to make such a differentiation, as the building use determines the objectives sought through the building's morphology and envelope design. More specifically, the domestic constructions usually hosted systems for satisfying the lighting needs and the visual comfort conditions, while the public buildings often employed more elaborated strategies in order to create special environments and achieve certain goals.

The term "public building" includes many different uses as it represents buildings that are used by the public for any purpose, such as assembly, education, entertainment, administration, or worship. For determining the association of daylighting with the public building characteristics, certain types of public buildings were selected, one for each historic period. The selection of the examined building types is based on a plethora of daylight strategies found in each function and the objectives that they served. Within this context:

- The Minoan Palace represents the prehistoric Greek architecture.
- The ancient temples epitomize the Classical or Hellenistic ages.
- The churches delineate the major buildings of the Byzantine Empire.
- The ottoman bathhouses – hamams outline the post-Byzantine or Ottoman period in Greece.

Before discussing the evolution of building forms and architectural elements and their association with daylighting, it is important to provide information on the factors that determined the building construction for each period, i.e., the available materials and technological level, as well as the social background. Moreover, the status of light in science, philosophy, and religion will contribute to the interpretation of the interconnection between the building configurations and the indoor visual environment.

2.1 Available Materials and Achieved Technology

In general, constructions of the past were usually built of locally available materials. In the Minoan Ages, the building techniques incorporated the use of ashlar masonry in the lower levels, made of limestone and a mortar of mud and usually supported by a timber frame. Upper stories were built entirely of mud-brick and roofs were flat, often constructed with wooden beams covered by reeds, cane, and mud [10].

Similar building techniques were also used in the Classical period. The walls of many dwellings were made of limestone bedded in clay and sun-dried bricks above, covered on both sides with three layers of hard stucco made of lime mixed with sand, gravel, and broken pottery. Roofing was formed either as flat roofs made of lime concrete or with ridged roofs consisting of timber elements and clay tiles [11]. Constructions of higher importance, such as public or religious buildings, were more elaborated and were usually built of more luxurious materials, such as marble. Through the ages, building techniques became gradually refined and allowed greater spans between columns and windows, more extended building plans, and higher constructions.

Bearing masonry remained the prevailing building technique during the Byzantium, but in this era, baked bricks and lime mortars with powder of crushed bricks were most often used. Wooden beams were also used in order to form a frame on different levels of the building envelope, i.e., walls, arches, openings of doors and windows, as well as in the basis of domes [12]. Similar building techniques were found in later periods as well as in the vernacular architecture. The main building materials were wood, stone, adobe, and bricks.

2.2 Social Factors

The social character of each civilization undoubtedly influenced the morphology of the constructions housing the people's activities.

In the Minoan Period, all archaeological evidence suggests that the Cretan states of the first half of the second millennium BC were monarchies. Although the government was dominated by priests and the king seemed to have some religious functions, the principal role of the monarch seemed to be that of "chief entrepreneur" of the Cretan state. Beneath the king was a large administration of scribes and bureaucrats who carefully regulated production and distribution both within the state and without. The palace was not only the residence of the royal family and the administrators but also housed many other citizens, and, thus, it was designed in order to cover multiple needs. Furthermore, despite the concentration of wealth, a strange phenomenon in the ancient world was created: social equality. In the excavated city of Gournia, the "poor" parts of town can be easily discerned; even there, however, people were living in four-, five-, and six-room houses [13]. Moreover, there is no evidence in the architecture of the individual units for the existence of a social hierarchy within the settlements. In addition, it seems that no inequality existed between genders, although the gender relations in Prehistoric Crete cannot be fully established due to limitations on interpreting the written works.

In ancient Greece, there were no houses that revealed the wealth, power, or good taste of the owner. This does not mean that rich and poor did not exist; simply the rich did not set themselves off by a particular type of house [14]. The competitive spirit and the basic values of the early Greek society were honor and fame. Learning became gradually a new value of that epoch. It was often practiced in private houses, which progressively were adjusted to this new way of life. Such a house provided a suitable setting for those wishing to "cultivate things of beauty and excellence at home," and thus the public character emulated in domestic space was developed.

Byzantium comprised the rational continuation of the Eastern Roman Empire, and, despite the geographical particularities and the ethnographical differences, the spread of Christianity had managed to unite the state. During this era, new architectural types were developed, mainly on the basis of Greek and Roman forms [15]. The focal point of the Byzantine building construction was undoubtedly the religious and administrative centers, which were built in an exceptional and splendid way. Apart from monumental buildings, the Early and Middle Byzantine city centers probably contained several categories of dwellings, ranging from elite residential complexes with courtyards and attached churches to individual low-status houses and shops. Indications of social grades could be discerned; in fact, it has been reported that the poorest inhabitants lacked any housing. Furthermore, the urban structure was defined by functional zones, although there was not an absolute spatial division between high- and low-status houses [16].

2.3 *Sun and Light in Science, Philosophy, and Religion*

The study of light has been a major topic in the subjects of mathematics and physics from ancient Greek times up to the present day. In fact, the early Greek ideas on natural philosophy and in particular on the nature of light would influence the world for two thousand years. Apart from the philosophical attempts to analyze the origin, the quality and the physics of light, mythological, and religious aspects were also attributed to sun and daylight. As in most polytheistic civilizations, sun was also worshipped in Greece. Homer quotes [17]: "As sun (Helios) rides in his chariot, he shines upon men and deathless gods and piercingly he gazes with his eyes from his golden helmet. Bright rays beam dazzlingly from him and his bright locks streaming from the temples of his head gracefully enclose his far-seen face; a rich, fine-spun garment glows upon his body and flutters in the wind, and stallions carry him. Then, when he has stayed his golden-yoked chariot and horses, he rests there upon the highest point of heaven, until he marvelously drives them down again through heaven to Ocean" (Fig. 1).

Christianity also associated light with divine, spiritual, and metaphysical powers. It is a frequently repeated theme both in the Old and the New Testaments and is often strongly associated with God's presence. In fact, God and the Holy Trinity could be revealed as the "true and eternal" light only to those that had confessed their sins and had thus acquired a purified heart. In that way, light represents truth, purity, and virtue (Fig. 2). The meaning and the merit of light in the Orthodox Church are often illustrated in the icons by highlighting those with higher spirituality and symbolizing life, joy, and the Resurrection of Christ.

Christianity was not the only religion in which light was associated with divine and spiritual powers and symbolized truth, purity, and life. In Quran, it is stated that "Allah is the light of the heavens and the earth; a likeness of His light is as a niche in which is a lamp, the lamp is in a glass, (and) the glass is as it were a brightly shining star, lit from a blessed olive-tree, neither eastern nor western, the oil whereof almost gives light, though fire touch it not. Light upon light. Allah guides to His light whom He pleases, and Allah sets forth parables for men, and Allah is Cognizant of all things" [18].

Fig. 1 Relief showing Helios, riding a chariot of four winged horses. From the North-West pediment of the temple of Athena in Ilion (Troy). Between the first quarter of the third-century BC and 390 BC. Marble, 85,8 × 86,3 cm. Found during the excavations led by Heinrich Schliemann in 1872, now in the Pergamon Museum in Berlin, Germany. (Source: Ident.Nr. LV 21, 1 (LG) Sammlung: Antikensammlung © Foto: Antikensammlung der Staatlichen Museen zu Berlin – Preußischer Kulturbesitz. Fotograf/in: Johannes Laurentius)

Fig. 2 Eighteenth-century icon of Christ the Great Archpriest, Holy Monastery of Agios Pavlos – Athos. In the inscription, it is quoted: "I am the light of the world; he that followeth me shall not walk in darkness, but shall have the light of life". (Source: M. Vasilaki, G. Tavlakis and E. Tsigaridas, Icons from the Holy Monastery of Agios Pavlos – Athos, Holy Mountain of Athos, 1998)

3 Identification of Daylighting Strategies Found in Residential Buildings

The diachronic analysis of daylight-related strategies in domestic buildings concerns the architectural elements that contributed to the formation of indoor daylighting and visual comfort conditions.

3.1 Prehistoric Era

The Minoan civilization bloomed on the island of Crete and on some smaller islands nearby. At the end of the Neolithic period, caves were still the main housing premises and slowly open settlements started to appear [19]. There are also traces of rectangular buildings and of small free-standing buildings on hilltops, perhaps seasonally occupied [20]. As centuries went by, the settlements became more organized.

More information regarding the housing premises exist for the mid-Minoan ages, thanks to artifacts found during excavations. More specifically, a series of mold-made faïence plaques representing building facades, known as the "Town Mosaic," shows that the houses of this era have two or three stories, while windows are common in the upper stories and rare on the ground floor (Fig. 3a). Moreover, a small (0.31 × 0.29 × 0.15–0.18 m. high) terracotta model of a two-story building having windows, columns, a lightwell opening onto a typical Minoan hall, a stairway, and a projecting balcony on the second story (Fig. 3b) provides more information for the domestic architecture of this prehistoric, yet not fully explored till today, times. Although these artifacts provide an invaluable outlook of the Minoan houses, little can be said for the actual configurations and prevailing conditions, mainly due to the fact that the written documents are scarce and still not decoded [21].

However, the excavated Minoan towns and villages show their organization with small houses, one or two story height set on either side of narrow streets [20]. The positioning of more windows on the upper story may be a result of the builders' wish to admit more daylight into the interior. As the clay house indicates, the plethora of openings, clerestories, and light wells could guarantee a well-lit indoor environment.

3.2 Classical Era

In the Hellenistic and Classical period, dwellings usually consisted of rooms arranged around an inner courtyard, which gradually acquired a "peristyle" (colonnaded portico) [22]. The principal openings for the admission of light and air faced the peristyle; windows overlooking the street are rare and appear to be positioned chiefly in the upper story (Fig. 4).

Fig. 3a The "Town Mosaic": models of house facades made of faience plaques from the temple repository from Knossos. Herakleion Archaeological Museum. Ministry of Culture/Archaeological Receipts Fund. (Source: Michailidou, A., Knossos. A Complete Guide to the Palace of Minos, Ekdotike Athenon, Athens 1993, p. 31, ill. 12–13. © Ministry of Culture, Ime.gr)

Fig. 3b Clay house model from Archanes. 23rd EPCA. Ministry of Culture/Archaeological Receipts Fund. (Source: Sakellarakis, Y. and Sapouna-Sakellaraki, E., Archanes, Ekdotike Athenon, Athens 1991, p. 61, ill. 36. © Ministry of Culture, ime.gr)

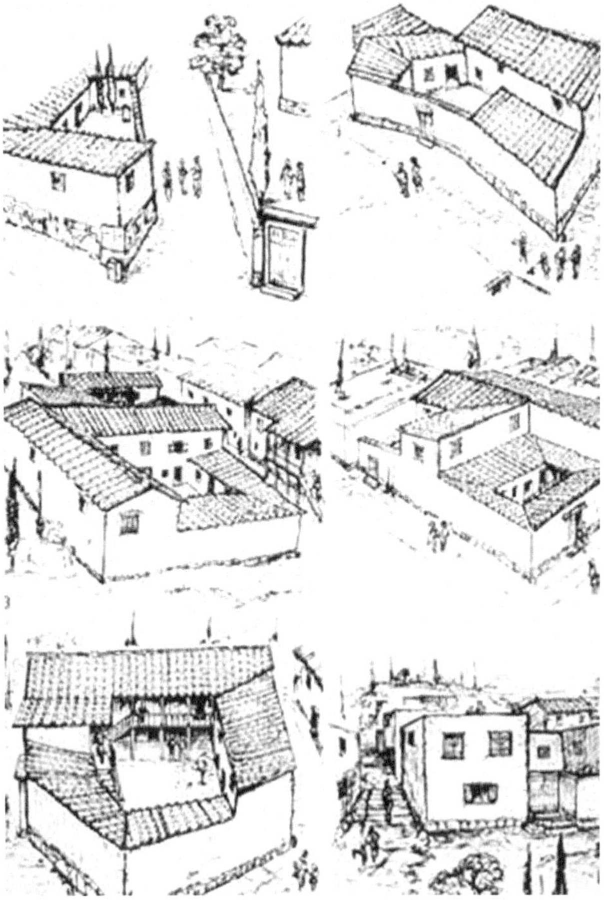

Fig. 4 Illustrations of typical Greek houses of the fifth-century BC in Athens. (Source: W.B. Dinsmoor, The architecture of Ancient Greece: an account on its historic development, Batsford, London, 1975)

Rooms of main use, such as the Men's Quarters, where the social life was taking place, were positioned on the northern part of the building unit, facing south in order to be warm in winter and cold in summer, as well as to capture as much daylight as possible.

This technique is vividly reported in Xenophon's works Memorabilia and Oeconomicus, in which Socrates, the great Greek philosopher, discusses with Ischomachus and Aristippus about the benefits of proper orientation [23]:

> In houses with a south aspect, the sun's rays penetrate into the porticoes in winter, but in summer the path of the sun is right over our heads and above the roof, so that there is shade. If, then, this is the best arrangement, we should build the south side loftier to get the winter sun and the north side lower to keep out the cold winds.

Similar references can be found in the work "De Architectura," written by Vitruvious. He describes the proper orientation of rooms according to their use with respect to the amount of required light [24]:

> Winter dining rooms and bathrooms should have a south-western exposure, for the reason that they need the evening light and also because the setting sun, facing them in all its splendour but with abated heat, lends a gentler warmth to that quarter in the evening. Bedrooms and libraries ought to have an eastern exposure, because their purposes require the morning light and also because books in such libraries will not decay. Summer dining rooms to the north, because that quarter is not, like the others, burning with heat during the solstice, for the reason that it is unexposed to the sun's course and hence it always keeps cool and makes the use of the rooms both healthy and agreeable. Similarly with picture galleries, embroiderers' workrooms and painters' studios in order that the fixed light may permit the colours used in their work to last with qualities unchanged.

He was also the first who expressed the "rule of thumb" for adequate daylighting [24]:

> On the side from which the light should be obtained let a line be stretched from the top of the wall that seems to obstruct the light to the point at which it ought to be introduced, and if a considerable space of open sky can be seen when one looks up above that line, there will be no obstruction to the light in that situation. But if there are timbers in the way or lintels or upper stories, then make the opening higher up and introduce the light in this way. And as a general rule, we must arrange so as to leave places for windows on all sides on which a clear view of the sky can be acquired, for this will make our buildings light.

3.3 Byzantine Era

In the bibliography, it is stated that in the Early and Middle Byzantine times, several categories of domestic structures existed, such as large elaborated residential complexes, private stone houses, multistory apartment blocks, and individual low-status houses [16].

The typical characteristic of ancient Greek houses, i.e., the organization of the spaces around courtyards, remains a common feature for many types of byzantine residences. More specifically, such examples are common in upper-class buildings, in which the internal courtyard often contained chapels and churches, but there were also other residential and commercial structures built around a communal courtyard [16]. In these complexes, the indoor visual environment was much better compared to the low-status buildings, which were built very close together with a pattern of narrow passages between them [25].

However, the urban form and main characteristics of the built environment (i.e., building heights, street widths, etc.) were basically defined by the Byzantine Building Codes concerning housing and neighborship. [26–28]. The content of the codes concerning the built environment mainly referred to the distances between buildings, so as to ensure the right of light and of view [27], as well as the construction of balconies, which gave access to the sun and view.

In written works of the period, a distinction is made between two types of windows [27]: the “panoramic” window, from which one can enjoy the view (i.e., the view to the sea was highly appreciated), and the light hole (clerestory), which was a high-level window (6 feet, i.e., more than 1.8 m above the floor), which provided views to the sky and diffuse daylight. The widths of the windows varied, but in the bibliography, it is reported that they ranged between 1.0 m and 1.4 m and had a continuous sill. In the cases that they were double, or even triple, they were most likely arched [29]. The daylight conditions were greatly affected by the presence of surrounding buildings, especially in areas with narrow street widths.

Based on the above, it is easily concluded that the main daylight strategies found in byzantine residential buildings are the openings toward the central courtyard, small windows in the upper stories facing the narrow streets, “view” windows overlooking the sea, and clerestory windows for admitting as much daylight as possible.

3.4 Post-Byzantine and Ottoman Period: Vernacular Architecture

After the end of Byzantine Empire, the morphology of domestic constructions evolved slowly through time and gradually shaped the vernacular or traditional architecture. Through its evolvement, it embraced the accumulated experience and practice; it was influenced by the economic, social, and cultural values of each society; and it responded to the topography, climate, and availability of materials of each different region [30].

Although the typology of vernacular architecture differs significantly within the country, the concept of addressing both the lighting needs and the control of the energy flows (heat losses and solar heat gains) is shared among all locations and climates, i.e., from the mountainous regions to the islands.

This concept is best expressed through the openings’ configuration, which was carefully chosen in order to form a comfortable visual and thermal environment in the living spaces. The size of the windows was determined by the space’s use, needs, and seasonality and ranged from large openings, which resulted in solar gains during the winter in northern parts of the country, to smaller apertures, which blocked the excessive intense solar gains during the hot summer months of southern regions.

The availability of daylight indoors naturally depended on the openings’ size, but it is characteristic that there was a special care in order to enhance the visual environment and reduce the glare formation. This was mainly achieved either by admitting additional daylight through clerestory windows or by attempting to distribute it in a more uniform way.

More specifically, a common characteristic among the traditional buildings found in the mountainous regions of Greece is the use of small clerestories placed above the view windows, which were designed to provide diffuse light on the upper part of the wall of the room, resulting in a smooth distribution of luminance on the

Fig. 5 The clerestory windows found in the traditional architecture of continental Greece, as a means for providing more uniform distribution of illuminances on the interior surfaces

Fig. 6 The inclined configuration of the sill and the jambs of the windows, as seen from the interior, in the case of small windows positioned on thick, stone masonries

interior surfaces (Fig. 5). The roof eave blocks the direct solar radiation during the summer months, while extrasolar gains are admitted during the winter [31].

On the other hand, in locations where the management of the solar radiation was more intricate, due to the local climate, the flat roof configuration or the thick stone masonry, the windows were small and they caused a significant contrast between the dark wall and the bright aperture [32]. For this reason, the window jambs and sills were often constructed with an inclination toward the interior, resulting in a broader window area as seen from the inside (Fig. 6). This broadened window area had an intermediate illuminance and resulted in the reduction of glare caused by the ultra-bright windows.

Inner courtyards are still present in many traditional buildings; together with balconies, summer-closed or semi-open spaces ("hagiatia") and projections constitute distinctive paradigms of daylight exploitation strategies [33].

4 Identification of Daylight Strategies Found in Public Buildings

As public buildings gained more attention, special daylight strategies were often employed with the purpose to create special environments. With respect to the position of light in the religion of each era, the commonly found techniques are identified and presented.

4.1 The Minoan Period

As is often stated in the bibliography, the Minoan palace at Knossos, Crete, is a building that encompasses the breadth and depth of its culture more eloquently than any other single building in the history of European architecture [34]. Though by far the largest among palaces of the same period, it incorporates all the distinctive elements of Minoan architecture. Many of them can be associated with the provision of daylight and adequate ventilation.

Firstly, there is a central court, orientated across the North-South axis [11], which has been variously explained as either to provide maximum daylight to the colonnades bordering the central court or to have the openings of cult rooms along the west side of the central court facing toward the rising sun.

Undoubtedly, the focal point of a Minoan palace is the central court; it is surrounded by a large number of architectural blocks, which tend to have specific and at the same time discrete functions [35]. The irregular set out of the building blocks around the central court as well as the complex system of the narrow corridors leading to it (assumed to form the labyrinth) necessitated the consideration of light and air circulation. This was achieved by placing numerous light wells, colonnades, and clerestories in areas, which usually served specific purposes, such as the Royal Quarters (Fig. 7), the throne room, and the administrative chambers.

The light wells extended from ground level up to the ceiling of the building units (usually consisting from more than one or two stories) and were unroofed. Light was transmitted to the adjacent rooms through colonnaded porticos (Fig. 8). Such porticos were commonly used as transient spaces, which separated the building units from the open spaces.

Undoubtedly, a common feature of most palatial complexes is that the provision of light and air was a priority for the builders. For that purpose, colonnades, lightwells, clerestories, and windows were employed.

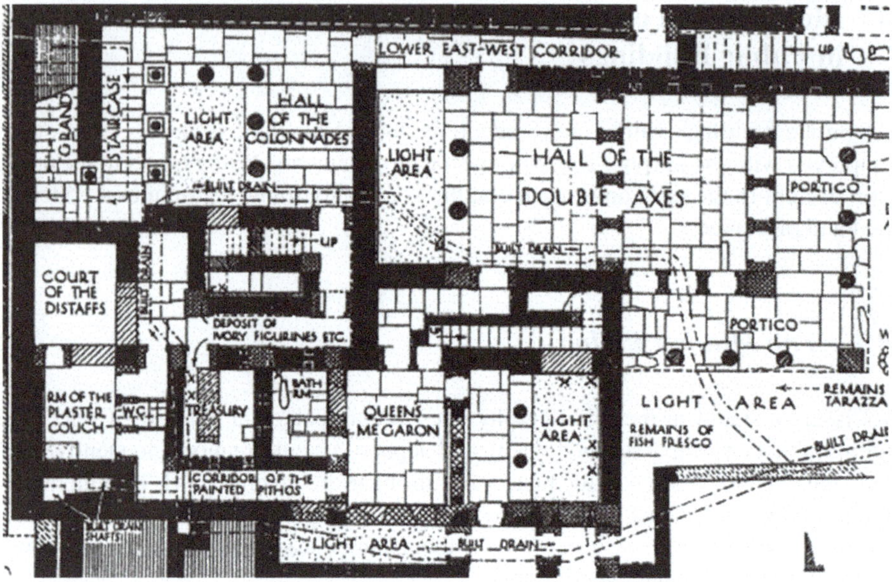

Fig. 7 Section of the plan showing the eastern part of Knossos Palace, which houses the Queen's Quarters and the Hall of the Double Axes with numerous light wells. (Source: Shaw, Joseph, & Arron Lowe. 2002. "The 'Lost' Portico at Knossos: The Central Court Revisited". American Journal of Archaeology 106 no 4 (October): 513–523)

Fig. 8 (Restored) light wells in the Hall of the Royal Guard in Knossos. (Source: K. Tsikaloudaki: "Shedding light on constructions of the past and today: the use of advanced daylight systems through the centuries", international symposium "Advanced Daylighting and Artificial Lighting in Architecture in 21st Century", Seoul, 18.06.2006, proceedings pp. 119–153)

4.2 *Classical Era*

Apart from the private houses, special attention was given to the lighting of the temples. The motion of the sun and the position of the planets were well known before the Classical period and often formed the focal point of their design. It is worth mentioning that more than 80% of the Greek temples run within the arc formed on the horizon between the sunrise directions at the summer and winter solstices [22]. It seems that most temples were orientated in order to face exactly the sunrise on the actual day of their foundation, which presumably coincided with the celebration day of the divinity. Temples were constructed without windows [36]. It is believed that the interior of the temple was accessible to privileged persons only and therefore the only view of the deity's statue was achieved through the open gateway on the eastern side of the building (Fig. 9). The awe and sense of mystery inspired among the congregation by such a view of the statue is reported in the work of Aeschylus, titled *Agamemnon*: "Gods who face the rising sun … with gleaming eyes" [37].

Daylighting in sacred and public buildings was often supported by small openings in the roof, known as "opaion" (oculus). It is also reported that marble from Paros was employed for roof tiles on account of its translucency, which would provide additional light to the interior of the temple through openings in the framed ceiling [22]. Again, emphasis was given by positioning such elements above and at the front of the divinity's statue.

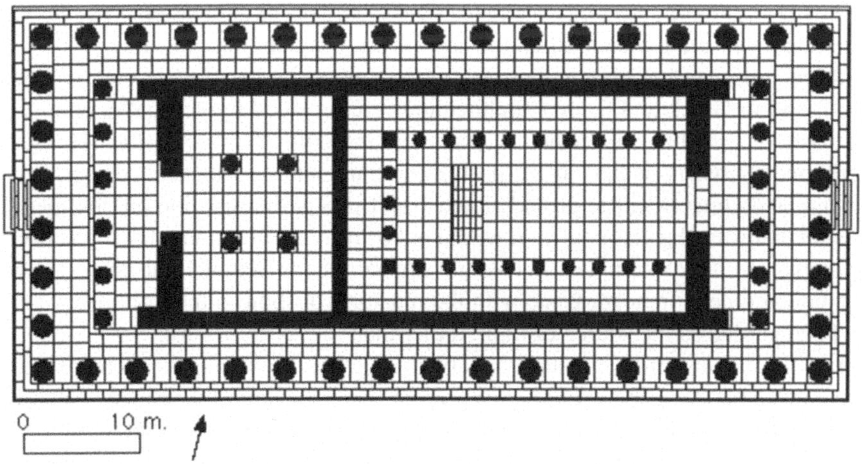

Fig. 9 Plan of Parthenon, Athens. The colossal chryselephantine statue of Athena – regarded as one of the seven miracles of the ancient world – was positioned among the pillars on the right section of the temple, facing the gate. (Source: C. Bouras, History of Architecture Vol 2, Melissa Ed., Athens, 1994)

4.3 Byzantine Era

The influence of ancient Greek architecture is evident during the Byzantine times. Undoubtedly, religion was the focal point of the Byzantine state and therefore the temples received high attention, particularly under the reign of Justinian. However, there are many crucial differences between the Classical and the Byzantine architectures, which emerged mainly from the dissimilarities in the religions' nature. In the Christian religion, the congregation participates in the ceremonies, which took place in the interior of the temple [15]; on the contrary, earlier forms of worship (Greek, Roman) were conducted on altars positioned outside the temples. Furthermore, the Christian churches were designed in proportion to the human scale, unlike the ancient temples, the form of which was determined by the harmony of its individual elements [15].

Light levels in churches are mostly related to the creation of a special environment to invigorate the faith, rather than to cover visual comfort needs [38]. Just as in previous eras, artifices were often employed, in order to cause sophisticated impressions and strengthen the religious spirit. Light was the main medium for achieving such an aim and its management was handled both with lateral and zenithal openings.

As in ancient Greek temples, the sanctuary of Byzantine churches is usually orientated due East; more specifically, the main axis of most Byzantine churches was adjusted toward the sunrise on the saint's celebration day. The morning daylight would be transmitted through the openings of the sanctuary's apse during the liturgy and would highlight the special character of the day. This has been also verified by measurements of the daylight levels prevailing in Byzantine churches existing in the area of Thessaloniki; the study indicated that regardless of the church type, the illumination levels were increased on the area of the sanctuary and the altar [39].

As regards zenithal illumination, the roof openings known as the "opaion" (oculus) found in Greek and Roman architectures are replaced by vertical openings on the basis of the dome. In orthodox churches, the dome represents the "House of God" and Christ is usually there portrayed; thus, among the main objectives of Byzantine architects was to provide intense and steady illumination on this specific area. In that way, the vision of God would not be directly illuminated, like the divinity statues in Greek temples, but would appear as the actual light-giver. Such a design outcome is in accordance with the meaning of light in the orthodoxy, since God is regarded as the source of light, rather than the receptor [40].

For the achievement of such an effect, ingenious techniques were used. In Byzantine churches, the artifices were more sophisticated and refined. Although there were no significant developments in science, there was remarkable progress in architecture, which is verified by the integration of already known scientific principles in the construction. The space configuration, together with a patchwork of light and dark areas, creates dramatic conflicts

The ingenuity of Byzantine architects reached its peak in Hagia Sophia of Constantinople (532 AD–537 AD). The initial dome of Haghia Sophia is assumed to be the application of Anthemios' treatise titled "On paradox mechanisms," which discussed the construction of a reflector able to direct the solar rays toward specific indoor areas, taking into account the different positions of the sun on an annual basis.

Historians of the period claim that the sill and the vertical sides of the openings surrounding the dome had an unusual elliptic shape, so as to guide the reflected on them daylight toward the center of the dome. These surfaces acted as reflectors, which ensured that the dome would be illuminated by as many windows as possible during the whole day, acquiring thus a uniform and steady luminance. Furthermore, the paradoxically low curvature of the dome, which was probably the reason of its collapse, would create multiple light reflections; the reflected light would not be directed downward, but it would be retained in the dome. The optical impression of such a design is best described through the words of Procopius: "The church is singularly full of light and sunshine; the place is not lighted by the sun from without, but that the rays are produced within itself, such an abundance of light is poured into this church... A spherical-shaped dome... does not appear to rest upon a solid foundation, but to cover the place beneath as though it were suspended from heaven by the fabled golden chain" [41].

4.4 Post-Byzantine/Ottoman Era

In many examples of Ottoman architecture, daylight flowing down from the main dome is used to emphasize the dome and to gather the people under it. This feature is also met in the bathing establishments, known as hamams, which served as one of the focal points of everyday life in the post-Byzantine/Ottoman period in Greece.

However, they were not only functional buildings; they were symbolic implications of status and offered an expression of existential meanings [42], as they became paired with mosques to satisfy the needs for hygiene and purification, which are significant in Islamic regulations [43]. To serve this role, the hamams were actively promoted by the elaboration of their design, techniques, and embellishments, and illumination through daylight served as a medium for achieving these goals.

Daylight in ottoman hamams is admitted through lateral and horizontal openings. The position and the size of the openings vary significantly with regard to the use of each individual space constituting the hamam, which are organized in a common layout and followed a plan related to temperature gradation, which corresponds to the ritual of a hamam visit [44]. More specifically,

- The cold room is the welcoming and disrobing area.
- The tepid room is the warmed space that links the disrobing area to the hot room and prepares the body for the increasing temperature and humidity that is supplied through the water vapor.
- The hot room is the destination space of the ritual, where the gathering and the activities take place. It is mostly a spacious room, usually enlarged with cells or

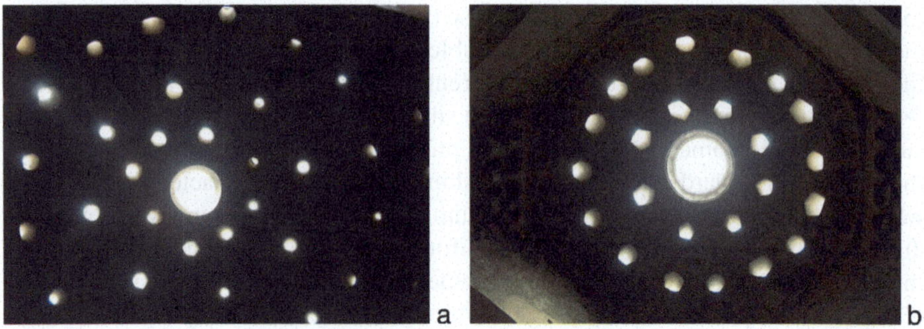

Fig. 10 The light openings on the dome of the tepid (**a**) and the hot (**b**) room. The former provides a more uniform daylight in the area, but the latter delivers more intense and directed daylight toward the central marble slab. (Source: Katerina Tsikaloudaki, Öget Nevin Cocen, Kyriaki Tasopoulou, Ioannis Milonas, Daylighting in historic bathhouses: the case of ottoman hamams, METU JFA (30:1) pp. 45–55, 2014.)

sections around the central area. It is mostly dominated by a central marble slab that is surrounded by basins where the water is supplied from. The central slab, as its seating, is the center of the activities, where massage, cleaning, or leisure activities take place.

The light distribution in the individual spaces follows the temperature gradation. The cold room is the only space of the hamam that would have windows at street level and/or at higher levels and it is moderately lit. In tepid rooms, the natural light levels are lower and are produced through small light apertures on the dome of the spaces. The hot room is mostly a moderately illuminated space with a mystical atmosphere created through the downward daylight from the domed superstructure. The character of the light mostly varies by the positioning, the geometry, and the color of the capping glass of the oculi (Fig. 10). However, in the hot chambers, special attention is given to the central area, where the marble slab stands [45]. The centrally oriented composition of the hot room is enhanced by a central lantern and/or the oculi around it. Taken into account the symbolic aspects of light and bathing in the Islamic culture, the high daylight levels prevailing on that part of the chambers could be associated with the presence of God, who is apparent as light from above to those taking part in the corporal cleanliness procedure. The transition to the darker areas leads to the completion of the ritual procedure; the bather feels relaxed, calm, and satisfied with fulfilling his religious duty.

The above-described visual environment has been verified by measurements in the interior of the Ottoman baths in Greece [46], which proved the interconnection between daylighting, the uses, and the objectives of each particular space.

5 Conclusions

From the above analysis, it is depicted that light has been employed in each era in order to achieve certain goals. In Minoan architecture, light wells were used primarily for illumination. The integration of the known principles of light transmission and the benefits of proper orientation in the domestic architecture of the Classical period, as well as the careful design of the Greek temples with regard to the daylight, which would intensify the power of the deity and invoke awe to the members of the religious community, are the most characteristic paradigms of light management in ancient Greece. The most sophisticated and refined techniques of light management are met in the Byzantine Architecture and are almost exclusively used for the enhancement of the experiential impact of the religious teachings and the invigoration of faith. Accordingly, special openings of the domes of the ottoman baths allowed daylight to serve once more as the medium for creating sophisticated impressions and strengthening the religious spirit by combining the corporal cleanliness achieved through the ritual bathing with the presence of God presented as light.

Transformations of the above principles and techniques are found in vernacular architecture. Paradigms of proper light management can be found both in the islands (widening of the window sides and sill toward the interior for glare control) and in mountainous areas (addition of openings above the conventional windows for exploiting sky luminance).

Unfortunately, such strategies are hardly met in contemporary architecture, especially as regards ordinary constructions. Although, nowadays, quantum physics has solved the mystery of light's nature, complicated computer programs enable the reliable simulation of indoor lighting conditions, and accomplishments on the technological sector have introduced numerous advanced systems for the exploitation of daylight, the former policy of implementing each era's technological and scientific achievements in building design has been abandoned [47]. However, it could be adopted even today, not in a way that would imitate former building techniques, but by utilizing the present knowledge, skills, and means in order to achieve well-lit environments and visual comfort with the minimum use of artificial energy sources. Daylight is almost everywhere available; what has to be done is to introduce it to our constructions by focusing on the needs of the users, the character of the space, and the desired outcome. Such an approach would form the best lesson from our history and an important step toward sustainability; after all, understanding the past has always been a decisive factor in designing the future.

References

1. H. Coch, Bioclimatism in vernacular architecture, Renewable and Sustainable Energy Reviews, 2(1–2), pp. 67–87, 1996.
2. M. Parker Pearson, C. Richards, (Eds.), Architecture and Order: Approaches to Social Space (1st ed.). Routledge, 1994.
3. R. Kittler, S. Darula, History of the daylight criteria conditions influencing new methods for the determination of interior daylighting, 2016 IEEE Lighting Conference of the Visegrad Countries (Lumen V4), pp. 1–8, 2016.
4. J. Manuel Monteoliva, L. Bellia, F. Fragliasso, A. Pattini, Ancient Romans and daylighting: the case of Villa of the mysteries in Pompeii, Journal of Cultural Heritage, 43, pp. 204–218, 2020.
5. C. Balocco, R. Calzolari, Natural light design for an ancient building: A case study, Journal of Cultural Heritage, 9, pp. 172–178, 2008.
6. M. Sibley, Let There Be Light! Investigating Vernacular Daylighting in Moroccan Heritage Hammams for Rehabilitation, Benchmarking and Energy Saving, Sustainability, 10, 3984, 2018.
7. S. N. Kamaruzzaman, R. Edwards, E. M. Ahmad Zawawi, A. I. Che-Ani, Achieving energy and cost savings through simple daylighting control in tropical historic buildings, Energy and Buildings, 90, pp. 85–93, 2015.
8. L. Ide, M. Gutland, S. Bucking, M. Santana Quintero, Balancing Trade-offs between Deep Energy Retrofits and Heritage Conservation: A Methodology and Case Study, International Journal of Architectural Heritage, 2020.
9. H. Elkadi, S. Al-Maiyah, Daylight for Strategic Intervention in Historic Towns: The Cases of Cairo and Edinburgh, Journal of Urban Technology, 14:3, pp. 83–102, 2007.
10. K. Tsikaloudaki, Harnessing the sun through the centuries: an analysis on the history of building construction in Greece, international conference Environmental Sustainability. The Challenge of Awareness in Developing Societies" – PLEA 2005, Beirut Lebanon, 13–16.11.2005, proceedings volume I, pp. 185–190.
11. C. Bouras, Lessons on the History of Architecture-Vol 1, Simmetria Ed., Athens, 1999.
12. A. Moropoulou, A.S. Cakmak, N. Lohvyn, Earthquake resistant construction techniques and materials on Byzantine monuments in Kiev, Soil Dynamics & Earthquake Engineering, 19, pp. 603–615. 2000.
13. R. Hooker, The History of the Minoans, A learning module of Washington State University, 1996.
14. E. Walter-Karydi, The Greek House, Archaeological Society at Athens, Athens, 1998.
15. C. Bouras, History of Architecture Vol 2, Melissa Ed., Athens, 1994.
16. K.R. Dark, Houses, streets and shops in Byzantine Constantinople from the fifth to the twelfth centuries, Journal of Medieval History, volume 30, issue 2, pp. 83–107, 2004.
17. Anon., The Homeric Hymns and Homerica, translated by H.G. Evelyn-White, Harvard University Press, London, 1914.
18. A. Ali, Al Quran, Princeton University Press, Princeton, 2001.
19. V-P Herva, The life of buildings: Minoan building deposits in an ecological perspective, Oxford Journal of Archaelogy, 2005.
20. https://sites.dartmouth.edu/aegean-prehistory/ Project sponsored by Dartmouth College. Site and content developed by: Jeremy B. Rutter, Professor of Classics, Dartmouth College; JoAnn Gonzalez-Major, Instructional Designer. Accessed 15 June 2021.
21. L. V. Watrous, Ayia Triada: A New Perspective on the Minoan Villa, American Journal of Archaeology, vol. 88, no. 2, pp. 123–134, 1984.
22. W.B. Dinsmoor, The architecture of Ancient Greece: an account on its historic development, Batsford, London, 1975.
23. Xenophon, Memorabilia and Oeconomicus, Harvard University Press, 1953.
24. Vitruvious, The Ten Books on Architecture, Dover Publications, New York, 1960.

25. F. Bougiatioti, A. Oikonomou, Bioclimatic analysis of streets and buildings of the Byzantine period, IOP Conf. Ser.: Earth Environ. Science, 410 012100 2020.
26. V. Tourptsoglou Stephanidou, The Roman-Byzantine building regulations, in B.S. Hakim (Ed.), Mediterranean Urbanism. Historic urban/Building Rules and Processes, Springer, pp. 127–170, 2014.
27. C. Saliou, The Byzantine house (400-912): rules and representations in Grünbart, M., Kislinger, E., Muthesius, A., Stathakopoulos, D. (dir.) Material Culture and Well-Being in Byzantium (400-1453), Wien, pp. 199–206, 2007.
28. B.S. Hakim, Mediterranean urban and building codes: origins, content, impact, and lessons Journal of Urban Design, 13, pp. 21–40, 2008.
29. F. Bougiatioti, A. Oikonomou, Architectural characteristics and environmental performance of byzantine houses and streets, Building and Environment, Volume 170, 106605, 2020.
30. T. Tassiopoulou, Thermal Behaviour of an Eighteenth-Century Athenian Dwelling, Applied Energy, 53, pp. 383–398, 1996.
31. A. Oikonomou, Daylighting issues in 19th century traditional buildings of the town of Florina, in North-Western Greece, Proceedings of the 22nd conference on PLEA, Beirut Lebanon, 13–16 November 2005, pp. 359–364.
32. N. Sakarellou-Tousi. B. Lau, The Visual Environment in the vernacular dwellings at Mount Pelion, Greece, PLEA 2011 – 27th Conference on Passive and Low Energy Architecture, Louvain-la-Neuve, Belgium, 13–15 July 2011, pp. 849–854.
33. A.-M. Vissilia, Evaluation of a sustainable Greek vernacular settlement and its landscape: Architectural typology and building physics, Building and Environment, 44, 1095–1106, 2009.
34. L. A. Hitchcock, Understanding the Minoan Palaces, Athena Review, Vol. 3, No. 3, pp. 27–35, 2003.
35. J. Shaw, L. Arron. The 'Lost' Portico at Knossos: The Central Court Revisited. American Journal of Archaeology, 106, no 4, pp. 513–523, 2002.
36. M. M. Azad, A. Barua, S. Sultana, A Review Analysis of Ancient Greek Architecture, Civil and Environmental Research, Vol.6, No.11, 2014.
37. Aeschylos, Agamemnon, Kaktos, Athens, 1992.
38. T. Antonakaki; Lighting and Spatial Structure in Religious Architecture: A Comparative Study of a Byzantine Church and an Early Ottoman Mosque in the city of Thessaloniki, Proceedings, 6th International Space Syntax Symposium, İstanbul, 12–15/06/2007.
39. G. Triantafillidis, Daylight Elements in Byzantine Churches (in Greek), Athens, 1964.
40. I. Potamianos, The Light in the Byzantine Church, University Studio Press, Thessaloniki, 2000.
41. Prokopios, On Edifices (in Greek), Georgiadis Ed., Athens, 1996.
42. I. Buyukdigan, A critical look at the new functions of Ottoman baths, Building and Environment, 38, pp. 617–633, 2003.
43. F. Yegul, Baths and bathing in Classical Antiquity, MIT Press, New York, 1992.
44. E. Kanetaki, The ottoman baths in the Greek region (in Greek), Technical Chamber of Greece, Athens, 2004.
45. E. Kanetaki, The still existing ottoman hamams in the Greek territory, METU JFA, 21, pp. 81–110, 2004.
46. K. Tsikaloudaki, Ö. N. Cocen, K. Tasopoulou, I. Milonas, Daylighting in historic bathhouses: the case of ottoman hamams, METU JFA (30:1) pp. 45–55, 2014.
47. K. Tsikaloudaki, S. Tsoka, T. Theodosiou, D. Tsirigoti, Exploring the symbolisms and techniques of daylight management in historic Greek constructions, international conference Places and Technologies 2019, Pecs Hungary, 09-10.05.2019, proceedings pp. 605–612.

Dr. Katerina Tsikaloudaki, civil engineer, has been working for the Civil Engineering Department, Aristotle University of Thessaloniki, Greece, first as a lecturer (2006), then as an assistant professor (2013), and now as an associate professor (2017). She teaches courses related to building construction, building energy performance, and sustainability at the undergraduate and postgraduate levels.

Her research interests are in the performance of the building envelope and building elements, building physics, and sustainability in the built environment. She has participated as a researcher and principal investigator in 26 projects funded by national (19) and international (7) sources. The results of her research have been published in peer-reviewed journals (23) and conference proceedings (more than 80).

She has an active participation in the development of the Greek Regulation of Building Energy Performance, as well as in the preparation of the relevant technical guidelines and the manuals for energy auditors. She has participated in the national committee for the recast of the energy performance legislation, as well as in the relevant committees of the Technical Chamber of Greece and the panel working for the definition of nearly Zero Energy Buildings (nZEB) framework in Greece.

She has also served as a member of the scientific committees of national and international conferences, reviewer of international scientific journals, and expert evaluator of research proposals at the national General Secretartiat for Research and Technology (GSRT) and European level (European Commission).

Passive Solar Design: The Influence of Building Geometry and Orientation on Solar Performance of Mosque in the Tropics

Mastura Adam and Norafida Ab Ghafar

1 Introduction

Mosque architecture has progressed excessively throughout the Muslim world. It reveals a wide range of design interpretations and adaptation to local context, tracing back over 1443 years where the first mosque has been built as part of Prophet Muhammad's house compound (p.b.u.h.). The mosque is a of religious building, a place for worshipping and functioning as a community for Muslims. This building typology has been a common sight in Malaysia because it immersed well with local culture and context which portrays Islam as the country's primary religion.

Besides functioning as the main congregational space, the mosque also serves as the center of religious education for the community. Mosque design has evolved through times; hence, the architecture styles are diversifying based on the current trend. Traditional vernacular mosque, for example, incorporates environmental responsiveness which equivalents to sustainable design concepts. The architecture has considered sustainable design strategies in making sure that the building has sound construction methods, use of green materials which are ecologically friendly and robust and flexible internal planning, hence contributing to the reduction of building energy consumption.

The influence of foreign architecture style during the colonial period is non-exceptional for mosque design, which changed the traditional vernacular mosque with pitch roof to dome type. The transformation of design also introduced the use of bricks and glass as new building materials. Nevertheless, those designs have been adapted to the local climate by applying passive design strategies such as verandah,

M. Adam (✉) · N. Ab Ghafar
Department of Architecture, Faculty of Built Environment, The Centre for Building, Construction & Tropical Architecture (BuCTA), University of Malaya, Kuala Lumpur, Malaysia
e-mail: mastura@um.edu.my

A. Sayigh (ed.), *Achieving Building Comfort by Natural Means*, Innovative Renewable Energy, https://doi.org/10.1007/978-3-031-04714-5_10

long eaves, ventilation block, clerestory windows, and long windows. The wide spread of these colonial mosque styles has perceived the dome as a compulsory design element for mosque architecture in Malaysia. Aesthetic aspects seem to supersede the most important value which is the thermal performance in the tropics where the sun path right on top of the head throughout a year. This leads to high solar radiation received on the roof surfaces. Therefore, this chapter aims to explore the solar radiation received on different mosques roof geometries: flat, tiered, and domes which are widely applied in mosque design in Malaysia.

2 Mosque Design Evolution in Malaysia

Islam has spread over the globe; mosque architecture and construction have begun in every country. As can be seen, many countries have different mosque architecture styles that are influenced by their culture and tradition and architectural heritage. The Peninsula of Malaysia has been under foreign colonial administration for about 446 years since the fall of Malacca in 1511. Malaysia gained independence in 1957 (84 years ago). Islamic architecture has been influenced by a variety of styles and techniques before and after independence [2].

Scholars have updated and developed a historical timeline of mosque architecture evolution from vernacular to contemporary era (Siti Dalila et al. 2019). Mosque architecture has been the subject of numerous studies, including architectural styles, classification, and typology. Mosque design in Malaysia was quite diverse and can be divided into three periods; pre-colonial, colonial, and post-independence [12], whereas Tajuddin (2000) had classified seven styles of mosque architecture language known as traditional vernacular, Sino-eclectic, colonial, north Indian, modern vernacular, modernistic expressionism, and post-modern revivalism. The evolution of architecture styles is mapped with a timeline, and influencing factors that formed the transformed mosque design in Malaysia are shown in Fig. 1.

2.1 Mosque Design and Its Architecture Styles

Mosque design and architecture style can be explained into seven categories. First, the traditional vernacular style where the context applies the practice of vernacular architecture within Malay Archipelago. There are three type of mosque: three-tier roof, double-tier roof, and gable roof [3, 8]. These three types of mosques did not incorporate with minaret structure in its design. Kampung Belimbing Mosque (Fig. 2) and Dato Dagang Mosque (Fig. 3) represent double-tier roof design style built during the pre-colonial period.

Second is the Sino-eclectic style, which is derived from the combination of Sino (Chinese architecture influence) and eclectic that denotes a combination of two or more architectural language. This style applies pyramidal of three- and double-tier

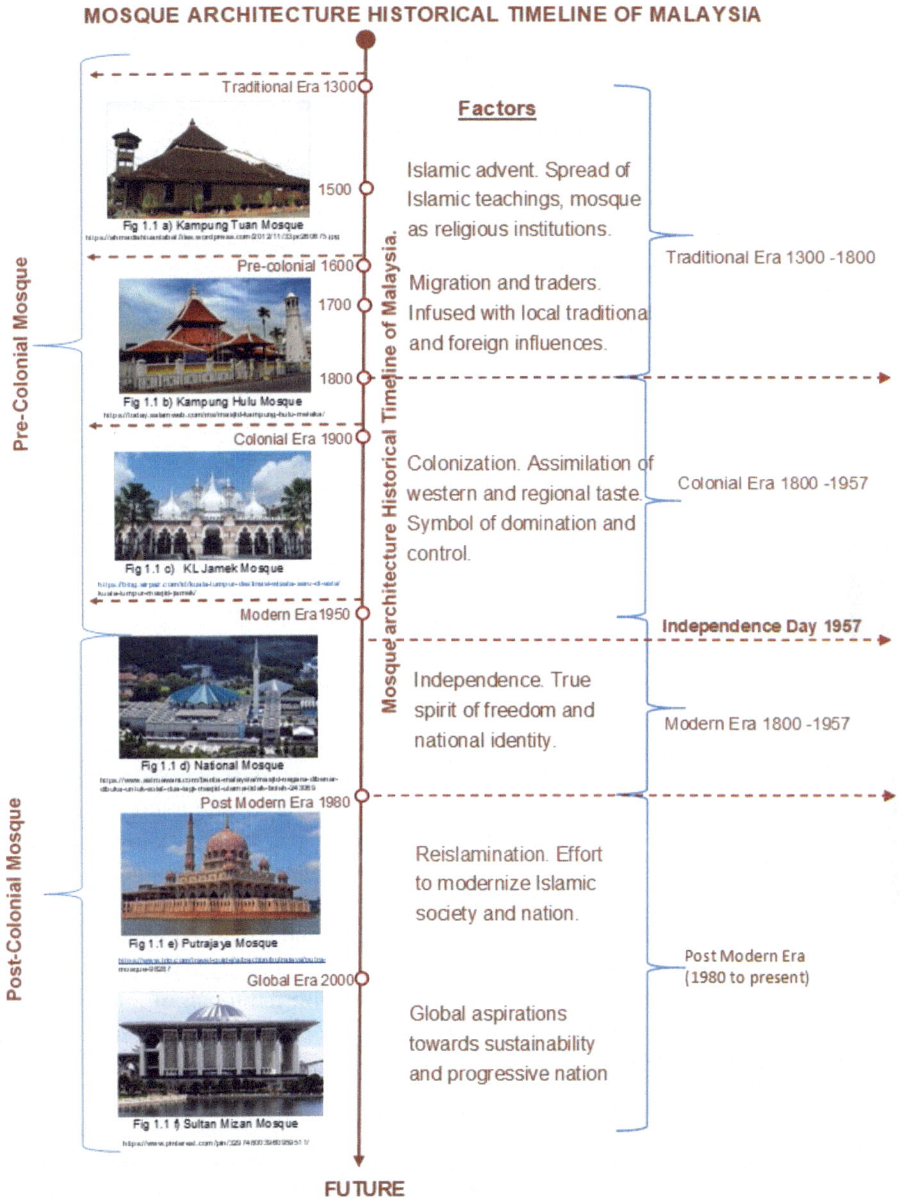

Fig. 1 Mosque design evolution and architecture style in Malaysia

roof with prominent curvature made of cement at the roof ridges. Kampung Hulu Mosque (Fig. 4) and Kampung Kling Mosque (Fig. 5) represent a cross-fertilization of culture between Chinese and traditional vernacular architecture with minarets that are influence from Chinese pagoda.

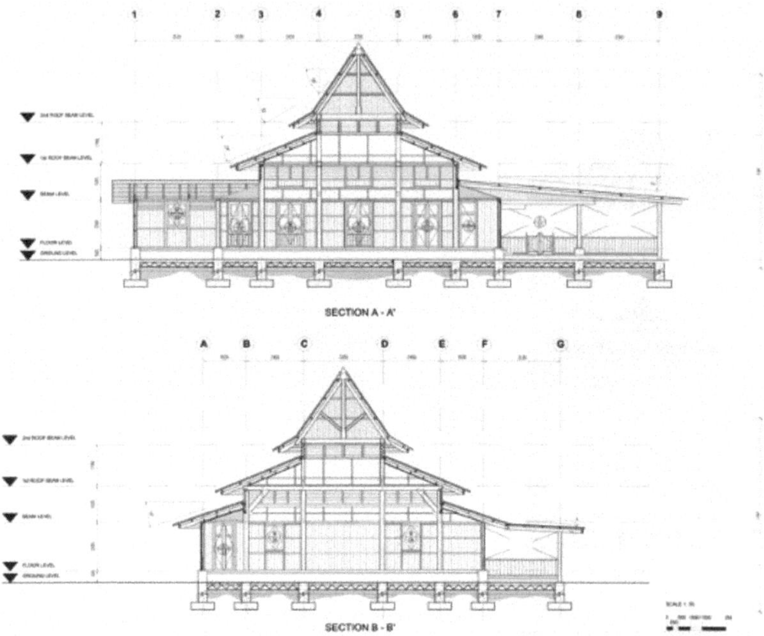

Fig. 2 Kampung Belimbing Mosque, Selangor. (Source: Measured Drawing, Dept. of Architecture, Universiti Malaya)

Third is the European classical style that refers to the High Renaissance architecture derived from Greco-Roman heritage. The main characteristic features, emphasizing scale, proportion, and order, enhancing three parts of building division, consist of base, middle, and top with double column supporting the semicircular arches or walls with pilaster. Sultan Alaeddin Mosque, Klang (Fig. 6) is one of the examples that applied these characters with a minaret and smaller domes crowning the main pointed dome. Sultan Sulaiman Mosque, Klang (Fig. 7) also uses the same division principles in elaborating the façade design with the touch of art-deco and emphasizes the large central half dome with the addition of four smaller half domes around its square plan. Both mosques articulate their minaret design to fit their function as a call tower for prayers.

The fourth is the north Indian style, with the influence of Moghul-style architecture widespread during the colonial era. The major characteristic is defined by central onion domes, also known as bulbous domes, and surrounded by smaller domes in various sizes, as shown in Jamek Mosque, Kuala Lumpur (Fig. 8). A series of pointed arches have been inserted into the façade design as well as along the verandah way.

Fifth is modern vernacular style that used that parallel-reinforced concrete structural frame with plaster brick infill as main construction. The use of new material and construction technique made it established as vernacular. The form used is

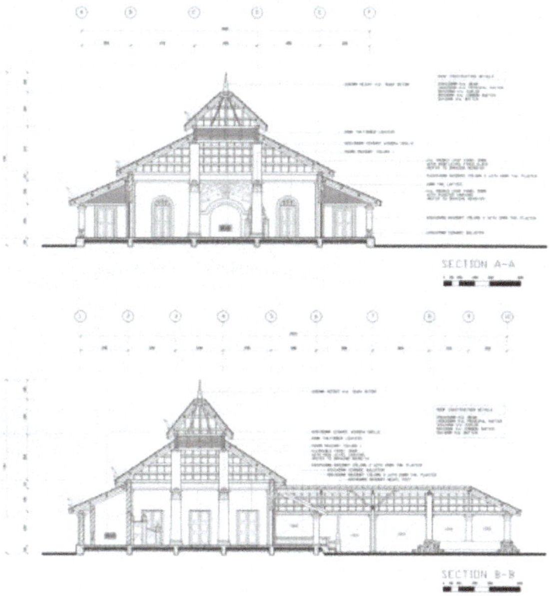

Fig. 3 Kampung Dato Dagang Mosque, Selangor. (Source: Measured Drawing Dept. of Architecture, Universiti Malaya)

Fig. 4 Kampung Hulu Mosque, Melaka. (Source: https://today.salamweb.com/ms/masjid-kampung-hulu-melaka/)

either gable or pyramidal roof with a small dome or huge-sized dome over the main prayer hall (Nayeem.A. et al. 2019).

Sixth is the modern style from the revolution of architectural and structural expression in Europe.

Fig. 5 Kampung Kling Mosque, Melaka. (Source: https://www.flickr.com/photos/gordontour/7096150831)

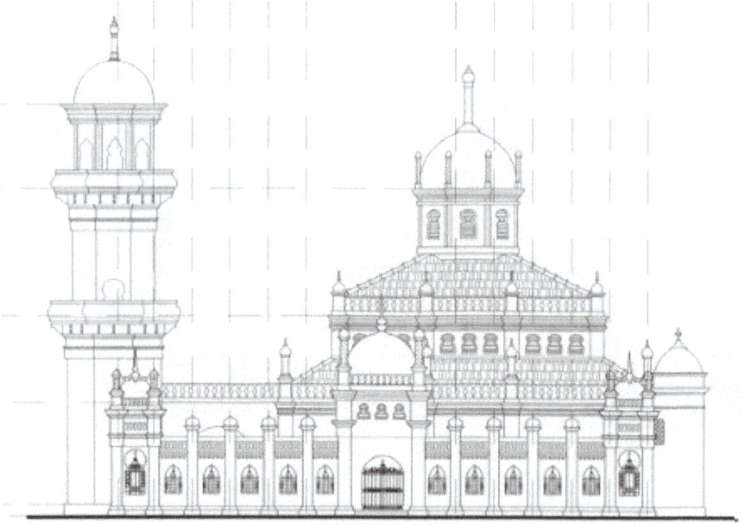

Fig. 6 Sultan Alaeddin Mosque, Klang. (Source: Measured Drawing, Dept. of Architecture, Universiti Malaya)

Expressionism derived from the metaphoric message through structurally expressive form, for example, National Mosque and Negeri Sembilan State Mosque. The use of verandah, serambi, long eaves, ventilation block, and water features in the National Mosque contributes to passive cooling of the building. The uniqueness of

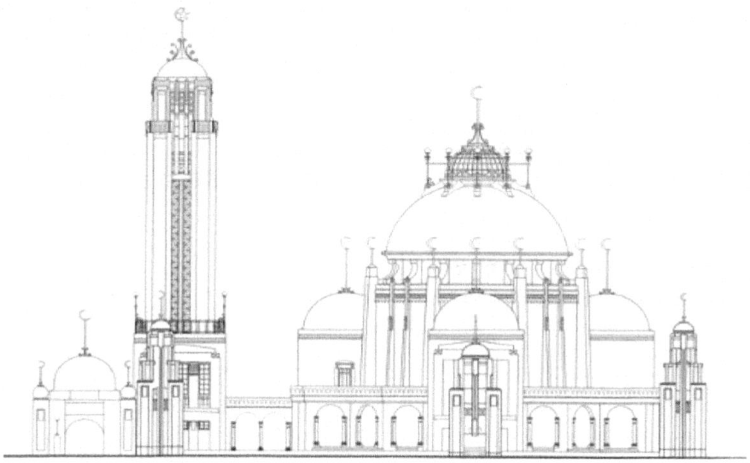

Fig. 7 Sultan Sulaiman Mosque, Klang. (Source: Measured Drawing, Dept. of Architecture, Universiti Malaya)

Fig. 8 Jamek Mosque, Kuala Lumpur. (Source: Measured Drawing, Dept. of Architecture, Universiti Malaya)

the roof portrayed the combination of modernist reinterpretation of traditional Malay architecture with the metaphor of royal umbrella. Structural expression in the National Mosque through the usage of concrete and steel as main construction materials.

Lastly, the seventh is post-modern revivalism style influenced by foreign revivalism and vernacular revivalism. Foreign revivalism adapted toward Iranian and Turkish dome, Egyptian and Turkish minarets, Persian gateway, lavish courtyard, and hypostyle planning composition from Arabian, whereas vernacular revivalism is less monumental in approach with the use of three-tier pyramidal roof from either concrete or timber.

Mosque evolution in Malaysia can be confirmed to have diverse architectural style and influences worldwide. The information gathered from mosque evolution is able to give an in-depth understanding on mosque roof geometries that have been applied in mosque design in Malaysia. The mosque roof geometry can be classified into three (3) types which will be examined, simulated, and evaluated on their thermal performance. The mosque roof's geometry types are as below:

(i) Flat roof
(ii) Pitch roof consists of single tier and double tier
(iii) Dome roof consists of half dome, pointed dome, and bulbous dome

3 Bioclamatic Architecture in Malaysia

The term "bioclimatic architecture" refers to the planning of buildings and facilities depending on the local climate, with the goal of delivering thermal and visual comfort with low conventional energy usage while utilizing various renewable energy sources. Bioclimatic architecture is extremely important, and its appropriate application benefits both humans and the environment.

Geographically, Malaysia is located near the equator and experiences hot, humid weather throughout the year. The average temperature is 27 °C (80.6 °F) with an annual rainfall of 250 centimeters (98 inches) (Fig. 9). Malaysia consists of two mainlands: Peninsular Malaysia and East Malaysia, which have quite different weather. In Peninsular Malaysia, the weather is influenced by the wind from the mainland, whereas in East Malaysia, the weather is dominated by the ocean. Climate change is expected to have a big impact on Malaysia, raising sea levels and rainfall, increasing the risk of flooding, and causing large drought (Marshall Cavendish Corporation 2008).

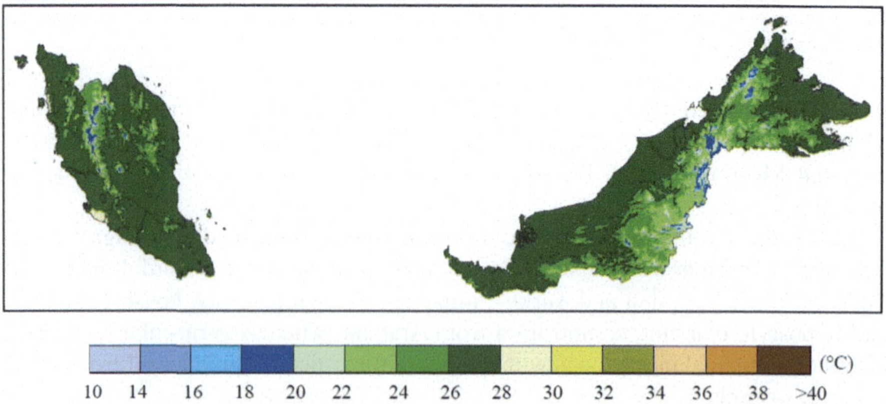

Fig. 9 Malaysia's Mean Monthly Temperature Tabulation in June 2015 (Malaysian Meteorological Department 2015)

The rainfall distribution patterns across the country are determined by seasonal wind flow patterns along with local topographic factors. Exposed locations such as Peninsular Malaysia's east coast, Western Sarawak, and Sabah's northeast receive considerable precipitation during the northeast monsoon season. Inland places or areas shielded by mountain ranges, on the other hand, are relatively free of its influence shown in Fig. 10.

The context of the building is important in determining the design strategies that need to be adapted to achieve indoor thermal comfort. In the tropics, architecture must incorporate passive bioclimatic design, which takes into account current weather, hydrography, and natural ecosystems to reduce the building's overall energy balance, which encompasses design, construction, use, and end of life.

3.1 *Passive Solar Design Strategy for the Tropics*

Passive solar design helps to minimize energy consumption in building and improve building thermal performance by considering building form, ventilation, building fenestration, and fabric (skin). This goal can be achieved by strategizing the building orientation, natural ventilation, sun shading devices, glass insulation, building fabric, and thermal mass as shown in Fig. 11. The passive design considers the local temperature and conditions to provide thermal comfort while utilizing the least amount of energy possible.

3.2 *Factors Affecting Solar Radiation*

The energy that is discharged from the sun when hydrogen diffuses into helium through nuclear fusion is fundamental to understanding solar radiation. The sun is a star that emits energy as its atoms combine with one another. Solar radiation

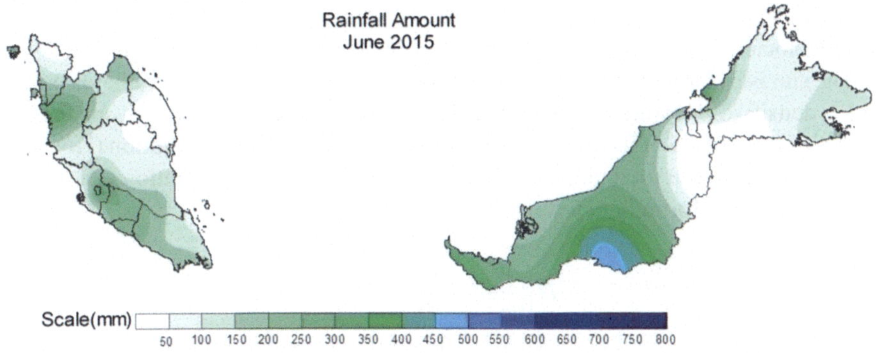

Fig. 10 Malaysia's Rainfall Tabulation in June 2015 (Malaysian Meteorological Department 2015)

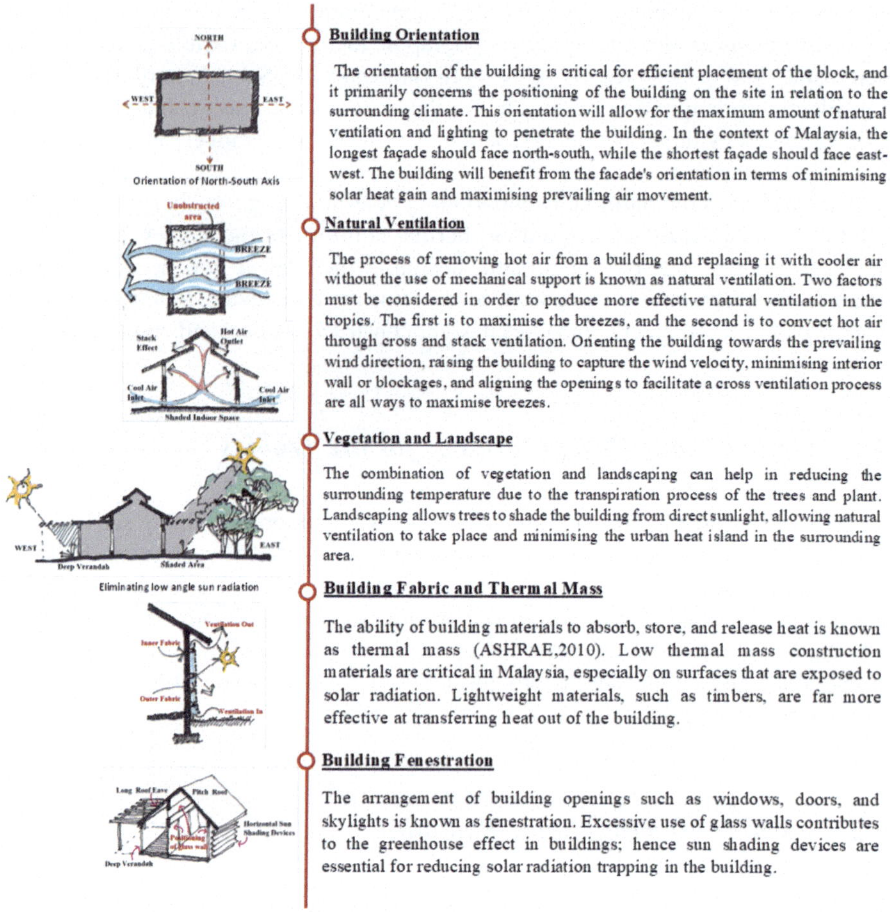

Fig. 11 Passive solar design strategy for the tropic

intensity was affected by several factors including water vapors, air particles, and sun altitude. Solar radiation is not a constant factor because it varies depending on latitude, location, season, cloud cover, pollution, sea levels, and solar elevation.

The sun is a star with a temperature of approximately 6000 K that radiates power with a density of H (sun) = 73 MW/m2. The energy then travels through space until it reaches the surface of the planet Earth, which is known as solar radiation (Honsbeg 2008). The sun has a total power of 4R2, and the density of its radiation can be calculated using the formula below:

$$H\left(\frac{W}{m^2}\right) = \frac{R_{sun}^2}{D^2} \times H_{sun}$$

There are two types of solar radiation received at ground level is either direct or indirect radiation. The term "direct radiation" refers to solar radiation that has not been reflected, dispersed, or diffused by the earth's atmosphere, whereas "indirect radiation" refers to solar energy that has been reflected, scattered, or diffused by the earth's atmosphere [7, 13]. The intensity of the sun is affected by a variety of components in the atmosphere, including dust particles, ozone levels, water vapor, and solar altitude as shown in Fig. 12. As a result, solar radiation is not a fixed factor on Earth; it fluctuates depending on latitude, location, season, cloud cover, pollution, sea levels, and solar elevation.

There are five (5) contributing factors for solar radiation that need to be considered for building design in the tropics such as the Earth's rotation on its axis, the Azimuth angle, solar altitude, solar radiation measurement, and terrestrial radiation (Honsbeg 2008) as explain in Fig. 13.

4 Simulation of Different Mosque Roof Geometries

The roof is the topmost of building components that receives approximately 70% of solar radiation, particularly in the tropics. Therefore, knowledge on the potential of roof geometry and orientation in reducing total amount of solar irradiance received on the roof surface is important for designer at the early stage of design. This will help the designer in deciding the appropriate mosque roof geometry according to

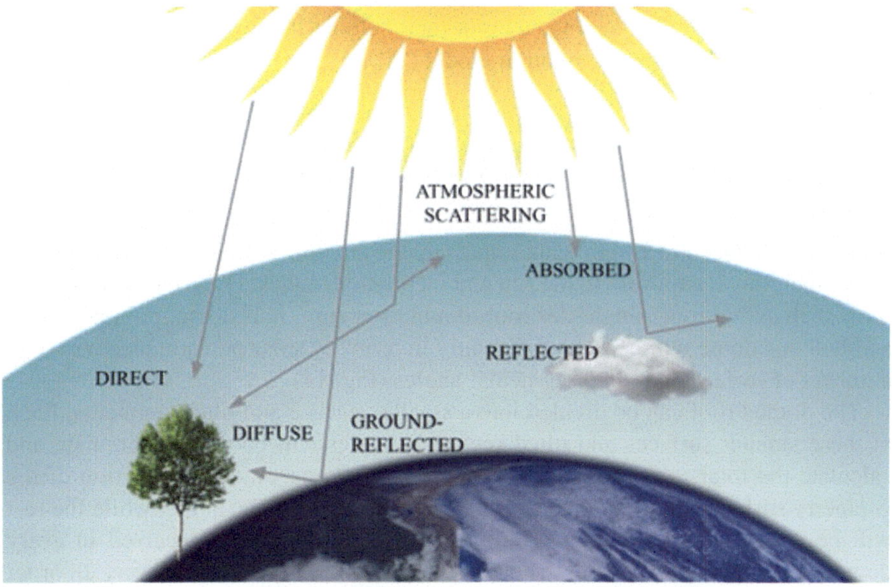

Fig. 12 Different types of solar radiation

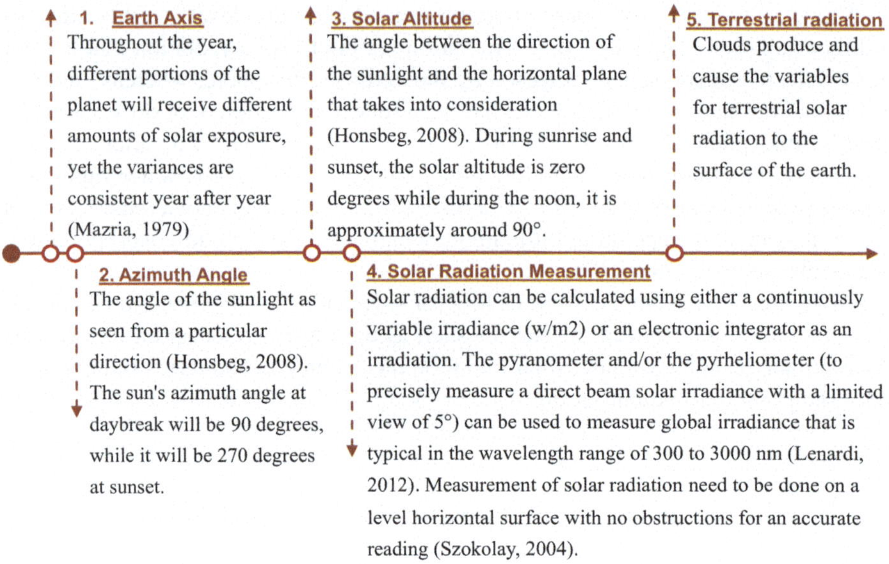

Fig. 13 Consideration of solar radiation's factors in building design approach for the tropics

specific context, location, and latitude as a solution for energy conservation and environmental responsiveness. This section will be examining the influence of solar altitude, surface azimuth angles, and the orientation of mosque roof geometries on the amount of solar irradiance I(HTCS) received on the roof surfaces. Solar performance of different mosque roof geometries, namely, flat roof, single-tier roof, double-tier roof, half dome, pointed dome, and bulbous dome were tested and simulated using solar radiation simulation model (SRSM).

4.1 *Solar Radiation on Roof Surface*

Tilted roof surface has a significant impact on the amount of solar radiation received due to their direct exposure to the sun and inclination angles. Different types of roof form such as flat roof, single-tier roof, double-tier roof, half dome, pointed dome, and bulbous dome will perform differently in terms of solar performance based on numbers of surface and tilted segments' angles (Fig. 14)

The domed roof can be divided into a smaller surface and tilted segments. Each of these smaller surfaces and tilted segments angles will be used to simulate and calculate the total solar radiation received by the dome. A smaller portion of the geometry surface area was in contact with the solar incidence angle, while the rest will be shaded from radiation. The amount of solar radiation received in every domed geometry varies during the day as the sun moves from east to west. In order to estimate the solar radiation, the latitude and longitude of the location are

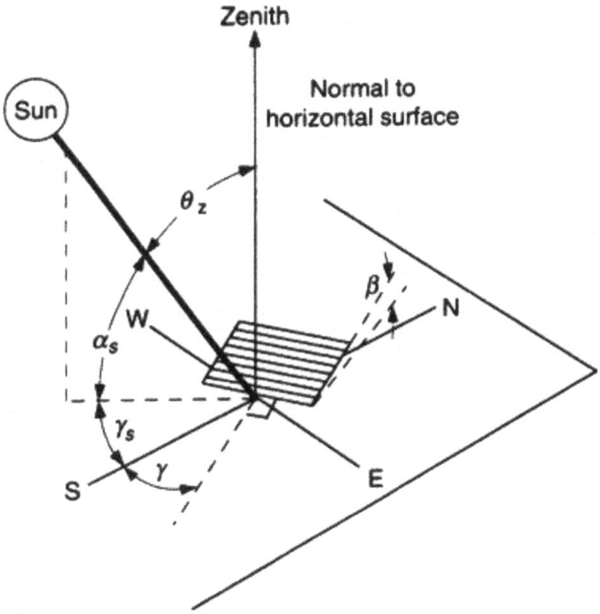

Fig. 14 The geometry of solar radiation on surfaces and tilted segments' angles. [5]

determined at chosen hours of the day and the day of the year. As for this experiment, Subang has been selected as the location and estimated on 20 March (equinox).

4.2 Sun Path

The considerable positional changes in the sun's yearly and hourly location as the earth rotates on its axis and orbits around the sun are referred to as the sun's path. The sun path can show us how the sun will affect our building over the course of the year. The sun path diagram can be used to determine the route of the sun at any location on the planet. The solar azimuth angle and the altitude for a specific location can be determined using the sun path diagram (Fig. 15).

4.3 Equinox and Solstices in the Tropics

Equinox can be defined in two events on earth when the position of the sun is exactly above the equator. During this moment, the sun moves across the celestial equator, and the Northern Hemisphere begins to experience spring, which occurs on 21 March, while on 23 September, the sun crosses the celestial equator to go to the south. It rises straight in the east and sets straight in the west. Furthermore, Solstices

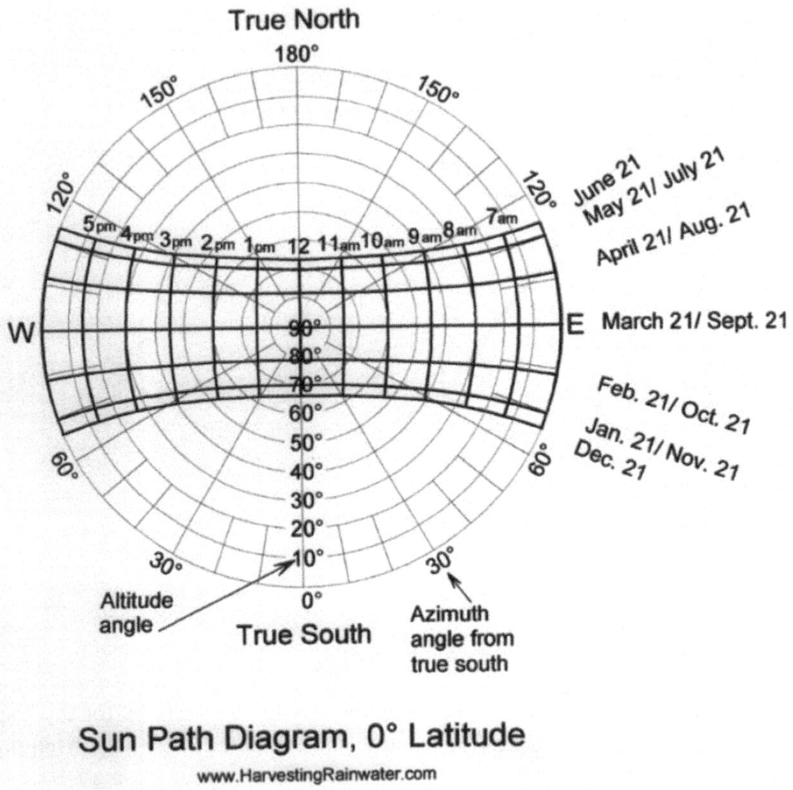

Fig. 15 The sample of sun path diagram. (Source: http://www.harvestingrainwater.com/wp-content/gallery/0-15degreess, retrieved July 2021)

occur twice a year, in June and December, and during these times, the sun reaches the northernmost point on the earth (Fig. 16).

4.4 Solar Radiation Independent Variable for Simulation

Solar radiation simulation model (SRSM) was chosen in determining the solar irradiance of the difference dome in Malaysia. SRSM is a mathematical model and highly reliable to calculate the intensity of received solar radiation I(HTCS) on different surfaces (horizontal and sloped) at any orientation; it has high accuracy for locations within the range of +25° north add −25° south (Muneer 1997). This simulator is a JavaScript program using the same properties as real solar radiations data in the tropical climate in generating numerical data. Initially, it was developed by Excel in 1986 and was upgraded in the year 1999. The software is able to calculate the total hourly clear sky irradiance, I(HTCS) W/m^2 on horizontal and different

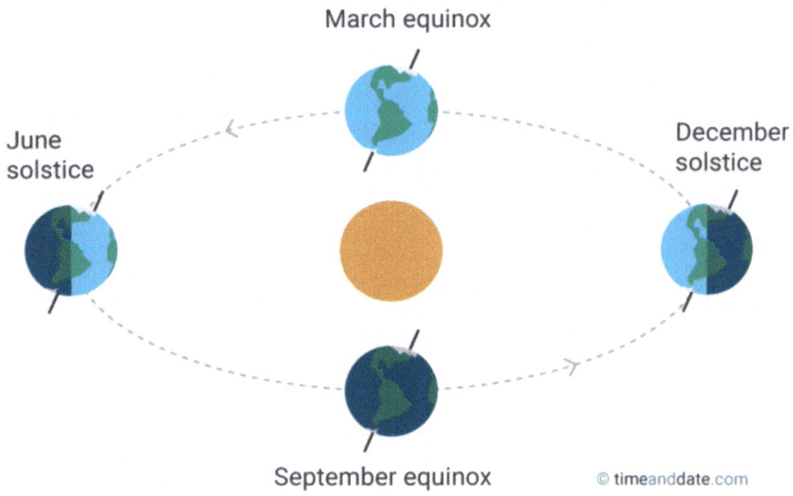

Fig. 16 Earth's tilt during the equinoxes and solstices. (Source: http://www.timeanddate.com/calendar/spring-equinox.html, retrieved 9 July, 2021)

sloped surfaces using the empirical formula suitable for the tropical climate. SRSM required the following input of independent variables:

(a) **Latitude (deg)** – SRSM model is only accurate in the tropics with the latitude within the interval +25° (north of the equator) to −25° (south of the equator). Therefore, the simulation for Subang, Malaysia, with the latitude 3° 12′ N is reliable and valid.

(b) **Mean daily solar irradiation on a horizontal surface each month (MJ/m2)** – The mean of solar irradiance I(HTCS) W/m^2 on horizontal surfaces in Subang (3.12°N) for clear-sky conditions is estimated at 75% of cloudy sky conditions.

(c) **Days of the Year** – The data in SRSM were computer programs generated for the whole 365 days of a year. For instance, 1 January is shown as Day 1, while 31 December is shown as Day 365.

(d) **Surface Slope Angle** – It is the angle between the normal angle of that surface and the zenith or the angle between the surface and the horizontal plane. It is also known as tilt angle (degree) (Fig. 17).

(e) **Orientation of tilted surface** – The Azimuth (degree West) is the angle between south (= zero degrees) and the direction in which the tilted surface is facing, measured toward the west. For example, west is 90°, north is 180°, and east is −90°. The tilt angle (degree) is the angle between the surface and the horizontal plane, which is the same as the angle between the normal to the surface and the zenith (Fig. 18).

All of the above variables must be inserted into the SRSM as an input to create the model as shown in Fig. 19.

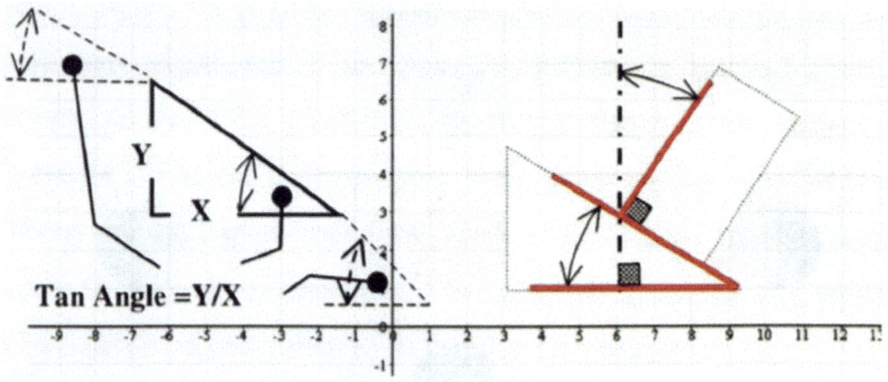

Fig. 17 Surface slope angle and geometrical relations [5]

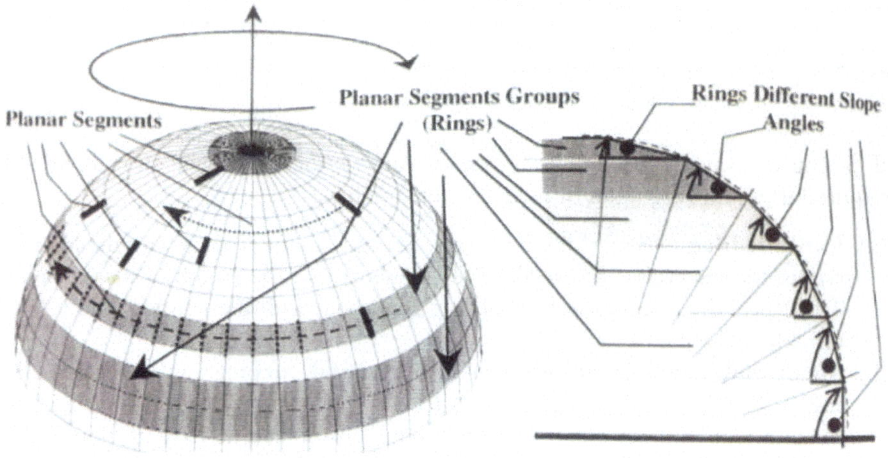

Fig. 18 Dome geometrical resemblance of surface segments [5]

5 Solar Radiation Performance on Mosque Roof Geometries

The solar behavior of all the specified mosque roof geometries including flat roof, single-tier roof, double-tier roof, half domes, pointed domes, and bulbous domes were studied and examined using SRSM software. All simulations had been tested on the same day which is 21 March (equinox) due to the highest average solar irradiance I(HTCS). Solar radiation simulation models on six mosque roof geometries emphasize on different surfaces and segments' tilted angles as shown in Fig. 20. Ring slope angles have been established in 14 rings altogether for all dome roof types (Table 1).

The solar performance of different types of mosque roof geometries will be examined and evaluated based on solar irradiance I(HTCS) W/m^2 received on the

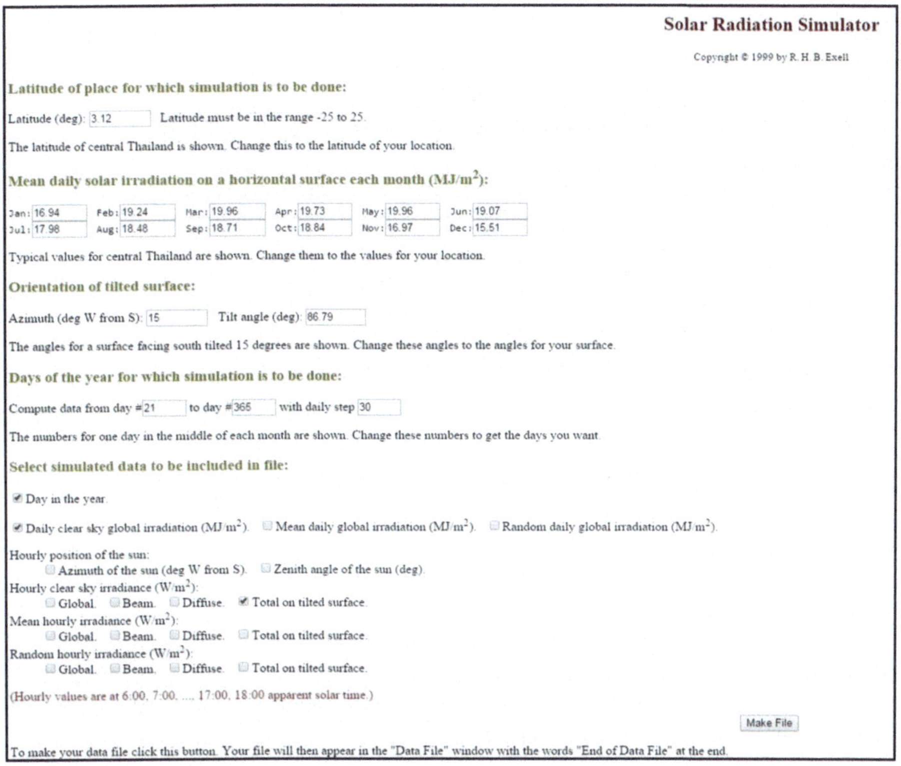

Fig. 19 Solar radiation simulator model (SRSM). (Source: http://www.jgsee.kmutt.ac.th/exell/Solar/SolradJS.htm, retrieved on 11 March 2017)

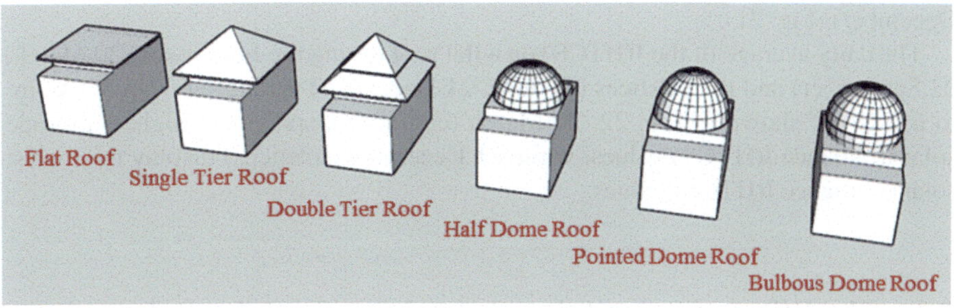

Fig. 20 Six mosque roof geometries simulation models emphasize on different surfaces and segments' tilted angles

roof surfaces. The data from each type of roof geometry were compared to one another to arrive at a conclusion on which type of mosque roof design has the most efficiency in receiving less solar radiation in the context of Malaysia.

Table 1 The ring segments tilted angle for different type of domes

Type of dome	Ring Segment Tilted Angle						
	R1	R2	R3	R4	R5	R6	R7
Half Dome	86.79	80.36	79.93	67.5	61.07	54.64	48.21
Pointed	90	87.33	80.99	74.98	68.97	62.96	56.96
Bulbous	113.86	105.42	96.99	88.55	80.12	71.69	63.25
	R8	R9	R10	R11	R12	R13	R14
Half Dome	45	35.36	30	22.5	16.07	9.64	3.21
Pointed	50.95	44.94	38.93	32.93	26.92	20.91	14.9
Bulbous	54.82	46.39	37.95	29.52	21.08	12.65	4.22

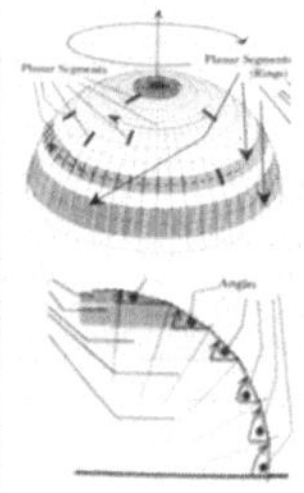

5.1 *The Effect of Solar Latitude on the Solar Behavior of Flat Roof*

The flat roof is used in this experiment to investigate the effect of solar behavior and solar latitude on the amount of solar irradiance received I(HTCS) W/m^2. This style of roof was chosen because of its low tilt angle, which means that the quantity of solar irradiance received by the roof's surface is comparable to that obtained by any horizontal surface.

The solar irradiance received I(HTCS) W/m^2 during both equinoxes on 20 March and 22 September as well as solstices on 21 June and 21 December were calculated. The roof is oriented in the East and West during the equinoxes and solstices. Based on the graph, similar behavior of I(HTCS) curves can be observed. The curve starts to rises differently after 6:00 in the morning and reaching its peak around midday or noon period before declining differently and overlapping at 18:00 in the evening. The amount of solar irradiance I(HTCS) on the flat roof during the equinoxes (20 March, 22 September) is greater than the I(HTCS) during the solstices (21 June, 21 December) (Fig. 21).

The daily average of the I(HTCS) on a flat roof during the Equinoxes (20 March, 22 September) and the solstices (21 June, 21 December) is ranging from 550W/m^2 to 617 W/m^2 shown in Fig. 22. 20 March (equinox) displays the highest average solar irradiance I(HTCS) values, while 21 December (solstices) display the lowest solar irradiance I(HTCS) values.

5.2 *The Effect of Solar Azimuth Angle on the Solar Behavior of Single-Tier Roof*

The effect of solar azimuth angle on the solar behavior of single-tier roof is being tested. On March 20, the graph below (Fig. 23) depicts the solar irradiance I(HTCS) on a single-tier roof hourly. It explains that solar behavior of a single-tier roof is similar to that of a flat roof. They progressively ascended around 6 a.m., reaching peak at 12 p.m. As the day progresses into nightfall, they steadily decrease.

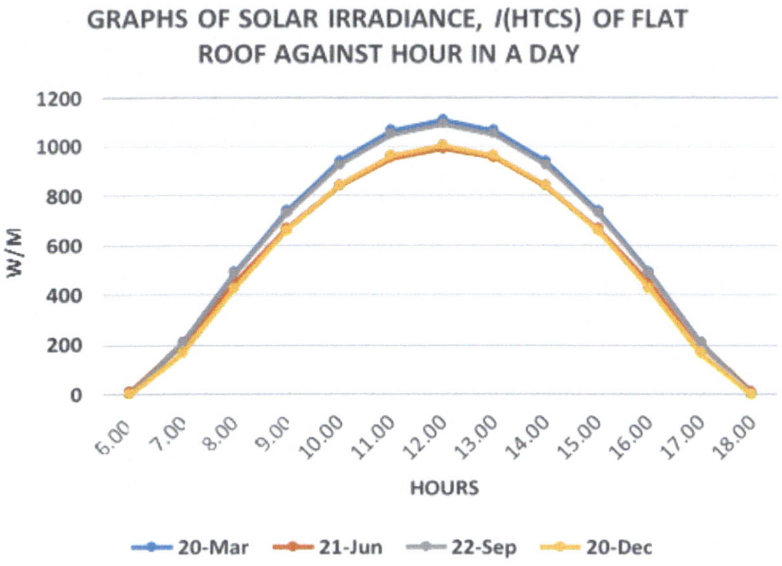

Fig. 21 I(HTCS) of flat roof against hour in a day: E-W facing in Equinoxes and Solstices

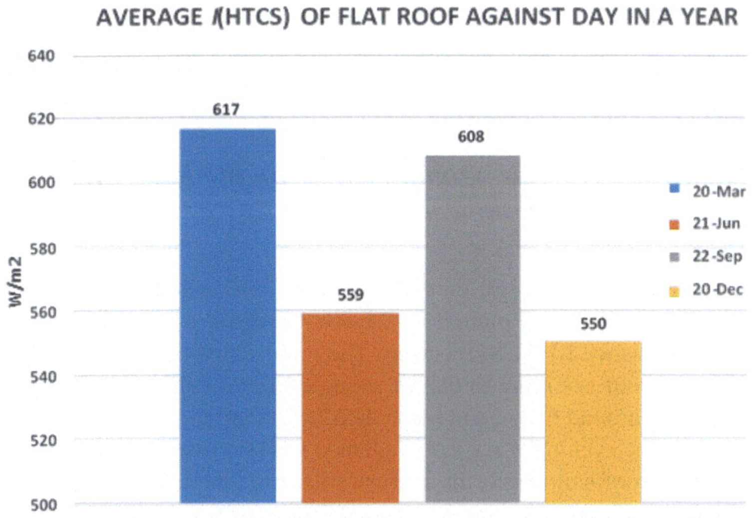

Fig. 22 Average I(HTCS) of flat roof against day in a year

The average solar irradiance I(HTCS), received by a single-tier roof was simulated based on each roof surface's azimuth tilted angle (44.54°). Throughout the month of March, the single-tier roof receives approximately 472 W/m^2 as shown in (Table 2).

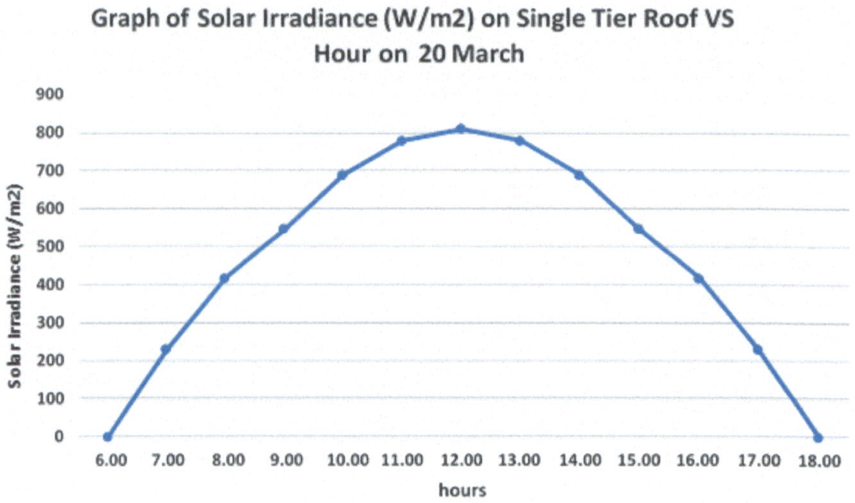

Fig. 23 Solar irradiance, I(HTCS) of single-tier roof

Table 2 The daily average of solar irradiance received, ***I***(HTCS).

Tilt Angle azimuth	6.00	7.00	8.00	9.00	10.00	11.00	12.00	13.00	14.00	15.00	16.00	17.00	18.00	Day Av.
0	0	163	379	567	718	812	843	812	718	567	379	163	0	471
90	0	56	83	100	353	601	810	959	1027	994	851	541	0	871
180	0	156	355	527	663	748	776	748	663	527	355	156	0	436
270	0	541	851	994	1027	959	810	601	353	100	83	56	0	110
	0	229	417	547	690.25	780	809.75	780	690.25	547	417	229	0	472

5.3 *The Effect of Solar Azimuth Angle on the Solar Behavior of Two-Tier Roof*

The effect of solar azimuth angle on the solar behavior of a double-tier roof is being investigated, calculated, and plotted for March 20 (equinox) due to the maximum average sun irradiance I(HTCS) received. The graph shows that the solar behavior of a double-tier roof is similar to that of a single-tier roof (Fig. 24). They begin to increase steadily around 6 a.m. and reach their height around lunchtime. As the day progresses into the evening, the solar irradiance I(HTCS) steadily decreases.

Based on the roof's tilt angle, the average solar irradiance, I(HTCS), received by the double-tier roof was determined (Table 3). The results revealed that the double-tier roof received approximately 480 W/m^2 (equinox), which is higher than the single-tier roof's average solar irradiance, I(HTCS).

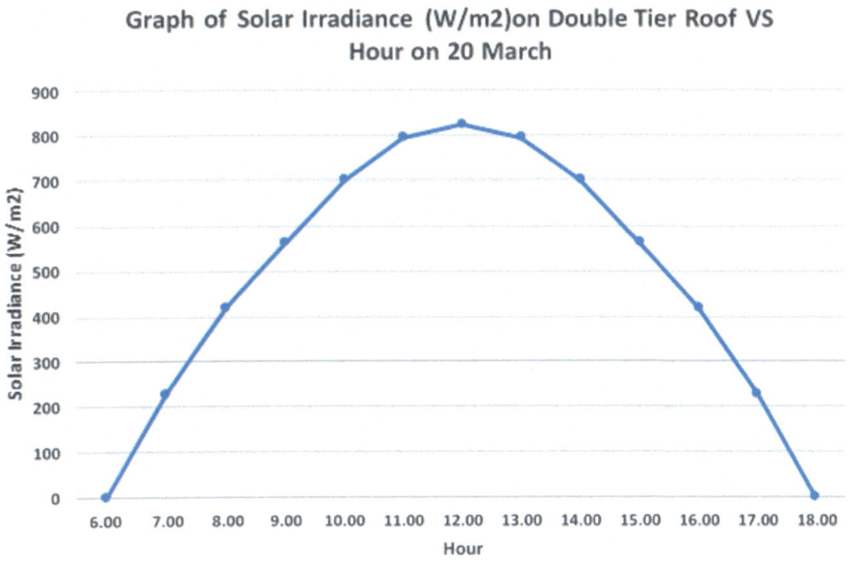

Fig. 24 Solar irradiance, I(HTCS) of Double Tier Roof

Table 3 The daily average of solar irradiance received, I(HTCS)

Tilt angle														
	6.00	7.00	8.00	9.00	10.00	11.00	12.00	13.00	14.00	15.00	16.00	17.00	18.00	Day Av.
38.15	0	228	428	598	756	855	887	855	756	598	428	228	0	509
48.2	0	229	409	528	649	734	761	734	649	528	409	229	0	450
	0	228	419	563	702	794	824	794	702	563	419	228	0	480

5.4 *The Effect of Solar Azimuth Angle on the Solar Behavior of Half Dome*

The solar behavior of the half dome revealed that the solar irradiance I(HTCS) displayed the curve's mirror pattern of morning to evening (Fig. 25). As can be seen in the graph below, the first three rings (R1, R2, R3) showed distinct behaviors than the rest of the rings. R1, R2, and R3 received less solar irradiance I(HTCS) at the midday compared to the other rings due to high value of tilted angle (86.79°, 80.36°, and 79.93°). It also demonstrates that the solar irradiance I(HTCS) values of solar irradiance are generally growing toward the top ring. These show that the peaks value increases with the decrease of the ring tilt angle. The value of each ring is varying from bottom part of the dome to the top surface. This difference is most noticeable between the hours of 11:00 and 1:00 in the afternoon. In contrast to the early morning and evening periods, when the difference in received, solar irradiance I(HTCS) from one ring to the next is insignificant.

Table 4 shows the daily average solar irradiation, I(HTCS), from R1 to R14, as well as the entire dome. The findings revealed that the lower half of the half dome (R1-R7) performs differently from the upper ring (R8- RHO) in terms of solar

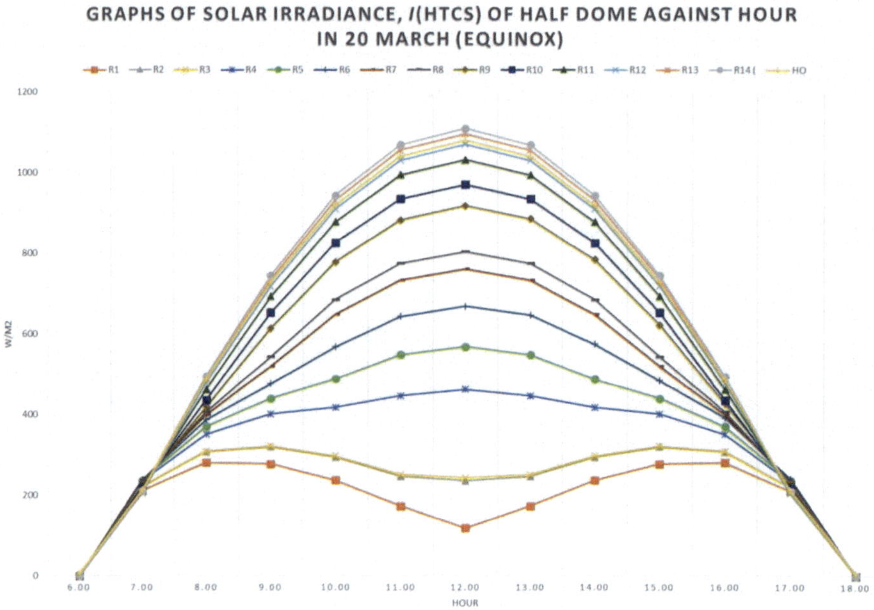

Fig. 25 Solar irradiance I(HTCS) of half dome

Table 4 Average solar irradiance I(HTCS) of half dome on 20 March (equinox)

Ring Tilt													
	6.00	7.00	8.00	9.00	10.00	11.00	12.00	13.00	14.00	15.00	16.00	17.00	18.00	Day Av
R1	0	211	281	278	238	174	119	174	238	278	281	211	0	191
R2	0	221	308	320	294	247	236	247	294	320	308	221	0	232
R3	0	222	310	323	298	252	244	252	298	323	310	222	0	235
R4	0	234	352	401	418	446	462	446	418	401	352	234	0	320
R5	0	237	371	440	488	549	569	549	488	440	371	237	0	364
R6	0	237	387	476	568	643	669	647	575	484	392	241	0	412
R7	0	235	399	520	649	733	761	733	649	520	399	235	0	449
R8	0	233	406	543	685	775	804	775	685	543	406	233	0	468
R9	0	224	418	614	778	882	918	886	784	622	427	229	0	526
R10	0	220	435	653	826	935	971	935	826	653	435	220	0	547
R11	0	211	461	693	878	994	1032	994	878	693	461	211	0	577
R12	0	205	478	719	911	1031	1071	1031	911	719	478	205	0	597
R13	0	207	488	735	932	1056	1097	1057	934	737	490	209	0	612
R14	0	211	495	745	944	1069	1110	1069	944	745	495	211	0	618
HO	7	213	487	728	920	1041	1081	1041	920	728	487	213	7	606
	0	221	405	546	655	722	743	722	656	547	406	222	0	450

performance. The difference in solar irradiance, I(HTCS), from one ring to another ring reaches its maximum value at midday and receives solar irradiance I(HTCS) approximately 450 W/m^2.

5.5 *The Effect of Solar Azimuth Angle on the Solar Behavior to Pointed Dome*

Pointed dome has more sharp profile at the top in comparison to half dome. The distribution of solar irradiance, I(HTCS) was received on 14 rings of the pointed dome as shown in Fig. 26. The majority of rings received maximum solar irradiance, I(HTCS), value at noon accept R1, R2, and R3. The pointed dome geometry or profile appears to be a factor influencing solar behavior, as the dome received less sun radiation intensity during the Equinox, which has similar behavior to the half dome

In addition, the value of solar irradiance I(HTCS) is varying from one part to other surfaces of the dome. The solar behavior has significant effect during noon period, while during the early morning period and evening period, value of solar irradiance, I(HTCS), received on one ring to another is insignificant. The table below shows the hourly solar irradiance, I(HTCS) value received by each ring of the

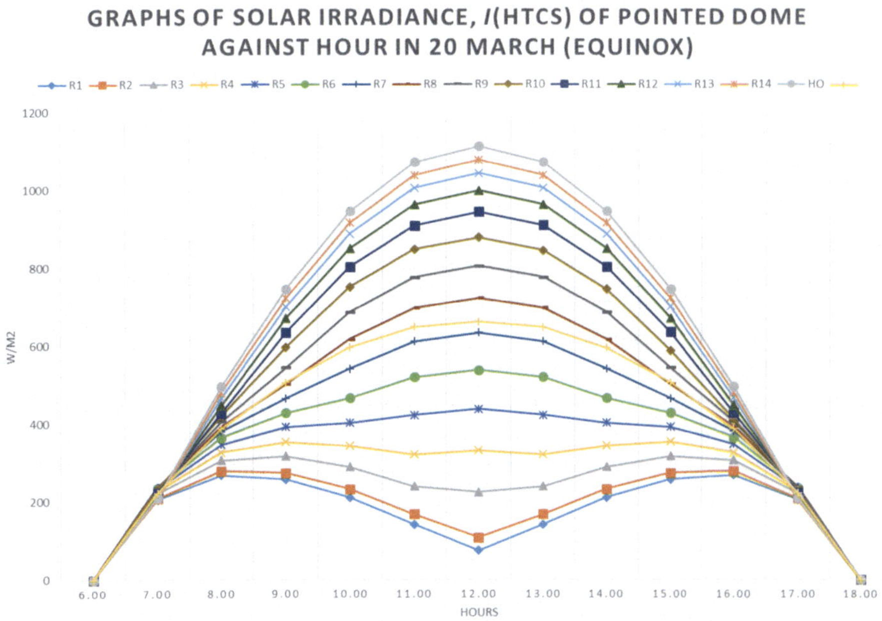

Fig. 26 Solar irradiance, I(HTCS) of Pointed Dome

Table 5 Average solar irradiance, I(HTCS) of Pointed Dome

Ring Tilt Angle	6.00	7.00	8.00	9.00	10.00	11.00	12.00	13.00	14.00	15.00	16.00	17.00	18.00	Day Av
R1	0	205	268	257	211	142	75	142	211	257	268	205	0	172
R2	0	210	279	274	233	168	109	168	233	274	279	210	0	187
R3	0	220	306	316	289	239	224	239	289	316	306	220	0	228
R4	0	228	328	354	344	322	332	322	344	354	328	228	0	268
R5	0	233	347	392	403	422	437	422	403	392	347	233	0	310
R6	0	237	365	429	466	519	538	519	466	429	365	237	0	352
R7	0	238	381	465	541	611	633	611	541	465	381	238	0	393
R8	0	236	394	502	617	697	722	697	617	502	394	236	0	432
R9	0	234	406	544	686	775	804	775	686	544	406	234	0	469
R10	0	233	421	596	751	848	878	845	745	588	412	228	0	500
R11	0	224	428	634	803	908	943	909	803	635	428	224	0	534
R12	0	216	447	671	849	961	998	961	849	671	447	216	0	560
R13	0	209	466	700	887	1004	1042	1004	887	700	466	209	0	583
R14	0	205	480	722	915	1037	1076	1037	915	722	480	205	0	600
HO	0	211	495	746	945	1071	1112	1071	945	746	495	211	0	619
	0	222	387	507	596	648	662	648	596	506	387	222	0	414

pointed dome was around 414 W/m^2 (Table 5) which is more efficient as compared to half dome that received 450 W/m^2.

5.6 *The Effect of Solar Azimuth Angle on the Solar Behavior to Bulbous Dome*

The curvature of the bulbous dome has a rounder profile than the half dome and pointed dome geometry which is commonly known as onion dome in Malaysia. The highest value of solar irradiance occurs during midday for rings R7 to R14 in contrast with R1 to R6. This scenario happened due to R1 to R6 receiving lower I(HTCS) value starting from 8 a.m. to midday (almost 0 W/M^2), increasing gradually from midday to 5 p.m., and drastically dropping till 6 p.m. (Fig. 27). Bulbous dome appears to have the best profiles for a dome-type design to receive solar radiation on 20 March (equinox) approximately 382 W/m^2 which is the lowest reading among dome types (Table 6).

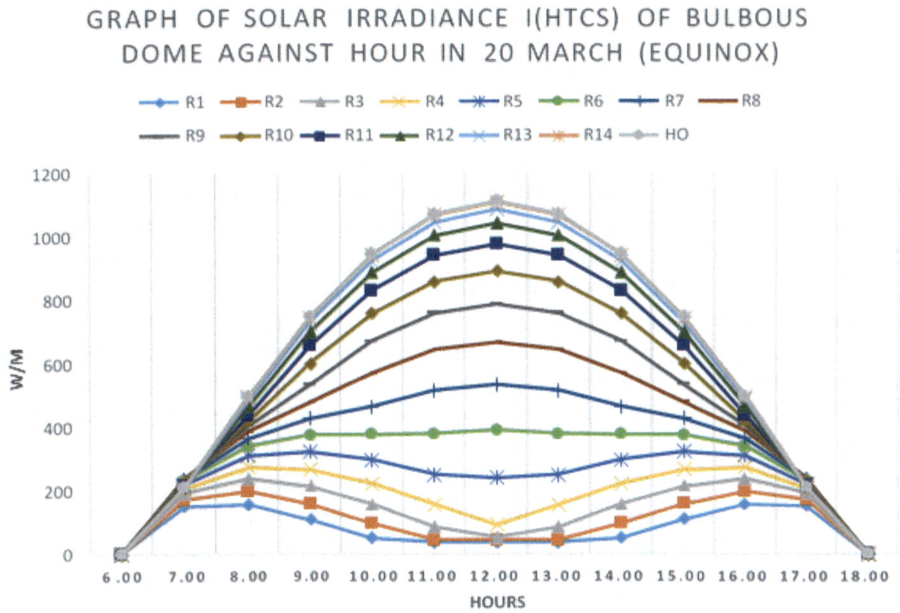

Fig. 27 Solar irradiance I(HTCS) of bulbous dome

Table 6 Solar irradiance I(HTCS) receive by each ring of bulbous dome

Ring tilt angle	6.00	7.00	8.00	9.00	10.00	11.00	12.00	13.00	14.00	15.00	16.00	17.00	18.00	Day Av.
R1	0	150	155	107	49	35	36	35	49	107	155	150	0	79
R2	0	172	198	158	97	44	44	44	97	158	198	172	0	106
R3	0	192	237	213	155	85	52	85	155	213	237	192	0	140
R4	0	207	274	267	223	156	91	156	223	267	274	207	0	180
R5	0	221	309	321	296	250	240	250	296	321	309	221	0	234
R6	0	231	339	375	375	377	390	377	375	375	339	231	0	291
R7	0	236	365	427	463	515	533	515	463	427	365	236	0	350
R8	0	237	386	478	569	642	666	642	569	478	386	237	0	407
R9	0	234	403	533	670	757	785	757	670	533	403	234	0	460
R10	0	229	418	599	758	857	889	857	758	599	418	229	0	508
R11	0	219	437	656	830	940	975	940	830	656	437	219	0	549
R12	0	209	465	699	886	1003	1041	1003	886	699	465	209	0	582
R13	0	207	485	729	924	1046	1086	1046	924	729	485	207	0	605
R14	0	210	494	744	943	1068	1109	1068	943	744	494	210	0	617
HO	0	211	495	746	945	1071	1112	1071	945	746	495	211	0	619
	0	211	364	470	545	590	603	590	545	470	364	211	0	382

5.7 *Solar Behavior of Five Selected Mosque Roof Geometries*

All solar data behavior of mosque roof geometries were tabulated and compared to determine the level of efficiency in receiving the low solar radiation I(HTCS) on the roof surface. From the graph, the maximum solar irradiance I(HTCS) amount on all roofs is during midday. Flat roof received the highest solar irradiance I(HTCS) and bulbous dome received the lowest. Although pyramidal roof such as single-tier and double-tier roof has a different roof geometry and profiles, they all have similar characteristics of solar radiation I(HTCS) receives pattern.

Flat roof received the highest average I(HTCS) with an amount of 617 W/m^2, follows by double-tier roof with 480 W/m^2, single-tier roof with 472 W/m^2, half dome with 450 W/m^2, pointed dome with 414 W/m^2, and lastly bulbous dome with 383 W/m^2. It can be noted that there is only a very slight difference between the results of double-tier roof and single-tier roof in terms of the average solar irradiance I(HTCS) received. Bulbous dome received an average solar irradiance of 383 W/m^2, which is the lowest value compared to other five roof geometries.

To surmise, the bulbous dome roof geometry has the best solar performance and characteristics, whereas the flat roof has the opposite effect (Fig. 28).

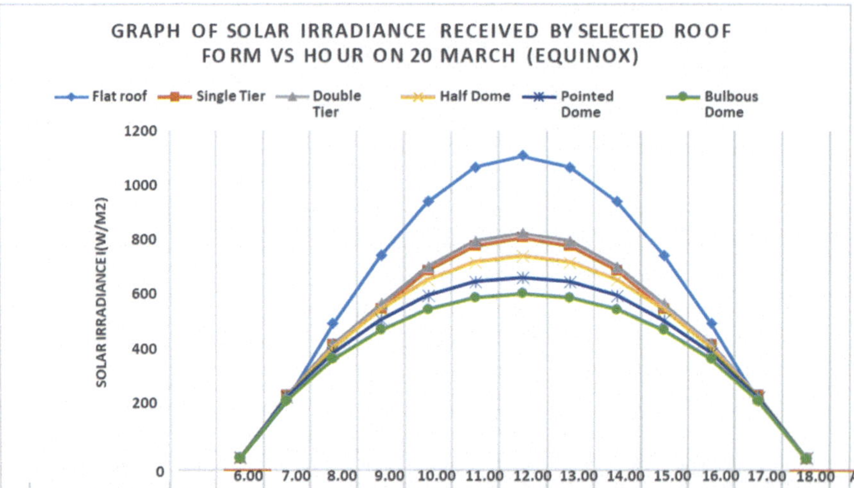

	6.00	7.00	8.00	9.00	10.00	11.00	12.00	13.00	14.00	15.00	16.00	17.00	18.00	Ave
Flat roof	0	210	493	743	942	1067	1108	1067	942	743	494	210	0	617
Single Tier	0	229	417	547	690	780	810	780	690	547	417	229	0	472
Double Tier	0	228	419	563	702	794	824	794	702	563	419	228	0	480
Half Dome	0	221	405	546	655	722	743	722	656	547	406	222	0	450
Pointed Dome	0	222	387	507	596	648	662	648	596	506	387	222	0	414
Bulbous Dome	0	211	364	470	545	590	603	590	545	470	364	211	0	383

Fig. 28 Cross tabulated of I(HTCS) of six selected mosques roof geometries

6 Conclusion

The main basis of the research was to provide an architectural scientifically proved of solar behaviors and solar irradiance I(HTCS) on different mosque roof geometries in Malaysia. This research conduct experimentation and simulation using Solar Radiation Simulation Model (SRSM) which is appropriate and reliable in calculating solar irradiance I(HTCS) in the tropic region. The SRSM enables to obtain the influence of the tilt surface angle and surface azimuth angle on the amount of solar radiation received. The input data needed to get the result was the latitude, the mean daily solar radiation, the tilt angle, the azimuth angle, and the day of the year.

This study portrays the architectural significance and importance of roof geometry, especially dome roof. The solar radiation does cause an overheating in tropical climate because the roof is the topmost part of the building that is highly exposed to solar radiation as compared to other parts of the building. It is also important to clarify that this study of the solar performances on the different form of roof mosque was conducted during 20 March (equinox). Equinox season is important than the rest of the year because on 20 March (equinox), the equator experiences the highest amount of solar radiation due to the position of the sun at a much steeper altitude; thus, the amount of solar energy fall onto the roof surface are greater.

In contrast, during the solstices, the solar radiation hits the surface of the earth at much lower angle, which in turn received less solar energy. Throughout this study, the potential of different mosques roof geometries as passive solar design strategy in Malaysia. Pitch roof such as single tier and double tier and dome room such half dome, pointed dome, and bulbous dome can be applied as this study revealed their significant performance on the amount of solar irradiance I(HTCS) received on the roof surfaces. Nevertheless, dome roofs have showed the most efficient result due to its curvature profiles that assist the dome in reducing the solar radiation received. Dome roofs contributes to minimal heat gain from the roof surface due to various rings or segment surfaces with different tilted angles.

Based on the evolution of roof design in Malaysia, the designer could choose appropriate roof geometry not only based on architectural style and local identity. Furthermore, this study can guide them in understanding on roof design could improve building thermal performance that contributing to low energy consumption in buildings. Other passive design strategies such as building orientation, ventilation, vegetation, and landscape; building fabric and thermal mass; and building fenestration should also be incorporated in the early stage of design

References

1. Abdul, H. N., Wan, H. W. T. (1996). *The Traditional Malay House*. Kuala Lumpur: Penerbit Fajar Bakti Sdn. Bhd.
2. Aziz, A. A. (2013) *Redefining Islamic Architecture through Sustainable and Progressive Approach*

3. Aziz, A.A. (2016) *Execution Of Contemporary Islamic Architecture Through Design: The Cyberjaya Green Platinum Mosque Project In Malaysia,* WIT Transactions on The Built Environment, Vol 159, © 2016 WIT Press
4. Asif, N., Utaberta, N. and Sarram, A. (2019). *Architectural Styles of Malaysian Mosque: Suitability In Compact Urban Settings*. MATEC Web of Conferences, 266, p.06001.
5. Ab Ghafar, N. (2014). Solar Performance on The Atrium Roof And Skylight Design In Warm-Humid Region: With Reference To Kuala Lumpur. Nottingham University, PhD Thesis.
6. Bremner, S. P., Levy, M. Y., Honsberg, C. B. (2008). *Progress in Photovoltaics: Research and Applications*, 16(3), 225-233.
7. Mazria, E. (1979). The passive SOLAR ENERGY book: A complete guide to passive solar home, GREENHOUSE, and building design. Emmaus, Pa: Rodale Press.
8. Mohamad Rasdi, M.T (2007). *Mosque Architecturre in Malaysia: Classificattion of Styles and Possible Influence, Journal AlamBina Mosque Architecture in Malaysia*: Penerbit Universiti Teknologi Malaysia
9. Szokolay, S. V., & Brisbin, C. (2004). *Introduction to architectural science: The basis of sustainable design*. Amsterdam: Elsevier, Architectural Press.
10. Sojak S. D. M. et al (2019). *Mosque Architectural Timeline In Malaysia: From Vernacular To Contemporary.* The European Proceedings of Multidisciplinary Sciences.
11. T Muneer, C Gueymard and H Kambezidis (2004). *Solar radiation and daylight models.* Amsterdam; Boston ; London: Elsevier Butterworth Heinemann.
12. Yeang, K. (1992). *The Architecture of Malaysia.* Amsterdam; Kuala Lumpur: The Pepin Press.
13. Duffie, J., and Beckman, W. (1991), Solar Engineering of Thermal Process. John Willey and Sons: New York.

Mastura Binti Adam is a senior lecturer in the Department of Architecture, Faculty of Built Environment, University of Malaya (UM), Kuala Lumpur, Malaysia. She has a PhD from the University of Salford, Greater Manchester, UK, awarded in 2013; MA in Urban Design from Oxford Brookes University, Oxford, UK, awarded in 2000; postgraduate diploma in Architecture from Oxford Brookes University, awarded in 1995; and diploma in Architecture from Mara Institute of Technology, Malaysia, awarded in 1992. Her 20 years' career as an academician in UM has contributed to several successes in the department, including the architectural school's recognitions from the Malaysian Architects Board (LAM), Royal Institutes of British Architects (RIBA I and RIBA II), numerous batches of Bachelor in Science and Bachelor in Architecture graduates, and Mubin Sheppard awards for Measured Drawings of Heritage Buildings. She has been awarded the Excellence Service Award University Malaya in 2017. Her area of expertise is architectural design and urban design, with emphasis on sustainable design, inclusive and universal design, sustainable urban mobility, outdoor thermal comfort in public urban spaces, and walkability in the pedestrian environment. She has had articles published in national and international journals, conferences, as well as books and book chapters. She has been appointed as an external reviewer of an architecture design course in University Kebangsaan Malaysia (Ukm), part of an external critique panel for architectural design studio presentations in several public and private universities in Malaysia, an associate editor for the *Journal of Design and Built Environment*, a research methodology expert, and an internal/external examiner for master's and PhD candidature. She has been organizing international seminars in urban design areas, several research workshops in multidisciplinary areas, and access audit training for people with disability (PWD). She recently collaborated with local and international experts on an endeavour to advocate Walkability in Kuala Lumpur City, Urban Transformation – Kuala Lumpur as a Liveable City, Inclusive and Universal design in Accessible Neighbourhood, Townscape Legibility and Cities for People. **Areas of expertise:** architecture design, architectural history, urban design, and inclusive and universal design.

Thermal and Visual Adaptive Comfort: Field Studies in Portugal

L. Matias, A. Santos, and M. Correia Guedes

1 Introduction

Personal well-being, health and productivity of building occupants depend to a large extent on available thermal and visual conditions.

Thermal and visual requirements have changed across centuries. Last century, as a result of the economic growth that occurred in developed societies, the satisfaction of those requirements gained importance. In a society and period where energy was abundant and cheap, the easiest way to provide desired indoor conditions consisted in the systematic use of artificial conditioning (AC) and electric lighting. As a consequence, resulting energy consumption became significant.

Actually, the growing tendency for the use of AC systems, particularly in office buildings, transports and, more recently, in residential buildings and the systematic and the permanent use of electric lighting proceed.

Although this fact may derive from a change in user comfort requirements and expectations, they may also reveal design and lifestyle trends conducting to unsustainable options concerning construction, rehabilitation and everyday use of buildings.

The original version of this chapter was revised: Modifications done to the order and photos of authors in the Author biography section. The correction to this chapter is available at https://doi.org/10.1007/978-3-031-04714-5_21

L. Matias (✉) · A. Santos
National Laboratory of Civil Engineering (LNEC), Lisbon, Portugal
e-mail: lmatias@lnec.pt

M. Correia Guedes
Superior Technical Institute (IST), Lisbon, Portugal

A. Sayigh (ed.), *Achieving Building Comfort by Natural Means*, Innovative Renewable Energy, https://doi.org/10.1007/978-3-031-04714-5_11

Fig. 1 Constructive solutions that cannot achieve thermal and visual comfort conditions without an excessive and unnecessary energy consumption

As a matter of fact, some contradictory aspects can be detected: a greater intolerance in relation to exterior/interior climate variations, the option of non-climate-conscious building solutions and the creation of controlled indoor climates which frequently do not provide the desired well-being conditions (in working, leisure or living places) and unnecessary and excessive energy consumption (Fig. 1).

As it is commonly accepted, energy use can hardly be kept at present-day levels and the dependency on non-renewable energy sources is going to continue. This reality justifies the European Community efforts to reduce energy consumption and increase the efficient use of energy in buildings, being an example the EC Directive 2002/91/EC (EPBD).

The first recast of the EPBD [1] requires that by 31 December 2020 all new buildings are nearly zero energy buildings. It also recommends that built-in electric lighting (mainly in the non-residential sector) shall be taken into consideration and recognizes the need to account for the positive influence of daylight.

The recently adopted second recast of the EPBD [2] integrates many requirements regarding the acceleration of deep energy renovation of buildings in Europe. The council has agreed on the establishment of a voluntary smart readiness indicator (SRI) promoting digitalization and smart technologies that improve capabilities to adapt its operation mode in response to the needs of the occupant in a user-friendly way, to maintaining energy efficiency and overall performance, including indoor comfort and health.

Beyond active and passive measures aiming at the improvement of equipment energy efficiency (air conditioning and lighting equipment) and building performance (thermal insulation, passive solar gains, daylighting), user attitudes and behaviour are determinant issues in the final, either positive or negative, results obtained.

Buildings, their indoor environment and their systems have been studied by a multiplicity of disciplines, from the most involved in its design and conception, such as architecture, engineering and physics, to the social sciences mostly engaged with the study of the dynamics of interaction between individuals and environment.

In this sense, an interdisciplinary research study aimed at the study of the conditions for comfort, well-being and productivity, reflecting objective and subjective regional and local conditions. This study has considered the influence of such relevant factors as the Portuguese temperate Mediterranean climate, the population diversity and the natural tendency of individuals to act and to adapt.

The National Laboratory for Civil Engineering (LNEC), a Portuguese public institute that covers a wide range of areas related to civil engineering, constituted a multidisciplinary team to carry out this study developed based on extensive field surveys on different types of buildings considering both physical (objective) and behavioural (subjective) factors related with thermal and visual comfort.

Building occupants have been requested to complete comprehensive questionnaires designed to study the influence of adaptive processes, namely, of behavioural, physiological and psychological nature.

Bivariate and multivariate statistical analysis has been used to describe and to analyse objective and subjective data. Special emphasis was put on structural equation modelling (SEM) approaches supporting in-depth-oriented interpretation and explanatory research in fields of work.

In the context of the perception of thermal and visual indoor environments, along with detailed measurements of relevant indoor environment parameters and questionnaires, an adaptive thermal comfort model and a comprehensive integration of the behavioural patterns and the energetic impacts in dynamic daylighting modelling have been developed [3, 4].

The results of these actions will be the basis for the development of an adaptive thermal comfort model, alternative or complementary to a conventional analytical approach in this area, which will be applicable to the specific characteristics of Portugal [3].

On the other hand, a model for the evaluation of the energetic impacts of daylighting shall be developed, considering the typical Portuguese Mediterranean climate [4].

This model shall be developed based on results of the following actions: (i) onsite evaluation of daylighting and artificial lighting systems of selected buildings, (ii) characterization of patterns of use of shading devices and artificial lighting systems and (iii) evaluation of the occupants' behaviour towards existing indoor environmental conditions including the possibility of its control.

The present research work was partially funded by a Portuguese national funding agency for science, research and technology (FCT – Foundation for Science and Technology), within the scope of the Project PTDC/ECM/71914/2006 – *Development of Sustainable Visual and Thermal Comfort Models*.

2 Thermal Comfort

2.1 The Adaptive Approach

Economic, cultural and social drivers and constraints often dictate current comfort standards. Global and indiscriminate trends tend to impose patterns and needs, regardless of regional and local specific characteristics and habits [5]. Most of the existing conventional regulations and comfort standards are based on complex equations resulting from laboratory research (EN ISO 7730). By the very nature of the laboratory derivation of these conventional standards, adaptive processes found in the real world have been limited or eliminated. This is largely responsible for the always growing demand and use of air-conditioned spaces, first in offices (productivity) and other service buildings and nowadays spreading to residential environments.

However, the implementation of adaptive comfort criteria and the recuperation of traditional passive concepts could contribute to a more rational and sustainable approach to building design and operation. Ultimately, this approach may lead to a reduction in the use of energy and to increase occupants' comfort perception and satisfaction. Nowadays, adaptive models are beginning to be incorporated in international thermal comfort standards such as ASHRAE 55:2020 [6] or EN 16789-1:2019 [7].

In Portugal, only a limited number of research studies were carried out on a small number of office buildings [8, 9]. The implementation of adaptive comfort criteria is dependent upon contextual objective and subjective factors specific to each region. There is still a need to validate the adequacy of the proposed models and further investigate the role of human adaptation and of new expectations created by the changing dependency on the use of active solutions to achieve thermal comfort.

The main purpose of this interdisciplinary research study was to develop a culturally oriented adaptive approach on the basis of extensive field surveys carried out on occupied buildings, in the sequence of the results achieved in a previous research.[1] The objective was to adapt or develop an adaptive approach oriented to the definition of indoor thermal comfort requirements applicable to Portuguese buildings.

These considered climate, social and cultural habits and adequate constructive solutions. A major attention was given to behavioural aspects in this study and deserved a detailed environmental psychology approach.

Unlike other research studies, besides office buildings, educational buildings, homes for the elderly and dwellings were also included. Geographical location has

[1] This field work was part of a Ph.D. research developed between 1996 and 2000 [9]. The research, involving monitoring in 26 office buildings during the two consecutive summers, aimed at assessing the relationships between the workers' adaptive behaviour and their comfort votes in each building type. This adaptive behaviour concerned both physical adaptation, i.e. the adaptive actions that the workers reported to carry out to make themselves more comfortable, and psychological adaptation processes, such as the workers' perceived freedom to control the thermal environment.

also been considered in order to take into account different Portuguese summer and winter climatic zones.

2.2 *Field Surveys*

Field surveys (2006–2008) involved the measurement of indoor environment parameters, namely, air temperature (T_a), operative temperature (T_{op}), airspeed (v_a) and relative humidity (*RH*), during hot, cold and transition periods of the year. All measurements were performed using sensors and probes (Fig. 2), in compliance with thermal comfort standard specifications [10], and the field studies can be classified as Class I [8].

Simultaneously, occupants have been requested to complete a comprehensive questionnaire including six groups of questions aimed at assessing the influence of multiple factors on human thermal sensation. Information was obtained about the identification of individual and technological adaptive opportunities, the easiness and degree of satisfaction in implementing them, the evaluation of respondents' expectations about thermal environment, the identification of relevant aspects of their daily life experience and the motivations towards mechanically controlled indoor environments.

Subjective opinion scales were defined [11, 12] which allowed the participants to vote on their thermal sensations (neutrality, preference, satisfaction level) regarding the surrounding thermal.

A total of 40 different buildings have been selected (Table 1) located in representative geographical and climatic regions of Portugal. The studied cases included

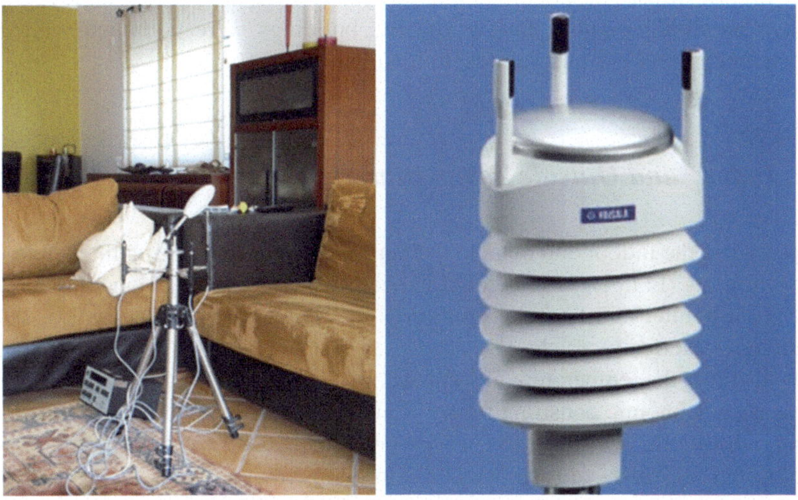

Fig. 2 Thermal comfort assessment equipment and weather station

Table 1 Distribution of the field survey sample (occupational profile)

Occupational profile	Buildings	Rooms	Surveys	Participants	Answers
Offices	9	36	130	242	690
Educational	6	23	38	853	945
Elderly homes	21	27	85	404	698
Dwellings	4	16	32	19	34
Total	**40**	**102**	**285**	**1518**	**2367**

Fig. 3 AC and NV buildings

both AC (mechanical heating/cooling systems) and naturally ventilated buildings (NV) (Fig. 3). Most NV buildings had passive features and technologies (greater adaptive possibility), and a few of the AC buildings were equipped with user-independent control systems (almost null adaptive potential).

2.3 *Indoor and Outside Temperatures*

A first look at the indoor and outside temperatures measured during the whole field surveys reveals (Fig. 4), as one could expect, that AC buildings have higher temperatures in winter and lower temperatures in summer and that control of indoor conditions is exerted more closely.

The fact that the temperature in some AC spaces was too high in hot periods may represent an economic constraint rather than a sustainable/adaptive option. On the contrary, during the winter period, a high percentage (~80%) of office rooms were being heated regardless of their actual indoor temperatures.

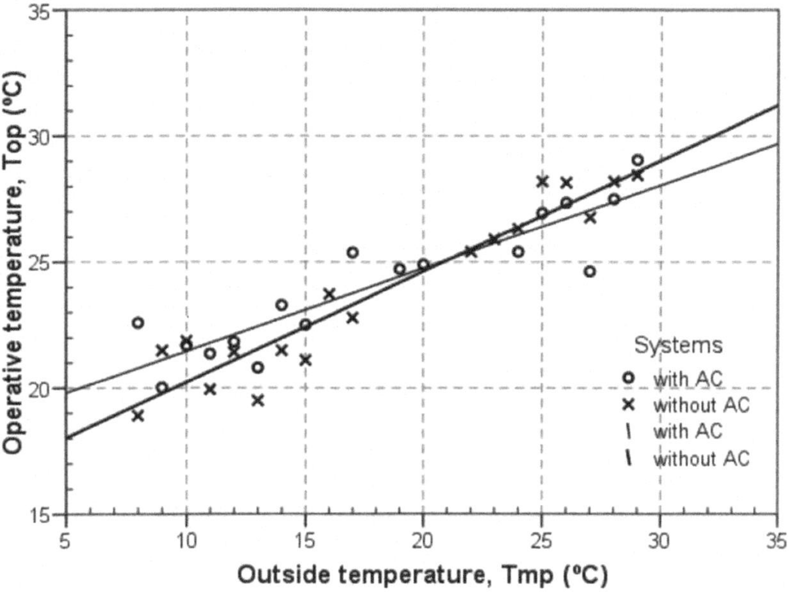

Fig. 4 Indoor and outside temperature measured in field surveys

More than 50% of the surveyed spaces in homes for the elderly and in classrooms were heated when the indoor temperature was <20 °C, but when temperatures were above 20 °C, the percentage of the use of the heating system was low (30%).

Another interesting conclusion drawn from winter surveys was that about 15% of office buildings were being cooled even when the indoor temperature was lower than 20 °C [13].

2.4 Thermal Sensation and Thermal Preference

Based on the answers to the questionnaires, a statistical linear regression was performed between indoor temperature (*Top*) and the mean thermal sensation (*Mts*) and, on the other hand, the mean thermal preference (*Mtp*). Assuming that when *Mts* or *Mtp* are null, the corresponding indoor temperatures indicate, respectively, the neutral temperature (*Tn*) and the preference temperature (*Tp*). Table 2 summarizes all available field data concerning *Tn* and *Tp* [3].

Unlike the conclusions from other important studies [8, 15], the present research study shows significant differences between the two calculated temperatures, *Tn* and *Tp*, mainly in the heating season (winter). Based on the distribution of the neutral temperature in the summer period (Fig. 2), it was also possible to conclude that

Table 2 Neutral and preference temperatures (°C)

	Neutral temperature (Tn) and preference temperature			
Season	Offices	Educational	Elderly homes	Dwellings
Summer	25.1–24.5	–	24.9–24.2	–
Mid-season	22.2–23.0	19.5–22.2	22.8–24.8	–
Winter	19.1–21.6	18.1–22.1	21.4–22.6	18.1–19.6

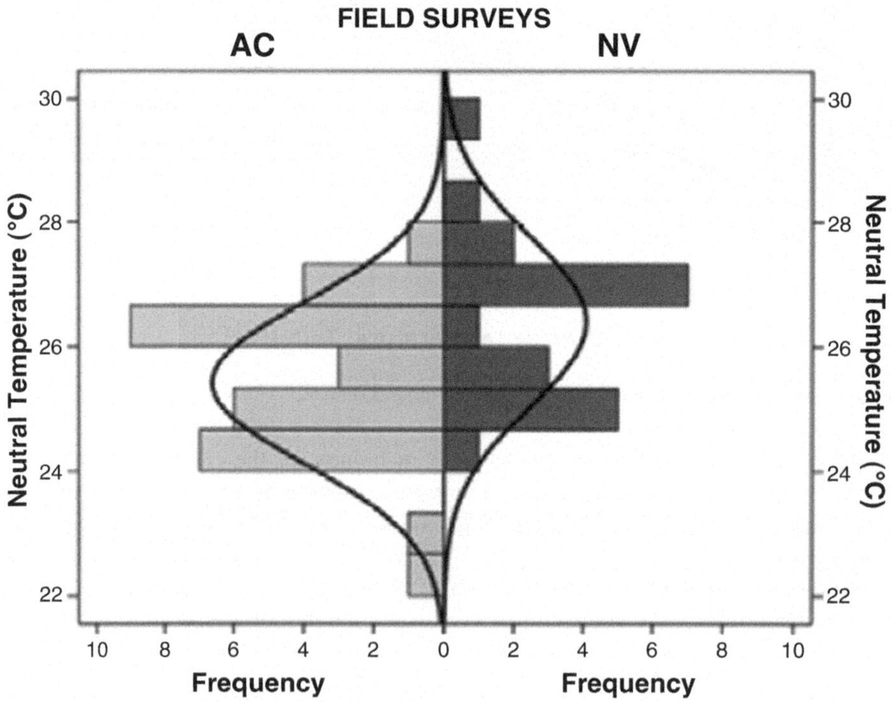

Fig. 5 Neutral temperatures for AC and NV buildings (summer period)

the occupants of NV spaces show a greater thermal tolerance, probably due to greater adaptive opportunities and different thermal expectations (Fig. 5).

Considering the differences between *thermal sensation* and *thermal preference*, this study presents a new index, *Comf,* that combines both the dimensions.

2.5 A Proposed Adaptive Comfort Model for Portugal

The adaptive model proposed in the current European standard EN 16798-1 [7] is based on the linear relation between neutral temperatures (*Tn*) and outside temperature (*Trm*). Figure 6 shows the acceptable temperature limits defined in the standard for a normal expectancy level [7] and ensuring that 90% of the occupants are satisfied.

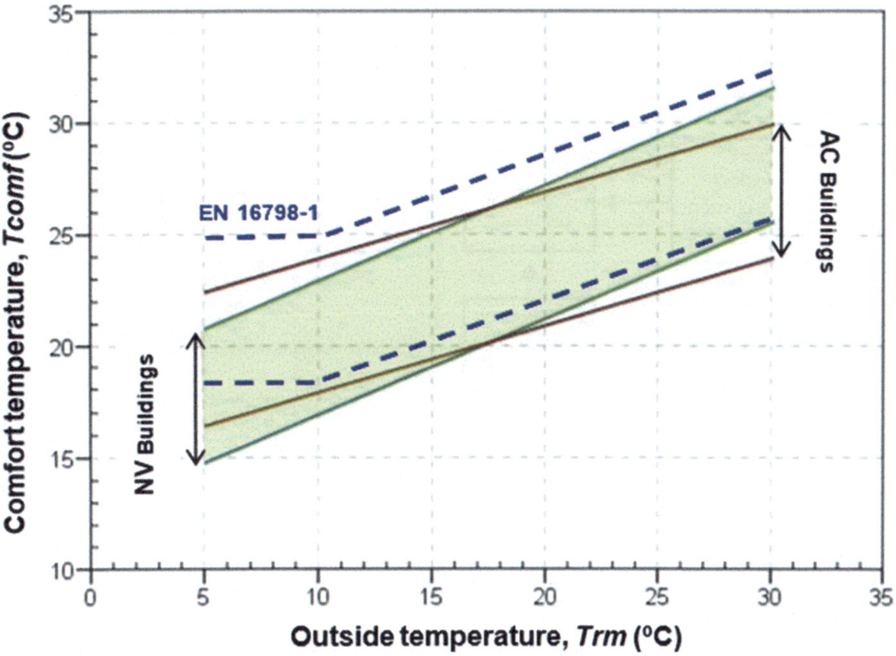

Fig. 6 *Neutral* (*Tn*) and *comfort* (Tcomf) temperatures according to EN 16798-1 and the proposed model

The same figure presents the comfort range temperature (*Tcomf*) developed and proposed in the present research study regarding NV spaces, as expressed in European standard, but also proposes a comfort range temperature for AC buildings that change in function of external conditions.

Like expected, AC range comfort temperature is more exigent (higher temperatures in winter and lower in summer) than for NV buildings, in accordance with different users' expectations.

Figure 6 denotes in the proposed model a different tendency compared to the European standard [7]. This is mainly because the model does not assume a "kink" point at lower outside temperatures, like in EN 16798-1. It may be argued that the model shows wider accepted comfort conditions as the results on which it is based are still dependent on prevailing cultural and local climatic specific conditions.

2.6 *Adaptive Thermal Comfort in Buildings: An Explanatory Model*

Using a structural equation model approach a comfort model (Fig. 7) was developed [13, 14] in order to understand the meaning of perceived thermal comfort in the Portuguese context, by establishing the way psychological, social and

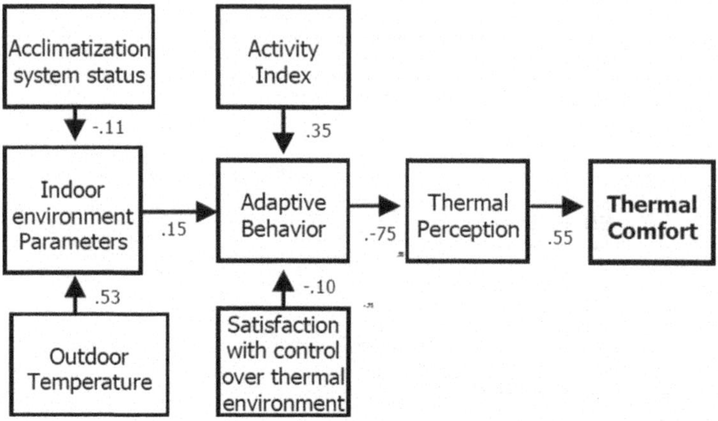

Fig. 7 Model of thermal comfort (Comf)

environmental variables are related and are influenced by each other. The dependent variable (*Comf*) was built by crossing the responses to thermal sensation and thermal preference, both measured by a 5-point Likert scale.

The final model stresses that thermal comfort is a subjective dimension directly explained by other subjective parameter (individuals' thermal perception) and indirectly related to objective comfort (indoor environment, acclimatization equipment and outdoor temperature), adaptive behaviour and satisfaction regarding control over thermal environment. A noticeable aspect of this model is the central role that is assumed by adaptive behaviour in the sense that is responsible for several and important links between all the other model components. In sum, this set of relations highlight that thermal comfort is mainly raised by social perceptions which, in turn, are based on the interaction between objective and subjective parameters.

3 Daylight Dynamics and the Luminous Environment

3.1 Modelling Daylight Dynamics

A similar multidisciplinary and objective/subjective context approach was adopted in the research carried out in the field of daylight modelling, its use in buildings and its impacts on visual comfort and energy efficiency.

Energy and sustainability issues must be invoked to justify technical empathy for such an approach in this field. However, the need to understand the relations between the dynamics of natural light, artificial light, behaviour patterns and satisfaction of the occupants towards daylighting and available control systems was a fundamental driver to the research presented hereafter.

A new model for the quantification of the dynamics of daylight, occupants' behaviour and energy-related impacts is being developed and will be calibrated and validated. The new model has four linked modules designated as i) exterior module, ii) transmission module, iii) interior module and iv) behavioural module [16].

3.2 Local Climatic Conditions (Exterior Module)

The energy-related aspects of daylight are particularly important where non-overcast skies conditions prevail, which is the case in most of the regions in southern Europe and in particular in Portugal. Due to the increased trend in the use of oversized glazed areas, either in office or residential buildings, sunlight is often a source of thermal discomfort (overheating) as well as unnecessary energy consumption in the cooling periods and excessive daylight.

The characterization and representation of the particular luminous climate uses the data collected and available under previous research projects [17–19]. The implementation of the exterior module was based on the methodology used in the representation of the luminous climate incorporated in the daylighting-hours indicators model [20, 21].

3.3 Glazing and the Effect of Shading (Transmission Module)

The global effect of glazing/shading devices was quantified considering daylight as well as thermal aspects. Besides the use of reference computational applications, an algorithm [22] for the determination of all illuminance and irradiance fluxes transmitted, reflected and absorbed within a multilayered glazing/shading system was developed and compared with experimental results. An experimental setup was used to validate the algorithm served also to test a net radiation method for determining solar optical properties of glazing with roller blinds or venetian blinds which are common shading devices in Portugal.

3.4 Daylight Availability and Distribution (Interior Module)

Quantification of daylight availability and conditions (work plane illuminances, uniformity and glare) and of the impact of electric lighting and shading systems (types, zoning, controls and patterns of use) is based on an improved version of the daylighting-hours indicators model [20, 21]. The indicator and the corresponding method of calculation are intended to be used in indoor spaces in regions where clear skies and sunlight effects prevail and, namely, in applications where sunlight is important to daylight performance.

This module uses information from the exterior and transmission modules and incorporates information and knowledge acquired in the behavioural module.

3.5 Understanding and Modelling the Occupants (Behavioural Module)

One of the underlying assumptions for this research project is that the conventional objective (physical) characterization of indoor luminous conditions is not able by itself to capture the subjective (sensorial) experience of the users [23]. This way, conventional established requirements may not respond to the occupant's effective needs. The unavoidable complex existing links and interrelationships influence, on the one hand, the performance of visual tasks, the overall satisfaction and the well-being of the users. On the other hand, energy and environmental performance of the buildings may be affected by behaviour patterns adopted by the users to achieve a better luminous environment.

The inclusion of subjective aspects, namely, perceptions, expectations and user behaviour attitudes and patterns towards the available control (shading devices, electric lighting) of the luminous environment will improve the interpretative and design capacities of conventional tools [23, 24].

The post-occupancy surveys performed in the scope of this research comprised a set of 13 selected multifunctional, office and educational (classrooms) buildings. These buildings presented different window orientations and geographical/climatic location. Daily and seasonal sunlight variations were also considered. Rooms were electrically lit, daylit or used a combination of both. Operation (control) strategies ranged from manual to user-independent control.

Surveys were based on a comprehensive structured questionnaire (Fig. 8) comprising 7 thematic groups of questions: working place, luminous environment, windows/views, shading options, electric lighting and related control, visual tasks and social characterization of the participants.

Subjective data was complemented and compared with in situ objective (physical) measurements (Fig. 8) conducted according to a predefined measuring protocol. Relevant parameters were observed, measured and registered for further analysis: room characterization, indoor illuminances either at work plane level or at

II – Lighting and Daylighting

6. Do you prefer to work with daylight, electric light or a combination of daylight and electric light?

☐ daylight ☐ Electric Light ☐ Combination of daylight and electric light

7. Do you have a desk lamp in your work place?

☐ **Yes** ☐ **No**

If your answer was **Yes**, say: **a.** When do you use it ....

b. How often do you use it: ☐ Always ☐ often ☐ Rarely

If your answer was **No**, do you think a desk lamp would improve your working conditions?

☐ **Yes** ☐ **No**

PLAN THE SPACE (Orientation, Level, Location, Occupants, Working Posts, etc...

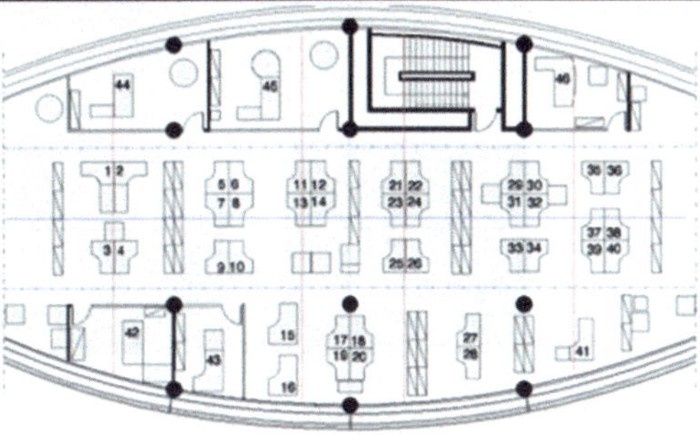

OBS:
- Location of Measurements
- Lines of Sight
- Direction of Views outside

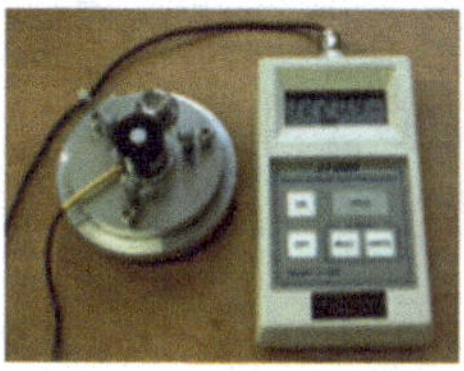

Fig. 8 Examples of questionnaire, room plan (incl. location of measurements) and measurement equipment

different vertical orientations and outdoor solar radiation and illuminances either due to the sky or affected by surrounding surfaces and obstructions [23].

Photographic records (Fig. 9) and complementary informal verbal contacts helped to complete and to detail other collected data.

All 584 participants (54% male, 46% female and ages ranging from 18 to >50) fulfilled a questionnaire providing an extensive and representative amount of valuable information about their perception of the luminous environment, availability and effective use of existing control systems.

More than 50% of the subjects worked at a distance less than 2 m from a window, thus increasing the potential use of daylight, but also the risk of glare (~84% used a computer for most of their working hours).

Objective performance indexes/scores were defined in order to enable a better comparison, calibration and validation of subjective data. These indexes result from the analysis and treatment of collected data and are being introduced in the dynamic daylight simulation model. One of these indexes – daylighting quality index (IDO_QUAL_IN) – was based on measured parameters (e.g. illuminances at the horizontal work plane level and at four vertical distinct directions) and was used to analyse and validate subjective data on the "quality of the luminous environment".

In the following items, some of the objective and subjective outputs of this study are presented and commented.

Fig. 9 Example of the type of interior spaces in one of the monitored office buildings

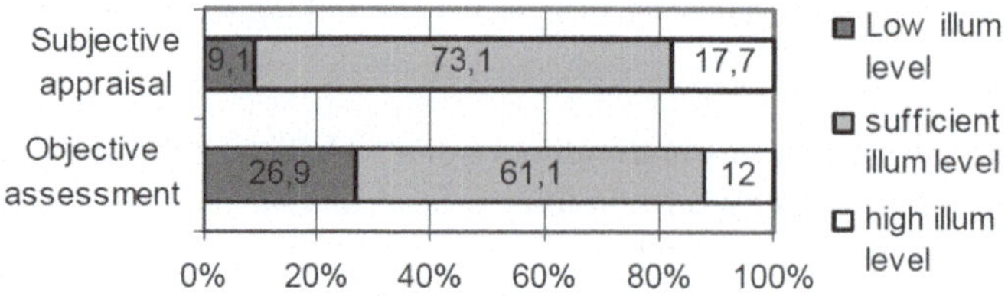

Fig. 10 Objective and subjective evaluation of electric lighting

3.5.1 Perception of the Luminous Environment

Illuminance Levels

Considering separately electrically and daylight environments, an expressive difference is found between objective assessment and subjective appraisal of illuminance levels. Figures 10 and 11 show that subjects under artificial lighting are much more tolerant to measured illuminance levels which are conventionally assessed as insufficient. This means that people are more demanding when working in daylight environments because there is an added value attributed to daylight beyond satisfaction of visual task performance requirements (Fig. 12).

The fact that occupants are more demanding when working under daylit environments is illustrated in Fig. 13, which details a subjective appraisal of luminous perception of daylit work plane levels. Although a significant percentage of measured illuminance levels were less than 500 lux, most of the participants considered that

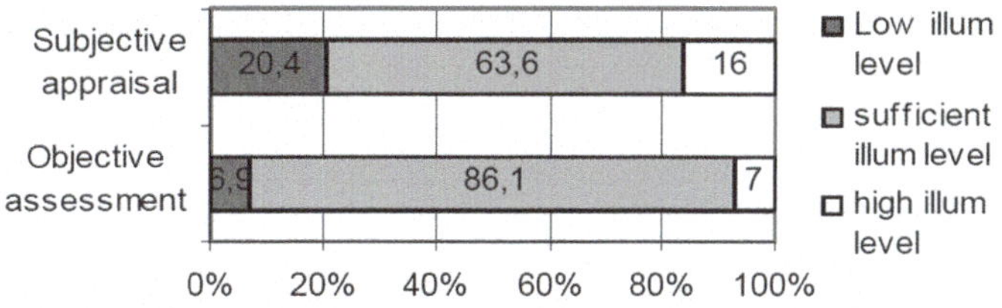

Fig. 11 Objective and subjective evaluation of daylight

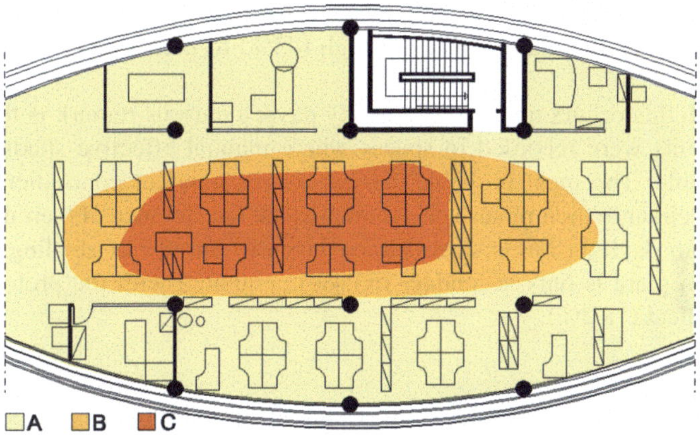

Fig. 12 Simplified daylight illuminance distribution, under overcast sky conditions in a semi-open plan office building. (**a**) – good to reasonable conditions, (**b**) – medium to almost insufficient conditions and (**c**) – inadequate conditions

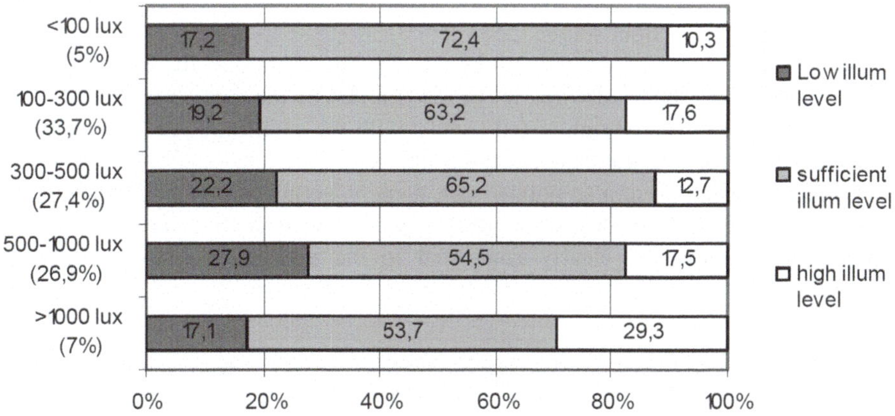

Fig. 13 Luminous perception as a function of daylight illuminance at wok plan level

they were working under satisfactory luminous conditions (sufficient illuminance levels) [25].

Subjective Visual Comfort

In the studied spaces, glare does not constitute a matter of complaint in electric lighting environments. Nevertheless, glare is a cause of discomfort for 30% of participants in daylit spaces, most of them working close (<1 m) to a window. The incidence of direct sunlight on a computer screen (70,7%) or a working plan (28,8%) are pointed out as the major causes of high (43%) to moderate (36%) degrees of discomfort.

Although this occurs mainly in clear sky days, a curious remark is that most of the complaints were recorded in spaces where manual effective shading devices were available. The most probable causes of this apparent contradiction can be pointed out either to their misuse or, eventually, the interference of their use with the general room daylight levels. Justification given to the use of shading devices to protect from glare is only secondary (63.4%) compared with the protection from solar gains heat (78.2%).

Expectations and Preferences About the Luminous Environment

Daylight ranks among the most named factors (multiple answers) that contribute to the creation of an "agreeable" environment. Daylight is clearly indicated as preferred (67%) to electric (0.5%) or "mixed" lighting (32.5%).

Nevertheless, the possibility of light control is essential (Fig. 14) for the users regardless the type of light source [25].

Windows are considered a "very important" (84%) feature of the working environment. Positive aspects of windows were identified by the users: a source of daylight and fresh air, a satisfying visual contact with the exterior, a sense of the time of the day and a distraction from current tasks.

Negative aspects were also pointed out, namely, acting as unwanted sources of heat in some periods (66%), noise (58%), glare and excessive daylight. These aspects are mainly and clearly connected to the prevailing non-overcast sky

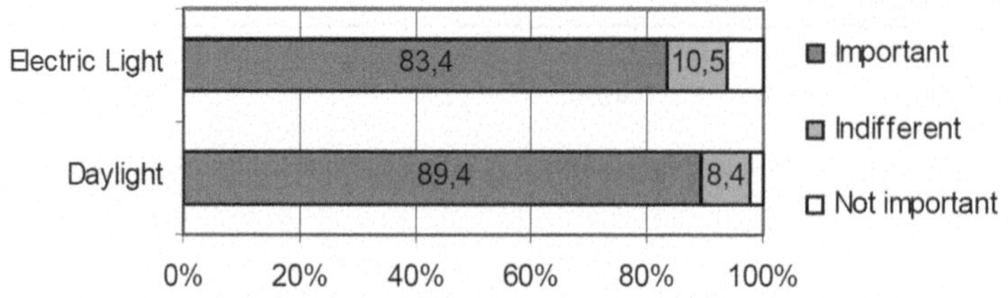

Fig. 14 Importance given to the possibility to control daylight and electric lighting

conditions and temperate climate of Portugal and other southern Europe regions. But one should also be aware that oversized and unshaded windows adopted in many new buildings must deserve an attentive look.

Satisfaction with the Luminous Environment and the Working Place

The study shows that subjects that usually perform their tasks under electric lighting conditions reveal indifference to daylight, once they are not dependent on it. Those who work in daylit spaces reveal a high degree of satisfaction but only if its control is possible, although as already mentioned before adequate control is often not exerted.

Current interior and exterior shading devices are perceived as moderately to highly efficient. Nevertheless, subjective appraisal and objective assessment of shading efficiency strongly diverge (Fig. 15).

Additionally, in situ observations showed that 61% of the users effectively had total freedom to control daylight admission in the working place but only 27.8% acknowledge that fact.

3.6 Linked Mechanisms Approach

The statistical analysis of the results of the surveys and of the on-site characterizations allowed the identification of important correlations between different variables (degree of control over shading devices versus satisfaction with the indoor environment, effectiveness of the controls versus adequate energy-saving attitudes, etc.).

The objective and subjective outputs of the field post occupancy surveys supplied the rationale to structure a hypothetical model establishing the links between users' perceptions and behaviours and luminous environment.

Figure 16 presents the first draft model representing the variables selected to measure all concepts in analysis relating indoor and outdoor lighting conditions, visual comfort and subjective appraisal of the luminous environment, sociocultural context building characteristics and users' behaviours and satisfaction.

Multivariate statistical analysis, testing and validation of this structural model will contribute to the improvement of the dynamic daylighting simulation model,

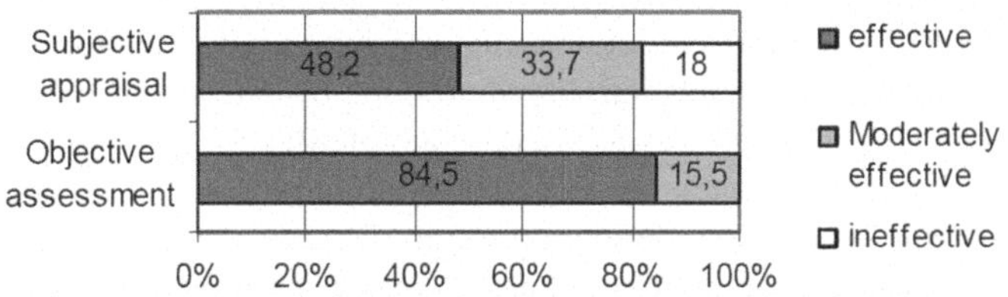

Fig. 15 Subjective appraisal and objective assessment of the efficiency of shading devices

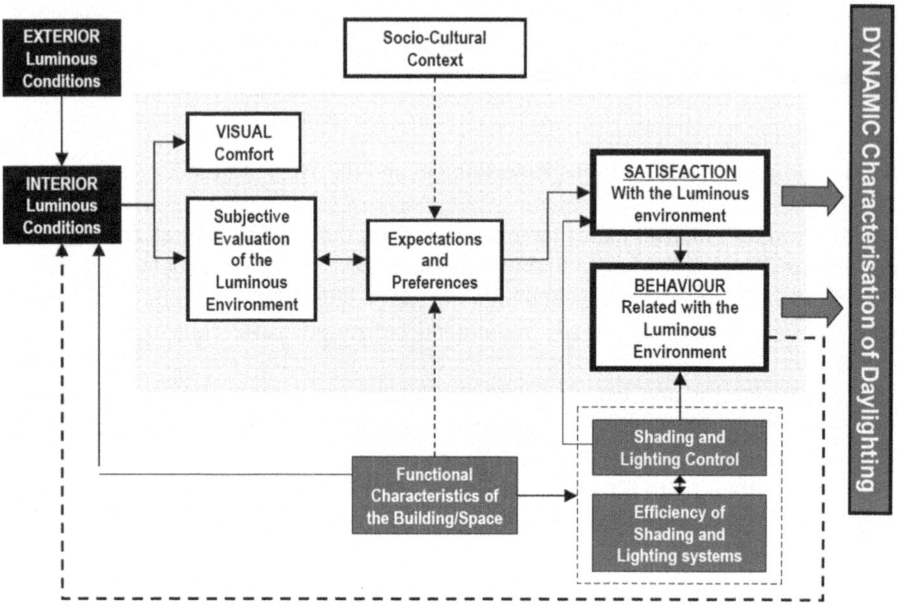

Fig. 16 Hypothetical structured individual-luminous environment relationships

namely, by supporting the definition of typical real-life patterns of behaviour related with daylighting, electric lighting and shading and respective control systems.

4 Conclusions

There is a recognized and even imposed need to reduce unjustified buildings energy consumption levels in most of Europe. In Portugal, due to climatic and economic specificities, the stress could be put on the need to limit growing energy use in buildings, namely, for indoor acclimatization (heating and cooling) and lighting, although ensuring the satisfaction of adequately formulated requirements.

The results of the research carried out showed that there is a trend towards energy consuming conventional "uniformity" (buildings, indoor environments, requirements). However, a deeper understanding of objective/subjective relationships reveals that it is possible to attain satisfaction "differently" and actually in a more sustainable way. Portuguese temperate climate, where predominantly non-overcast skies prevail, and traditional adaptability should not be extraneous to that premise.

In this sense, this research was focused on an interdisciplinary approach considering specific Portuguese characteristics and supported by extensive objective measurements and subjective data obtained through questionnaires on how occupants' sense and perceive and react to thermal and luminous indoor environments.

The following paragraphs summarize some of the most relevant results of the extensive field surveys and analysis carried out in the scope of the research project:

- Clearly the occupants' perception votes show that they accept temperatures beyond rigid indoor conventional (reference) comfort conditions.
- However, the present field study shows significant differences between the neutral temperature, *Tn*, and the preference temperature, *Tp*, mainly in the heating season (winter).
- A new adaptive thermal comfort model was proposed reflecting the extensive Portuguese objective and subjective data obtained and the elaborated analysis performed.
- Daylight ranks among the most named factors that contribute to the creation of an "agreeable" indoor environment.
- An expressive difference is found between objective assessment and subjective appraisal of illuminance levels.
- People are more demanding (less tolerant) when working in daylit environments because there is an added value attributed to daylight beyond satisfaction of visual task performance requirements.
- In the studied spaces, glare does not constitute a matter of complaint in electric lighting environments but is a cause of discomfort for respondents in daylit spaces.
- Daylight is also an increasing cause of complaints about undesired heat gains and discomfort.
- Although daylight control (shading) is important to those who work in daylit spaces, it is either often not adequately exerted or corresponding subjective appraisal and objective assessment strongly diverge.
- The better understanding acquired about the patterns of behaviour of the occupants supported the definition of objective performance indexes that will be integrated in a new dynamic daylight simulation model.

In the fields covered by this research – thermal comfort, luminous satisfaction and dynamic characterization of daylighting – the elaborated statistical analysis revealed the interest and clear advantages of the integrated study of the multiplicity of relevant physical and psychosocial factors and highlighted the complexity of intervening phenomena.

Future research will extend the present work. Maintaining this interdisciplinary approach, there are two domains that will deserve the attention of the multidisciplinary research team. On the one hand, the inclusion of other subjective variables of the indoor environment, namely, the perception of acoustic comfort (noise level) and indoor air quality; on the other hand, the study of residential environments, which have a growing impact on energy consumption in Portugal.

As a final remark, this study showed that regulations and standards may have a role to play in the sustainability of the working and living built environments by addressing indoor comfort from a wider perspective, without the need to sacrifice productivity and well-being.

References

1. "Directive 2010/520/EU of the European Parliament and of the Council of 19 May 2010 on the energy performance of buildings (recast)". Official Journal of the European Union (OJEU), L 153, pp 13–35.
2. "Directive 2018/844/EU of the European Parliament and of the Council of the Council of 30 May 2018 amending Directive 2010/31/EU on the energy performance of buildings and Directive 2012/27/EU on energy efficiency". Official Journal of the European Union (OJEU), L 156, pp 75–91.
3. Matias, L. "Development of an adaptive model for defining the thermal comfort conditions in Portugal". PhD Thesis. Lisbon: LNEC (in Portuguese).
4. Santos, A. J. "Development of a Method for the Dynamic Characterisation of Daylighting in Mediterranean Climates". PhD Thesis. Lisbon: LNEC (in Portuguese).
5. Correia Guedes, M. "Bioclimatic Design and Thermal Comfort Criteria". In *Bioclimatic Architecture in Warm Climates: A Guide for Best Practices in Africa*. Ed. Correia Guedes, M.; SPRINGER, Cham, Switzerland, ISBN 978-3-030-12035-1, 2019.
6. ASHRAE "ANSI/ASHRAE 55:2020. Thermal Environmental Conditions for Human Occupancy".
7. CEN "EN 16798-1:2019. Energy performance of buildings - Ventilation for buildings - Part 1: Indoor environmental input parameters for design and assessment of energy performance of buildings addressing indoor air quality, thermal environment, lighting and acoustics - Module M1-6".
8. McCartney, K. & Nicol, F. "Developing an adaptive control algorithm for Europe". Energy and Buildings, 34: pp. 623–635.
9. Guedes, M. "Thermal Comfort and Passive Cooling Design in Southern European Offices", PhD Thesis, Cambridge University, Cambridge.
10. CEN "EN ISO 7726:2001. Ergonomics of the thermal environment. Instruments for measuring physical quantities (ISO 7726:1998). Brussels, European Committee for Standardization (CEN)".
11. Matias, L.; Santos, C.; Rebelo, M.; Silva, D. & Guedes, "Constant or variable indoor environments? Sustainable option of adaptive thermal comfort". Conference Portugal SB07 – Sustainable Construction. Materials and Practices, Challenge of the Industry for the New Millenium. Lisbon, September 12–14 2007. pp. 403–410.
12. ISO "ISO 10551:2019. Ergonomics of the physical environment — Subjective judgement scales for assessing physical environments".
13. Matias, L.; Almeida, S.; Santos, C.; Rebelo, M. & Guedes, M. "Adaptive Thermal Comfort for Buildings in Portugal Based on Occupant's Thermal Perception". PLEA 2009 – 26th Conference on Passive and Low Energy Architecture, Quebec, June 22–24. pp. 504–509.
14. Rebelo, M.; Almeida, S.; Matias, L. & Santos, C. "Adaptive Thermal Comfort in Buildings". An Explanatory Model, Poster, 8th Biennial Conference on Environmental Psychology, Zurich, September 6–9.
15. Dear, R.; Bragger, G. & Cooper, D. "Developing an Adaptive Model of Thermal Comfort and Preference. Final Report ASHRAE RP-884".
16. Santos A. J.; Carvalho L. C.; Moret Rodrigues A. & Santos C. "Sustainable Daylighting Design in Southern Europe". Proceedings, PLEA 2008 – 25th Conference on Passive and Low Energy Architecture, Dublin, October 22–24.
17. LNEC "LNEC Contributions to JOULE program: vols. I – V. Lisbon: National Laboratory of Civil Engineering", 1983.
18. IDMP – International Daylight Monitoring Programme, [Online], Available: http://idmp.entpe.fr/ [May 2008].
19. Satel-light [Online], Available: http://www.satel-light.com [April 2008].
20. Carvalho, L. C., "Influência da iluminação natural nas necessidades de iluminação eléctrica". Revista Energia, 45, pp. 18–24, 1997.

21. Carvalho, L. C. & Santos, A. J. "Daylighting Availability Characterisation". Proceedings, PLEA 2000: The Millennium Conference - The 15th International Conference on Passive and Low Energy Architecture. Cambridge, U.K., July 2–5, 2000.
22. Gomes, M. G.; Santos, A. J. & Rodrigues, A. M. "Experimental and Numerical Determination of Radiative Properties of Fenestration Systems". Proceedings, PLEA 2008 – 25th Conference on Passive and Low Energy Architecture, Dublin, October 22–24, 2008.
23. Santos, A. J.; Almeida, S.; Rebelo, M. & Santos, C. "Subjective Aspects of Visual Comfort in Buildings: Perceptions and Expectations". Lisbon: LNEC (in Portuguese), 2000.
24. Santos A. J.; Carvalho L. C.; Moret Rodrigues A. & Santos C. "Sustainable Daylighting Design in Southern Europe". Proceedings, PLEA 2008 – 25th Conference on Passive and Low Energy Architecture, Dublin, October 22–24, 2008.
25. Santos, A.J. (2014) "Daylighting: A Perspective in the Context of The Environmental Comfort and Energy Efficiency in Buildings". ITE 57. Lisbon: LNEC (in Portuguese), 2014.

Luis Matias is a Research Officer at Building Finishes and Thermal Insulation of Building Department of National Laboratory of Civil Engineering (LNEC), Portugal. He works in Research & Development (Programmed Research, Research by contract and participation in national and international R&D Projects) in the domains of thermal performance and energy efficiency in buildings, indoor thermal comfort, and nondestructive techniques for thermal assessment of buildings. Dr Luis Matias is the coordinator of the Thermal Insulation Testing Laboratory of LNEC. Since 1991, he has been participating in various national and international research projects, in thermal area, namely 'ESBU – Energy Efficient Sportive Buildings – Leading the way towards a sustainable environment' (4.130/Z/1-137); 'Net Zero Energy School – Reaching the community' (MIT-Pt/SES-UES/0037/2008); 'Development of sustainable thermal and visual comfort models' (PTDC/ECM/71914/2006) and 'Highly efficient cladding eco-panels with improved nano-insulation properties (GELCLAD)'. He participated in various international and national conferences, seminars and workshops. His PhD was on thermal comfort. Based on a significant set of field studies performed on buildings, an adaptive model that defines thermal comfort conditions for buildings in Portugal.

António Santos is a research officer at the Acoustics, Lighting, Building Components and Facilities Unit of the Buildings Department of National Laboratory of Civil Engineering (LNEC), Portugal. He works in Research & Development (Programmed Research, Research by contract and participation in national and international R&D Projects) in the domains of daylighting, artificial lighting, solar radiation, visual comfort, energy efficiency, indoor environmental comfort and quality and other related areas Dr António Santos is the Coordinator of the Photometry and Colorimetry Testing Laborator within LNEC. He is also a member of several WG and TCs within national and international standardization organizations. Since 1991, he has been participating in various national and international research projects, in domains of daylighting, electric lighting and energy efficiency, such as: EU-financed JOULE R&D Projects, 'ESBU – Energy Efficient Sportive Buildings – Leading the way towards a sustainable environment' (4.130/Z/1-137); 'Net Zero Energy School – Reaching the community' (MIT-Pt/SES-UES/0037/2008); 'Development of sustainable thermal and visual comfort models' (PTDC/ECM/71914/2006) and 'Highly efficient cladding eco-panels with improved nano-insulation properties (GELCLAD)'. Dr António Santos participated in various international and national conferences, seminars and workshops and is the author of over 200 publications. The main themes of his PhD were daylighting and visual comfort in buildings. His PhD research was based on a significant set of field studies performed on buildings, which led to the development of a behavioural model used in the dynamic characterization of daylight in buildings in Mediterranean Climates.

M. Correia Guedes is a professor in the Department of Civil Engineering and director of the Architectural Research Centre, Instituto Superior Tecnico (ICIST – Group 8), Portugal. He is responsible for several disciplines of the courses of Architecture, Civil Engineering, and Territorial Engineering. He is supervising several PhD and MSc students. **Participation in various research projects**: chief coordinator of a COOPENER EU project (SUREAFRICA), national coordinator of an ASIA-LINK project. Dr Guedes has participated in various international and national conferences, seminars, and workshops. His PhD was on 'Environmental Design in Southern European Office Buildings', covering sustainability, passive design, energy efficiency, and thermal comfort. Both his PhD and MPhil were done at Martin Centre, Cambridge University, UK. The MPhil thesis was sponsored by the EU 'Praxis' scholarship in Environmental Design, 'Aspects of Environmental design of Office Buildings in Portugal'.

Since 1985 he has participated in various projects, namely the Portuguese Pavilion in Seville's EXPO 92, two residential buildings in Vila Real, the competition for the National Assembly building, and the building of the Agronomy Faculty (UTL). He worked as an architect in several Portuguese architectural companies. He has had many articles and papers published in the fields of bioclimatic architecture and the built environment.

Rethinking Building Habitat for Comfort and Human Well-Being: Digital Technologies for Nature-Based Design

Antonella Trombadore and Gisella Calcagno

1 Well-Being in Buildings: How the Pandemic Experience Has Changed the Domestic Habitat

In Italy, the slogan #Istayathome was the refrain of the first lockdown when, to protect us from the risk of contagion, we were all forced to stay indoors. In this cloistered time, we changed our habits and, above all, reassessed the importance of living in a comfortable environment, desiring wider green and open spaces like our balconies, terraces and gardens.

The pandemic has in fact changed the use of domestic habitat: It has become a place to be protected, a multifunctional study and work environment. When we were forced to remain inside our home, however, the architectural quality of our homes and, even more so, the comfort levels of those interior spaces showed many weaknesses. Being inside 24 hours a day in close contact with family members and begin connected to the outside world only digitally due to smart working and remote learning was challenging (Fig. 1).

This experience has led us to rethink the housing model: our homes have suddenly turned into office, classrooms and virtual meeting places, combining new work needs with those of ordinary home management. It has also shown us that the contact with nature is necessary for our psychological, physical and relational well-being.

The consequences of discomfort caused by #Istayathome were obviously not the same for everyone. It reflected the dimensional characteristics of the houses, their location and neighbourhood, the engagement with neighbours, the presence of available local services and, above all, the possibility to relate with natural elements, such as fresh air, sunlight and vegetation. Yet, housing and building

A. Trombadore (✉) · G. Calcagno
DIDA Department of Architecture, University of Florence, Firenze, Italy
e-mail: antonella.trombadore@unifi.it; gisella.calcagno@unifi.it

A. Sayigh (ed.), *Achieving Building Comfort by Natural Means*, Innovative Renewable Energy, https://doi.org/10.1007/978-3-031-04714-5_12

Fig. 1 Smart working during the Covid-19 pandemic

regulations have traditionally focused just on dimensional aspects: simplifying the rationalism tradition and undervaluing the ecological objectives; today's building practice is still mainly oriented to satisfy basic human needs in minimal spaces.

2 From the Existenzminimum to the Human-Centred and Sustainable Building Design

In Germany of the early 1900s, illustrious architects gave life to the concept of *Existenzminimum*, based on the idea of houses able to meet human primary needs by optimizing spaces and reducing expenses. That was also the title of the II CIAM (*Congrès Internationaux d'Architecture Modern*), held in Frankfurt in 1929, focusing on the organization of domestic spaces, analysing the morphological, dimensional and distributive characteristics of spaces in order to design houses, with the words of Ernst May, "*made in a way to satisfy material and spiritual needs of their inhabitants*".

In this period, foundations were laid not only for what becomes the study and analysis of ergonomics but also for a new attention to more sensitive aspects; opening the road towards energy-saving, bioclimatic architecture; the reuse of construction materials; the attentive and conscious use of resources along the building lifecycle and themes emerged just in the post-1970s energy crisis. A lot of experimentation in the Weimar period led to the creation of residential complexes in the outskirt of many German cities. Designed to address the productive and organizational needs of cities: the Siedlungen represented a significative result, under the rationalism, of researches conducted by architects as Gropius and Mies Van der Rohe, characterized by essentiality and formal schematization, wide openings towards the outside and strictly functional layout of the roads and general services. Houses featured by modular typing, prefabrication in reinforced concrete and the

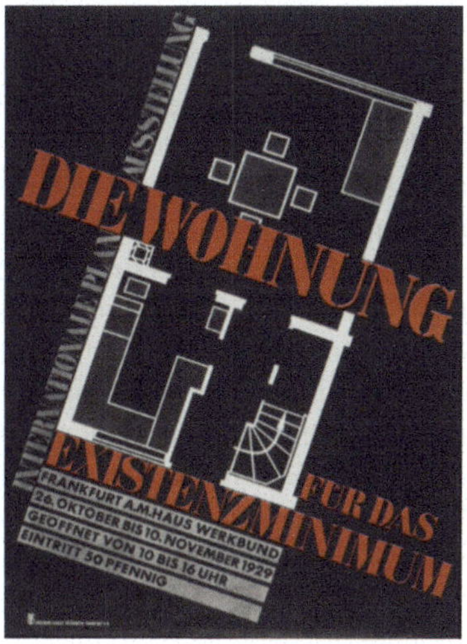

Fig. 2 Hans Leistikow. The Dwelling for Minimal Existence, 1929 – with Ernst May's floor plan for houses in Frankfurt

"minimum standardized cell", a minimal living unit to be used as a standard element for the construction of different housing solutions (Fig. 2).

Across the twentieth century, a design approach based on rationalism drives new housing models based on scientific parameters (typologies, functional schemes, taxonomies of values and hierarchies of behaviours) [1]. As explained by Maurizio Vitta, "*a spatial and interior design inspired by the pure structural, ergonomic and industrial calculations, which were framed by more vanishing, but not less normative, representations of ethical and social order. Even the aesthetical result (the image of the "bella casa") has generally agreed to this conception, founding the topic of mass housing on formal protocols elaborated in the light of the same productive criteria*".[1]

This pandemic has accentuated even more the obvious signs of crisis that already appeared in the twentieth century, drawing attention to a deeper reflection on the environmental perceived quality in living spaces, starting from a radical rethinking of the relationship between the living built space and nature. According to the World Green Building Council[2] in fact, "harmony of the built environment with nature" is a key principle for a heathy and sustainable built environment, interrelated with the objectives of taking climate action, protecting health, prioritizing comfort, facilitating healthy behaviour and creating social value [2]).

Our homes today lack the component of nature, in particular green spaces and elements of vegetation, which should not only be a prerogative of villas or luxury apartments but be applicable to family residential buildings. We need now to bring the concept of *urban jungle* into cities in order to restore balance and, perhaps,

[1] Vitta M., *Nuovi modelli dell'abitare - XXI Secolo*, Treccani, 2010.

[2] World Green Building Council, *Health & Wellbeing Framework: Six Principles for a Healthy, Sustainable Built Environment*, November 2020, online at https://worldgbc.org/health-framework

reduce the effects of pollution and increase biodiversity. Gardens and terraces are not a new frontier in architecture, since they already belong to the historical tradition and culture of Mediterranean living: patios, porches, interior gardens and open-air domestic environments are in clear opposition to the new dense models that deny a tangible relationship with the outside or the simple opening of a window for natural ventilation (Fig. 3).

We must now focus on the ecological quality of our home environment and on more clever use of building technologies. We are seeing experimentation of the relationship between architecture, technology and nature where buildings are designed to breathe, with innovative materials and a wiser use of kinetic components including natural light and energy from the sun, rain and wind.

We need a change of pace: The concept of a natural space for living must be further expanded. The necessary reduction of emissions, reduction of energy consumption and the design of intelligent buildings need to be integrated into the new approach of healthy living. Above all, it is necessary to move from the functional concept of "habitation" to the cultural concept of "living", placing the inhabitant in a more human-centric approach, understood in his corporeity and behavioural models, such as in his ecological consciousness and social interaction.

This in turn requires to go beyond the logic based on need-performance, moving towards a human-centred and sustainable building design. It is necessary to emphasize the real immersive and perceptive experience of well-being of the inhabitants in an anthropized habitat, their adaptive capacity and conscious and sustainable interaction with the natural context: an inhabitant taking care of his own habitat, as an amniotic fluid in which he is immerged and that guarantees, in return, nourishment, health and well-being.

This means beginning a regeneration process of our cities and homes and rediscovering the beauty of our buildings with a sensitivity to culture, where architecture becomes the art of living in harmony with nature.

Fig. 3 Reconstruction of the Pompeian "Casa Vettii" (Firenze, Giardino di Boboli, 2007)

3 The Building Habitat and the Lack of Nature

The human being, without his own habitat, has transformed the whole world into anthropic environment, through design and technique, as an artificially recomposed nature placed at his service. Agricultural areas have been urbanized and natural habitats anthropized. The anthropogenic activities, in this geological time of the Anthropocene, have profoundly altered the metabolism of the biosphere superorganism, and the main symptoms are illness and discomfort, global warming and the loss of biodiversity, which is the main indicator of the health of ecosystems. As Federico Butera points out very well in the introduction of his book,[3] "*in facing complexity to govern the ecological transition, everything is complicated by the fact that the two phenomena are connected to each other. To repair the damage we must first be clear that the environmental problems we face are systemic: a mixture of physical, chemical, biological and social changes that interact and reinforce each other. If we continue to think that we can tackle the various problems without connecting them, there is no hope of success. We need to combine the pessimism of intelligence with the optimism of the will, where the pessimism of intelligence derives from the knowledge of the complexity of the problem and the optimism of the will is the answer, that is, roll up your sleeves and find solutions*" [3].

The necessity to rethink the ways of living in the light of the growing global complexity, complicated by the pandemic explosion, is addressed by the ongoing Architecture Biennale in Venice, curated by the Lebanese architect, Professor and Dean of the School of Architect and Planning at MIT Hashim Sarkis, and exemplary titled: How will we live together? According to the curator, such a question cannot have a unique answer and cannot be answered by the political and administrative bodies: it requires the contribution of architects, as design experts capable to attract a wide range of stakeholders, such as in the design and construction process.

If we focalize the five thematic areas composing the Biennale, it is possible to capture the challenges for the present and future architects: Among Diverse Beings, As New Households, As Emerging Communities, Across Borders and As One Planet.

There is a strong desire for architecture, and above all, we wonder how architecture can respond to the new challenges that climate change urgently poses to us, the role that public spaces can play in the urban revolution due to the pandemic and the new reconstruction techniques.

To use the metaphor of the symptom of this virus that Covid 19 really "*took our breath away*", the pandemic forced us to reshape the consolidated familiar dynamics which marked our daily lives: the lockdown and social distancing measures created a spatial-temporal distortion and forced us to change the way we inhabit the city.

The most important distortion was precisely the changing of perception of the urban space, which has been progressively reduced. Having the freedom to go out

[3] Butera, F., *Affrontare la complessità. Per governare la transizione ecologica*, Edizioni Ambiente, 2020.

just to shop, take a walk and look for some greenery within 200–300 metres from home-made us discover the urban space that surrounds our homes: we rediscovered the beauty of living in a neighbourhood.

In the last century, we saw the frenzy of concentration in urban centres contaminating large swathes of populations. They were pursuing the utopia of a better life, living in cities increasingly full of services, infrastructure, work and opportunities. Yet, according to UN Habitat's World City Report 2020,[4] the pandemic was first and foremost an urban phenomenon: more than 90% of cases confirmed were concentrated in big cities like Wuhan, Milan, Madrid, New York City and Mumbai. During the pandemic, the new need for social distancing brought back to life the quality of a slower and more sustainable lifestyle linked to the rhythms of nature, restoring attractiveness to the smaller historic centres with low population density.

The challenge of improving the quality of life in our cities is certainly not based on a nostalgic return to the past: We actually need to reactivate the balance between man and nature. We need to learn from the crisis and activate sustainable urban systems that are capable of amplifying community life without consuming natural, energy and environmental resources. The historic districts of many Mediterranean cities have survived thanks to the strong osmosis between workplaces, agricultural areas, public socialization space and private housing. The contemporary urban layout with the list of separate functions has not worked. Only by applying an ecological approach we can trigger a true and therapeutic regeneration process for the city and its inhabitants.

The real post-Covid-19 challenge has to focus on the actual level of sustainability we will achieve in our cities, restoring the balance between urban areas and agricultural territories in order to increase the green areas for better air quality. The design of buildings and green areas towards reduced environmental impact via ecological footprint, pollution, GHG emission, using recycled or recycling material and with production logistic based on the green and circular economy.

4 The Green and Ecological Approach for Well-Being

The solution that is pervading many cities is to reinvent the built environment through nature-based solutions and green infrastructure, aiming to re-naturalize cities (urban jungle effect) and to accelerate the ecological transition through green architectures, reducing energy consumption and the effects of pollution, consciously managing natural resources and increasing biodiversity. The dominant idea, however, is to think that the integration of simple hanging gardens and climbing plants can restore a new balance and a new relationship with Nature. The challenge should be more ambitious, addressing the processes of urban transformation in a systemic

---

[4] UN-Habitat, *World Cities Report 2020: The Value of Sustainable Urbanization*, 2020, online at https://unhabitat.org/World%20Cities%20Report%202020

way, as claimed by the architect and ecologist Ken Yeang (Fig. 4). According to the bio-philic thinking[5] (*Biomimicry Thinking process*), we need to deepen the knowledge of the systemic relationships between man-nature and addressing the criticalities of the anthropic habitat by applying creative approaches of biophysics and biomimicry projects, developing innovative solutions to increase the levels of sustainability and improve the well-being of people [4].

We need a change of pace starting from the cognitive approach: we have the opportunity to focus attention on the ecological quality of the built environment, further expanding the concept of a natural space for living. To the necessary reduction of emissions, reduction of energy consumption, to the design of smart buildings, we need to integrate a new approach to *well-being*, the ability to make us live better.

This means not only restarting, in a healthy and virtuous way, the regeneration processes of our cities and our buildings but above all rediscovering the beauty of our architectural heritage, the sensitivity of a culture in which architecture becomes art to live in harmony with nature.

Combining the holistic and ecological approach in architecture means the art of integration different competencies, technologies and materials, to meet project requirements with a sustainable strategy. In this new approach, the value and the lesson from vernacular architecture are evaluated as an important root to understand how to reach to bio-climatic condition and how to use local materials (0 km materials) to reduce unuseful energy use. They are the set of strategies of design and construction, through which it seeks to achieve the realization of a sustainable architecture, an architecture that is using its formal configuration, technologies, components, materials and equipment to establish an optimal relationship with the

Fig. 4 Ken Yeang & TR Hamzah Architects, EDITT Tower, Singapore, 2008

[5]Yeang K., *Saving the Planet by design. Reinventing our world through ecomimesis*. Ed Routledge, 2020.

surrounding environment, so as to reduce energy consumption and provide the best comfort to the occupants, using as far as possible natural systems and reducing the use of mechanical systems of the building [5].

We can apply this design approach to valorize the transformation of the natural and built environment at different operational scales. From the size of regional and urban planning to town and district design, architectural concept, building detailed design, mechanical plant, up to orient the choice of building component and to sustainable building materials and even applied to rules setting and post realization management.

4.1 The Adaptive Design of Green Envelope

Adaptive design of green facades and natural-based solutions for resilient architecture and environment are considered effective tools for dealing with the cities' challenges and make buildings' envelopes efficient and green. Introducing vegetation elements as decorative or productive plants into architecture is the newest tendency that integrate previous recommendation of sustainable architecture implementing holistic design principles hollistic design principles with the benefit associated to vegetation that many people emphasize with the name of eco-design or eco-friendly architecture. Vegetation in architecture means health contribution, through oxygen production and air purification, as well psychological and social well-being. The good feeling of being surrounded by natural vegetal elements reduces the stress of living in full artificial environment, made with artificial materials, in artificial controlled atmosphere without c*onnection to nature and natural processes.*

Green architecture strives to minimize the resources consumed and the energy balance embedded in materials and in construction processes, both for buildings and cities, where green architecture usually symbolizes the possible sustainability of modern cities, improving environmental quality of the natural and built environments.

Vertical greening in architecture provides a cooling effect on the building surface, which is very important during summer periods in hot and temperate climates. The cooling effect of green facades has also an impact on the inner climate in the building by preventing warming up the walls (Fig. 5).

Greening the building envelope especially in the Mediterranean climate and in the warm season reduces the peak temperature on the wall surfaces with climbers' plants. Some studies demonstrate that greenings buildings technology, for roofs or facades, can increase the dynamic thermal characteristics of the wall surfaces temperature to reach a good thermal behaviour of the building envelope. The use of vegetation in the building envelope is an interesting sustainable strategy to save energy. Green roofs, façades greened with climbing plants or living wall systems using modular pre-vegetated panels or vertical greening systems, are solutions that can be used as energy-saving methods for the built environment.

Fig. 5 One Central Park – two residential towers in Sydney, Ateliers Jean Nouvel, 2008

Greening on roofs determines better climatic and microclimatic conditions. Green roofs perform many functions: insulate the building, protective function from atmospheric agents; increase the thermal inertia, better indoor climate comfort and reduction of energy costs. In winter, heat losses are considerably limited. In summer, on the other hand, thanks to the shading and evapotranspiration, lower temperatures are obtained.

The presence of greenery in the building complex increases the well-being of the inhabitants by influencing the psychological enjoyment of the surroundings. The initial costs to implement this system are amortized by the decrease in energy [6].

4.2 *Greening the Public Areas*

Beautiful trees make great cities because a dynamic urban forest supports a healthy community, economy and environment. In addition, as sustained since the 1980s by Arnold,[6] trees are integral to the urban design of any city or town: Information on best practices needs to be more widely disseminated [7].

There is an urgent need to develop and apply larger strategies for sustainable urban forests and street tree plantings. In this field, the competences of architects, botanic experts need to be integrated to define better standards for site planning and suitable tree species. The design of streets has to be defined in order to maximize their benefits and avoid potential conflicts with traffic and utilities and tree quality standards are key to achieving this goal.

There are many reasons to justify the new trend to implement the use of green and nature-based solution in the cities that we may summarize in the following two main aims:

[6] Arnold. H. F., *Trees in Urban Design*, Van Nostrand Reinhold Company, New York, 1980.

- *Reducing heat island during summer*: All over the world, the growing of heat pick in town is not only a consequence of climate change, but the primary result of urbanization (narrow street, uncontrolled sprawl of construction, high density, parking, urban high ways) but also the reduction or the absence of green areas, parks and trees bordering. All these aspects may be moderate with the introduction of vegetation, trees, gardens, vertical greening in the urban tissues.
- *Improving thermal comfort and human behaviour:* Water that plants soak into the soil from irrigation or rain ultimately returns as water vapor through direct evaporation or by transpiration through plant leaves. This loss cools the surface and plant canopy just like the evaporation of sweat cools our skin. This improves microclimate condition both in open and interior spaces and human comfort.

5 People Acknowledging Buildings for an "Ecological Well-Being"

Even if the integration of nature in the built environment is widely recognized as an absolute benefit for both the human beings and the planet, its conscious and comprehensive consideration, and valorization, in the architectural design phase is still an ongoing effort.

Beyond general ideas and conceptual statements in architectural projects, the design process of nature-based solutions needs the support of technological elements and the scientific assessment by validated methodologies and tools. In this way, it is possible to fully capture and demonstrate the real impact of natural elements (sun, water, air and soil/vegetation) in living spaces and their contribution on improving human well-being and on reducing the ecological footprint.

In order to sustain the efficacy of nature-based technological solutions in the design phase, we require considering a wide range of quantitative and qualitative data. From a side, benefits can be quantified in terms of reduction of the energy demand and improvement of indoor environmental quality (physical parameters); from the other side, the perceived quality generated by natural elements in built spaces can be assessed only through qualitative and human *soft* data (real-life experience).

Nowadays, this data challenge can be addressed by the digital technologies (intended as electronic tools, systems, devices and resources to generate, store or process data) specifically evolved for the building sector. They allow the virtual replication of the built environment in order to advance its knowledge and analysis for a more conscious decision-making across the building lifecycle, since the very early phases of the architectural design process.

Grounded in the scientific approach, the computation of large amount of qualitative and quantitative data in the design of future building and cities has to be intended not to define standard solutions but as a means to collectively acknowledge and foresee a wide range of possible scenarios, in which society can reflect its diversity, aspirations and desires.

6 The Digital Twin as Predictive Tool for Ecological Design

The need to support an ecological approach involving nature in the design of future buildings and cities can be addressed by a more conscious exploitation of digital technologies. As recognized by the European Union,[7] the spreading diffusion and evolution of digital technologies represents a huge opportunity for the AEC sector to reach a more sustainable built environment.

Green and digital transitions are blended in a virtuous circle, with the digital domain enabling data-rich and predictive processes of co-planning and co-design, which also intercept end-user's feedbacks and behaviours, supporting the definition of more ambitious scenarios in terms of environmental quality and sustainability.

Since the introduction of computer-aided design (CAD) in the 1960s, many advancements in the field of informatics have been gradually integrated in architecture, with software applications permitting a more accurate and controlled management of the large amount of building data, from the design phase to the whole building lifecycle.

The design phase has been enriched by the development of simulation tools, whose computation permits to predict the behaviour and the performances of buildings according to different aspects (i.e. structural, thermo-hygrometric). In the 1970s, the necessity to cope with the energy crisis gave the impulse to the experimentation of software applications able to calculate the energy behaviour of buildings before their actual implementation in the real-life context. Yet, a complex data management and non-intuitive interfaces characterized first generation software, which prevented their widespread application in the design field, remaining in the engineering one.

In the most recent years (late 1990s), the most influential digital innovation for the AEC sector has been the introduction of building information modelling (BIM), a methodology to organize the high quantity of building information within a structured set of data (models). Beyond the three-dimensional representation of the building's geometry, the core of BIM is its relational and semantic database, consenting to collect, exploit and share also non-graphic data on building technological components, their proprieties and functioning.

Nowadays, the turn on Industry 4.0 is enriching the BIM approach with the contribution of the most advanced digital technologies, from sensors to IoT (Internet of Things) applications, from Big Data to Semantic Web and Machine Learning, foreseeing the most advanced possibilities of Artificial Intelligence [8].

At the forefront of this next evolution, there is the concept of Digital Twin. It becomes a virtual replication of the built environment, where it is possible to combine technical static data (BIM models) with real-time environmental data fed by sensors located in the real-life contexts (buildings and urban spaces) and connected

[7] The Directive (EU) 2018/844, amending the Directive 2010/31/EU on the energy performance of buildings and Directive 2012/27/EU on energy efficiency, introduced the role of digitalization to reach environmental sustainability in the building sector.

with IoT systems, which also consent to take into account, and interact with, the end-user experience.

The competitive advantage of Digital Twins is the fully and comprehensive acknowledgement of the built space. It is a very strategic aspect (i.e. energy) to predict future scenarios, thanks to the availability of technical data (i.e. geometry or materials) combined with real-time environmental monitored data (i.e. IEQ). Besides, the further possibility to integrate even more softened digital data (i.e. end-users' feedbacks or big data) is the basis for a more conscious decision-making not only in the design phase but across the whole building lifecycle (i.e. operational phase) [9].

This approach appears particularly profitable for supporting a design process oriented to the integration of nature-based solutions in the existing built environment: DTs can be exploited to measure the human and environmental benefits deriving from green technologies, making people aware of the positive contribution of ecological solutions in the built living spaces (Fig. 6).

DT will represent a new "design environment" where to strategically predict future sustainable buildings by even more reliable simulations and where the exceptional contribution of nature-based solutions can be analyzed, designed and managed in a more efficient and scientific way [10].

Embracing the *New European Bauhaus*[8] objectives under the ambitious EU *Green Deal,*[9] a digitally based integration of nature-based solutions in the regeneration of the existing built environment (main EU challenge for the building sector) has the potential to address all the three dimensions of "sustainability, quality of experience and inclusion". Moving "beyond buildings" to firstly rethink human behaviours, Digital Twin can effectively transform the design process into a human-centred, positive and tangible experience [11].

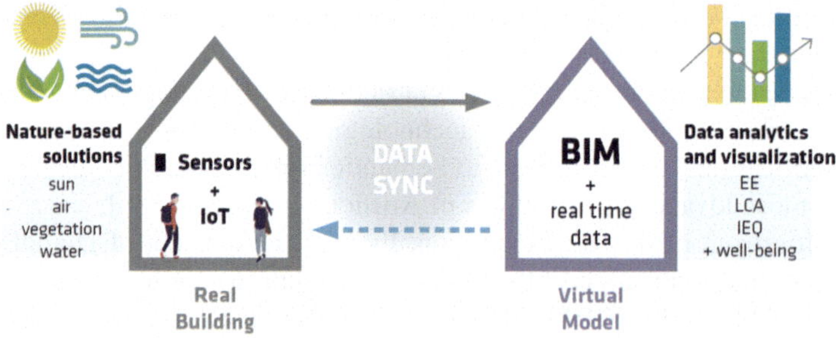

Fig. 6 Digital Twin functioning for nature-based solutions

[8] https://europa.eu/new-european-bauhaus/index_en

[9] https://ec.europa.eu/info/strategy/priorities-2019-2024/european-green-deal_en

7 Conclusions and Next Challenges

The pandemic experience is suggesting us that naturalizing the built environment is a required urban evolution to create healthy living spaces, also matching the overarching objective of environmental sustainability to contain the climate change effects. Yet, nature-based solutions need to be sustained in the architectural design domain by a new capability of active listening to the great benefits they bring in the real world: only a scientific-based common awareness can support effective human and eco-centric design approaches, oriented to harmonize the relationship between the built and the natural environment.

This challenge can be addressed by tackling the digital revolution. The newest possibility of digitally twinning the built environment opens to the opportunity to look at our buildings and cities as experimental spaces to *reimagine how to live better together*, where people can be aware. Thanks to the data support, it will be possible to preview the real impacts of living spaces both on the planet, in terms of energy efficiency and conscious use of resources, as well as on people, in terms of comfort and wellbeing, in order to support nature-based regeneration processes.

The naturalization, through digitalization, of the built environment can be addressed only by a cultural evolution of current design practices, which should start from the public buildings sector and based on open-source data and technologies. The challenge is to be resilient to the crisis: the lessons learned from the 1920s housing crisis was the rationalism, and the ecological turn was a reaction to the 1970s energy crisis. Today, the post-Covid-19 architecture needs to overcome the ongoing ecological effort of reducing the negative impact of buildings on the natural environment, to new attention on human health, comfort and well-being in living spaces. Sharing this vision, the integration of nature-based solutions appears as the most appropriate and promising strategy for sustainable, inclusive, but also beautiful, future buildings and cities.

References

1. Brysch, S., *Reinterpreting Existenzminimum in Contemporary Affordable Housing Solutions.* Urban Planning, 4(3), pp 326–345, 2019.
2. World Green Building Council, *Health & Wellbeing Framework: Six Principles for a Healthy, Sustainable Built Environment*, November 2020, online at https://worldgbc.org/health-framework.
3. Butera F., *Affrontare la complessità. Per governare la transizione ecologica*, Edizioni Ambiente, 2020.
4. Yeang K., *Saving the planet by design. Reinventing our world through ecomimesis*, Routledge, 2020.
5. Balocco C., Leoncini L., *Energy cost for the effective ventilation and air quality for healthy building. Plant proposals for a historic building school under reopening in the Covid-19 era*, Sustainability International Journal 2020, 12, 8737, Special Issue Energy Efficiency of the Indoor Environment, pp 1–16, 2020.

6. Trombadore A., *Mediterranean Smart Cities. Innovazione tecnologica ed ecoefficienza nella gestione dei processi di trasformazione urbana*, Altralinea Edizioni s.r.l., Firenze, 2015.
7. Arnold H. F., *Trees in Urban Design*. Van Nostrand Reinhold Company, New York, 1980.
8. Khajavi S., Motlagh N., Jaribion A., Werner L., Holmstrom J., *Digital Twin: Vision, Benefits, Boundaries, and Creation for Buildings*, IEEE Access, vol. 7, pp 147406-147419, 2019.
9. Tagliabue, L.C., Re Cecconi, F., Maltese, S., Rinaldi, S., Ciribini, A.L.C., Flammini, A., *Leveraging Digital Twin for Sustainability Assessment of an Educational Building*, Sustainability 2021, 13, 480. https://doi.org/10.3390/su13020480, 2021.
10. Trombadore A., Calcagno G., Pierucci G., *Advance smart cities through Digital Twins expanding the knowledge and management capacity of public buildings stock for energy efficiency rehabilitations,* Contesti – Città Territori Progetti, 1/2020, pp. 129–139, University Press, Firenze, 2020.
11. Trombadore A., Calcagno G., *Mediterranean University as Catalyst for Eco-Sustainable Renovation: the experience of Med EcoSure Cross Border Living Lab*, in A. Sayigh (ed.), Proceeding of WREC 2020 International Congress in Lisbon, Springer, 2021.
12. Butera F., *Dalla caverna alla casa ecologica: storia del comfort e dell'energia*, Edizioni Ambiente, 2014.

Antonella Trombadore is a contract professor at the University of Florence, a senior researcher with a PhD in Architectural Technology from the University of Florence and an MSc in Energy Management, and an architect. From 1999 she works at the Department of Architecture and ABITA interuniversity research centre, University of Florence.

She works as a senior expert in the field of project management, integrated projects, sustainable urban design, energy-conscious design, low-energy buildings, sustainable architecture and bioclimatic design, and communication and dissemination strategies.

She is a team leader of several European research projects in the field of architectural integration of innovative solutions for energy retrofitting action, nearly zero buildings as well as integrated smart process for Abitaremediterraneo, multi-issues platform as interactive network for technical regulations management, technologies' databases, and best-practice dissemination activity. She is also a collaborator for the management of the European master's course 'Sustainable Design and Technologies for Built Environment' and postgraduate courses in Bioclimatic Architecture and Energy Saving in Buildings at the University of Florence.

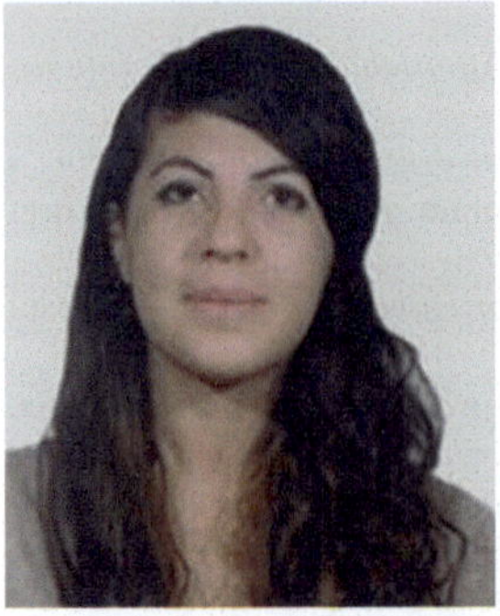

Gisella Calcagno is an architect. After her graduation (MSc) in Architecture at the University of Florence (2013), she attended the II level Master Course at the research centre ABITA (University of Florence) on Bioecological Architecture and Innovative Technologies for the Environment (2014). Focusing on the most social aspects of architecture, she completed a PhD thesis on urban and housing solutions for the reception of refugees in the Italian context (2019), by deepening the topics of temporary uses, urban regeneration, and social innovation. On this occasion, she had the opportunity to approach interdisciplinary research (advanced university course) and to actively participate in the international debate (conferences, publications, and workshops). She is actually a research fellow at the beXLab (building environmental eXperience) of the DIDA Department of Architecture – University of Florence, working in the field of EE for the sustainable renovation of Mediterranean buildings, by investigating approaches and methodologies to empower the decision-making across the building lifecycle thanks to Digital Twin applications.

Renovating Heritage Buildings into Daylit Enjoyable and Visually Comfortable Museums/Galleries

Khaled A. Al-Sallal

1 Introduction

When people in the past constructed what we highly regard today as "the Arab Gulf heritage buildings", they followed a sort of agreed-upon building rules/directions acquired as a result of many historical real-life experiences and knowledge learning until it proved to work well and produce these amazing buildings. In this process, many factors were taken into consideration that resulted in making these buildings not only appealing and beautiful but also highly functional from the socially, culturally, and climatically points of view. Respecting the traditions, considering the availability and suitability of building materials, and adapting the buildings to the harsh desert climate of the region were some of these factors.

In the UAE, like other places in the Arab Gulf region, heritage sites were somehow neglected in the early 1980s amid new pressing demands of modernizing the society and upgrading the way of people's living. This led to removing large areas of heritage sites (Figs.1 and 2) and replacing them with modern roads, highways, and settlements and to the introduction of new building materials and methods and neglecting the traditional ways of building. Yet, this was followed soon by another awakening period that started in the late 1980s and continued until now. This period has witnessed high appreciation of these treasures of the past by the authorities who developed plans to revive them and adapt them to museums, galleries, and social places capable of serving the country's agenda for touristic and cultural economy.

One should keep in mind that a museum building cannot function properly and effectively unless many environmental and technical factors are considered in its design; and the heritage buildings converted to museums/galleries were not designed or built as museums/galleries from the start and thus concern about their

K. A. Al-Sallal (✉)
Professor of Architecture & Consulting Architect, Concreto, Canada
e-mail: Khaled@Concreto.ca

A. Sayigh (ed.), *Achieving Building Comfort by Natural Means*, Innovative Renewable Energy, https://doi.org/10.1007/978-3-031-04714-5_13

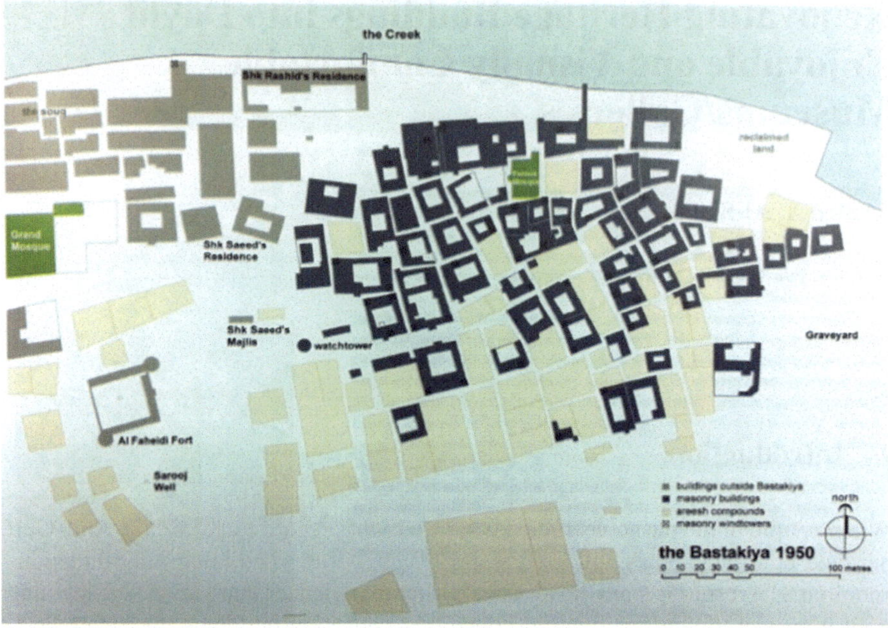

Fig. 1 Shows a top view of how heritage buildings looked like in the Bastakiya area in 1950 (left) [1]

Fig. 2 Sheikh Saeed House in Dubai, before and after restoration

performance can be raised. One of the most important challenges in designing a museum is how to design its lighting so that it is sufficient to enjoy seeing the artifacts and their details; make the exhibits' gallery exciting and vibrant, effective in showing a true experience of its architecture and the colour of the artifacts (as artificial lighting might result in deceiving effects); and minimize any visual discomfort coming from glare or direct sun.

The architectural configuration of the heritage building in the Arab Gulf region was originated around the concept of having a central courtyard to give access to natural light, in addition to satisfying other needs such as circulating fresh air and passively cooling the building and occupants and providing a private outdoor place furnished by greenery for the family to enjoy and mingle when the weather conditions are suitably comfortable. In many of these beautiful heritage buildings when they were renovated to museums or galleries, access to daylighting was blocked and their designs were modified to rely totally on artificial lighting. Although no clear answer was found from the authorities about the main reason for taking this approach, an assumed justification could be to control exposure of the artifacts to light, which of course can be done a lot easier in the case of the artificial lighting than that of the natural lighting. Daylight is dynamic and changing over the course of the day, while artificial light is constant; and hence a gallery space lit by daylighting would be much more complex to design than another one lit by artificial lighting. Choosing the easiest solution though does not necessarily mean it is the best one when considering the visitors' experience and enjoyment. For instance, in the Tate Modern Museum case of London, it has been observed that daylight areas are the most crowded mainly because daylight does not only provide a better colour rendering but also have better luminous quality that cannot be achieved by artificial lights. Visitors of the Tate always argue that natural light is more pleasurable and enjoyable [2]. Artificial light on the other hand is static and monotonous, which may lead to visual efficiency, but also can result in "museum fatigue". Museum fatigue can be experienced by a visitor when his/her interest initially reaches a high level and remains constant for about 30 min then later decreases to a lower level [3].

According to [4], the visual quality of viewing any artwork mainly depends on the field of view, which should have both aesthetic qualities and a certain degree of interest. The use of natural light, which changes with time, can help in providing relief from museum fatigue. There are several examples of well-known museums and top-notch galleries considering natural light as an important factor for ambient light, like the following: the Louvre in Paris, Lisson Gallery in London, Sackler Galleries, Royal Academy of Arts, Tate Britain, the Solomon R. Guggenheim Museum in New York, and Louis Vuitton Foundation in Paris. These all factor in a bit of natural light spilling into their galleries so that the viewing of artwork can be experienced in a holistic more exciting manner [2].

The approach to block natural light and rely on artificial lights only can lead to undesirable results. One of which is the impression the visitor will develop about the heritage place under the provided lighting conditions. In other words, if a heritage museum/gallery is visited when it is lit by artificial lights only (i.e. daylight is blocked), this will indeed provide a very different experience than the real historical one when it was lit mainly by natural lighting. One shouldn't overlook that the architecture of the heritage building is in fact the most valuable piece in the whole heritage museum/gallery. Its high value comes mainly from its unique courtyard form and how its indoor spaces were configured to enclose the central courtyard and take advantage of the natural light captured by it. Hence, the design of the heritage building and its lighting should facilitate for the visitor to obtain a true impression of how people in the past lived in these houses and experienced its spaces under natural lighting conditions. Another important issue is related to the way visitors perceive the true colour of the artifacts under a certain provided lighting. Colour rendering index or CRI, devised by the International Commission on Illumination which as CIE R_a, is a scale from 0 to 100 (100 is the best) that assesses the effectiveness of light sources in making the objects appear true or closer to true to their real colour. Numerically, the highest possible CIE R_a value is 100 and would only be given to a source identical to standardized daylight.

This chapter describes the challenges in adapting these heritage buildings as museums or galleries from the daylighting point of view and explains the adopted process. The focus is to investigate how differently configured exhibit spaces affect the luminous environment with regard to their artifacts' safety and visual comfort and what are the possible solutions to improve their performance while in the meantime achieves more pleasurable and enjoyable visitor's experience.

2 Daylighting as Formgiver of Architecture

Masters of architecture like have always been fascinated by natural light. The architecture of Louis I.Kahn, Alvar Aalto, Le Corbusier, Jorn Utzon, and Johannes Exner showed to what extent natural light was considered an essential form giver of architecture. Some of Louis I. Kahn's quotes were as follows:

- "No space, architecturally speaking, is a space unless it has natural light".
- "Every architectural form, every building or group of buildings, regardless of the practical purpose or the expressive need that formed it is a visible form built from differences of light qualities, ... Within our perception of these patterns of light, our distance sense, our appreciation of the qualities of a wider space would completely disappear".
- "We were born of light. The seasons are felt through light. We only know the world as it is evoked by light, and from this comes the thought that material Is spent light. To me natural light Is the only light, because it has mood – it provides a ground of common agreement for man – it puts us In touch with the eternal. Natural light Is the only light that makes architecture".

Alvar Aalto knew exactly how to treasure and maximize the use of daylight. He understood how to maximize the use of daylight and distinguish between the lighting requirements for the changing seasons of a year. The spaces designed by him provided carefully and properly distributed amounts of light and revealed at the same time the beauty of the interiors. To achieve his goals, he accurately chose the suitable architectural language and elements including a variety of skylights, clerestories, screen-glazed openings, vaults, and lighting scoops. He also used baffles, polished brass, louvers, and white-painted surfaces. All these elements were always an integral part of his design idea and structure. His carefully designed architecture and the accurate selection of elements helped to denote movement from place to place, to crown or accent spaces, and to evenly spread the most precious warm, diffused light.

3 Architecture of Heritage Buildings

Many of these heritage buildings used to be houses for rulers or rich merchants. Others included schools and souqs (traditional markets). Some of the highly sensitive holdings in these heritage buildings are historical documents such as letters between rulers, contracts, deeds, or itineraries describing the pearl industry. Therefore, strict action is needed to ensure the conservation and protection of these vital documents. The heritage building usually takes an introvert courtyard form configuration, and the most common shape of the courtyard is the rectangle (Fig. 3). The courtyard is the main source of natural light and air, and most windows in the building are opened to it. Fewer windows are opened to public alleys, but these are limited to guests' reception rooms that do not necessitate high level of privacy. The rooms in these buildings usually take the shape of a rectangle for the large rooms and a square or semi-square for the small ones, with a width of around 3 m and a length that ranges from around 3 m for the small rooms to reach to more than 8 m for the large ones. The rooms are connected to each other and to the courtyard by a beautiful-arcaded verandah that encloses the courtyard from all or most sides. The verandah with its arcade helps provide substantial shading on the room windows.

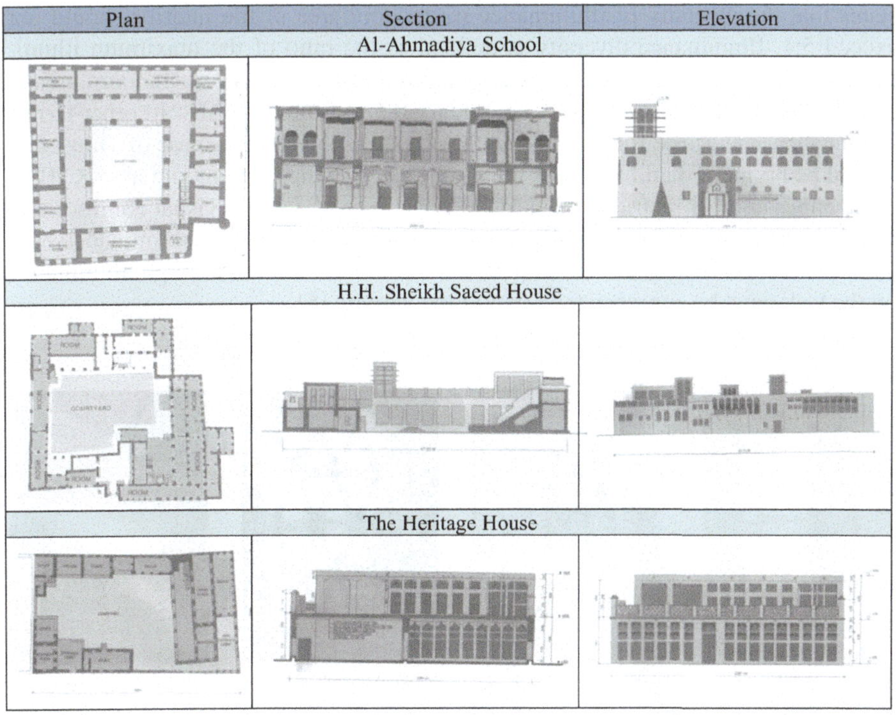

Fig. 3 Architectural drawings of selected heritage buildings' cases in Dubai [5]

The heritage openings (Fig. 4) can be classified into four categories: Dreesheh (double panel window), Mesbah (lantern window), Barjeel (wind tower), and Masgat (air puller). Dreesheh in the local dialect means simply a window. It is basically the main type of the windows and the largest in the heritage buildings and is used for providing daylighting, ventilation, and views (Fig. 5). It is made of local wood, measures about 110 × 70 cm, and is covered by iron rails for security. It has wooden shutters on the interior that act as control elements for fully/partially opening/closing the window. It can be found on the exterior of the house or the interior, looking into the courtyard. Three different components of the traditional windows were used to maintain privacy and security and control of daylight levels; these were gypsum ornaments, iron bars, and wooden shutters. Glass was introduced to the heritage windows at a later time when maritime trading rose up. The windows in the current heritage museums are usually constructed with double panel glass. Since Dreesheh is the main opening for providing daylighting to exhibit rooms in the current heritage museum/gallery buildings, the research presented here focused on it.

4 Theoretical Background

The code for lighting of the Chartered Institution of Building Services Engineers [6] states that the diversity of illuminance in the core area of the interior should not exceed 5:1. Illuminance diversity is defined as the ratio of the maximum illuminance to the minimum illuminance at any point in the core area. The core area is area of the working plane having a boundary 0.5 m from the walls. In other terms, illuminance at any task, for example, at the object displayed in the case of a museum, should be no more than five times the illuminance of the surrounding area. CIBSE also defines another metric to be used with diversity to evaluate visual comfort, that is, the illuminance uniformity. It is defined as minimum to average illuminance over the task area, and it should not fall below 0.7, provided that the average illuminance on the task must be appropriate to that of the activity [6].

Dreesheh (double panel window) | Barjeel (wind tower) | Mesbah (lantern window) | Masgat (air puller)

Fig. 4 Four categories of the heritage building openings

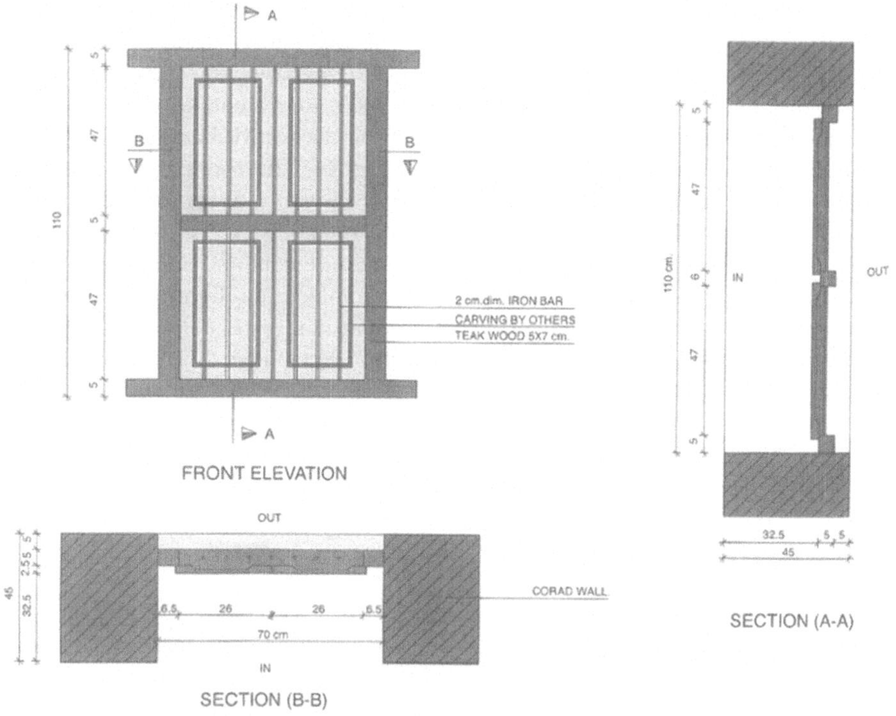

Fig. 5 Architectural drawings and dimensions of the traditional Dreesheh window in Dubai (Dubai Municipality, 2000) [5]

The luminance distribution in the field of view is another important metric that can be used to evaluate to what extent visibility at the task is good or bad. A good task visibility depends on easy adaptation level of the eyes at the scene points; and these are all controlled by the luminance distribution in the field of view [6]. CIBSE [6] states that a well-balanced adaptation luminance is needed to increase visual acuity (sharpness of vision), contrast sensitivity (discrimination of small relative luminance differences), and efficiency of ocular functions (such as accommodation, convergence, pupillary contraction, eye movements, etc.). The human visual system can comfortably perceive only a narrow range of brightness at one time as several issues should be avoided: (A) luminances that are too high, which may give rise to glare; (B) luminance contrasts that are too high, which will cause fatigue because of constant readaptation of the eyes; and (C) luminance and luminance contrasts that are too low, which may result in a dull and non-stimulating working environment. Contrasting luminances should not exceed the following recommendations:

- 3:1, task-to-immediate surround [6]
- 10:1, task-to-general background [6]
- 20:1, between the light source and the surrounding surfaces [7]

The author with his research team has been investigating daylight performance in heritage museums for quite some time. One of their earlier studies [8] relied on the daylight metrics, *Illuminance level (Lux), Light Exposure (Lux.Hrs), Light Distribution and Uniformity, and Luminance Levels,* to evaluate a relatively small exhibition room (2.73 m × 3.65 m), located in the Sheikh Mohammed Centre for Cultural Understanding (SMCCU) of Dubai (Fig. 6). The room has a Dreesheh window on the east sidewall (a total of window-to-wall-ratio or WWR = 12%), looking upon the public alley. The window has no shading except for the operable window shutters. The study depended on field measurement and computer simulation using Desktop Radiance 2.0 Beta (DR) to simulate daylighting at 9:00 am, 12:00 pm, and 3:00 pm on the solar summer and winter solstices (June 21 and December) and spring and fall equinoxes (March 21/September 21). The findings of the study can be outlined as follows:

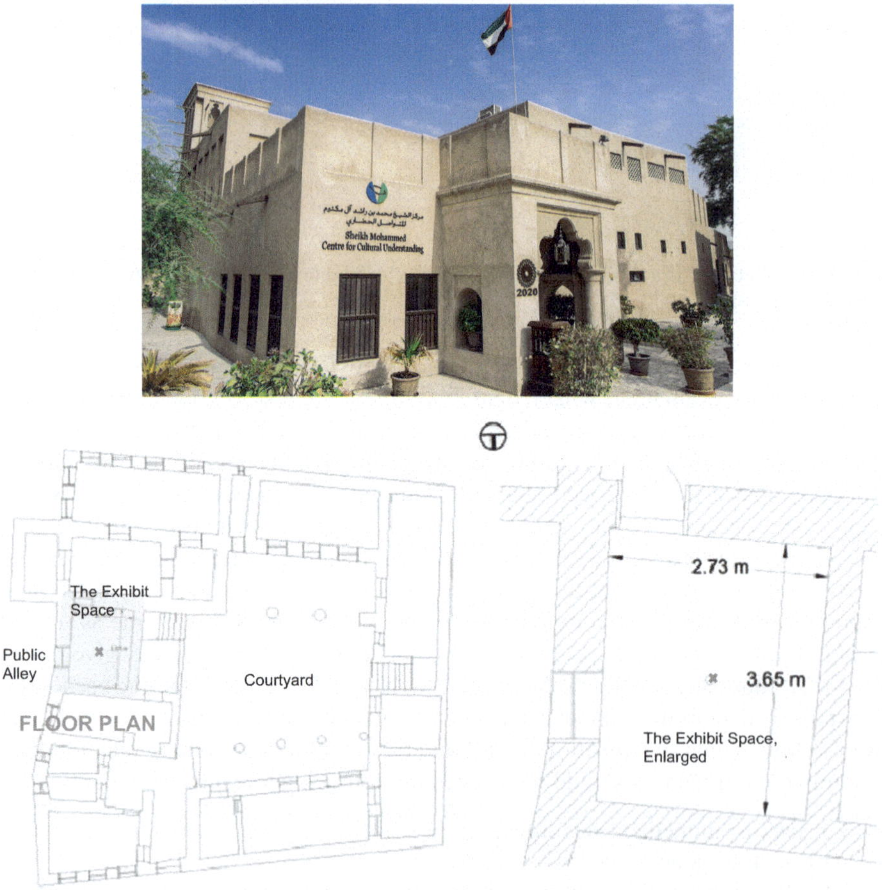

Fig. 6 Photograph and a floorplan of the SMCCU case and the selected exhibition room

1. The difference in illuminance results between the actual measurements and the simulated values was around ±4%. These values were seen as reasonably sufficient to validate the calculations of the simulation engine in the investigated environment.
2. The illuminance values were higher than the recommendations for a museum or gallery use, which can lead to increased light exposure and serious risks of artifacts' deterioration.
3. The results for diversity in the exhibit room showed values of 7:1 for most of the tested hours and 13:1 as the highest recorded diversity recorded on December 21 at 9:00 am. The results showed also that all the recorded values of uniformity fell below the minimum recommended value (0.7) in spite of the average illuminance on the horizontal displays fell within the acceptable level of 300 lux. These high variations in illuminance levels can lead to visual discomfort.
4. The simulated value for contrasting luminances on the window and the surrounding surfaces on June 21 at 09:00 am was 16:1; which is acceptable. This date and time were judged as extreme time since the tested room takes the East orientation for its windows.
5. It was recommended to add a filtering element that helps to uniformly distribute the admitted daylight, most preferably adding shade plants or trees on the outside. Such a solution can help to mitigate the high variations of light in the indoor space and the excessive brightness of the sky seen through the window. It is also an architecturally desirable and safe solution since it will not harm or ruin the architectural character of the heritage building like perhaps changing the design of the façade or window design. The contrasting luminances can be improved further by using surrounding surfaces with higher luminance values around the window. Planning the indoor display furniture so that the window location is off the line of sight was also recommended.

Daylight performance needs to be considered over time. Climate-based daylight metrics (CBDM) like daylight autonomy (DA) and useful daylight illuminance (UDI) are far more useful in daylight calculations than simple metrics such as daylight factor (DF) or illuminance because daylight illumination levels are dynamic and the effect of local climate conditions and daily and seasonal balance of daylight, provided from direct sunlight versus the sky and clouds, on daylight availability is critical. DA calculates the percentage of the occupied hours when the minimum illuminance level is delivered by daylight only for a point in the room. UDI divides the year into three bins based on lower and upper thresholds of 100 lux and 2000 lux [9–10]. The upper bin (UDI > 2000 lux) represents the hours when there is an oversupply of daylight that can lead to visual or thermal discomfort, the lower bin (UDI < 100 lux) represents the hours when there is a lack of sufficient daylight, and the intermediate bin (UDI100–2000 lux) represents the useful daylight. Daylight safety (DS) is another metric derived from DA, which was developed before by the author of this chapter [11]. It calculates the percentage of the occupied hours when the delivered daylight does not exceed a maximum illuminance level, above which the risk of artifacts' deterioration could exist. When DS is applied for a whole room, sDS term is used. sDS calculates the percentage of the room area when the delivered daylight does not exceed a maximum illuminance level based on a certain number of exposure hours per year (e.g. 500 Hr/year), above which the risk of artifacts' deterioration could exist. The volumetric aspect ratio of the courtyard (CVAR) was introduced by the author in a previous study [11] as an effective parameter since it expresses the full relationship between space volume and lighting penetration to the indoors. It was defined as CVAR = $(W.L)/H^2$, where W is the courtyard width, L is the courtyard length, and H is the courtyard height.

Excessive light and improper distribution of luminance seen by a person in the field of view causes glare and visual discomfort [12]. Assessing glare is difficult compared to other simple metrics like illuminance since it involves subjective judgement and relies on the position and view direction of the person viewing a scene [13–14]. It also requires to consider the vertical illuminance that falls within the observer's field of view. One should distinguish between the two types of glare, discomfort and disability glare; the first is generally irritating while the latter impairs vision. Discomfort glare can turn into disability glare if the cause of glare is acute. Most of the existing methods in calculating glare predict the probability of experiencing discomfort rather than visual disability [12, 15]. They are based on the ratio of the size, location, and luminance of glare sources in a field of vision when compared to the average luminance not inclusive of the glare source.

Elimination of glare and control of surrounding brightness in museums can serve to reduce the need for additional light on the artifacts. The human eye has greater ability to tolerate glare from the sky, seen from the windows, than to tolerate glare from a comparable artificial lighting situation. That is because a window as a light source is usually much larger than artificial light. To demonstrate this idea, the limiting value for the glare index (GI) in museums is 16, while for daylit spaces, this could translate to a daylight glare index (DGI) of 20 and can be considered as "just acceptable" glare criterion [7]. Thus, previous research found that it is more important to find out how to decrease the DGI than to calculate its exact value [4]. The major controlling factor is the glare source luminance and its contrast with the surrounding surfaces; it should not exceed 20:1 (Baker et al. 1993). It was also suggested to reduce the luminance by a factor of 4 in order to obtain a significant decrease of the DGI value (−3 units) [4]. In a previous study on classrooms in the UAE, Al-Sallal and Al-Rais [16] demonstrated how the use of local shade trees helped reducing the windows' glare luminance from >2000 cd/m^2 to 350 cd/m^2 and the luminance contrast with the surrounding surfaces from 1:50 to 1:8.75. Movable shading systems like movable louvers and light shelves can also be effective.

In a simulation study, Jakubiec and Reinhart [17] compared between five glare metrics under 144 clear sky conditions in three spaces to investigate their ability to predict the occurrence of discomfort glare and hence support the design of comfortable spaces. The analyzed metrics were daylight glare probability (DGP), daylight glare index (DGI), unified glare rating (UGR), visual comfort probability (VCP), and CIE glare index (CGI). The results showed that daylight glare probability (DGP) was the most robust metric and the least to produce misleading predictions. The DGP glare metric was defined by the following formula:

$$\mathrm{DGP} = 5.87\times10^{-5}E_{\upsilon} + 9.18\times10^{-5}\log\left(1+\sum_i \frac{L_{s,i}^2\omega_{s,i}}{E_{\upsilon}^{1.87}P_i^2}\right)$$

where:

L_s is the luminance of the glare source (cd/m^2)
ω_s is the solid angle of the glare source (sr)
P is the weight factor based on positioin in a viewing hemisphere, the position index
E_{υ} is the total vertical eye illuminance (lux)

In a study by Berardi and Wang [18], the authors used several glare metrics (i.e. DGP, DGI, UGR, VCP, and CGI) to evaluate glare caused by daylighting in an atrium-type high-performance house. They considered the DGP as the main metric in their analysis since its formula accounts for the vertical eye illuminance (Ev) in its input, and thus it can predict discomfort even without visual contrast in exceedingly bright scenes. More importantly, the results of the DGP matched the visual perception in the house and showed indications to resolve some of the problems found in the other metrics [15, 17, 18].

5 Methods

The research work in this chapter was conducted in two stages. The first stage was simple and depended on limited fund. The focus was to define the research problem, to identify its limits (scope and variables), and to collect needed information that will feed the simulation runs in the next stage. Stage 2 was more comprehensive and involved wider and deeper investigations. The focus was to reach to a full understanding of the problem' variables, to examine the variables under different design and climate-based and sky models' conditions, and finally to suggest effective solutions. All these necessitated to rely on multiple daylight metrics and methods.

Stage 1 – Historical and technical information about several cases of heritage buildings were collected from several resources including published materials, architectural drawings, and historical documents. Site visits were made to obtain photographic information and experience the spatial configurations and the availability and quality of the admitted daylight in the visited cases. The photographs focused on the spaces and their arrangements, the windows and other architectural elements that affect daylight performance such as the arcaded verandah, and the exhibited artifacts. The photographical survey of the artifacts was accompanied with meetings with the museum curators to obtain information about each one with regard to its kind of material, its age and value, and its sensitivity to light exposure. Figure 7 shows typical artifacts in the heritage museums. Based on the information collected from the site surveys and technical information collected from the literature, the research classified the artifacts into groups according to their sensitivity of light exposure (Table 1). These are as follows:

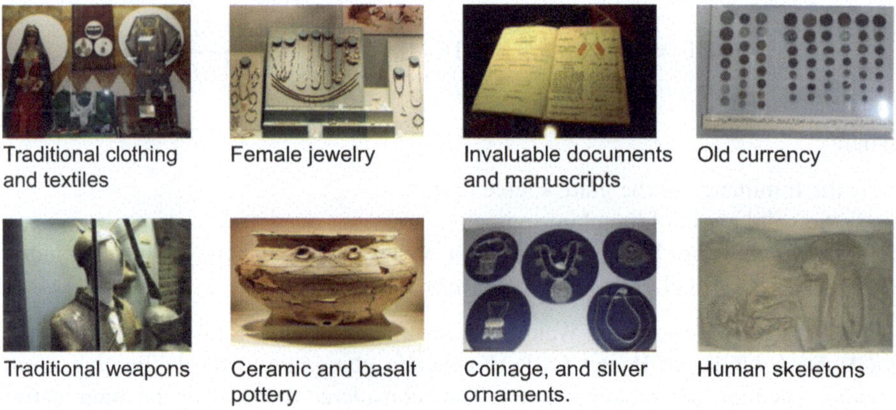

Fig. 7 Typical displays in the heritage UAE museums

Table 1 Recommended illuminance (Lux) and cumulative light exposure (Lux.Hr/year) limits, as adopted from IESNA [19–21]

	IESNA		CCAHA[c]
Type of material	Max illuminance (Lux)	Light exposure (Lux.Hr/year)	Light exposure (Lux.Hr/year)
High sensitivity to light Books, botanical, specimens, costumes, cotton, drawings, dyed, leather, feathers, fugitive dyes, fur, gouache, insects, manuscripts, miniatures, paintings in distemper media, paper, prints, silk, skins, some minerals, some photographs, stamps, tapestries, textiles, wallpapers, water colours, wool, and writing inks	50	50,000[a] 150,000[b]	50,000
Low sensitivity to light Bone, horn, ivory, lacquer, leather, oil paintings, some plastics, some photographs, tempura painting, textiles with stable dyes, and wood finishes	200	480,000[a] 600,000[b]	100,000
Less light sensitive Paper-based artifacts	–	–	300,000
No light sensitive Ceramics, enamel, glass, jewels, metal, most minerals, stone, and wood	1000	–	–

[a]Adopted from IESNA [19]
[b]Adopted from IESNA [20]
[c]Adopted from CCAHA [21] CCAHA specializes in the treatment of art and historic artifacts on paper

High sensitivity:

- Significantly valuable manuscripts and documents such as treaties, letters, maps, decrees, and agreements that were vital in shaping the country's history
- Old currency notes, stamps, and postal stationery

Low sensitivity:

- Traditional clothing and textiles.
- Female jewelry with a large selection of bead necklaces agate, bronze, and soft stone. Some of this jewelry is considered prehistoric.
- Human skeletons and bones that were excavated from graves that date back to the third millennium BC.

No sensitivity:

- Some of the earliest coinage, silver ornaments, and costume accessories.
- Traditional weapons such as rifles and guns. Additionally, bronze daggers and arrowheads that date back to the first, second, or third millennium BC.
- Ceramic and basalt pottery, of which some are considered prehistoric, consisting mostly of vessels and plates.

Stage 2 – Stage 2 aimed in achieving a more accurate and in-depth investigation of the factors that affect daylighting performance and its effect on artifacts exposure to light versus occupants' visual comfort. This required to conduct additional site visits of the heritage buildings to obtain photographic and building information. The site visits helped also to take new on-site measurements to find characteristics of the interior surfaces needed for the simulation process and produce actual reference readings of lighting levels useful in simulation results' checking and correlation.

The need to perform more accurate and in-depth investigation required to depend on daylight autonomy metrics (DA) and use advanced methods of calculations/simulation. Stage 2 depended on Diva-for-Rhino program to perform the needed CABD-detailed simulations. The tested metrics included Lux levels, DA, UDI, light exposure Lux.Hr/year, DS, sDS, DGP, and contrasting luminances. DIVA-for-Rhino is a highly optimized daylighting and energy modelling plug-in for Rhinoceros, a NURBS modelling software [22]. The plug-in was initially developed at the Graduate School of Design at Harvard University and is distributed by Solemma LLC. It uses RADIANCE and DAYSIM as its basic daylight simulation engines. Both RADIANCE and DAYSIM have been validated in many previous studies [23–30].

The simulations used a large number of sensors (160 sensors) in order to ensure covering the entire space and areas that could be potentially shadowed by the solid portions of the external walls or the structural columns of the verandahs. These sensors were equally distributed on a horizontal grid, placed at a height of 0.81 m to match the common heights of display. To produce useful analysis, the room was divided along its depth into three zones: (1) far from windows zone (FZ), (2) middle zone (MZ), and (3) near windows zone (NZ). To give a full picture of how daylight will perform under different design configurations, many cases based on combinations of WWR and verandah projections were modelled and tested.

Stage 2 presented also a comparative analysis between the SMCUU case, investigated earlier by the research, and an additional case study, a newly added case study, which is the Heritage House Museum (HHM) in Al-Ras District of Dubai (Fig. 8). Like the case study SMCCU, the HHM was also seen as a good representative of the traditional buildings in Dubai and as a good example of the coastal architecture of the Arab Gulf in general. This heritage museum is a two-story building with a courtyard configuration. The HHM includes several exhibit spaces in the ground and first floors. The exhibit space selected for the daylight investigation in this case study measures 8 m length × 3 m width × 4 m height. It has 3 Dreesheh windows on the southwest sidewall with WWR of 12%, looking upon the courtyard. This room is shaded with deep arcade/verandah (3m) in addition to operable window shutters. The CVAR values of the SMCCU and HHM cases are 1.19 and 3.52, respectively.

Stage 2 evaluated glare risks and visual discomfort inside different design variations of the exhibit room. The lack of established methodologies for determining which times and scenes to use, among a multitude of potential daylight glare situations, confronted the researcher with a real challenge. The worst case was imagined when a person is observing a displayed artifact while high luminance windows appear in the background. It represents a very critical situation with regard to the potential of glare risks. Such a situation can be identified during the times when the direct sunlight could penetrate into the exhibition room at low solar altitude angles (e.g. afternoon time for a southwest-oriented room). This scene can also reveal

Fig. 8 Photograph and a floorplan of the HHM case and the selected exhibition room

issues during other times when the sun shines brightly on the courtyard wall opposite to the room (the morning time as predicted). To determine the times of critical glare, the study relied on two types of Diva simulations: the annual DGP analysis and the luminance-based visualizations. Figure 9 shows how the camera was positioned for the DGP analysis in the tested room (i.e. a room with size measurements and luminous characteristics similar to the exhibit room in the HHM case).

6 Results and Discussion

6.1 Daylighting Autonomy (DA)

The visual results in Figs. 10 and 11 use a colour scale (on the right) that represents the DA percentage for each module of the evaluation grid in the two simulated exhibition rooms of the SMCCU and HMM cases. For each room, the thresholds of the DA calculations were set to 50 Lux, 200 Lux, and 1000 Lux to represent high sensitive, low sensitive, and no sensitive materials, respectively; based on the IESNA lighting handbook recommendations [19, 20]. The spatial DA value was also calculated, which represents the percentage of the space in which the illuminance levels exceeded the Lux threshold. The results can be outlined as follows:

- The SMCCU case (Fig. 10):

  The $DA_{(50\text{ Lux})}$ simulation revealed that the room of the SMCCU case is not safe for high sensitive materials; almost all the room area will experience illuminance levels are greater than 50 Lux threshold, for 50% of the occupied hours.
  The $DA_{(200\text{ Lux})}$ simulation revealed that the room of the SMCCU case is not safe for low sensitive materials; almost all the room area will experience illu-

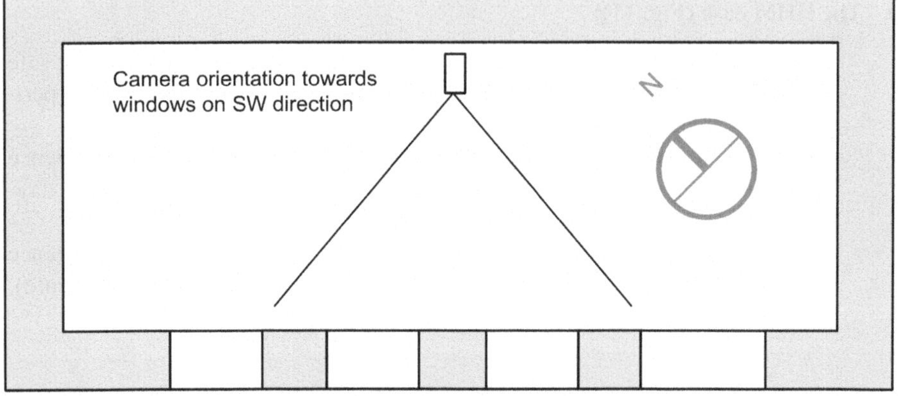

Fig. 9 Camera positioned in the room for the DGP analysis

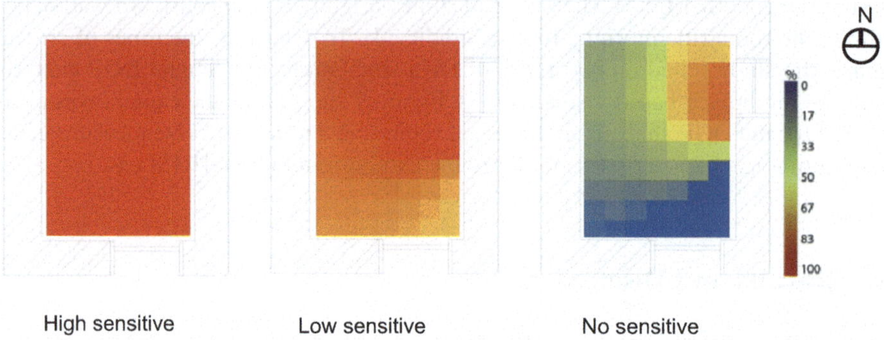

Fig. 10 Daylighting autonomy (DA) results for the exhibit space in the SMCCU case, shown for the high, low, and no sensitivity categories

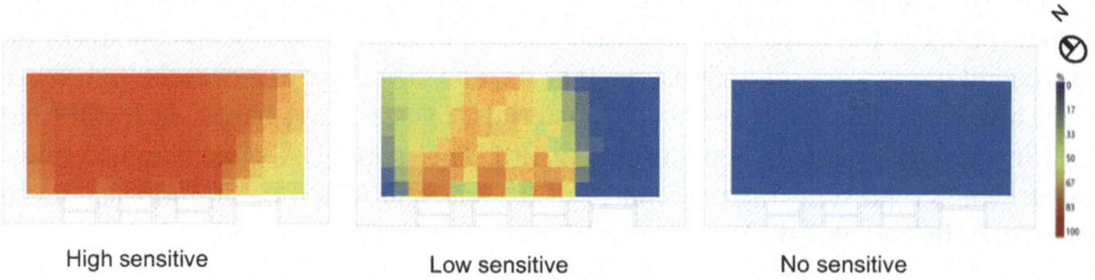

Fig. 11 Daylighting autonomy (DA) results for the exhibit space in the HMM case, shown for the high, low, and no sensitivity categories

minance levels are greater than 200 Lux threshold, for 50% of the occupied hours.
The $DA_{(1000\ Lux)}$ simulation revealed that 23% of the room area will experience illuminance levels ≥ 1000 Lux, for 50% of the occupied hours (77% daylight safety).

- The HHM case (Fig. 11):

The $DA_{(50\ Lux)}$ simulation revealed that the room of the HHM case is not safe for high sensitive materials, because (almost) all of the room area will experience illuminance levels ≥ 50 Lux, for 50% of the occupied hours.
The $DA_{(200\ Lux)}$ simulation revealed that 52% of the room area will experience illuminance levels ≥ 200 Lux, for 50% of the occupied hours (48% daylight safe).
The $DA_{(1000\ Lux)}$ simulation revealed that 0% of the room area will experience illuminance levels ≥ 1000 Lux, for 50% of the occupied hours (77% is safe).

7 UDI

The common method when performing the useful daylight illuminance (UDI) calculations considers the values that are less than 100 Lux as "too little" light, the values that are between 100-2000 Lux as "useful daylight", and the values that are more than 2000 Lux as "too much" light [19, 20]. The UDI results are expressed as the percentage of the room area that detected illuminance levels for at least 50% of the time. Since in museums, conservation of valuable artifacts is at the uppermost priority, illuminance values that are less than 100 Lux (but not less than 50 Lux) can also be useful or even recommended. The UDI results can be outlined as follows:

- SMCCU case:

  < 100 Lux is 0%
  between 100-2000 Lux is 94%
  > 2000 Lux is 4%

- HHM case:

  < 100 Lux is 21%
  between 100-2000 Lux is 79%
  > 2000 Lux is 0%

These results indicate that the SMCCU room is not as safe for the exhibited artifacts as the SMCCU room, especially for the high sensitivity category. With regard to sunlight exposure, the low value of the $UDI_{(>2000\,Lux)}$ (4%) in SMCCU room indicates a reasonably high level of artifacts' safety and also visual comfort but the HHM room has lower values, which indicates a higher level of artifacts' safety and visual comfort than the SMCCU. Figures 12 and 13 show the UDI visual results for the exhibition spaces of the SMCCU and HHM cases, respectively. These visual results can help museum curators to choose the best locations for the artifacts' safety and visual comfort.

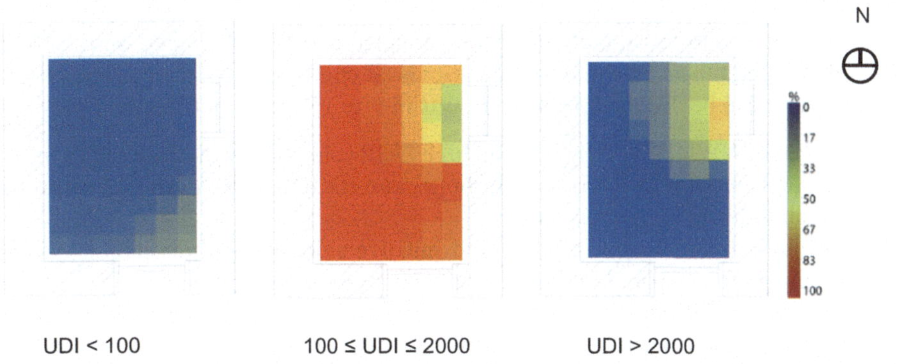

Fig. 12 Useful daylight illuminance (UDI) results for the exhibit space in the SMCCU case

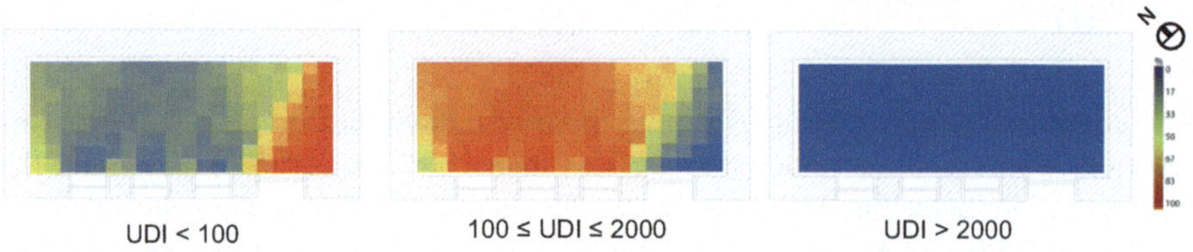

Fig. 13 Useful daylight illuminance (UDI) results for the exhibit space of the HHM case

Figure 14 shows the $UDI_{<100\ Lux}$ and $UDI_{100\text{-}2000\ lux}$ (averaged values) as a function of distance from window. The use of verandahs improves the daylighting distribution throughout the room, with minimal impact on the UDI levels at the back zone. The increase of the verandah's size (from 1 to 3m) has more impact on the improvement of daylight distribution and quality than the reduction of the WWR (from 18% to 5%). The results show that the lower WWR with the larger verandah depth improves the $UDI_{<100\ Lux}$ performances (Fig. 14). Thus, the best option for the HS and LS categories is to use WWR-5% with a verandah that has a depth of 2 m or 3 m. The performance improves in the MZ and FZ zones of the room (at a distance of 1.5-3m from windows). Figure 14 also supports the adequacy of these same two cases for the LS category, as suggested by their low results of $UDI_{100\text{–}2000\ lux}$, especially at a distance of 1.5–3 m from windows.

8 Annual DGP Distribution Analysis

The two cases that were simulated for the DGP distribution in the exhibit room of the HHM case were as follows:

- The room with 18% WWR without verandah (WWR18%-V0)
- The room with 18% WWR with 3 m verandah (WWR18%-V3)
- The room with 5% WWR without verandah (WWR5%-V0)
- The room with 5% WWR with 3 m verandah (WWR5%-V3)

The aim was to test the effect of the verandah and WWR change on the DGP distribution and their potential to improve visual comfort performance. The observations that can be deduced from the results in Fig. 15 can be outlined as follows:

1. The case (WWR18%-V0) results in high DGP values that can cause either intolerable glare (DGP $\geq$ 0.45) or disturbing glare (DGP $\geq$ 0.40) during most of the operation hours of the museum.
2. The days of the fall and spring months involve more hours of intolerable or disturbing glare compared to those in the winter and summer months.
3. The lower DGP values of the winter (especially during December) can be rationalized by the capacity of the courtyard design configuration to shade the low

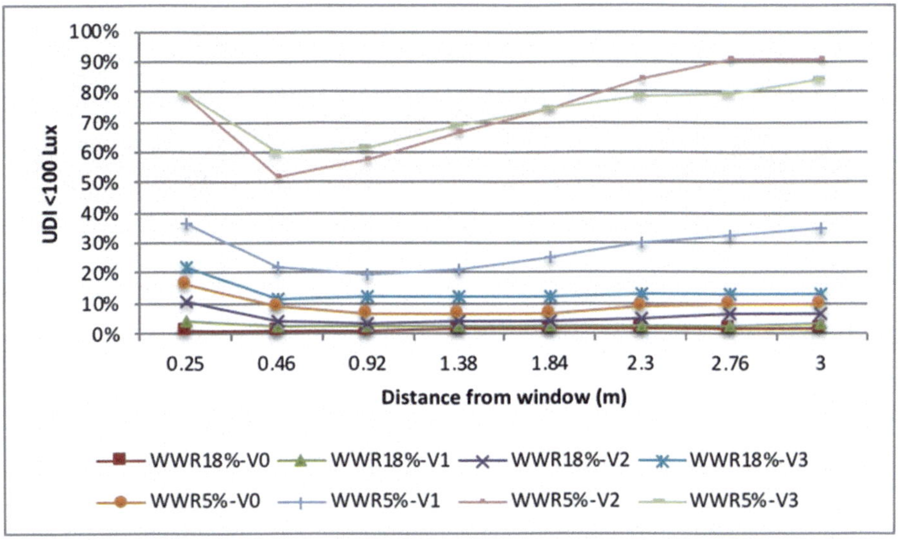

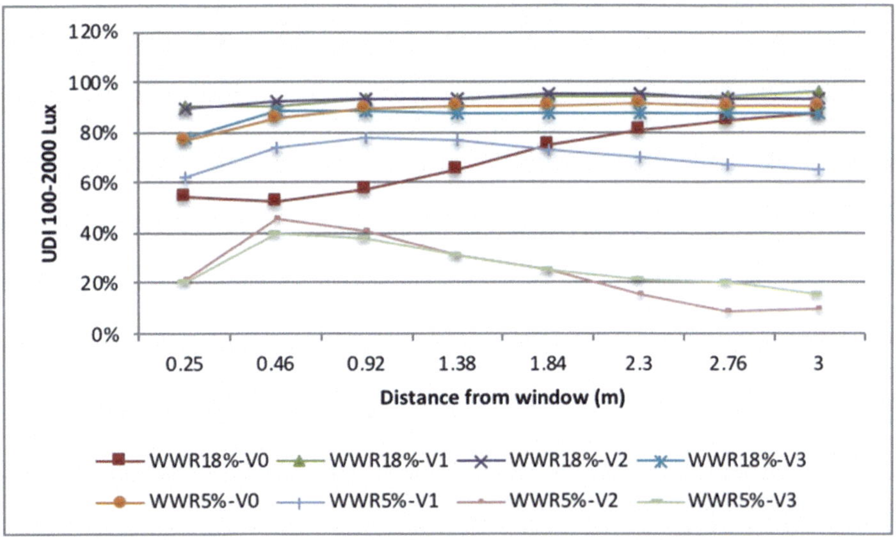

Fig. 14 $UDI_{<100\ Lux}$ and $UDI_{100–2000\ lux}$ as a function of distance from window for WWR = 5% and 18% with different sizes (0, 1, 2, and 3 meters) of verandahs on SW orientation

altitude sun of the winter especially during the early morning hours and late afternoons.

4. The lower DGP values of the summer can be rationalized by the high position of the sun in Dubai's summer. That is because during the summer solstice of Dubai, the sun is nearly perpendicular to the building roof due to the fact that Dubai (25.20° latitude) is very near to the tropic of cancer (23.5°) and hence the

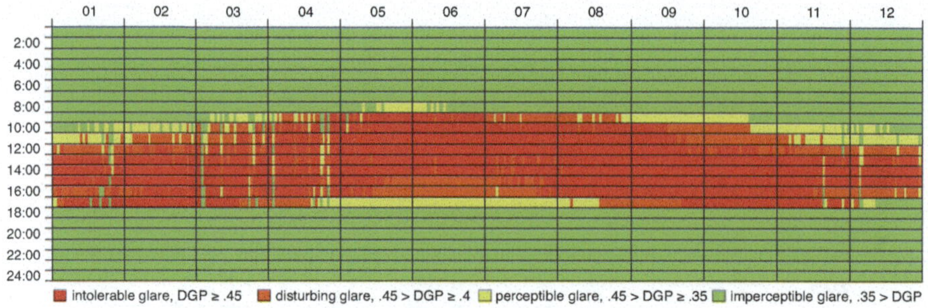

Fig. 15 WWR-18% without verandah – hourly DGP values through the year of the SW orientation

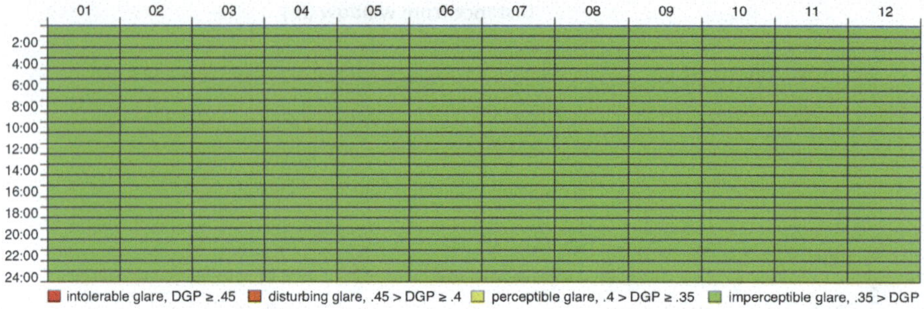

Fig. 16 WWR-18% with 3m-verandah – hourly DGP values through the year of the SW orientation

potential of direct sunlight penetrating through the windows and the risk of glare is minimized.

When the DGP distribution analysis was applied to the cases WWR18%-V3, WWR5%-V0, and WWR5%-V3, the results showed great improvements. As can be seen in Fig. 16 for the case WWR-18% with verandah, there are zero instances of glare levels above DGP of 35% (no disturbing or intolerable glare issues). Hence, reducing the WWR of the windows and/or adding the traditional arcaded verandah on the courtyard side to shade the windows can help greatly overcome any possibility of disturbing or uncomfortable glare.

9 DGP Cumulative Numbers of Hours

The cumulative number of hours when the DGP is above a certain benchmark (imperceptible, perceptible, disturbing, and intolerable) was also simulated and the results are shown in Fig. 17. For the 18% WWR without verandah, the number of annual hours (and % of total hours) is as follows:

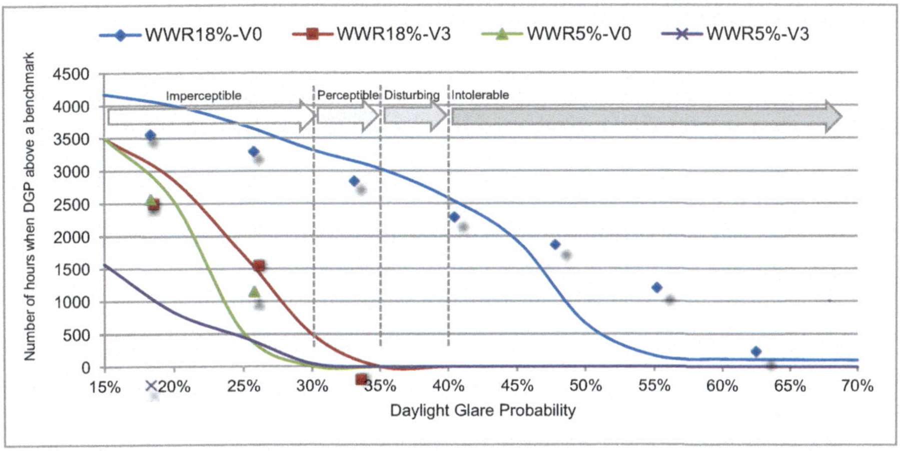

Fig. 17 Cumulative numbers of hours when the DGP is above a certain benchmark, divided into four ranges

Table 2 Improvement of DGP values after adding 3m-verandah and reducing the WWR to 5%

	Hr (% of operating time)		
Case	Perceptible, DGP > 30	Disturbing, DGP > 35	Intolerable, DGP > 40
WWR18%, without verandah	3321 (37.91%)	3032 (34.61%)	2583 (29.49%)
WWR-18%, with 3m-verandah	497 (5.67%)	0	0
WWR-5% without verandah	10 (0.11%)	0	0
WWR-5%, with 3m-verandah	49 (0.56%)	0	0

- 2583 (≈30%) when the DGP is above 40% (i.e. intolerable glare)
- 3032 (≈35%) when the DGP is above 35% (i.e. disturbing or intolerable glare)
- 3321 (≈38%) when the DGP is above 30% (i.e. perceptible, disturbing or intolerable glare)

When the 3m-verandah is added to the case 18%-WWR, the simulation results showed huge improvements in DGP reductions: from 30% to 0% (i.e. no danger of intolerable glare), from 35% to 0% (i.e. no danger of disturbing or intolerable glare), and from 38% to 6% (i.e. only limited perceptible glare). Further improvements were achieved by reducing the WWR to 5% either with or without verandah (Table 2).

10 Luminance-Based Visualizations

Three times were judged to represent critical conditions of potential glare problems: 21 September at 9:00, 21 December at 12:00, and 21 March at 15:00. To quantify the risk of glare, point-in-time DGP simulations for four cases were performed:

1. Base case: WWR-18% without verandah
2. Improvement Case 1: WWR-18% with 3m-verandah
3. Improvement Case 2: WWR-5% without verandah
4. Improvement Case 3: WWR-5% with 3m-verandah

The aim was to quantify the capacity of the verandah element and the smaller-sized windows (smaller WWR) to control the penetrated direct sunlight and minimize possibilities of glare.

The visualization results of the base case in Fig. 18 showed a DGP range of 43-45% (i.e. disturbing to intolerable glare). One can also notice the penetrations of the direct sunlight into the room (on the display stands) on December 21 at 12:00 and on March 21 at 15:00. The simulation results of the case WWR-18% with

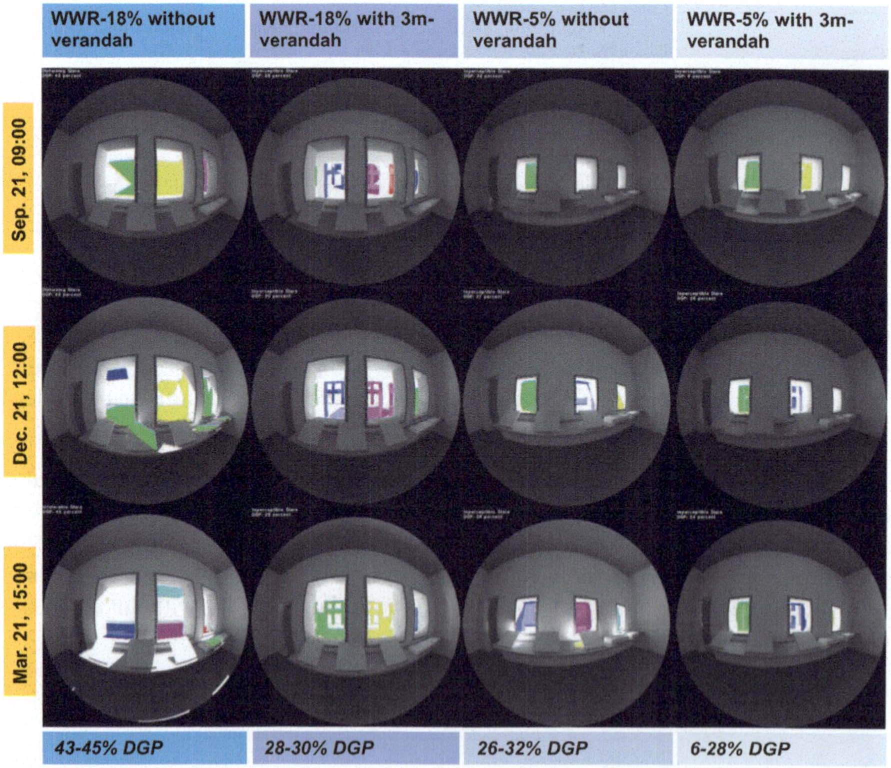

Fig. 18 Point-in-time DGP analysis of the four simulated cases

3m-verandah showed a significant improvement; the DGP range went down from 43–45% (disturbing to intolerable glare) to 28–30% (imperceptible glare). The simulation results of the cases WWR-5% without verandah and WWR-5% with 3m-verandah showed further improvements; the DGP range went down to 26–32% and to 6–28%, respectively (imperceptible glare in both cases). In addition, all improvement cases did not involve any dangers of direct sunlight.

11 Conclusion

The chapter raises the question "if the architectural configuration of the heritage building in the Arab Gulf region was originated around the concept capturing daylight from a central courtyard, why should we block it (instead of controlling it) and replace it with artificial lighting" that could lead to undesirable issues such as:

- Giving the visitor an untrue impression of how people in the past experienced the interior spaces of these buildings due to a completely different luminous environment
- Creating a static and monotonous environment that might operate better in terms of visual efficiency but it could also result in "museum fatigue"

The traditionally thought claim that "artificial lighting is the best solution and one should block daylight to protect artifacts from light exposure" is seen as an oversimplification assumption of the problem, and hence it is challenged here with presented proofs that one can reach better solutions if we control daylighting rather than blocking it – solutions that offer high levels of artifacts' safety – but with better visual comfort and more pleasurable and enjoyable visitor experience. Stage 1 implemented preparatory and data collection work.

The first part of Stage 2 focused on how daylight could affect the artifacts and possible measures to mitigate this effect. The most important findings of stage (2) are as follows:

1. The $DA_{(50\ Lux)}$ and $DA_{(200\ Lux)}$ results showed that the highly shaded exhibit room of the HHM case performed better than the no-shading room of the SMCCU case for the low sensitivity category (48% difference in spatial DA between the two cases); while for the high sensitivity material category, both cases did not perform well.
2. The $DA_{(1000\ Lux)}$ simulations revealed that 23% of the SMCCU room area (77% daylight safety) while 0% of the HHM room area (100% safe) will experience illuminance levels ≥ 1000 Lux, for 50% of the occupied hours.
3. The UDI results indicated that the highly shaded room of the HHM case was safer than that of the SMCCU safe for both the high sensitivity and low sensitivity categories ($UDI_{<100Lux}$ of 21% versus $UDI_{<100Lux}$ of 0%).
4. The zero value of the $UDI_{(>2000\ Lux)}$ in the highly shaded room of the HHM case also indicated high level of safety with regard to sunlight exposure.

5. For the low sensitivity category, the results indicate that the exhibit space in the SMCCU is unsafe from the window till at least the centre of the room, while the exhibit space in the HHM case is safe from at least the centre of the room and the back wall.
6. The results generally suggest to keep the artifacts with high sensitivity to light exposure in separate exhibit rooms as this will facilitate controlling dangers of light exposure. Daylight might be blocked totally for this category only. To control direct sunlight, one should keep the shutters of the windows closed during the direct sun hours, especially during the morning time. The use of veiling structures such as fixed shading devices, movable shading devices, arcaded verandahs, or landscape elements on the interior/exterior of the openings must be used to control daylight access to the space. Using local shade trees is a more desirable option sometimes since they will not affect the architectural character in any manner and will add environmental benefits as well as aesthetics to the area.

The second part of Stage 2 focused on potential visual discomfort and glare risks based on annual DGP analysis and luminance-based visualizations. It identified three critical times of potential glare risks in a model room similar to the one in the HHM case with WWR = 18%. The annual DGP analysis revealed high DGP values for the case WWR-18%, which can cause either intolerable glare (DGP $\geq$ 0.45) or disturbing glare (DGP $\geq$ 0.40) during most of the operation hours. After adding the 3m-verandah to the case WWR-18%, the annual DGP results showed great improvements (zero instances of glare levels above DGP of 35%, meaning no disturbing or intolerable glare issues). That was also confirmed by the produced luminance visualizations that showed how the verandah element managed to overcome the problems of direct sunlight and reduce bright window luminance.

Acknowledgement This research is funded by the UAE University under UAEU Program for Advanced Research (UPAR), Grant # G00001939-UPAR (9) 2015.

References

1. Source: Dubai Culture, Government of Dubai (2020), accessed June 26, 2021. https://dubai-culture.gov.ae/en/Live-Our-Heritage/Pages/AlAhmadiyaSchool.aspx
2. Stir World, https://www.stirworld.com/think-opinions-here-comes-the-sun-the-importance-of-daylight-in-galleries-and-museums
3. Falk, J., Koran, J., Dierking, L., & Dreblow, L. "Predicting Visitor Behaviour", Curator, 28(4), 1985, 249–257.
4. Baker, N., Steemers, K. "Daylight Design of Buildings" James & James, 2002.
5. DM. "*Elements of traditional architecture in Dubai*". Dubai Municipality, 2000.
6. CIBSE "CIBSE code for interior lighting". London: Chartered Institution of Building Service Engineers, 1994.
7. Baker, N., Fanchiotti, A., & Steemers, K. "Daylighting in Architecture a European Reference Book" (1st ed.). London: James & James, 1993.

8. Al-Sallal K., Dalmouk M. "Indigenous buildings' use as museums: Evaluation of day-lit spaces with the Dreesheh double panel window". Sustainable Cities and Society. 2011;1(2):116-24.
9. Nabil A, Mardaljevic J. Useful daylight illuminances: A replacement for daylight factors. Energy and buildings. 2006 Jul 1;38(7):905–13.
10. Berardi U, Anaraki HK. The benefits of light shelves over the daylight illuminance in office buildings in Toronto. Indoor and Built Environment. 2018 Feb;27(2):244-62.
11. Al-Sallal KA, AbouElhamd AR, Dalmouk MB. UAE heritage buildings converted into museums: Evaluation of daylighting effectiveness and potential risks on artifacts and visual comfort. Energy and Buildings. 2018 Oct 1;176:333-59.
12. Rea M. Lighting handbook. 9th ed. New York. Illuminating Engineering Society of North America, 1999.
13. Jakubiec JA, Reinhart CF. The 'adaptive zone' - a concept for assessing discomfort glare throughout daylit spaces. Light Res Technol 2012;44:149-70.
14. Wienold J. Dynamic daylight glare evaluation. In Proceedings of Building Simulation 2009 Jul 27 (pp. 944–951).
15. Wienold J, Christoffersen J. Evaluation methods and development of a new glare prediction model for daylight environments with the use of CCD cameras. Energy and buildings. 2006 Jul 1;38(7):743–57.
16. Al-Sallal, K., Al-Rais, L. "Improving natural light in classroom spaces with local trees: Simulation analysis under the desert conditions of the UAE". In Proceedings of Building Simulation 2007 Beijing, China: International Building Performance Simulation Association; September 03–06, pp. 1168–1174, 2007.
17. Jakubiec JA, Reinhart CF. The 'adaptive zone' – A concept for assessing discomfort glare throughout daylit spaces. Lighting Research & Technology. 2011;44(2):149-70.
18. Berardi U, Wang T. Daylighting in an atrium-type high performance house. Building and Environment. 2014 Jun 1;76:92–104.
19. IESNA, "The lighting handbook (10th ed.)". DiLaura, D. L., Houser, K. W., Mistrick, R. G., & Steffy, G. R. (eds.). New York: Illuminating Engineering Society of North America, 2011.
20. IESNA, "Lighting handbook. (9th ed.)". Rea, M. (ed). New York, USA: Illuminating Engineering Society of North America, 1999.
21. CCAHA. "Artifacts, Guidelines for exhibition light levels for photographic materials". Philadelphia, Pennsylvania. CCAHA.org. Retrived from http://www.ccaha.org/uploads/media_items/hbcu-guidelines-for-exhibition-light-levels.original.pdf, 2007.
22. Solemma LLC. DIVA-for-Rhino. Harvard's Graduate School of Design – Sustainable Design. 2018 [20 Feburary 2018]. Available from: http://www.solemma.net/Diva.html.
23. Grynberg A. Validation of RADIANCE. Berkeley, USA: Lawrence Berkeley Laboratory, Berkeley LBID, Applied Science Division, Lighting System Reserch Group; 1989. Contract No.: 1575.
24. Reinhart CF, Walkenhorst O. Validation of dynamic RADIANCE-based daylight simulations for a test office with external blinds. Energy and buildings. 2001;33(7):683-97.
25. Jakubiec JA, Reinhart CF. The 'adaptive zone' – A concept for assessing discomfort glare throughout daylit spaces. Lighting Research & Technology. 2011;44(2):149-70.
26. Reinhart CF, J. W. The Daylighting Dashboard-A simulation based design analysis for daylight spaces. Building and Environment 2011;46(2):386-96.
27. Reinhart C, Breton P-F. Experimental validation of Autodesk® 3ds Max® Design 2009 and DAYSIM 3.0. Leukos. 2009;6(1):7-35.
28. Reinhart CF, Mardaljevic J, Rogers Z. Dynamic daylight performance metrics for sustainable building design. Leukos. 2006;3(1):7-31.
29. Reinhart CF, Andersen M. Development and validation of a Radiance model for a translucent panel. Energy and Buildings. 2006;38(7):890-904.
30. Reinhart CF. Lightswitch 2002: a model for manual and automated control of electric lighting and blinds. Solar energy. 2004;77(1):15-28.

Khaled A. Al-Sallal is a full professor of architecture and a specialist in sustainable design, with many years of experience in architectural education, research, and consultancy. He currently works as a consulting architect and project manager in Concreto in a wide range of projects that includes schools, commercial, office, and residential buildings in the greater Vancouver area, Canada. Professor Al-Sallal has been very active in producing research supported by substantial funds from governmental and private sector organizations. His research focuses on the interface between building technology and design of ecologically responsible environments. He gave talks in numerous prestigious conferences and wrote more than 100 research papers published in international high impact journals and contributed chapters in edited books produced by leading publishers. He worked for many years at UAE University where he was promoted to a full professor of architectural engineering and appointed as the Director of University Research Support. He also served as the Technical Manager of the Daylighting Simulation Laboratory, a state-of-the-art project that he built at UAEU from the ground up. He assumed major roles in supervising and examining doctoral and master's students several universities in the Arab Gulf. He is an active member of International Building Performance Simulation Association (IBPSA) and founder and president of IBPSA-UAE, an active member in the steering committee of the World Renewable Energy Network (WREN) and the World Renewable Energy Congress (WREC), a member of Canadian Green Building Council (CaGBC), a full member of American Society of Heating and Air-Conditioning Engineers (ASHRAE) and has served in the Board of Governors and as the Vice President, Chapter Technology Transfer (CTC) Chair and Research Promotion (RP) Chair of the ASHRAE Falcon Chapter. He holds a B.Sc. of engineering from Ain Shams University, an M.Sc. from Arizona State University, and a Ph.D. from Texas A&M University.

Assessment of Indoor Heat Gain Using Overall Thermal Value (OTTV) in the Rural Houses of Andhra Pradesh, India

J. Vijayalaxmi and Priyanka Jaiswal

1 Introduction

Factors such as building envelop material, air conditioning systems, water heating system, the dynamic equipment and illumination play an important role in performance evaluation of a building. However, the most important factor in terms of energy efficiency is the building envelop [6]. It has a significant role to regulate indoor temperature and to lower the amount of energy used in heating and cooling [4]. Hence, good energy savings can be achieved if the building envelop properties are improved [6].

Globally, thermal insulation standard is used to measure thermal performance of building envelop in cold climate, and overall thermal transfer value (OTTV) building regulations are used in hot climates. This regulation is mostly applicable to mechanically cooled buildings, as the OTTV is used as a measure to design building envelope and control indoor heat gain. OTTV is seldom used to calculate heat gain through less energy-intensive rural buildings, as the envelop design of these buildings is not studied as much as the energy-intensive buildings. Nevertheless, it is important to study the same, as indoor thermal comfort of a large number of socio-economically challenged rural populace is affected due to indoor heat gain through envelop. Building envelops get influenced by the factors like climate, building and user-related characteristics, building services and operation, occupant's activities and behavior, socioeconomic factors, and indoor environmental quality [4].

This study is carried out for the rural houses in hot-humid climate of Andhra Pradesh. These rural houses are made up of locally available building materials with minimal window openings to the exterior. The inhabitants cannot afford air-conditioning system due to their economic reasons. The people in this area are

J. Vijayalaxmi (✉) · P. Jaiswal
School of Planning and Architecture, Vijayawada, India
e-mail: vijayalaxmij2@gmail.com

A. Sayigh (ed.), *Achieving Building Comfort by Natural Means*, Innovative Renewable Energy, https://doi.org/10.1007/978-3-031-04714-5_14

compelled to stay indoors as it is a heat prone area and during heat time, they need to protect themselves from the heat waves. The disaster mitigation manual of the government suggests that people should stay indoors during the heat waves. Therefore, heat gain through the walls becomes an important factor to be taken into consideration. Building envelops of an unconditioned building mitigate the heat stress in hot climate [7]. Therefore, to minimize the heat transmission, to reduce the excessive energy consumption, to maintain the comfortable temperature in a building, and to control the heat flow through the building envelop, the most efficient way is to choose the appropriate materials used in envelop design [4, 7].

Thus, the building envelop heat gain can be reduced by the selection of appropriate material and enhancement in the design strategies which include (1) envelop designing, (2) the use of appropriate materials, (3) the use of insulating materials, (4) proper ventilation system, (5) designing of shading devices and enhancement in surrounding microclimate, (6) ideal building orientation, and (7) ideal window areas [7].

Al-mulla Hwaish et al. [7] discusses the heat exchange and energy conservation through the building envelop in hot climate. He also discusses the response of building envelop to climatic factors as it is a major factor to calculate the amount of energy required to maintain the indoor comfort condition. In his study, he explains the importance of material selection based on low U-value and high thermal storage capacity and construction technique implemented to get the desired time lag. A study by Tammu et al. in the hot climate of Thailand analyzes the OTTV of enclosing spaces of building envelop under bedroom function [5]. Residential characterization study states the importance of OTTV other than energy performance index (EPI) to calculate the thermal performance of a building envelop and also defines the factors leading to the high energy consumption in a building [3]. Netaphra et al. [4] discuss the acceptable thermal performance values of building envelops for designing as compared to different materials through BIM-based approach by analyzing it in a typical office building model (rectangular plan) in Southeast Asia (Thailand, Hong Kong, and Singapore). The thermal behavior of hotel building façade made up of double brick wall through performance-based evaluation in the tropical climate of Malaysia is studied by Nasir and Hassan [2]. A similar study was carried out for an office building [9]. A study conducted by Sheng et al. [1] assesses the effect of OTTV legislations on electricity consumption of residential and commercial buildings in Hong Kong using econometric energy epidemiology model [1].

All the previous studies have been conducted on commercial buildings to make the building energy efficient. No such studies are carried out for residential areas, specifically for the rural houses, in heat-prone areas of Andhra Pradesh. Therefore, in this study, OTTV is used as a tool to measure heat gain through the building envelop made up of different materials to analyze the indoor comfort conditions.

1.1 What Is OTTV?

In hot climate, the heat transfer through the building envelop can be calculated by means of OTTV. The initial calculation method of OTTV was suggested by ASHRAE in 1975. The concept of OTTV has been illustrated in the USA with the aim to ascertain the heat gain and loss through a building envelop. However, the concept of OTTV is applied to mechanically cooled buildings only. Conduction and radiation are the two principles of heat transfer to assess OTTV. The transfer of heat flow through the solid walls and glazed window due to the temperature difference between the two sides of the envelop is called conduction. The heat waves transmission through the window glass in the form of electromagnetic heat radiation energy is called as radiation. While selecting the material for building construction, the following properties should be considered to control the heat conducted into the building:

(i) The thermal transmittance coefficient (U-value) is used to measure the heat transfer rate through building envelop material. Low U-value provides good insulation and low energy consumption.
(ii) R-value is inversely proportional to U-value; it influences the heat flow resistance ability of a material. Higher R-value reflects the good thermal resistance.
(iii) Absorption coefficient expresses the absorptivity of the external surfaces which depend on the color of external envelops.
(iv) The solar gain that passes through the glass is measured as shading coefficient (SC); the low value indicates the low solar transmittance through the glass.
(v) The solar control through shading devices to reduce radiant and re-radiant solar gain can be achieved through external solar coefficient.
(vi) The solar heat gain transmitted through the glass is measured as SHGC on a scale of 0–1. Lower value indicates no heat transmitted, while the higher value indicated the amount of heat energy transmitted through the glass.
(vii) The building location, orientation, and WWR ratio play an important role in minimizing the amount of solar radiation received through a particular orientation which also affects the ambient temperature and ventilation [4].

The calculation method of OTTV as per ASHRAE standards is based on the empirical formula as mentioned below – [8]:

$$OTTV_{\alpha} = \frac{Q_{wc} + Q_{gc} + Q_{sol}}{A_i}$$
$$= \frac{\left(\left(A_w\right) \times U \times \Delta T_s\right) + \left(A_g \times U \times \Delta T\right) + \left(A_g \times I \times \theta\right)}{A_i} \qquad (1)$$

where $OTTV_a$ = overall thermal transfer value of outer walls (W/m^2)
Q_{wc} = conduction through opaque wall
Q_{gc} = conduction through window glass

Q_{sol} = solar radiation through window glass
A_i = total wall area
A_w = area of wall + area of glass (m^2)
U = transmittance value (W/m^2. °C)
$\Delta T_s = T_s - T_i = (T_o + Ix\ a/f_o) - T_i$
T_o = outside air temperature (°C)
I = radiation intensity (W/m^2)
a = absorbance of the surface
f_o = surface conductance outside (W/m^2. °C)
T_i = inside air temperature (°C)
A_g = area of glass in m^2
ΔT = difference between internal and external air temperature
Θ = solar gain factor of window glass

2 Methodology

Field studies were carried out to identify rural residences which are typical to the villages in Andhra Pradesh, India. Besides, all the five houses are representative of the houses in rural Andhra Pradesh in terms of their floor plan and building materials. They were chosen from the same location to avoid the influence of microclimatic variations.

These houses are made up of different locally available materials as mentioned below:

1. House-01 – Reed with mud-plastered walls and reed roof
2. House-02 – Random rubble masonry with mud-plastered walls and reed roof
3. House-03 – Brick with cement-plastered walls and reed roof
4. House-04 – Cement block walls and reed roof
5. House-05 – Brick with cement-plastered walls and RCC slab

The first four houses are made of different walling materials, but the roofing material is the same reed grass. This reed is a local variety known as "Rail Gaddi" as it is the type of reed grass grown along the railway tracks.

The fifth house (House 5) is a conventional house made of brick walls and reinforced cement concrete roof.

2.1 Study Area

The area taken for the case study is the Vallabhapuram Village, 26.9 kms away from Vijayawada. It is located in Kollipara Tehsil of Guntur district in Andra Pradesh state at 16.35°N latitude and 80.72°E longitude having elevation of 11 meters (Above sea level). Vallabhapuram falls under in hot and humid climate. The population count of Vallabhapuram is 6753 people and the number of houses in Vallabhapuram are 2231. The five houses are chosen from the same location as shown in Fig. 1.

The outdoor temperature can reach up to 45 °C during the summer months of April and May. The outdoor temperature and humidity for 21st of each month is depicted in Fig. 2. The 21st of each month is chosen for analysis due to its proximity to the solstice date.

The surface radiation value is high for the North-East and North-West wall during the summer months from April to September, while the surface radiation of the south-east and south-west wall is high from September to April (Fig. 3).

The surface radiation during the summer months is almost the same in all directions from April to August.

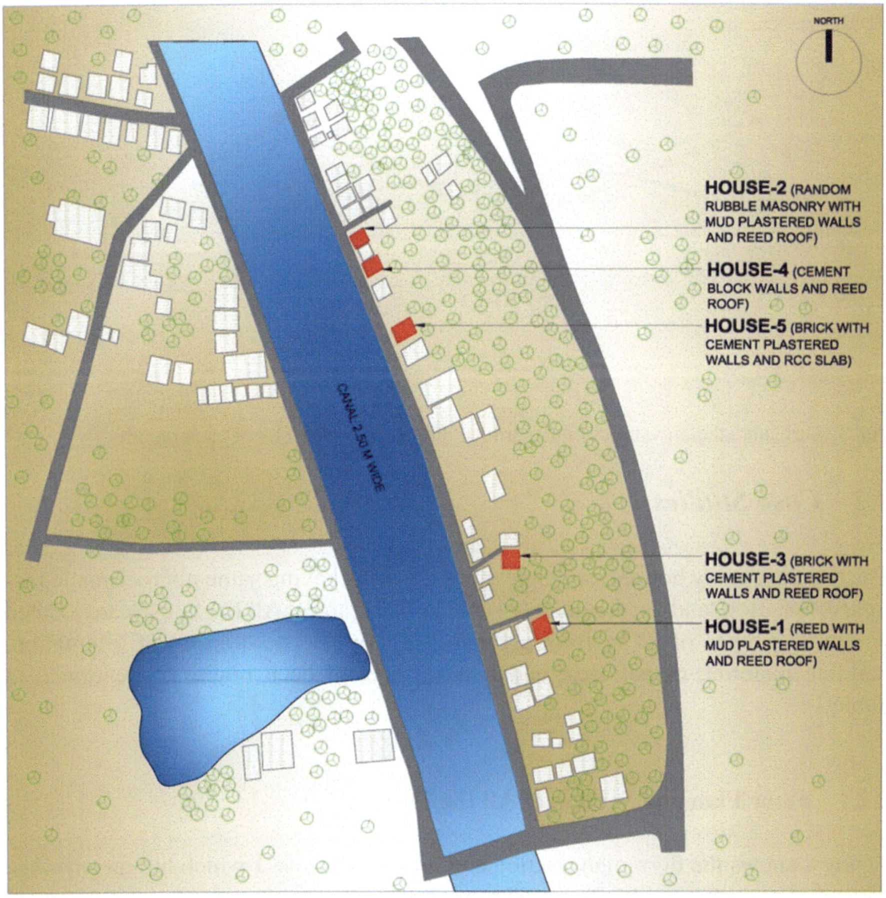

Fig. 1 Site plan of house location for field studies

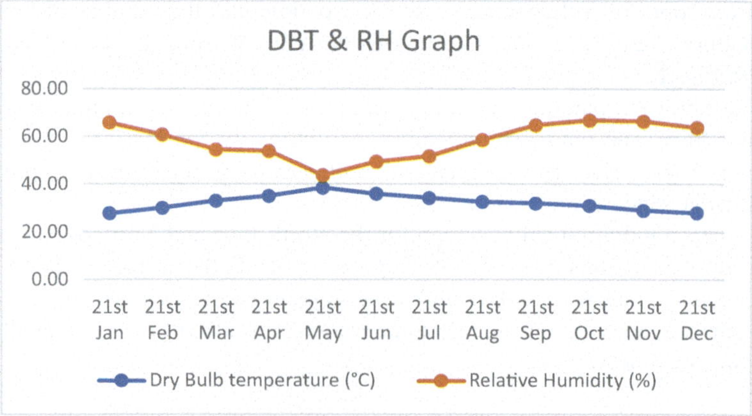

Fig. 2 Dry bulb temperature (DBT) and relative humidity (RH) in Vallabhapuram village

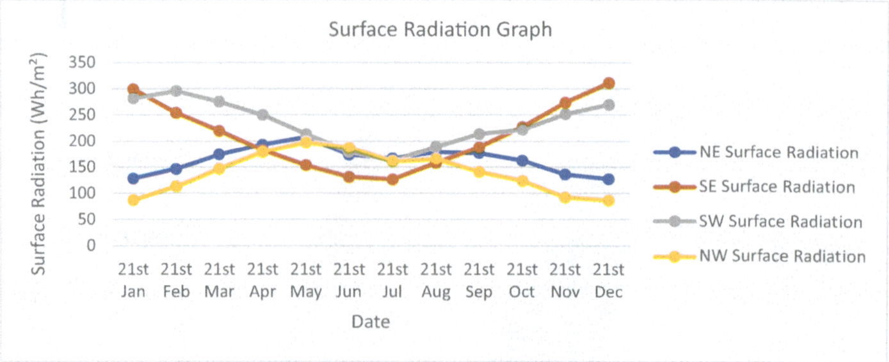

Fig. 3 Surface radiation values of intercardinal directions on the 21st of each month

2.2 *Case Studies*

The five case study houses are single-storied houses having same microclimatic factors, orientation, and approximately same built-up areas. All the houses are located within a distance of 24 m from a canal. The detailed description of orientation, building material, carpet area, and built-up area of each house is mentioned in Table 1.

2.2.1 Floor Plan and Section of All the Houses

Figure 4 shows the floor plan, section and image of House 1 which has an envelope of reed plastered in mud with a reed roof (Table 1).

Table 1 Details of all the houses

	House-1	House-2	House-3	House-4	House-5
Orientation	66° N	66° N	66° N	66° N	66° N
Material	Reed with mud-plastered walls and reed roof	Random rubble masonry with mud-plastered walls and reed roof	Brick with cement-plastered walls and reed roof	Cement block walls and reed roof	Brick with cement-plastered walls and RCC slab
Carpet area	32.85 sq. m.	16.50 sq. m.	30.06 sq. m.	22.55 sq. m.	31.60 sq. m.
Built-up area	34.55 sq. m.	20.93 sq. m.	36.87 sq. m.	32.24 sq. m.	43.86 sq. m.

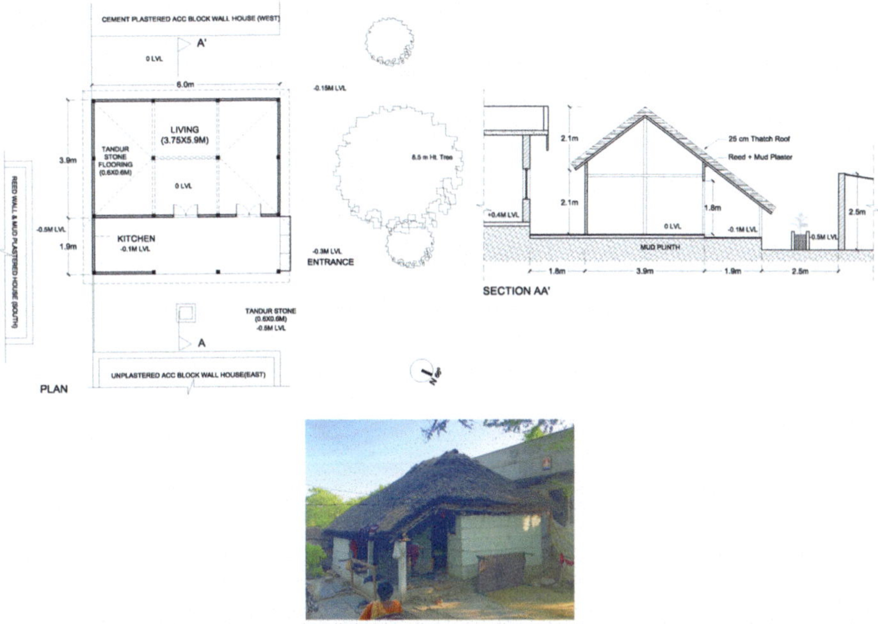

Fig. 4 Floor plan, section, and image of House 1

Figure 5 shows the floor plan, section, and image of House 2 which has an envelope of random rubble masonry with mud-plastered walls and reed roof (Table 1).

Figure 6 shows the floor plan, section, and image of House 3 which has an envelope brick with cement-plastered walls and reed roof (Table 1).

House 4, which is made of cement block walls and reed roof, is shown in Fig. 7.

Figure 8 shows the floor plan, section, and image of House 5 made up of brick with cement-plastered walls and RCC slab. This is the conventional envelop material in suburban and urban areas.

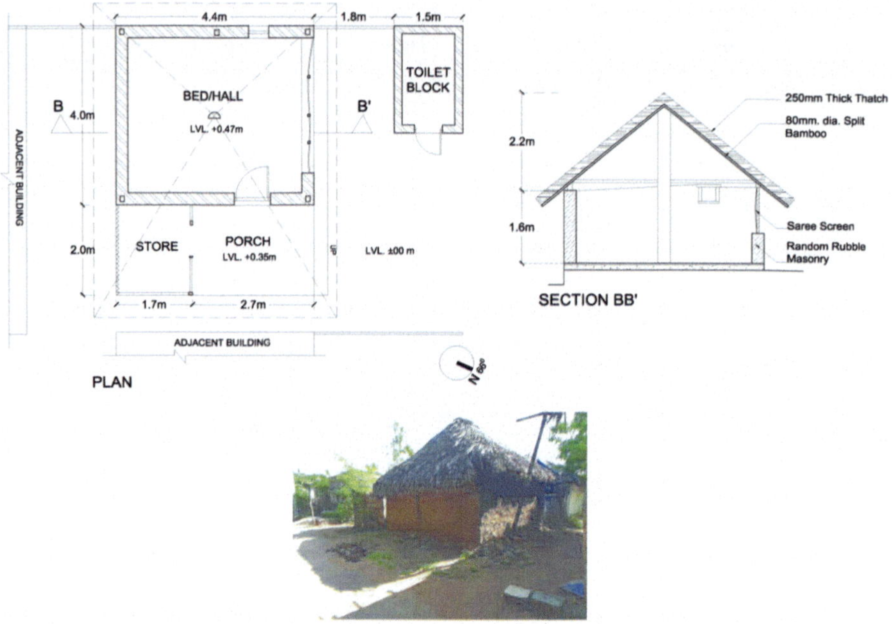

Fig. 5 Floor plan, section and image of House 2

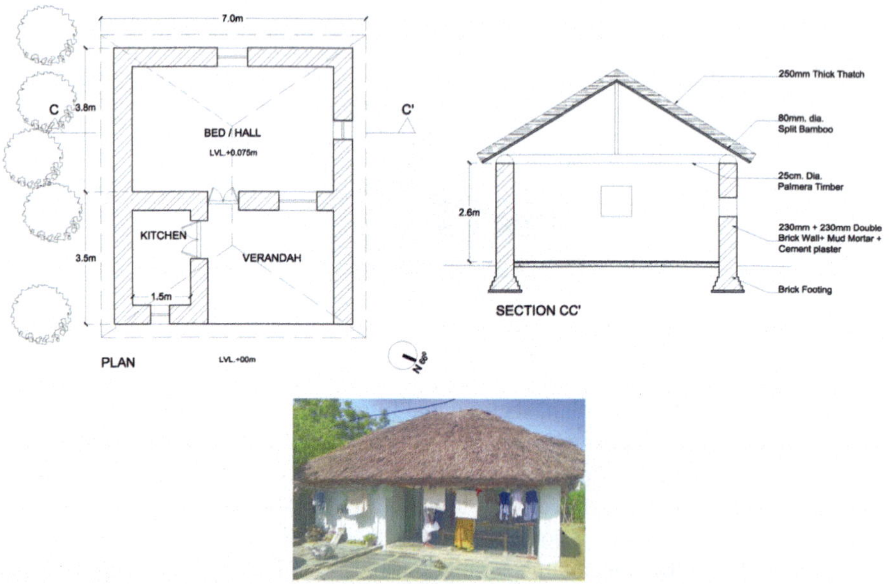

Fig. 6 Floor plan, section, and image of House 3

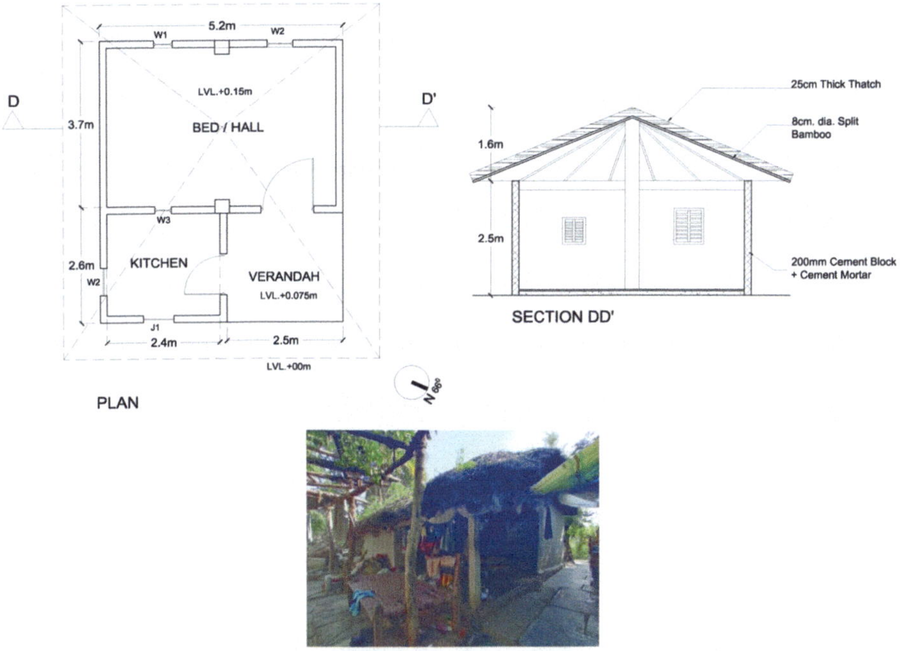

Fig. 7 Floor plan, section, and image of House 4

3 Findings: OTTV Calculation of All the Houses

The OTTV through walls of all the houses are calculated for the 21st of each month, and Figure 9 shows the graphical representation of OTTV through NE, SE, SW, and NW direction walls of all the houses, respectively. The calculations are done considering the most extreme cases when there are no windows on the external façade, as is the case in most rural houses. The graph shows the variation in per square meter OTTV through the walls when only material and indoor temperature vary.

The comparative average OTTV values of all the houses for a year in each façade is shown in Fig. 10.

From Fig. 9, it can be seen that the OTTV of House 2 made up of random rubble masonry with mud plaster is much higher than the OTTV of all the other houses in all the facades followed by House 5 made up of brick with cement plastered walls. Also, the OTTV value of all the houses in each façade is least in June and July months as these are monsoon months; hence surface radiation will be less due to cloudy condition. As compared with all the houses, House 4 with walls made up of cement blocks without plaster has less per square meter OTTV values in all the facades.

In NE and NW wall OTTV calculation graph, the highest per square meter value of OTTV is seen in April as it is a summer month and the radiations will be higher due to clear sky conditions. While in SE and SW OTTV calculation graph, the values are higher in December and January months. These months are winter months, and in these months, the sun is at a lower angle in the south direction.

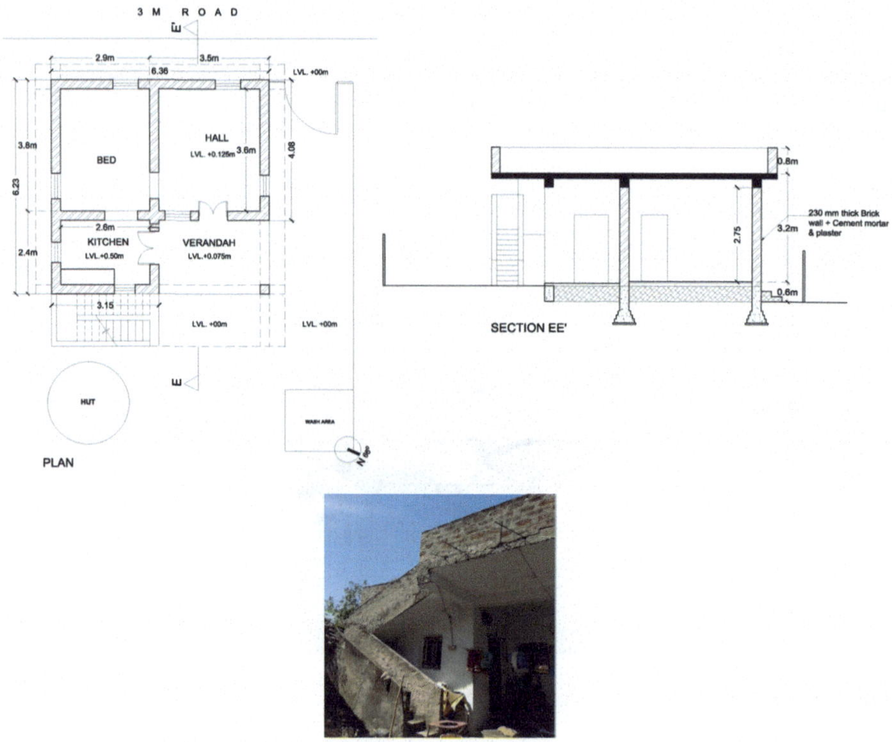

Fig. 8 Floor plan, section, and image of House 5 made up of brick with cement-plastered walls and RCC slab

From Fig. 10, it is seen that the average OTTV values for a year of House 3 made up of brick with cement plaster and House 4 made up of cement blocks are similar within a range of 0.10–0.14 W/m^2 having a variation of 2% only. While House 2 (random rubble masonry with mud plaster) has the maximum OTTV values as compared with all the houses because the U-value of random rubble masonry is higher than other walling materials; hence it is increasing the OTTV values in the house.

Of all the houses, the house with cement blocks and reed roof (House 4). The average OTTV of all the surfaces of House 4 is 0.11 W/m^2, and it is 85% lesser than House 2 which has an average OTTV of 0.76 W/m^2. The house made up of bricks with cement plaster has an average OTTV of 0.13 W/m^2, which is 83% less compared to House 2. Similarly, House 1, made up of reed with mud-plastered walls and reed roof, and House 5, made up of brick with cement plaster walls and RCC roof, have the average OTTV values of 0.18 W/m^2 and 0.34 W/m^2, which are 24% and 45% less than House 2, respectively.

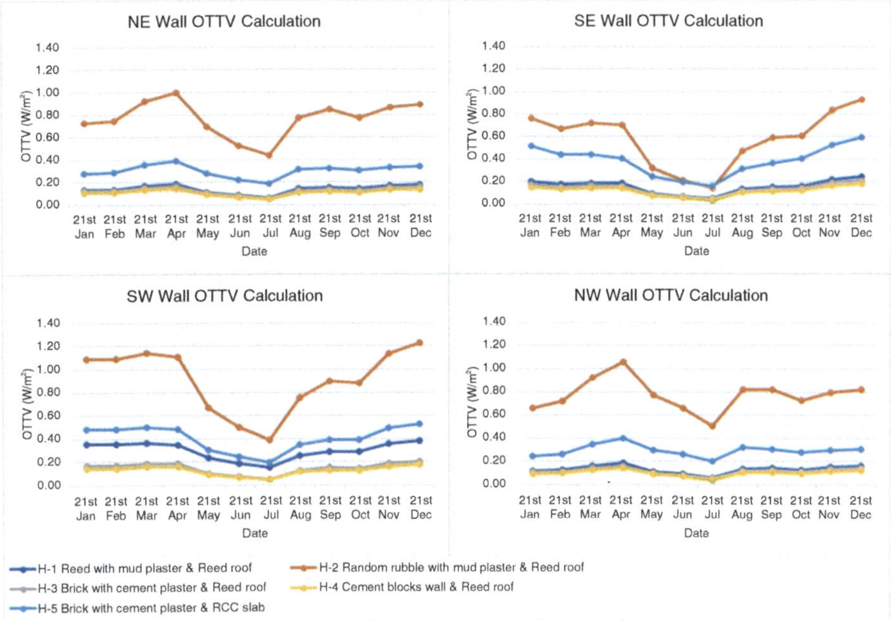

Fig. 9 OTTV of intercardinal directions of all the houses for the 21st of each month

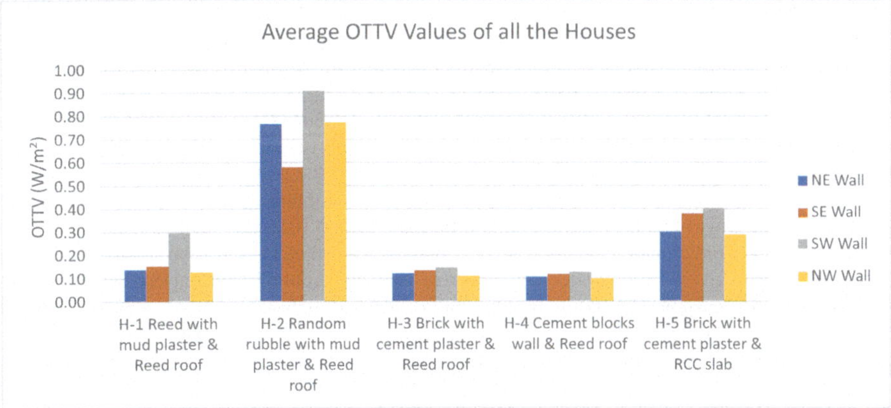

Fig. 10 Average OTTV values of all the houses in W/m^2

4 Discussion and Conclusion

For all the houses, maximum OTTV is found for SW wall; hence shading is recommended on this façade to reduce the surface radiation. House 3 and House 5 both have same wall material (brick with cement plaster), but House 3 has less OTTV values as compared to House 4. This is due to the variation in roofing material. In

House 3 the roof is reed roof and in House 5 roofing material is RCC slab. Due to the variation in roofing material, the indoor temperature varies; and hence it reduced the OTTV values to almost 50%. Reed thatch roof performs best and is recommended as compared with the RCC slab in hot and humid climate. To reduce the OTTV for the houses made up of random rubble masonry, there is a need to reduce the U-value of walls which can be possible by applying plaster and some insulating materials.

This study uses OTTV as a tool to measure indoor heat gain inside rural houses. The values of OTTV are less than $1W/m^2$. This is not comparable to the OTTV values of energy-intensive buildings, whose values range between 45 and $60W/m^2$, when designed with utmost precautions of following passive design strategies. This shows that the OTTV of rural envelope materials are nearly 50 times lesser than energy-intensive commercial and office buildings. Hence the results of this study can be used to understand that an alternative material based on the rural building materials, as studied in this chapter, for envelope in energy-intensive buildings can reduce the indoor heat gain drastically.

References

1. Sheng, W., Zhang, L., & Ridley, I. (2020). The impact of minimum OTTV legislation on building energy consumption. Energy Policy, 136, 111075. doi: https://doi.org/10.1016/j.enpol.2019.111075
2. Nasir, M. H. A., & Hassan, A. S. (2020). Thermal performance of double brick wall construction on the building envelope of high-rise hotel in Malaysia. Journal of Building Engineering, 101389. doi: https://doi.org/10.1016/j.jobe.2020.101389
3. Residential Characterization (2017–2018) – Centre for Advance Research in Building Science & Energy, CEPT University.
4. Natephra, W., Yabuki, N., & Fukuda, T. (2018). Optimizing the evaluation of building envelope design for thermal performance using a BIM-based overall thermal transfer value calculation. Building and Environment, 136, 128–145. https://doi.org/10.1016/j.buildenv.2018.03.032
5. Tummu, P., Chirarattananon, S., Hien, V. D., Chaiwiwatworakul, P., & Rakkwamsuk, P. (2017). Formulation of an OTTV for walls of bedroom in Thailand. Applied Thermal Engineering, 113, 334–344. https://doi.org/10.1016/j.applthermaleng.2016.10.055
6. Egwunatum, S., Joseph-Akwara, E., & Akaigwe, R. (2016). Optimizing Energy Consumption in Building Designs Using Building Information Model (BIM). Slovak Journal of Civil Engineering, 24(3), 19–28. https://doi.org/10.1515/sjce-2016-0013
7. Al-Mulla Hwaish, Akeel & Hwaish, Almulla. (2015). Impact of heat exchange on building envelope in hot climates. https://doi.org/10.13140/RG.2.1.3535.7284.
8. ASHRAE/IES Standard 901.-1989, Energy Efficient Design of New Buildings Except Low-rise Residential Buildings, American Society of Heating, Refrigerating and Air- Conditioning Engineers, Atlanta, 1989.
9. J. Vijayalaxmi. (2010). Concept of Overall Thermal Transfer Value (OTTV) in Design of Building Envelope to Achieve Energy Efficiency. International Journal of Thermal and Environmental Engineering. 1. https://doi.org/10.5383/ijtee.01.02.003.

Iyer Vijayalaxmi Kasinath has completed her BArch from the School of Architecture and Planning, Anna University, Chennai, India, in first class with distinction. She went on to pursue the MArch programme from the School of Architecture and Planning, Anna University, in 1997. She completed her PhD as an AICTE Quality Improvement Programme (QIP) Scholar in the year 2008 on the topic 'Thermal Performance of Naturally Ventilated Residential Buildings with Varying Opening Sizes and Orientation in Hot-Humid Climate – A Case of Chennai City' from Anna University. Subsequently, she was awarded the Erasmus Mundus Fellowship for Academic Profile, and completed her higher research at Politechnico di Milano, Italy, in 2010 in the area of 'Smart Materials and Nano Technology in the Construction Industry'.

She has over 25 years of experience in teaching Architecture and specializes in sustainable built environment and climate-responsive architecture. She is a professor and Dean (Research) at the School of Planning and Architecture, Vijayawada, an Institute of National Importance under the Ministry of Education, Government of India, and was also the Head of the Department i/c. She has had many technical papers published in peer-reviewed international journals and has presented technical papers at many international conferences, including at Harvard and Politechnico di Milano. She has travelled widely in her academic quest to countries such as Greece, France, Canada, Singapore, Malaysia, the US, and Italy.

Ms. Priyanka Jaiswal has completed under graduation from School of Architecture, COET, Akola (MH) as a gold medalist from Amravati University in the year 2015. She has 6 years of work experience and has worked as a senior architect in a renowned firm in Amravati, (MH). She is currently pursuing M. Arch. in Sustainable Architecture from School of Planning and Architecture, Vijayawada (A.P.)

Climate-Sensitive Architecture, Is Natural Comfort Possible?

Carolina Ganem-Karlen

1 What Do We Mean by Comfort?

The image of Adam by Vitruvius, the first writer of western architecture, wisely rebuilt by Filarete in 1490, offers a tormented humanity that does not willingly commune with the environment to which it seems to be unswervingly thrown (Fig. 1a). Arms overhead would not so much indicate a mere reactivity deposited in the gesture of despair, but rather the tracing of a fine line that must separate a chaotic and untamed nature from the borders of that human interiority.

Indeed, the so-called 'primitive hut' of Vitruvius is nothing but the mimesis of that first body gesticulation, not so much because they try to replicate the action of the cover, but because, above all, it stands under the firm conviction of drawing the difference between an interior and an exterior in search for comfort (Fig. 1b).

The architect designs the limit between an inside and an outside. Interior and exterior are not fixed concepts in architecture. Sometimes buildings are barriers to rain and wind and sometimes subtle filters for light and heat [1]. There are spaces in architecture that cannot be considered interior or exterior and sometimes change in response to changing climatic conditions. If we were to represent interior and exterior environments as a positive-negative image, where the black surfaces represent the building and the white surfaces the exterior space, we would have to envisage grey areas, which appear as a blurring of the line that separates black and white [2] (Fig. 2).

These grey interior–exterior areas were developed over time in very diverse forms in several cultures and are keys in the search for natural comfort (Fig. 3).

C. Ganem-Karlen (✉)
CONICET – UNCUYO, Mendoza, Argentina
e-mail: cganem@mendoza-conicet.gov.ar; carolinaganem@gm.fad.uncu.edu.ar

A. Sayigh (ed.), *Achieving Building Comfort by Natural Means*, Innovative Renewable Energy, https://doi.org/10.1007/978-3-031-04714-5_15

Fig. 1 From left to right. (**a**) –Adam by Vitruvius in Filarete's *Libro Architettonico* [3] (**b**) – Primitive hut by Vitruvius in Laugier's *Essai sur l'architecture* [4]

Surrounded by variable environments, where day and night, heat and cold, wind and calm, rain and sun change, buildings become havens of artificial conditions, like islands of tranquillity in an uncomfortable world [1].

Then, the question arises once again: What is comfort? Comfort in architecture is a subject that has been widely studied, although not always successfully. Many factors influence the appreciation of space. The simultaneous existence in time and space of different types and quantities of energy makes it very difficult to study them in an integrated manner, which would be the most appropriate approach in architecture. It is important to consider up to what point it is possible to study the concept of comfort objectively.

For Rybczynski [6], the simplest answer would be that comfort confines itself to human physiology. But that would not explain why – although the human body has not changed – our idea is different from that of a hundred years ago. But if comfort were subjective, one might expect a greater diversity of attitudes on the matter; yet, the opposite is true because in any specific historical epoch, there has always been demonstrable consensus.

While comfort is a subjective concept, it is also an objective fact. The most interesting point in all of this is that both assertions seem to be true at once, which leads

Fig. 2 Space analyses of St. Pietro Cathedral by Bruno Zevi in *Saper Vedere l'Arquitettura* [5]

Fig. 3 From left to right. (**a**) – *In a courtyard in Pompeii* by Luigi Bazzani (1878) (**b**) – *Coffee House in Cairo* by Konstantin Egorovich Makovsky (1870). (Source: Getty Images)

us to think that comfort may be both subjective and objective, without there being any contradiction in the statement [2].

Thermal comfort is defined in ISO 7730 [7] as '…that condition of mind which expresses satisfaction with the thermal environment'. In practice, the achievement of comfort is the result of a complex phenomenon in which objective space parameters – such as thermal, visual or acoustical – coexist with subjective physiological, sociological, cultural and psychological factors.

The comfort parameters are those objective characteristics of a given space which can be valued in energy terms and which summarize the actions that the people who occupy it receive in a given space. As such, these parameters can be analyzed independently of the users and are the direct object of environmental design in architecture.

The comfort factors, on the other hand, are those characteristics that correspond to the users of the space. They are therefore conditions external to the environment, but which influence the appreciation of said environment by these users. The comfort offered by a given environment will depend, in each case, on the combination of the objective parameters and subjective factors.

There is an analogy between comfort and an onion [8]. Like an onion, the notion of comfort has many layers, which were added historically. The image of the onion also conveys the elusive nature of comfort: If you take it apart and look at the layers one by one, you lose the overall shape, and yet the layers are still visible one beneath the other. These layers consist of ideas such as privacy or intimacy, convenience and physical ease. Each generation has added something to the definition, has added a layer without necessarily contradicting what came before.

2 Climate-Sensitive Architecture in Temperate Climates

In temperate climates, there are marked changes in conditions throughout the year. In these climates, architecture becomes more complex, having to be adaptable, even for short periods of time, to the entire spectrum of basic types of climate, such as hot humid, hot dry and cold climates.

The basic problem of these climates is not their harshness, but the fact that nearly in any period of the year and time of day, conditions of the opposite sign can occur. Cold in winter and heat in summer, and both can be almost as intense as in extreme climates, and also the problem of variable weather that, in the intermediate seasons, can generate cold or heat separated for short periods of time, even in the same day.

Although each individual constriction is not really critical, together they make the architecture of temperate climates have this greater degree of complexity, which makes it more difficult from a design point of view [1].

Air temperature, humidity, radiation and air movement all produce thermal effects and must be considered simultaneously. Each of these factors influences in some way the heat exchange processes between the human body and its environment; each one can favour or prevent the dissipation of superfluous heat from the body [9]. In the search to facilitate the understanding of the climate as a whole, analysis methods have been devised that relate climate variables in accordance with perceptual aspects of well-being.

Victor Olgyay [10] was one of the first to propose a systematic procedure to adapt the design of a building to human requirements and climatic conditions. His

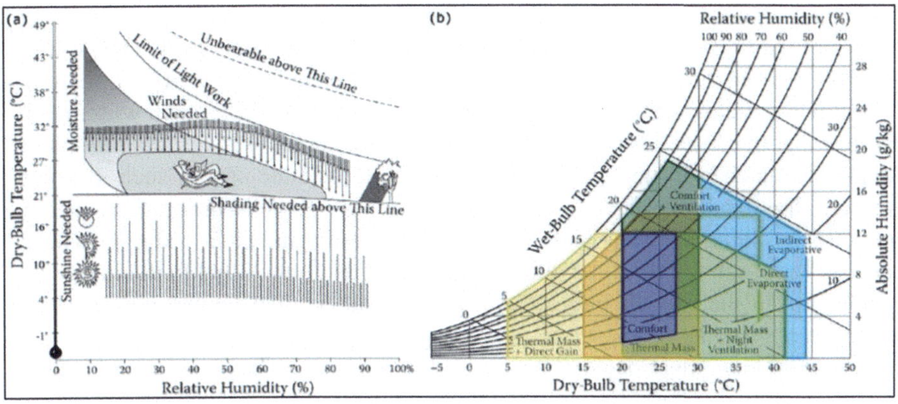

Fig. 4 From left to right. (**a**) – 'Bioclimatic Chart' by Olgyay in 1963 [10] (**b**) – 'Psychrometric Chart' by Givoni in 1988 [13]

method is based on a 'Bioclimatic Chart' showing the human comfort zone in relation to the conditions that surround it, such as ambient air temperature and humidity, average radiant temperature, air speed and solar radiation (Fig. 4a).

Another well-known diagram is the 'Psychrometric Chart'. In 1923, Richard Mollier [11] was the first to use enthalpy and moisture content as coordinates in the 'Mollier i-x' (Enthalpy–Humidity Mixing Ratio) diagram. The *ASHRAE Handbook of Fundamentals*, 1988, defines it as a graphic representation of the thermodynamic properties of humid air [12]. This means that each point on this chart will be defined by a value of the air dry bulb temperature (DBT), by a wet bulb value (WBT) and, therefore, by the ratio of both readings by a value of relative humidity (RH). The dew point temperature (DPT) is the temperature at which air with certain humidity becomes saturated and begins to condense the excess water contained in it.

Baruch Givoni [13] drew his bioclimatic diagram by incorporating the temperature and relative humidity of a place in the psychrometric chart. In this graphic, it is easy to recognize the relationship between climate and comfort and the demands to which architecture must respond to in a given climate. It is possible to define winter and summer comfort zones (Fig. 4b).

Givoni suggested the expansion of comfort limits from architectural performance through the application of bioclimatic strategies. These include passive heating, ventilation, thermal inertia with and without ventilation and evaporative cooling.

In winter, the effect of solar radiation on interior surfaces must be used in a certain way to balance the extreme minimum air temperatures. We can feel comfortable at low temperatures if the loss of heat from our body is counteracted by solar radiation.

In a climate with a majority of sunny days, the application of passive profit strategies is feasible. Direct Gain systems work from transparent surfaces oriented to the Equator (South for the Northern Hemisphere and North for the Southern Hemisphere). In this way, it is possible to capture the greatest number of hours of sunshine and those with the highest radiation intensity. These systems use the living spaces of the house to collect, conserve and distribute the solar gain. The building requires thermal mass to conserve heat during the day and to re-emit it at night.

Indirect systems work from the principles described for direct gain with the difference that the space into which the solar radiation enters through the transparent surface is attached to the space to be heated. Intermediate spaces such as greenhouses, glazed galleries and glazed balconies are very good architectural resources, and also specially designed walls such as trombe walls and solar walls. This space reaches very high temperature values, which are then transmitted to the interior by conduction in the case of indirect systems and by conduction and convection in semi-direct systems.

These systems must be combined with the presence of accumulation elements inside. That is, by building with materials with high thermal inertia like adobe, brick, stone and concrete. Thermal inertia is necessary because it allows maintaining higher indoor temperatures than outdoor temperatures during night-time periods in winter and achieving lower temperatures during daytime periods in summer.

To attend to the thermal conformation of the appropriate envelope is necessary to avoid losses by adding thermal insulation to the envelope, such as: cane, cork, cellulose, glass wool, expanded polystyrene and polyurethane. Also, to achieve an appropriate environmental regulation, the solar gain must be related to the reduction of the convective exchange of the glass with the outside. For this reason, the use of mobile night insulation on transparent surfaces is recommended, some examples are: shutters, blinds, lattices, roller blinds and sliding curtains.

In summer, the simplest strategy referred to passive cooling is in relation to the breezes of the place and the location of the practicable openings. The ideal position of the openings to favour ventilation in a space or in a sequence of spaces is to place them on opposite facades, that is to say, facing each other, in the direction of the prevailing breezes. This strategy is called cross ventilation.

Night cross ventilation is used in cases of continental climate, where the daily outdoor temperature differences are greater than 10 °C. The isolated thermal mass of the building must be kept closed during the day and avoid, by means of barriers and moderating filters, the entry of direct or indirect solar radiation. Given the construction characteristics, the insulation will prevent the entry of heat energy by conduction in the walls and the temperature will remain stable inside.

During the night, when the outdoor air temperature is lower than the indoor air temperature, the openings must be opened to promote nocturnal cross ventilation that will 'sweep away' – by convection from the inside – the heat accumulated during the day due to internal gains of its own related to the space use. If the openings are closed before the start of the sunny hours, the accumulated 'cold' will remain during the day, until a new cycle begins.

When high temperatures and low moisture content prevail in the air, the most viable design response is to raise the water vapour content of the outdoor air through the presence of moving water such as a fountain. This will produce an increase in the absorption of heat from the immediate surroundings, and thus reducing their temperature. The decrease in temperature caused by the evaporation of added moisture will restore the comfort temperature. This phenomenon will be influenced and modified by air speed, which will increase evaporation rates and with it the feeling of comfort.

Even though there is a consensus that climate-sensitive architecture is possible in temperate climates, the incessant search for what we call 'comfort', be it physical or psychological, has led to the evolution of the demands in all buildings. This process began to accelerate in the eighteenth century with the appearance of the first industries, massive production and new materials. Control systems, at first considered ingenious singularities created to provide some immediate benefit in the twentieth century, started to play a predominant role in architecture. Nowadays, comfort can be achieved by using auxiliary energy, usually from non-renewable and pollutant sources, that powers energy-intensive mechanisms with almost immediate response.

Comfort was repositioned and redefined as a 'product' sold by the HVAC industry [14]. The HVAC industry, therefore, needed to define 'comfort' in terms of the physical variables that could be controlled using the HVAC system. In such calculations, it is assumed that a thermal balance is needed between the environment and an 'average person', and this thermal environment is assumed to be constant. It, therefore, answers the needs of the engineering community in a way that allows them to size their plant. In doing so, it also dissuades them from addressing the shortcomings of the approach.

By inverting the original terms of what was understood as architecture (Fig. 1), it now seems that the building concept can be formulated as a 'support structure and enveloping shell of a set of facilities'. From ancient architecture we have left the most resistant parts, the fixed parts and the flat representations of them. In contrast, service and conditioning facilities are present in today's architecture in an increasingly conspicuous way and help it to function better, but most of the time they are hidden from the users' eyes (Fig. 5).

The volume that these mechanisms need occupies an increasing percentage of the built surface, but it is not only space that they are occupying in the building. The

Fig. 5 From left to right. (**a**) – Florida Gallery Building (1964 – Bonta and Sucari Architects, Buenos Aires) with air conditioners equipment added in the facade after occupancy in search for comfort. [15]; and (**b**) – Bank Macro Building (2019 – Cesar Pelli Architects, Buenos Aires) with all the high technology conditioning equipment hidden from our eyes [16]

decision-making power over them is increasingly moving away from the human being and is delegated to central offices that manage the control of the building.

Therefore and despite the existing richness of solutions and possibilities, it is more often to find poor architectural skins with rigid envelopes in the trust to achieve comfort through mechanical devices. It is very important to take into account that just as the climatic action changes during days and seasons, so do the demands of users regarding interior habitability, especially in changing and complex climates such as temperate [17].

Building comfort by natural means should be our main goal in the twenty-first century to protect the environment for future generations. In this holistic approach, well-designed climate-sensitive architecture will play a very important role, but also occupant's expectations and willingness to adapt to architectural possibilities and its response times will be key.

3 Occupant's Expectations Meet Architectural Possibilities

The overall mechanization of architecture has led to a disconnection between the occupants and the building. The mentioned and widely acknowledged historical transformation over the past century, where technological innovation led to a shifting of design responsibility in comfort provision from architects to mechanical engineering consultants, shifted the control responsibility from occupants to technology. Increased faith in technologically sophisticated environmental control systems meant that building occupants played little or no role in shaping the interior comfort conditions or indeed had little awareness and understanding of the systems that did and the energy required to operate the building.

The terminology used by comfort researchers is that of engineering and physics: temperature, humidity and airspeed, clothing insulation and watts of metabolic heat/m^2 [18]. The predicted mean vote (PMV) is the best known of such thermal indices [19]. Based on a simple steady-state physiological model, it predicts the mean comfort vote on the ASHRAE scale of a group of building occupants from a value derived from the four physical variables of radiant temperature, air temperature, humidity and air movement, along with the insulation of their assumed clothing level and their metabolic rate.

Acceptable thermal comfort in buildings is attained when 80% of occupants are satisfied with the provided conditions. The recognition of variation in comfort levels of the remaining 20% of occupants suggests that the notion of 'absolute' comfort is a privilege [20].

'Absolute' in this sense relates not only to thermal comfort but also to a range of possible comfort determinants including indoor air quality, visual and acoustic conditions as well as important psychological, cultural and behavioural aspects.

When referring to how individuals experience comfort, rather than the collective, having the ability to choose, for example, in a home or personal workspace where adaptive opportunities tend to be higher, represents a different state of individual comfort than in a shared or group situation.

Adaptation is central to current comfort discourse and those with more opportunities to adapt themselves to the environment or the environment to their own requirements will be less likely to suffer discomfort [21].

The notion of 'liberty to choose' is, in part, related to this ability to engage in adaptation strategies and behaviour changes and tolerances, and these may need to become much greater in an increasingly carbon-constrained world. Moreover, if comfort is considered a privilege, then the provision and expectation of its delivery are not a constant – even assuming 'a liberty to choose'. In a similar manner that limitations on the availability of natural resources can influence and limit consumption patterns, there will be times that an individual or group will be uncomfortable – and this may have to become an 'acceptable condition' [20].

Variability is generally thought of as a 'bad thing' in centrally controlled buildings because occupants are adapted to a particular temperature. In buildings where the occupants are in control, variability may result from people adjusting conditions to suit themselves. A certain amount of variability then becomes a 'good thing' [21].

The process of building systems and inhabitants dynamically responding to changing conditions and needs has been described by Cole, Robinson, Brown and O'Shea [22] as 'interactive adaptivity' and refers to the ongoing, bidirectional dialogue between building and inhabitant in which the outcome is not predetermined by building design parameters or performance metrics, but is rather an evolving process.

In this evolving process, results are not instantaneous. There is a time-lapse between the user's action and the building's response. This is a very important concept to bring into discussion because occupants (and there is a significant difference in the commitment between the term occupant and the term user) are used to push a button and obtain the desired change in environmental parameters.

Bordass and Leaman [23] have demonstrated that there is more 'forgiveness' of buildings in which occupants have more access to building controls. By forgiveness, they mean that the attitude of the users to the building is affected so that they will overlook the shortcomings in the thermal environment more readily. This can be explained as a function of who is in control.

The flexibility in modifying the comfort conditions of the space allows variations so that the last adjustment can be made by the same user. In this way, we will attend to your individuality. According to the adaptation principle of Humphreys and Nicol, what leads to a reaction in people's behaviour is lack of comfort: If a change occurs that leads to lack of comfort, people react in ways that they try to re-establish comfort [24].

As time passes, the temperature that people find comfortable (the 'comfort temperature') approaches the average temperature they have experienced. This implies that the conditions that occupants find comfortable are influenced by their thermal experience and that they can be adapted to a wide range of conditions.

The concept of acclimatization refers to the user's comfort factors. Although there are as many perceptions as there are users, generally particularities of the culture and climate of the region affect the expectations of comfort and therefore its standards. People put in place adaptation mechanisms and acquire more tolerance towards the most stressful aspects of the region's climate.

Popular architecture has always incorporated flexible solutions and systems, which components that can easily change their action according to climatic circumstances. Examples include mobile shading systems, which can prevent the access of solar radiation (hot weather in summer) or let it in completely if it is convenient (cold weather in winter); movable isolations in the openings, to allow night isolation; the same openings must be practicable for total ventilation; intermediate spaces located between interior and exterior, to generate favourable microclimates and be usable only in certain periods of time, among others. The components that make those changes possible can be an integral part of the envelope of the building (Fig. 6).

Therefore, the architecture in which natural comfort is possible takes time and space and engages the user as a part of its correct functioning, and that is a very different path to take when designing a building.

By accepting the adaptive hypothesis, it is argued that comfort is a 'social construction' – different societies, historically and geographically, have had different comfort temperatures. Therefore, previous experiences and cultural tolerance to change and education, among other factors, begin to participate in the equation of how architectural possibilities meet user's expectations – as the role of inhabitants

Fig. 6 Different shading devices. Sometimes buildings are barriers to Rain and wind and sometimes subtle filters for light and heat [25]

has shifted from occupants to users – engaged in a bidirectional dialogue with their buildings, reinforcing the idea that comfort is subjective and objective, without there being any contradiction in the statement.

4 Achieving Comfort by Natural Means: The User's Key Role from Theory to Practice

The adaptive model of thermal comfort rests on field-study research [26]. This research takes place in real buildings in everyday conditions, with the participants continuing their normal activities.

To assess the user role in achieving comfort by natural means, a case study of a free-running building in the city of Mendoza, Argentina is presented. A free-running building does not make any use of mechanical heating or cooling [27–28]. Its indoor temperature depends on the outdoor temperature and the total heat gains (from sun, occupants, lights and so forth) and the ability of users' 'interactive adaptivity' to adjust architectural devices to suit themselves.

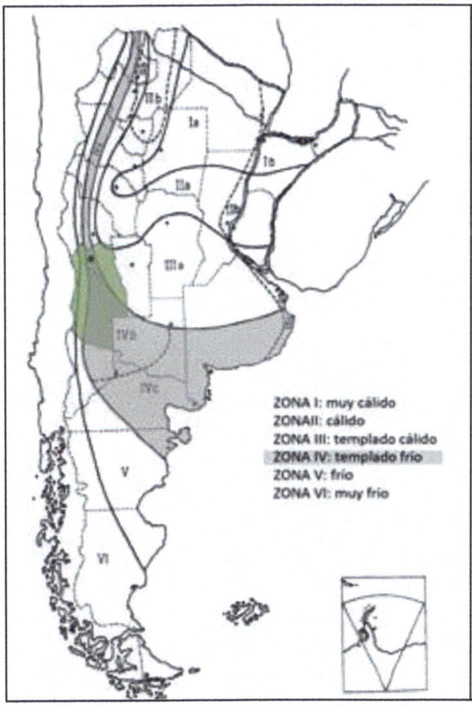

Annual values	
Maximum mean temperature	24.5 °C
Minimum mean temperature	9.6 °C
Mean temperature	16.5 °C
Global horizontal irradiance	18.4 MJ/m²
Relative humidity	56%
Relative heliophany	63%
July	
Maximum mean temperature	15.7 °C
Minimum mean temperature	0.8 °C
Mean temperature	7.3 °C
Global horizontal irradiance	10.2 MJ/m²
Relative humidity	63%
Relative heliophany	58%
January	
Maximum mean temperature	32.3 °C
Minimum mean temperature	17.4 °C
Mean temperature	24.9 °C
Global horizontal irradiance	26.1 MJ/m²
Relative humidity	49%
Relative heliophany	66%
Heating degree-days (T_b = 18 °C)	1384 °C
Cooling degree-days (T_b = 23 °C)	215 °C

Fig. 7 From left to right. (**a**) – Map: Location of the Province of Mendoza (green) in Argentina. The city of Mendoza is in climatic zone IV: temperate cold [29]. (**b**) – Table: Climate data for Mendoza, Argentina [30]

The city of Mendoza (32° 40′ South Latitude; 68° 51′ West Longitude and 827 masl) is characterized by an arid temperate continental climate, with strong thermal amplitude, low relative humidity, scarcity of rains and a high index of solar radiation and high heliophany (Fig. 7).

To obtain information about the thermal experience of real building occupants, a survey methodology was developed which simultaneously recorded room environmental conditions using conventional data loggers and sensors and the subjective feeling of warmth and preference for an individual together with a range of recent actions was recorded using a series of questionnaires.

The measurement period was of 59 days in the months of January and February, which is summer in the Southern Hemisphere. Temperature and humidity were recorded under pre-set conditions, testing the effect produced by different alternatives of occupation and envelope management by users. The information considered in the chart were: date, exterior and interior temperatures, rain, heliophany, hours of occupancy, closing/opening of windows, use of fan (the only device which required energy to function) and day/night comfort perception [31].

During the first measurement period, users opened all the windows during the night and closed them during the day. (Fig. 8) In a second period, users opened the window during the day and night (Fig. 9). Then in a third period, data were recorded for the dwelling without occupation, with all elements of the enclosure completely closed (Fig. 10).

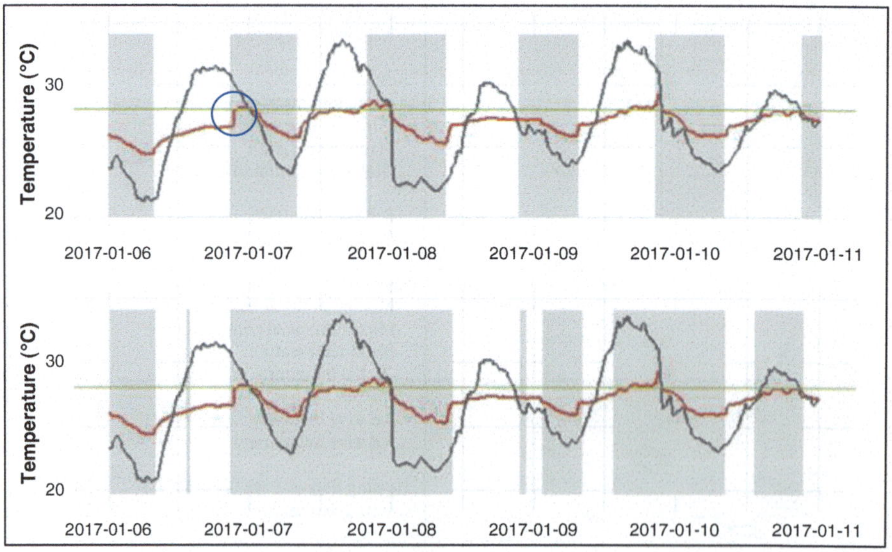

Fig. 8 Audit with occupation. Red: indoor. Black: outdoor. Green: 28 °C reference. From top to bottom: (**a**) – In grey: Windows opening to favour night-time natural ventilation. (**b**) – In grey: Space occupation [32]

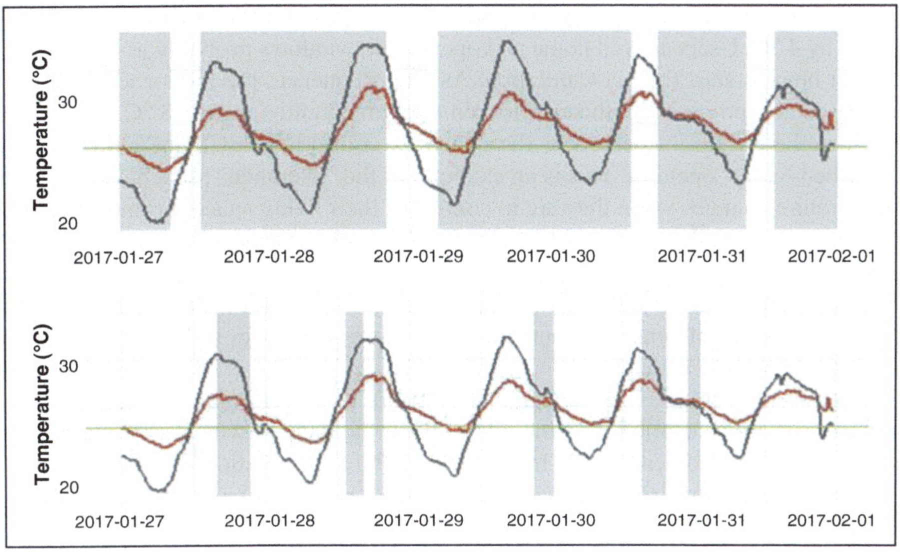

Fig. 9 Audit with occupation. Red: indoor. Black: outdoor. Green: 28 °C reference. From top to bottom: (**a**) – In grey: Windows opening day and night without any criteria. (**b**) – In grey: Space occupation with the use of a fan [32]

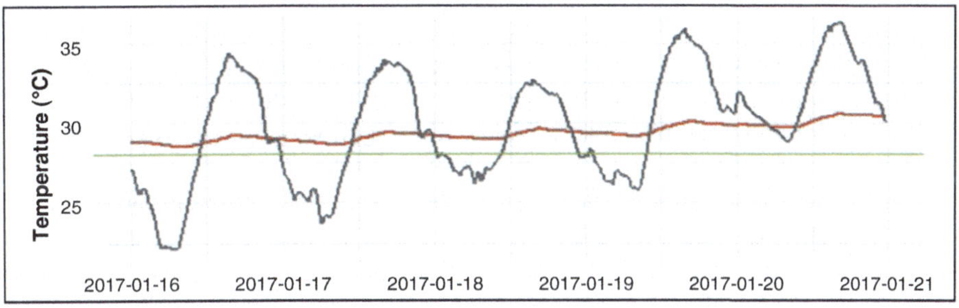

Fig. 10. Audit without occupation. Red: indoor. Black: outdoor. Green: 28 °C reference. Windows are closed during the day and at night [32]

Thermal performance was analyzed in relation to the compromised management of windows that users carried out to promote natural night ventilation, as well as the occupancy hours (Fig. 8).

The data from the record of occupation and management of openings graphed in grey bars (open/occupied) and in white bars (closed/unoccupied) in Fig. 8 show a general coincidence. We can assume that users are willing to manage the envelope of their homes while they are in them.

Note that the days in which the temperature exceeded the range of comfort correspond to an early opening of the windows (grey bars), which did not coincide with the drop in temperature outside. This can be clearly observed on 01/06 at 9:00 pm (marked in a blue circle in Fig. 8a), when the outside temperature exceeded the inside by 4 °C. Users arrived home and opened the windows probably assuming the correct hour to start (night) ventilation. As a consequence, the indoor temperature increased abruptly to meet the exterior temperature slightly above 28 °C.

Nevertheless, users perceived the space as comfortable without noticing the described ‘early’ opening. This is an example of the ‘tolerance’ users have towards temperature changes when they are in control of their living space. From this experience, users think their home is very comfortable and that they do not need to install any mechanical thermal conditioning.

In the second measurement period, users opened the window during the day and night. The direct influence that outside temperatures have over those of the interior was observed (Fig. 9).

Interior temperatures are above the comfort range throughout the observed period, with the exception of some night hours, the days in which minimum temperatures descended near 20 °C. In most cases, the accumulation of temperature in the interior mass materials during the day cannot be dissipated with natural night ventilation.

The lack of comfort manifested by users is evidenced throughout the period that they make intensive use of the fan, mainly in the hours of daily maximum temperatures. From this experience, users are willing to acquire energy-consuming mechanical equipment in order to condition their living space. Users do not know how to control their living space and therefore they think they need a machine to do that for them.

In the third measurement period, data were recorded with the dwelling without occupation, with all elements of the enclosure completely closed (Fig. 10).

The interior temperatures remain stable with a slight variation between 29°C and 30°C, staying above the range of comfort throughout the observed period.

Mass materials accumulate heat during the day, and by not ventilating inside spaces at night, heat builds up inside, which is reflected in a slight upward trend in interior temperatures. It is evident that even if temperatures are above the reference of 28 °C, they are stable. The fixed parts of architecture (material´s thermal mass and insulation) are keeping interior temperatures at a medium range.

Closed interiors won’t be at risk that daytime temperatures would rise to meet exterior temperatures. But this also implies that they won’t benefit from the night drop in exterior temperatures, also jeopardising the health and well-being associated with ventilation.

Opening windows are essential. All buildings should have opening windows and be habitable in natural ventilation mode [33].

Opening windows, along with all the architectonical devices and elements of the envelope in which the user can make changes and adjustments, lead to the fact that users add a level of uncertainty and unpredictability [34–36] and that it varies in relation to climate and customs. This fact must be seen as an advantage and not just a weakness – it is the means by which we manage risk and take advantage of opportunities by deviating from business as usual [37].

User behaviour has become key to reducing energy consumption. Many authors are starting to consider the education level of the user a differentiated point introducing management guides to give information on how to operate efficiently their homes, stating that a higher awareness and a better knowledge reinforce hazard prevention [38–39].

5 Building Natural Comfort for the Future

The existence of so many factors that influence the appreciation of architectural space makes the task of understanding space difficult. The simultaneity in time and of different types and amounts of energy makes it more complex to study this subject in an integrated way, which would be the most convenient in architecture. If we also consider the user interacting with architecture, the complexity rises even more. To these considerations we also need to add the future tense and a climate that is changing. Buildings are built not only for the present climate and the present user, but they will last over time. And even if the user remains the same, probably his preferences will evolve and change over the years.

Architectural design has an important impact on energy demand and efficiency. Decisions taken in the early stages of building design will influence and restrict the solutions for heating, ventilation and air conditioning in the present and also in the future. In this sense, the usage of the term 'resilient' has been increasing over the last decade, reflecting anticipated changes in our climate and resulting in necessary changes in our energy system and design practices.

Wilson [40] proposed a definition of resilience as the capacity of a system to absorb disturbance and reorganize while undergoing change to still retain essentially the same function, structure and identity. Resilience is measured by the size of the displacement the system can tolerate and yet return to a state where a given function can be maintained.

While referring to Wilson's definition, Roaf [41] states that it implies that the system should 'bounce back' to an original state and/or function and questions: Why not design systems that can 'bounce forward' to more failsafe states and functions? By designing systems that are enhanced by additional adaptive opportunities.

In the same direction, Trebilcock-Kelly, Soto-Muñoz and Marín-Restrepo [42] understand resilient buildings as designed and operated flexibly, so that the buildings can adapt to their occupants' requirements and promote the adaptability of the occupant.

Schweiker [43] establishes relationships between parameters and definitions of resilience and definitions and paradigms of comfort in the trust that connecting building resilience, human resilience, and paradigms of comfort opens several opportunities to scrutinise building design and operation practices together with research on thermal comfort (Fig. 11).

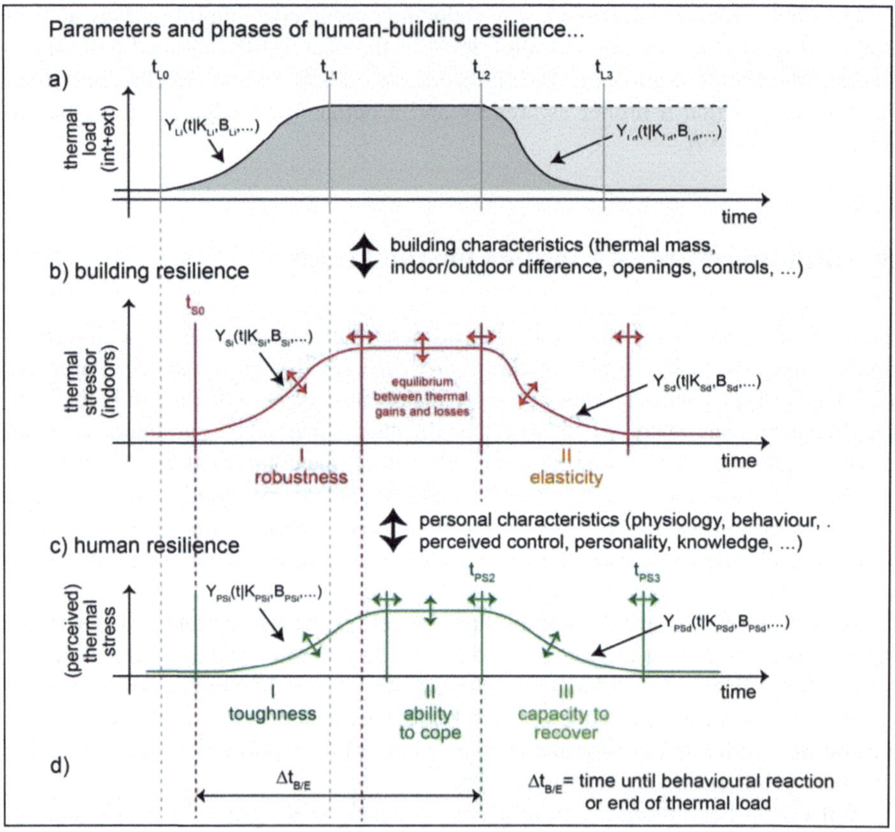

Fig. 11 Parameters and definitions of resilience in relation to comfort definitions and paradigms. Red and green arrows indicate that the position of the curve depends on building and personal characteristics [43]

Figure 11a presents the time course of a thermal load. This thermal load could be internal, meaning related to a high number of occupants or high equipment load, or external, such as high outdoor temperature or solar radiation. At times, the thermal stressor could be removed either by user action, for example switching on/off the equipment, closing blinds, or external conditions such as clouds or sunset.

Figure 11b shows the time course of a physical indoor thermal stressor, for example an increased operative temperature, as a result of the thermal load shown in Figure 11a. The relationship between thermal load and thermal stressor depends on the type of load and largely on the buildings' characteristics, such as thermal mass or window-to-wall-ratio. Different building concepts vary according to their robustness (increase) and elasticity (decrease). The values of these parameters can differ between robustness and elasticity. However, so far it hasn't been considered the effect on the thermal resilience of the user.

Figure 11c shows the time course of perceived thermal stress as a reaction to the thermal stressor shown in Figure 11b. While only one curve is drawn for reasons of clarity, note that thermal stress (e.g. physiological) can differ largely from perceived thermal stress. The relationship between thermal stressor and perceived thermal stress depends on personal characteristics such as the physiological constitution (e.g. level of fitness), behaviour (e.g. reducing workload to adjust metabolic rate) and perceived control, which was found to be related to perceived thermal stress and physiological reactions, personality or knowledge (e.g. that the stressor will end soon) [44].

The time course is split into three phases, each presenting a different parameter of human resilience: (I) toughness, (II) ability to cope and (III) capacity to recover. While the toughness related to thermal stress largely depends on physiological processes, the toughness related to perceived thermal stress might depend on other psychological influences. Phase II can be non-existing in case the occupant removes the thermal stressor before the thermal stress reaches its upper asymptote.

Thermal comfort is crucial in keeping building occupants safe, healthy, productive and happy. An unvarying environment may not just be psychologically boring, but may also reduce the ability of individuals to physiologically cope with environmental change. People adapt to a wide range of temperatures. Everybody is different. People habituate to the thermal environments they occupy over a day and year, and also their behaviours and expectations vary. Therefore, climate-sensitive architecture must vary too.

To achieve thermal comfort without a machine is the key to the success and resilience of a building. Architectural resources that users can use to adjust their spaces can collectively be described as adaptive opportunities, and their presence in a building may prove to be a key factor in the thermal satisfaction of the occupants.

The challenge posed is how to connect the work of comfort researchers to tangible improvements in the real world. In a future scenario that makes possible a more sustainable climate-sensitive architecture, comfort is not a given: Architectural adaptation possibilities must be conceived from the design, and users must be involved in the process of achieving comfort by interacting with their building over time.

References

1. Serra, R. (1999) *Arquitectura y climas*. Barcelona: GG.
2. Coch Roura, H. (2003) *La utilitat del espais inútils. Una aportació a l'avaluació del confort ambiental a l'arquitectura dels espais intermedis*. PhD Thesis. Barcelona: ETSAB-UPC.
3. Filarete, Antonio Averlino (1460 circa). *Libro Architettonico*. Von Oettingen version (1890) *Tractat über die baukunst*. Vienna: Verlag Von Carl Graeser. University of Michigan Libraries. https://openlibrary.org/
4. Laugier, Marc-Antoine. (1755) *Essai sur l'architecture. Nouvelle Edition*. Paris: Chez Duchesne. https://openlibrary.org/
5. Zevi, B. (1948) *Saper Vedere l'Arquitettura*. Enaudi

6. Rybczynski, W. (1986) *Home: A Short History Of An Idea*. London: Penguin.
7. ISO (2006) Standard ISO 7730. *Ergonomics of the Thermal Environment – Analytical Determination and Interpretation of Thermal Comfort using Calculation of the PMV and PPD Indices and Local Thermal Comfort Criteria*. Geneva: International Organization for Standardization.
8. Spronk, B. (1990). *A House is Not a Home: Witold Rybczynski Explores the History of Domestic Comfort.* Aurora Online. http://aurora.icaap.org/
9. Koenigsberger O.H., Ingersoll T.G., Mayhew A., Szokolay S.V. (1977) *Viviendas y Edificios en zonas Cálidas y Tropicales*. Buenos Aires: Paraninfo.
10. Olgyay, V. (1963) *Design with climate: Bioclimatic approach to architectural regionalism.* Princeton University Press.
11. Mollier, R. (1923) *Ein neues diagram für dampfluftgemische*. Z. Ver. dtsch. lag. 61, 869–872.
12. ASHRAE (1988) Handbook of Fundamentals. SI Edition. Atlanta: ASHRAE.
13. Givoni, B.(1988) *Climate Considerations in Buildings and Urban Design*. New York: Van Nostrand Reinhold.
14. Nicol, J.F. and Roaf, S. (2017) *Rethinking thermal comfort.* Building Research & Information, 45 (74): 1–5.
15. Villafañe, D. (2018) *Las tres primeras torres porteñas de cristal cumplen casi 60.* Revista ARQ Arquitectura. https://www.clarin.com/arq/arquitectura/
16. González Montaner, B. (2019) *César Pelli, un arquitecto que hizo más bellas las ciudades.* Revista ARQ Arquitectura. https://www.clarin.com/arq/arquitectura/
17. Ganem Karlen, C. (2006) *Environmental rehabilitation of the envelope of buildings.* PhD Thesis. Barcelona: ETSAB-UPC.
18. Fanger P.O. (1970) *Thermal comfort.* Copenhagen: Danish Technical Press.
19. ANSI/ASHRAE Standard 55-2017 (2017) *Thermal Environmental Conditions for Human Occupancy*. Atlanta: ASHRAE.
20. Cole, R.J., Brown, Z. and McKay, S. (2010) *Building human agency: a timely manifesto.* Building Research & Information 38(3): 339–350.
21. Nicol, J.F. and Humphreys, M.A. (2002) *Adaptive thermal comfort and sustainable thermal standards for buildings*. Energy and Buildings, 34(6): 563–572.
22. Cole, R.J., Robinson, J., Brown, Z. and O'Shea, M. (2008) *Re-contextualizing the notion of comfort.* Building Research & Information, 36(4): 323–336.
23. Bordass, B. and Leaman, A. (2007) *Are users more tolerant of "green" buildings?* Building Research and Information 35(6): 662–673.
24. Humphreys, M.A. and Nicol, J.F. (1998) *Understanding the adaptive approach to thermal comfort.* ASHRAE Transactions, 104 (1).
25. Belgrano, M. (2018*) Protectores solares, opciones de protección solar.* Apuntes. Revista Digital de Arquitectura. http://apuntesdearquitecturadigital.blogspot.com/2018/08/protectores-solares-opciones-de.html
26. Nicol, J.F. and Roaf, S. (2005) *Post-occupancy evaluation and field studies.* Building research and Information 33(4):338–346.
27. Ghiaus, C. (2003) *Free-running building temperature and HVAC climatic suitability.* Energy and Buildings 35 (4): 405–411.
28. Baker, N. and Standeven, M. (1996) *Thermal comfort for free-running buildings.* Energy and Buildings 23 (3): 175–182.
29. IRAM 11603/1996/2012. (2012) Acondicionamiento térmico de edificios – Clasificación bioambiental de la República Argentina. www.iram.org.ar.
30. National Meteorological Service of Argentina. www3.smn.gob.ar
31. Andreoni Trentacoste, S.E. and Ganem Karlen, C. (2017) *Influencia del uso y gestión de la envolvente en el comportamiento térmico de verano de una vivienda en la ciudad de Mendoza, Argentina.* Revista Hábitat Sustentable 7 (2): 64–75.
32. Andreoni Trentacoste, S.E. and Ganem Karlen, C. (2021) Influence of user behaviors on the thermal performance of a dwelling in the city of Mendoza, Argentina, ESTOA 10 (20): 1–19.

33. Roaf, S. (2020) *The Windsor Conferences. Past, Present and Future.* Windsor 2020 Resilient Comfort Conference.
34. Andersen, R. K., Fabi, V., and Corgnati, S. P. (2016). *Predicted and actual indoor environmental quality: Verification of occupants' behaviour models in residential buildings.* Energy and Buildings 127: 105–115.
35. Cuerda, E., Guerra-Santin, O., and Neila González, F. J. (2017). Definiendo patrones de ocupación mediante la monitorización de edificios existentes. Informes de la Construcción, 69 (548): e223.
36. Wagner, A., and O'Brien, W. (2018). *Exploring Occupant Behavior in Buildings.* Springer.
37. Jones, R. (2004) *Incorporating agency into climate change risk assessments.* Climatic Change, 67: 13–36.
38. Fernandez, Milan, and Creutzig (2015) *Reducing urban heat wave risk in the 21st century.* Environmental Sustainability (14): 221–231.
39. Guillén Mena, V., Quesada Molina, F., López Catalán, M., Orellana Valdés, D., and Serrano, A. (2015). *Energetic efficiency in residential buildings.* Estoa, 4 (7): 59–67.
40. Wilson (2012) Community resilience and environmental transitions. London: Routledge.
41. Roaf, S. (2018) Building resilience in the built environment (Ch.) in: Architecture and Resilience: Interdisciplinary Dialogues (Ed. Kim Trogal, Irena Bauman, Ranald Lawrence, Doina Petrescu) London: Routledge.
42. Trebilcock-Kelly, M., Soto-Muñoz, J. and Marín-Restrepo, L. (2019). *Adaptive thermal comfort for resilient office buildings.* Journal of Physics 1343:12148.
43. Schweiker, M. (2020) *Rethinking resilient comfort – definitions of resilience and comfort and their consequences for design, operation, and energy use.* Windsor 2020 Resilient Comfort Conference.
44. Schweiker, M. and Wagner, A. (2015). *A framework for an adaptive thermal heat balance model (ATHB).* Building and Environment, 94 (1): 252–262.

Carolina Ganem-Karlen is Professor and Senior Lecturer in the School of Design at the National University of Cuyo. (FAD-UNCuyo), Mendoza, Argentina and a Senior Researcher at the National ¬¬Research Council of Argentina (INAHE-CONICET). She is also the Academic Director of the Master in Sustainable Architecture at the Congress University (UC). Prof. Ganem-Karlen studied architecture at the University of Mendoza (FAU-UM) where she obtained her degree in 2001. That same year, she started her post-graduate studies in ABITA Centre at the Universitá degli Studi di Firenze (UNIFI), Florence, Italy with Professor Marco Sala. She holds a Ph.D. (2006) from the Universitat Politécnica de Catalunya (ETSAB- UPC), Barcelona, Spain under the guidance of late Professor Rafael Serra Florenza and Professor Helena Coch-Roura. Prof. Ganem-Karlen's research focuses are on energy efficiency; indoor environment and well-being; micro urban environments, occupants' thermal comfort and adaptive behavior, and energy assessment and certification of buildings. She has completed numerous research projects by a wide variety of resources. Prof. Ganem-Karlen has authored or co-authored many book chapters and more than 150 papers and research reports mainly presented in international journals or at international conferences with referee. Prof. Ganem-Karlen has been a member of the scientific committee and organization committee for many International Conferences. Referee for many international journals and international conferences in the field of energy and building. Prof. Ganem-Karlen is a Chartered Architect; a Fellow of the International Society of Building Science Educators (SBSE); a Fellow of the International Building Performance Simulation Association Latin-American Chapter (IBPSA-Latam) and a Fellow of the Argentinean Association of Renewable Energy (ASADES).

Design Elements of Building Comfort in Arid Zone of Arabian Countries – The Significance Role of Orientation and Courtyard

Falah AlKubaisy

1 Introduction

Human thermal in building comfort is defined as a condition of mind that expresses satisfaction with the surrounding environment. High temperatures and humidity provide discomfort sensations and sometimes heat stress (i.e., reducing the body's ability to cool itself). There are many factors and design elements that affect human comfort in the internal built environment. Human comfort is always affected by thermal, physical, and personal factors [1]. Another factor that can affect human comfort is the surrounding environment noise. Human comfort can also be affected by the visual of the space and the light intensity.

The issues of thermal comfort have become a global point of interest to many researchers around the world. This fact is attributed to the realization that people are spending most of their time (more than 90%) in an indoor environment. Passive methods of achieving thermal comfort in buildings in several countries have been commonly investigated. These passive methods are the solution to provide the comfort of the indoor environment, healthy lifestyle, and energy efficiency. Building comfort is considered an essential and necessary aspect in the urban sector, especially in desert climates where the outdoor temperature reaches 50 °C in the summer season and freezing point during winter [2].

The pre-occupation with privacy exerted another fundamental control. The roof was used for sleeping, for approximately six months of the year and the privacy of a family was fundamental, as such, roofs of houses should never allow to look down upon on its neighbor nor could one look into the court of another. This proscription effectively limited most houses to two stories above ground. Such self-imposed social restraint was more than mere good neighborliness. Buildings, particularly dwelling units, were therefore a culmination experience of influences from the

F. AlKubaisy (✉)
Architect and Planner, Manama, Bahrain

A. Sayigh (ed.), *Achieving Building Comfort by Natural Means*, Innovative Renewable Energy, https://doi.org/10.1007/978-3-031-04714-5_16

tradition, the climate, and the social conventions. The vernacular solutions emerged through an evolutionary phase over several centuries before they became the norm. It, therefore, becomes even more important which deserves attention to the traditional form of the built environment, always, for projects aimed at revitalizing or modernizing the Arabian traditional towns and particularly their historic background.

Since the 1950s, human thermal comfort under indoor and outdoor conditions has been discussed exhaustively in many reports and studies. These studies concerned the thermal comfort and/or heat stress for worldwide regions other than arid regions. Different scales for thermal comfort and heat stress have been produced in the form of numerical relations or graphs [3]. A survey of building comfort studies revealed that most of these studies focused on the environmental conditions for human occupancy either indoor or outdoor to evaluate the human thermal comfort and heat stress potential in different areas worldwide. However, the comfort conditions under arid climate in the Arabian Peninsula have never been evaluated deeply. Accordingly, evaluation of heat stress and human thermal comfort in an urban setting under arid environment–such as in the Arabian Peninsula–is urgently needed.

2 Building Comfort on Houses in Hot Arid Climates of Arabian Countries

Traditional designs of houses in Arabian countries allowed a comfortable environment for occupants in high-heat desert conditions; modern examples are far from this. Design elements of courtyards and the perfect orientation for the dwelling rooms are found to have adapted to desert environments, whereas the modern equivalent is by large ignoring this heritage. The designs of buildings with traditional construction and materials played a significant role in moderating the indoor environment when the outdoor climate of the Arabian Desert was unbearable. Unintentionally, energy for cooling was saved, creating a significant impact not only on the economy both at the national and global level but on the environmental scale as well.

2.1 Features Influencing the Built Form

In a manner like the effect of Islam on a city's form and structure, the religion has had its profound impact on the built form of buildings, particularly dwelling units. Also influencing the development of the ultimate form of buildings were the severe climatic conditions and to a considerable extent the established social conventions. It is useful to understand the social and spiritual implications of Islam on architecture. Islamic urban and cultural organization is also the physical manifestation of equilibrium between social homogeneity and heterogeneity in a system requiring both the segregation of domestic life and participation in the economic and religious life of the community [4].

A social law called the *Urf* [5], besides the faith beliefs, required the organization of a house, such as to provide maximum privacy and protect the residents from outsiders' sight. Added to this, the required segregation between males and females in Islamic life lead the layout of the traditional house to be developed on a "double circulation system," or the division of the house into two parts. Larger houses were planned with two courtyards, one for the men, *Diwan-khana,* and the other for the women, *Haram.* In smaller houses with a single court, the *Diwan-khana* was raised to the first floor with a separate staircase from the entrance, as in the many examples found in the traditional towns of Arabian countries, like Najaf, Iraq [6]. One of the main characteristics of a traditional house is the bent entrance leading to the courtyard, which for many centuries remained the dominant element in the plan of Arabic houses. For many centuries, the residences have evolved indigenous systems to sustain the harsh climate's discomforts.

Most attempted to create domestic micro-climates of which courtyard houses were the most common. Houses were laid out in appropriate orientation, and the fenestrations of rooms opened into the courtyard facing the direction of cool breeze. Environmentally, the courtyard has gone a long way to mitigate the hot and dry climate of Arabian countries. In most regions, there is a high percentage of sunshine and a high summer diurnal range (i.e., the difference between day and night temperatures) accompanied by low relative humidity. The traditional houses had an underground level, either cellars or semi-basements. The cellars were known as *Sirdabs* (Darkness) and the semi-basements as *Neems (Sleeping).* Both types in Iraq were built by brick-vaulted rooms, and both gained light as well as ventilation mostly from the courtyard, the *Sirdabs* having a central oculus and the *Neems* clerestories [7]; this is below and identical in plan dimensions to the family room upstairs, it is also here that the residents take their summer afternoon rest [8]. Because it is below ground level and naturally ventilated by cold air, the basement remains cool most of the day. The heat lost during the night to the clear sky by radiation allows the courtyard to remain cool most of the day. The covered terraces, usually on two or three (occasionally four) sides of the courtyard, and the identical first floor covered gallery immediately above, help to reduce the quantity of heat gain during the day by obstructing the direct solar radiation (see Fig. 1).

2.2 *Courtyards, Wind Tower, and Shading Devices Created a Sustainable Microenvironment*

Courtyards The height of the courtyard being greater than any of its plan dimensions, the area exposed to this radiation is reduced to a minimum, leaving adequate space in the shade, even at mid-day when the summer sun is near the zenith [9]. By means of a fountain, plants, or both, the very low relative humidity of the air is

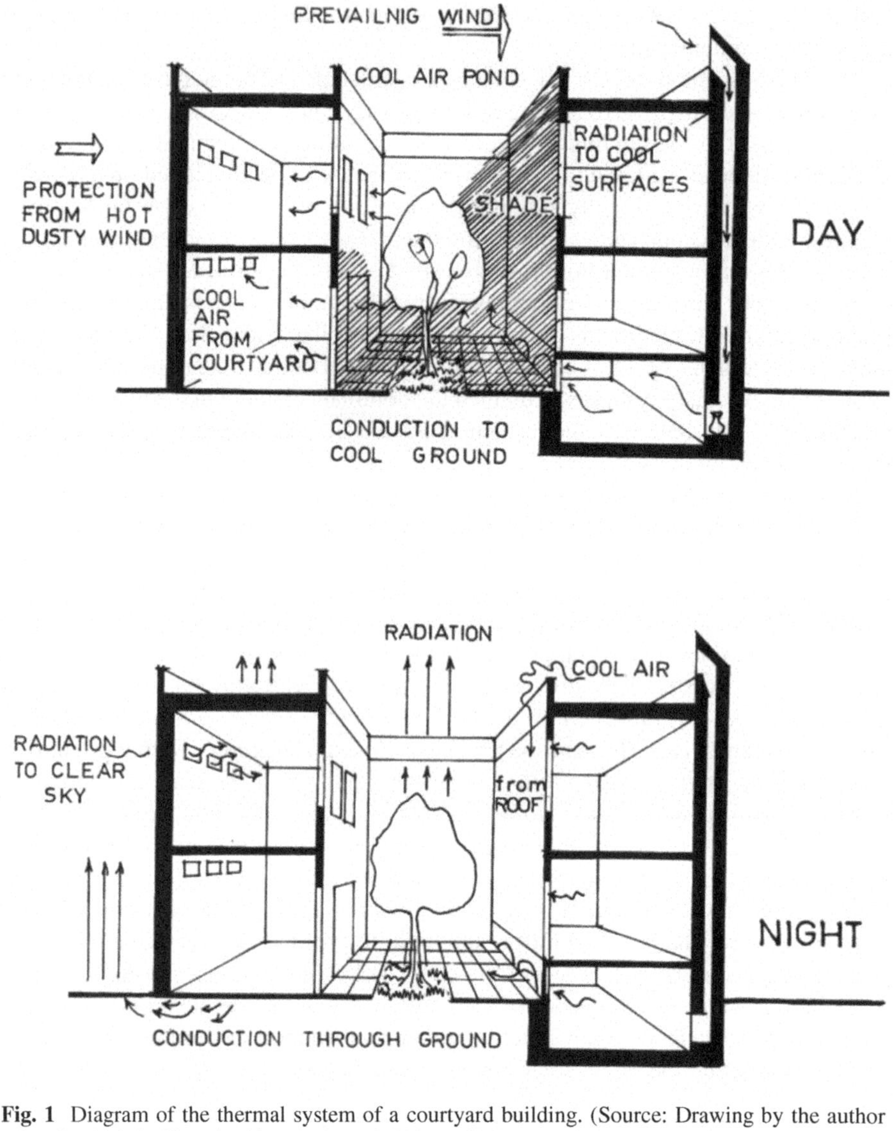

Fig. 1 Diagram of the thermal system of a courtyard building. (Source: Drawing by the author (also in Najaf) The Architectural and Urban Heritage of Iraq's Holiest City. page 34. KDP. Amazon. 2008)

raised to a comfortable level. In addition, the courtyard is usually washed at least once a day and showered a few times daily. All this is aimed at raising the relative humidity. Due to its position within the house, the courtyard is much quieter than the alleyways. The enclosing rooms which were built with thick walls also act as an effective buffer against noise. The social convention of providing privacy to each family was another major design consideration, which gave traditional houses an inward orientation. The openings were constructed in such a way as to prevent anyone intruding unseen into the intimacy of his neighbor's life.

Wind Towers Many towns of the Arabian Gulf have a humid climate, particularly in summer and other days of the year. Thick walls contain mid-wall wind catchers like wind towers, called Badgirs or Barajils. These towns used to have many wind towers when they are located close to the sea (see Fig. 2). They were designed to catch the faster flowing upper air stream and channel it down to ventilate and cool the rooms below for their occupants. Badgirs ventilated traditional basements to remove adores from heavily used areas with a few windows, providing refreshed air cooler than the rooms below. The convective techniques worked best when air temperature was around 35°C. Additionally, they would act as labyrinth cooling mechanisms, capturing the cool and saving it for the day [10]. Often the Badgirs were kept above water features to allow the cool air to pass over the water and provide refreshing moist air. Vernacular architecture was described as an expression to call traditional buildings that matched the local climate, culture, and economy.

There are several lessons that can be learned from vernacular architecture. The knowledge of creating passive low energy buildings, architecture has shown the way to achieve acceptable indoor environment for occupants with the least amount of energy consumption and materials. The usage of renewable energy sources in traditional architecture created a sustainable micro-environment [11]. Replacing unrenewable sources of energy with renewable natural sources is one of the most vital lessons. The usage of building physics helps to understand the best choice of materials that can provide better environmental properties [12]. Another type of popular shading device is the claustrum. The claustra are small vents that are made of plaster. These help to create a uniform distribution of air entering the room. In addition to security and privacy, it provides an aesthetic value. They are usually positioned on the upper section of the wall to allow the escape of hot air, see Fig. 1.

Fig. 2 In Abu Dhabi during the 1940s, several Badgirs or Barajils have demolished with the modernization era of the twentieth centuries

Shading Devices Another feature of traditional architecture in Arabian countries' dwellings is the excessive use of shading devices. These shading devices are used as privacy screens that have a climatic advantage, since they regulate the climate in the inner spaces. One of the well-known shading devices is the Mashrabiya, which is a widely used wooden screen (as shown in Fig. 3). The Mashrabiya had various functions, which include controlling the light passage; the airflow which helps to reduce temperature and plays an important role in creating privacy. A Mashrabiya, also called either Shanshūl or Rūshān, is an architectural element with a characteristic of traditional architecture in the Islamic World. It is a type of projecting oriel window enclosed with carved wooden latticework located on the upper floors of a building, sometimes enhanced in stained glass.

3 Re-inventing Traditional Elements for Building Comfort

There are two main lessons that can be derived from vernacular architecture; the first one is that traditional architecture provides many buildings which can be studied to understand the passive techniques and the environmental designs. The second reason behind considering this architecture is the best available lesson for us today, that it never ceases to appeal only to architects or students of architecture. The main lesson derived from traditional architecture is harmonizing the various conflict requirements that help to achieve a pure sense of community and timelessness.

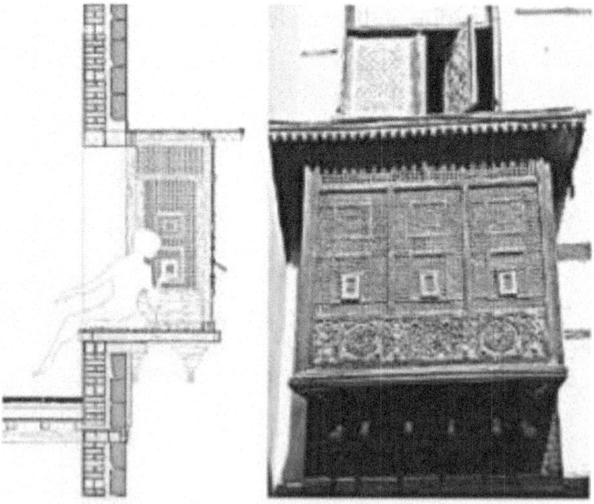

Fig. 3 The ladies behind the Mashrabiya. (Sources Ficarelli, 2008)

3.1 International Trends and Identity Charters

The Arabian Gulf countries, however, were not alone in the Arab World with such attempts to establish national identity through architecture. Pursuant to the oil boom and in the context of increasing Westernization that came along with the oil-based developments, there was in fact a larger soul-searching movement that swept through the entire Arabian Gulf region. Many similar nations also wanted to establish cultural uniqueness; distinct differences from others while re-affirming their associations with European cultures. From the 1950s of the last century, two interrelated trends emerged in the region: One was the use of European Classicism, subtly modified to create a national character to arise through an international form (see Fig. 4). The other was to look to one's own cultural heritage to find a truly national style in architecture. Fuccaro points out that they can be broadly organized around two main themes: the city as a recipient of modernity and the city as the focal point for the reclaiming of an Arab-Islamic identity [13].

The latter (the city as the focal point for the reclaiming of an Arab-Islamic identity) in Bahrain and Iraq, led to the discovery of a vernacular-influenced architecture with beliefs stemming from the ancient past that prevailed predominantly in Bahrain and traditional towns of Iraq. See Figs. 5 and 6. Of parallel significance are the practices that have evolved in the domestic architectural scene of Iraq and Bahrain, which have been driven by individual fascinations, market forces, and popular perceptions [14].

Fig. 4 Houses in the middle of the twentieth century built in Iraq by local materials of thick brick walls, with Art Nouveau style

Many of the villas and residential buildings imitated Palladian architectural forms and symbolism in preference to the vernacular. In fact, symmetry in form, elaborate domes, pediments, and porticos reminiscent of the Italian Renaissance were legitimate forms that could establish the newly acquired status as a wealthy nation.

3.2 New Era of Architecture Toward Sustainable Environment

The Bahraini wind tower is an architectural element that has now become a marker of its national identity, which has gained popular currency. A rising tower with openings orientated toward the good winds act as a funnel, catching the breeze and drawing it down into the cavities below, where the living spaces are located, while releasing the hot air like a chimney; the wind tower is one of the early forms of "air conditioning" by natural means. It is a built element that provides for iconic imagery, expresses local ingenuity in responding to the harsh climate, and therefore can be hailed as a unique symbol of Arab-Islamic-Bahraini identity for all times (see Fig. 7). Equally reproduced are the unique, wooden-poled palm-mat ceilings that had helped construct a flat mud roof for Bahraini traditional houses to combat the scorching sun [15]. Most present-day newly built "wind towers" are purely decorative identity advertisements that make a mockery of the past rather than celebrate it. Thus, the production of this form, devoid of functionality, raises serious questions about the engagement of elements of traditional urbanism to recall identity from the past.

Siyadi House, built by the pearl merchant Ahmed Bin Qassem Siyadi about a hundred years ago, is a fine example of vernacular architecture that represents a balance between sensitivities to climate and needs for privacy with fine, exquisite ornamentation to counter the barren desert; traits that Bahrainis would cherish and celebrate as uniquely theirs. This house does not have a window on the ground floor

Fig. 5 Different traditional windcatchers in Arabian countries. Windcatchers function during daytime and nighttime. **Left:** decorative Barjeel in Qatar; **Middle:** Barjeel in the Bastakiya Quarter of Dubai-UAE; **Right:** an example of Windcatcher in Bahrain
Reference: Jomehzadeh et al. Renewable and Sustainable Energy Reviews – A review on windcatcher for passive cooling and natural ventilation in buildings, Part 1: Indoor air quality and thermal comfort assessment of Energy Reviews journal

Fig. 6 Traditional courtyard house in Baghdad where the family realm with full privacy and comfort

Fig. 7 A traditional house with a wind tower converted to a restaurant in the middle of Manama in Bahrain

toward the streets, all spaces are open toward the courtyard, as well Khalaf House in Manama (see Fig. 8).

The building comfort of traditional buildings exists in areas that traditionally were used to open spaces inside the building for better ventilation. Creating

Fig. 8 **Right:** Siyadi House in the heart of Muharraq old town, one of the best examples of building comfort; all these magnificent screens are from Gypsum and wooden windows. **Left**: Courtyard inside Khalaf House in Manama

environments conducive to comfort and respite from the blazing sun. [16] Present-day architects may take the vernacular form to combine ideas of local material use and dimensions for the design of places between the indoors and outdoors. The examples used in traditional ecology will stand as a basis for future standards for building comfort design. The use of vegetation in micro-climates can provide a good basis for further investigation to reduce the Urban Heat Island (UHI) effect and improve thermal comfort for people entering the buildings, thus, reducing the cooling load [17] (See Fig. 6). Further research is needed to identify how vernacular use of design can reduce discomfort before entering the design concept of a building. Masdar in UAE provides a good example of this and can therefore shed light on modern construction. Other means of green building concept that can help to draw a decent courtyard and find the finest orientation to get the best building comfort and reduce the cost of energy in buildings.

3.3 Retrogressive Western Influence on the Concept of Building Comfort

Urbanization has long been a familiar social phenomenon in the Near East: Islam itself is largely urban in concept and is closely associated with urban-based culture. What is new and alien, however, is the strong tendency to link urbanization and modernization with Westernization. The incessant and indiscriminate importation of Western culture and its technology has unfortunately, but inevitably, resulted in the rejection of tradition. Thus, the old and indigenous aspects have now become inappropriate words falsely associated with reaction and conservatism. In physical terms, this blind rejection was quickly expressed in the neglect or outright destruction of historic areas, and much of what was valuable has now been lost forever [18]. Modernization does not have to be based on the Western model, and the supreme irony today is that while the Western society is going through a process of self-reappraisal and discovering some values and wisdom in its past; "follower" societies are busy trying to emulate and eventually catch-up with the West. However, it could also be argued that this race is not wholly self-imposed and that other deeper forces such as economic and technological colonialism are at work. While this is partly true, it should also be recognized that a cultural time gap and climate conditions between different societies and places will perhaps always exist and, therefore, the end of this race is fundamentally illusory. International culture, technological cooperation, and borrowing are necessary but must be selective and adaptive. Indeed, Arabian countries can take advantage of the Western experience and avoid their costly mistakes in this one good way (see Fig. 9).

Urbanism in Arabian countries is a product of the West—encouraging professionals, planners, architects, and the like that the Arab-Muslim city will evolve into this model of city development. There is no alternative. The fully recognized method of *Taqlid*—imitation architecture, abounds. Despite four decades of building development, the architectural style has been persistently reduced to a process of mimicking external designs; most large projects have been conducted by foreign consultants who are mostly not giving much attention to the climate nor social-economic issues that their design product is dominated by the idea to make something different from the surrounding, assuming that modernization is building façades of glass and aluminum. The return on investment of these high risers is approximately 10 years. Once the investments are paid off, the owner of the land may decide to renovate and build something higher and more commercially viable. With zero interest on loans and a very easy method for gaining commercial

Fig. 9 The first law court building in Bahrain in 1937. The designer succeeded to adapt the courtyards in public building. It is a style transition between tradition and modern

mortgages, the government provides ample opportunity for owners to demolish and rebuild [19]. Embodied energy becomes a real issue if these buildings are replaced so frequently.

However, during this process, particular interest could be paid to building comfort and green building concept for long-term costs of energy-efficient buildings to be minimized. There is a lesson we can take from traditional ecology and lifestyles to incorporate in modern construction. Higher UHI effect will only increase the use of AC consumption of power and exacerbate problems of the impact's fossil fuel energy costs on economy and environment as well as AC-related health issues. The increase in UHI will reduce the discomfort experienced by occupants as they enter the building, therefore, giving freedom to building owners to increase the temperature inside the in-between area of buildings and reduce energy bills, subsequently.

4 Building Comfort in Contemporary Architecture in the Arabian Countries

4.1 Imitating Traditional Forms of Building Comfort

The existing challenge is the absence of a conceptualization source to generate new patterns and design forms. Imitating old forms once fitted traditional houses, but this is no longer the case. The introduction of air conditioning and new manufacturing makes it possible for many products to be upgraded if thoughtful design consideration is articulated during the design and manufacturing phase. A true understanding of architectural element construction details, like the Mashrabiya, should be documented, understood, and tested with new materials and manufacturing before being used in modern interiors [20].

The Mashrabiya responds passively and positively to its surrounding environment, which proved the influence of the screen depth and perforation on decreasing the heating, cooling, and lighting energy loads in its environmental surroundings [21]. The wooden or gypsum screens were also found to reduce the penetrating level of light causing glare. Historians' studies have shown the Mashrabiya and Roshan to cool water jars in Islamic cities; some have classified these products as having more of a climatic character than a privacy role. The importance of the design that the old Mashrabiya allowed the inhabitants of the house to view the street from the upper floor but not to be subject to being viewed by outsiders (see Figs. 10 and 11).

The great Arabian architect Hassan Fathy states that the Mashrabiya is the best solution for thermal comfort issues in the region. The screens also allow a good passage of air and breeze, and their design delays the flow of heat into a building. At the same time, it has a positive consequence in enhancing a cooling effect.

Scholars investigated air movement in various areas of a traditional house in Jeddah City, Saudi Arabia, indicating an increased air velocity of about 0.3–1.3 m/s. Airflow simulation and shadowing evaluation gave the ability to determine the

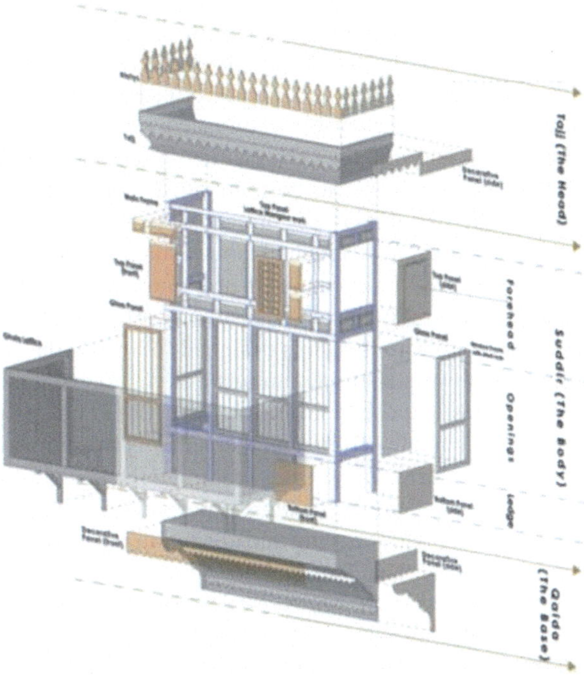

Fig. 10 Mashrabiya and Roshan construction. (Source: Mortada (2014) and Almerbati. N.)

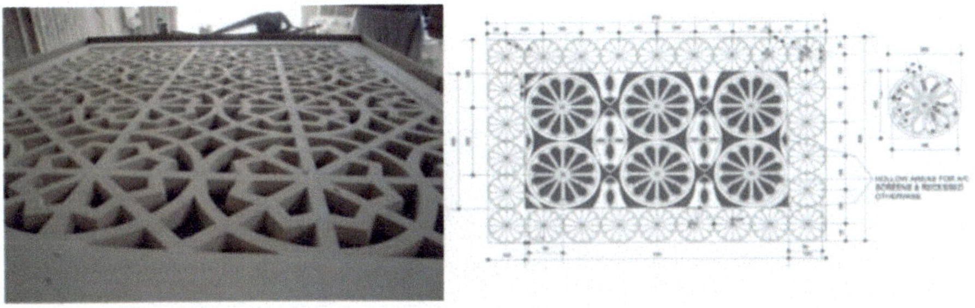

Fig. 11 A typical gypsum screen pattern. (Source: Bahrain Ministry of Housing, also Almerbati. N. page 66)

relation between the cooling effect and the geometric aspect of Mashrabiya concerning natural ventilation and solar control. However, the illuminance values of the screens proved the visual comfort level gained from Mashrabiyas, (see Fig. 11). Passive cooling can be argued as being inadequate for Bahraini new houses. The consequence of climate changes the high level of humidity and the introduction of air conditioning systems in modern houses because the availability of cheap power has placed an emphasis on the aesthetic function of Mashrabiya rather than environmental ones (see Figs. 12 and 13).

Respectable designs for large projects in the Arabian Gulf countries adapt to a traditional building element in contemporary design either for the identity reason or building comfort purposes. In terms of the environmental aspect, the traditional Mashrabiya plays a role in cooling and humidifying houses. The Mashrabiya's wood absorbs, retains, and releases water when faced with an air current. Once the wood fibers get heated by sunlight, they release their retained humidity. Wood is an expensive material to use for needs to be maintained continuously. The modern smart adaptive skin of the AlBahar Tower project wins the contest due to its intelligent Mashrabiya design and environmental context interaction. It uses parametric design and simulation as well as specialized programming methods and thermal actuators that open the Mashrabiya facade like an origami fold [22] (see Fig. 14).

An innovative honeycomb structure was designed following the analysis of high-efficient load paths. The towers also accommodate sky gardens at the top to reduce solar heat gain on the most exposed elevation of the roof. The designer described their project as a design generated from a mathematically prerationalized form that derives from Islamic design principles. A key feature of the design is the application of a diaphanous screen that envelopes the most exposed aspect of the building in the form of a dynamic "Mashrabiya," opening and closing in response to the sun's path, significantly reducing the solar heat gain and providing a more comfortable internal environment. The two honeycomb structures developed in the city are a decent example of the traditional Mashrabiya. These are intelligent facades that block the sun from heating the building, but still allowing for daylight penetration. The Abu

Fig. 12 Decent Mashrabiyas used for decoration and privacy elements covered the ladies' prayer room in Bahrain's City Center

Fig. 13 Al Ghurair Center in the middle of old Dubai, UAE; combining multi-use buildings to rewards the high price of land by using the concept of low-rise high density. Imitating traditional forms like the wind tower to be used as building lifts and landmark as a vertical element. Residential blocks on the left side consist of a large courtyard on third floor where all the amenity activities take place

Dhabi Investment Council's headquarters will be an example of energy reduction by design.

Within interior design, the Mashrabiya screen was used in houses, offices, and embassies to reflect cultural Arabic Islamic identity. It was also utilized to reflect the complexity of Islamic architectural patterns and its aesthetic endorsement of interior space quality. It was used as a window screen and sometimes as a room partition. Material and weather conditions regulated the use of Mashrabiya in an interior space. In Almerbati's thesis, she has classified as the contemporary of Mashrabiya in the context of modern architectural projects that have been used in three categories, which include adaptive skin, structural pattern, and cultural value [23] (see Fig. 15).

4.2 *Applicability of Green Building to Achieve the Building Comfort*

Building comfort and green building concept can be performed in an initial stage of architectural design development. It can involve all new buildings, additions, expansions, and restorations of buildings, particularly in town center areas. The next diagram can show that the design process has been introduced to develop the building

Fig. 14 Abu Dhabi Investment Council Headquarters Towers designed by Aedas – Arup, 2010. (Source: https://www.evolo.us/abu-dhabi-investment-council-headquarters-towers-aedas-arup/)

concept which must include building comfort criteria as well as matching the green building concept (see Fig. 16).

Design Process Any new design building should have the following:

A. Design aspects: Which include site analysis and conceptual design
B. Design process: Following a logical methodology
C. Economic aspects: Where the initial cost is high to save energy

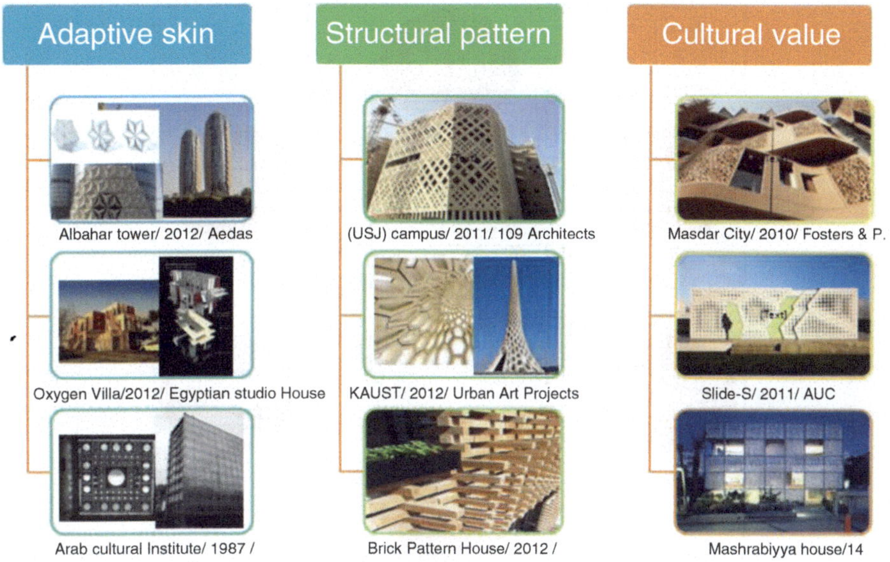

Fig. 15 Mashrabiya in the context of modern architectural projects. (Source: Almerbati et al. (2014) (page 72))

The design process was created a long time ago by the Chicago Architecture Center called Discover Design to help designers formulate a design concept accepted by community and suiting the micro-climate condition of that area. The design process is an approach for breaking down a large project into manageable chunks. Architects, engineers, scientists, and other thinkers use the design process to solve a variety of problems. Use this process to define the steps needed to tackle each project and hold all ideas as well as sketches throughout the process [24]. The design process consists of six steps; they are:

1. **Define the problem:** A solution cannot be found until having a clear idea of the issue.
2. **Collect information:** Collect sketches, take photographs, and gather data to start giving you inspiration.
3. **Brainstorm and analyze Ideas:** Begin to sketch, make, and study what can be started to understand how all the collected data and information may impact your design.
4. **Develop solutions:** Take the preliminary ideas and form multiple small-scale design solutions.
5. **Gather feedback:** Present an idea to as many people as possible: friends, teachers, professionals, and any other trusted persons to offer insightful comments.
6. **Improve:** Reflect on all the feedback and decide if or to what extent it should be incorporated. It is often helpful to take solutions back through the design process to refine and clarify them.

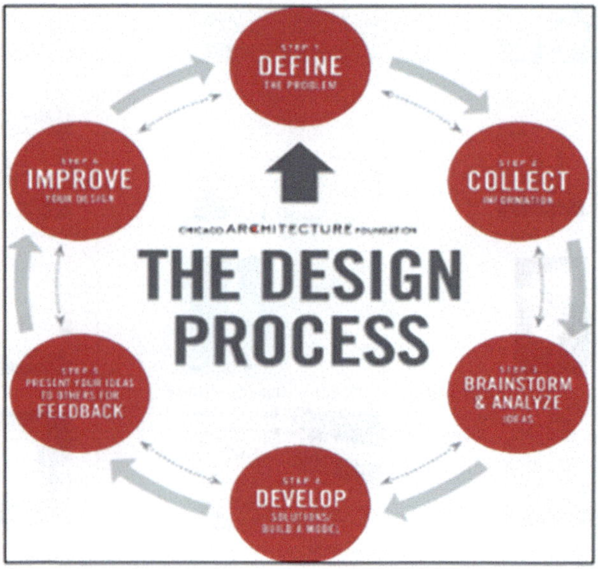

Fig. 16 The design process diagram. (Source: Chicago Architecture Center (2012–2019), https://discoverdesign.org/handbook)

Design Quality Features in Building Comfort

These are some design quality features that can be adapted to accomplish a building comfort in Arabian countries:

1. The spatial, temporal environment and privacy factors are clear and important in traditional designs.
2. Continuity in the development of architectural designs and inspirations of urban heritage according to the region or the city.
3. Continuous adjustments and development toward the quality of the design product according to the environmental, social, and spatial conditions in the city.
4. The response of architectural designs to the climatic environment, saving energy consumption and providing design rigor in placing buildings in healthy climatic directions.
5. Integrated and thoughtful architectural designs are put in place for all government businesses by architecture makers that seek reputation and workmanship and are far from business.

4.3 Building Comfort Requirements in the Arab Town Neighborhoods

There are at least 10 aspects of elements that are required to be provided for the design or rehabilitation of neighborhoods, which will affect the individual buildings in the context of building comfort. These aspects can be considered carefully to achieve new building designs. They are summarized as below:

1. The obligation to maintain a group of buildings that have advantages and characteristics of the urban heritage of the old city; they are the best example of building comfort and saving energy (see Fig. 17).
2. Renovation and Rehabilitation process is required for buildings that have damaged parts of the urban building fabric and contain heritage values.
3. Urban infill process to fill the empty and distorted spaces of urban scenes integrated with the urban fabric. This process will enhance the adjacent building and increase the building comfort in all buildings (see Fig. 18).
4. Developing urban designs inspired by the urban heritage of the city and determining the type of morphology of the Arab and Islamic city as well as not utilizing plans for modern prefabricated residential neighborhoods used in the outskirts of the city that do not fit with the compact urban fabric of the traditional city and not accomplish a building comfort.
5. Stressing on the principle of separation between vehicles and pedestrians in designing streets or alleys and making them shaded with evergreen and fruitful trees which can plant in arid zone or having a landscape of desert nature with water fountains (see Fig. 21).

Fig. 17 Restoration of the historical *Khan Margan* as a restaurant in Baghdad is the best example of saving energy and building comfort

Fig. 18 Solidere Downtown of Beirut- Lebanon shows real examples of creating building comfort by the process of urbanism to fill the gap between street buildings

6. Emphasizing the diversity in land subdivision schemes and layout alleys moving away from stereotypes in the designs of modern neighborhoods within the urban fabric of the traditional towns. Setting regulations and not encouraging the construction of tower buildings in the centers of historical towns that proliferate in the tribal way while not being environmentally sustainable, like Towers Area Zone in Doha, Qatar and in some areas in Dubai.
7. Setting a flexible urban requirement to improve the urban environment and building comfort while preserving the heritage values and architectural features of the urban fabric of the Arabian towns. Detailed planning and urban controls of reinforcing regulations that are compatible with the distinctive urban composition of the town. A clear written document should be available, binding and becoming the basis of any proposed development process toward building comfort (see Fig. 19).
8. Working to develop an urban design concept in developing new areas that are available in large plot areas and have been invested by joint-stock companies from the public and private sectors. Land subdivision of small plots should be halted. Small plots of land with the setbacks regulation for each building from four sides are difficult to organize and may only make small gardens, resulting in having at least three sides exposed to the sun and heat generation is increased inside the building: Other issues include construction wastage materials and spaces. A large plot of land subdivision easily creates successful design concept by having appropriate spaces for gardens, shaded walkways, livable amenities, and the best building orientation that can be achieved and would be suitable for the building comfort.
9. The necessity to place building designs in the accurate directions consider the movement of the sun and wind and not use exotic shapes, materials, and glass extensively over blind walls with glazed ceilings to obtain bright shapes that cannot respond to the sustainable environment and increase the heat load as well as energy consumption (see Fig. 20).
10. Encouraging to provide the concept of high density with low rise buildings in the center of the Arabian towns, using the local materials with insulation and reducing the opening sizes toward the bad orientation of buildings. Essentially,

Fig. 19 Muharraq Municipality building in Bahrain (when shaded, a shadow inner courtyard is the key to provide a successful sustainable urban environment). (Source: Yousif Dawood AlSayagh Office, Bahrain 2007)

providing the principle of privacy in all residential neighborhoods and separating the movement of vehicles from the dwellings. Low-rise housing developments for reasons of money, efficiency, simplicity and conceptuality, repetition seems difficult to avoid. When one unit or a group of units are designed, identical copies are stacked next to each other until a neighborhood is created, often with monotonous results (see Fig. 21).

Fig. 20 Development of the new Souq in the traditional town of Manama-Bahrain; using extensive skylight of steel glazed ceiling, increase the heat load and energy consumption with no care of sustainable environment building comfort (2006). This should be not acceptable with the increase of power consumption

5 Conclusion

Evaluation and discussion of building comfort and heat stress under arid climatic conditions was presented. The main conclusions from this chapter could be summarized that the arid zone climate of Arabian Countries can be responded to in building comfort from the initial stage of planning and designing such as:

1. Land-subdivision planning of a neighborhood should have as much of the land in rectangular shapes orientated toward the North-South direction as the prevailing wind is North-Western; such a direction can be easily designed in the concept of having the best orientation to the South for the sunlight and North for the light and the breeze. In addition, it can utilize the design plot to have a garden at the front or the back of the house. It is recommended to avoid the West-East direction for plots as much as possible; the East direction creates intensive heat during mornings, while the West direction has long hours of time exposed to direct sunlight at high temperatures from afternoons until the evenings.

   This design layout depends heavily on the size of the land plot from the land subdivision, where a building can be designed with a courtyard layout or as a U-shape where all major required spaces of the building have the best orientation toward the courtyard; meanwhile, if the plot is medium-sized, it can make an L- shaped design of the building and every two adjacent buildings can make a large courtyard (semi-detached style). It is often that land-subdivision planning departments struggle to subdivide small plots of land to increase the population density or planning for low-income groups. North-South orientation is still significant for this type of layout planning. However, regulations should be tolerated or abandoned by not providing a setback from two neighboring sides of the buildings, which will look like a row of house style. Good planning layout of a group of small plots can be created to generate a nice group of buildings with a

Fig. 21 Abu Nuase Street Development by Ammant Al Assama, Baghdad-Iraq. The project was completed back in 1986 to make a good example of low-rise high density in the centre of the capital of Baghdad. Solar energy had been used for 278 dwellings in different size areas. Providing the principle of privacy in all residential neighborhood as well as separating the movement of vehicles from dwellings. But the project as a group of buildings results with monotonous outcomes. (Source: Poul Mork Skaarup & Jespersen with Iraqi Architects Planner: Abad Al Radthi and Nazar Othman)

common open space, such as a large courtyard with vegetation of trees and arid zone landscape.

2. Using sold walls with small window openings toward the bad orientation will help build comfort, save energy, as well as enrich privacy.
3. One of the most harmful factors to building comfort is building a skydome made of glazed materials, which will increase the HVAC system in the building as well as face difficulty in maintenance. Such practice in buildings, shopping centers, or any function having a skydome as a design element is harming the sustainable environment and is considered a failure in architecture design and should be prohibited in regulations within the Arab countries.
4. Using strange shapes of buildings, shielded with glass and aluminum in the arid zone within the Arab countries, it is a matter of thoughtless ideas to the sustainable environment approach in architecture design and building comfort. This product will be classified in assumption of commercialism and does not follow the sense of place and the urban context. This kind of approach should be rejected and can be corrected in educating stakeholders within the field of architecture and construction.

References

1. UK Essays. (November 2018). Factors Affecting Human Comfort in Buildings. Retrieved from https://www.hse.gov.uk/aboutus/timeline/index.htm
2. Noor Aziah Mohd Ariffin, Amira Behaz, Zuraini Denan: Dept. of Architecture, International Islamic University Malaysia: Thermal Comfort Studies on Houses in Hot Arid Climates, Conference Series: Materials Science and Engineering, Published in Ser.: Mater. Sci. Eng. 401 012028 IOP, 2018.
3. Abdel-Ghany. A.M. and other "Human Thermal Comfort and Heat Stress in an Outdoor Urban Arid Environment: A Case Study". Academic Editor: Harry D. Kambezidis; Volume 2013 |Article ID 693541 | https://doi.org/10.1155/2013/693541
4. Akbar J.A:" Court Houses: a case study from Riyadh-Saudi Arabia" in the Arab City: Its Character and Islamic Cultural Heritage; Edited by Seragelden, Ismail 1981. p.164). Also, in AlKubaisy. F. "Najaf: The Architectural and Urban Heritage of Iraq's Holiest City", BookSurge. www.KDP.Amazon.com. 2008 page 33.
5. The ***Urf*** -"action or belief in which persons persist with the concurrence of the reasoning powers and which their natural dispositions agree to accept as right". Al-Hathoul," Tradition, Continuity and Change in the Physical Environment". The Arab Muslim City" Ph.D. Thesis (Boston 1981 p.139).
6. AlKubaisy. F. "Najaf: The Architectural and Urban Heritage of Iraq's Holiest City", BookSurge. www.KDP.Amazon.com. 2008 page 35
7. Warren, J and Fathi, I. Edited "Conservation of Traditional Houses" Amanat Al-Assima-Baghdad-Iraq; 1984; p. 10.
8. Fathi. I." Urban Conservation in Iraq with special references to Baghdad" Ph.D. thesis submitted to University of Sheffield (England 1977) p. 159.
9. Zuaiter Lina: Vernacular Passive Cooling Strategies in Hot- Dry Regions of Middle East and North Africa. Sustainable Residential Buildings in U.A.E. 2015
10. Zuaiter. L. op.cit. page 18
11. Jomehzadeh et al. Renewable and Sustainable Energy Reviews A review on windcatcher for passive cooling and natural ventilation in buildings, Part 1: Indoor air quality and thermal comfort assessment Energy Reviews journal.
12. Ibid. page 3.

13. Fuccaro. N." The Cities of the Gulf: a Challenge to Modernity?" in Arab Crossroads Studies, pp. 64–70
14. Dayaratne. R. "Landscapes of nation: Constructing national identity in the deserts of Bahrain" National Identities, 2012, 1–19, I First article. Page 6.
15. Dayaratne. page 9.
16. Kansara. T. page 601.Transitional Zone design in the desert environment of Abu Dhabi: Vernacular vs. Modern
17. The urban heat island (UHI) effect is a common environmental problem occurring in metropolitan areas in which the air temperature is significantly higher than in suburban areas. The UHI effect also leads to a smoggy climate.
18. Alkubaisy. F. edited "URBAN RENEWAL IN ARAB WORLD" KDP Amazom.com page 24.
19. Kansara. T. page 7.
20. Almerbati. N." Hybrid Heritage: An Investigation into the Viability of 3D-printed Mashrabiya Window Screens for Bahraini Dwellings. Ph.D. thesis published by De Montfort University, Leicester- UK. 2016.page 64.
21. Op. Cit. Pages 68–69.
22. Kansara. T. page 71.
23. Almerbati.N, Op. Cit page72.
24. UK Essayes.

Providing Thermal Comfort for Buildings' Inhabitants Through Natural Cooling and Ventilation Systems: Wind Towers

Alireza Dehghani-Sanij

Abbreviations

AC	Air-Conditioning
ASHRAE	American Society of Heating, Refrigerating and Air-Conditioning Engineers
CFD	Computational Fluid Dynamics
CTES	Cold Thermal Energy Storage
GHG	Greenhouse Gas
HVAC	Heating, Ventilation, and Air-conditioning
IAQ	Indoor Air Quality
IEQ	Indoor Environmental Quality
IIR	International Institute of Refrigeration
ISO	International Organization for Standardization
PCM	Phase Change Materials
RH	Relative Humidity
UAE	United Arab Emirates

1 Introduction

Since many people living in urban environments spend most of their time indoors [1, 2], more than 80% [3, 4], buildings of different sizes, types, and uses play a substantial role in the quality of human living and working environments in today's world [5]. In other words, buildings have a direct influence (short- and long-term) on inhabitants' health, performance, productivity, well-being, and thus satisfaction. In fact, a major contributor to occupants' health and happiness is indoor environmental quality (IEQ) [6–8]. Poor or bad IEQ can lead to physical, mental, and social

A. Dehghani-Sanij (✉)
Waterloo Institute for Sustainable Energy (WISE), University of Waterloo, Waterloo, ON, Canada
e-mail: a7dehgha@uwaterloo.ca

A. Sayigh (ed.), *Achieving Building Comfort by Natural Means*, Innovative Renewable Energy, https://doi.org/10.1007/978-3-031-04714-5_17

health problems for buildings' inhabitants [9, 10], such as asthma, respiratory allergies, shortness of breath, chronic lung diseases, eye infections, impaired vision, degraded attention, mental fatigue, elevated negative emotions, increased stress, depression, and amplified aggressive behaviours [11–14].

The IEQ is mainly evaluated by measuring four primary features of indoor comfort, that is, the air, thermal, visual, and acoustic qualities [15–17]. However, other factors such as water quality and interior design are less important [10]. In exploring the topic of this chapter, indoor air quality (IAQ) and thermal comfort are briefly discussed next. The IAQ, referring to the quality of air—encompasses the parameters of temperature, relative humidity (RH), and air pollutant concentrations (such as CO, CO_2, O_3, volatile organic compounds, and particulate matter)—within and around buildings and enclosed spaces [18–20]. In other words, the IAQ is influenced by a wide range of contaminants from various indoor and outdoor sources/activities, plus building-related factors (e.g., building materials and ventilation) [21, 22].

In accordance with the standard definition provided by the International Organization for Standardization (ISO) [23] and the American Society of Heating, Refrigerating and Air-Conditioning Engineers (ASHRAE) [24], thermal comfort is meant as "the state of mind that expresses satisfaction with the thermal environment in which it is located." In another definition, thermal comfort can be characterized as an individual's state of satisfaction—physiological, psychological, and physical—with his or her surrounding environments [25]. To determine the conditions of indoor thermal comfort, various criteria, including environmental parameters (air temperature, RH and velocity, plus mean radiant temperature), metabolic rate, and clothing insulation (or thermal resistance of clothing), must be considered and assessed during the building's design phase [26–28]. Additionally, a variety of factors such as gender, age, race, time of year, geographical features, weather conditions, building characteristics [29–32], and even culture [33] and individuals' expectations [34] can influence the thermal comfort felt by people. There are also direct correlations between thermal comfort, IAQ, and a building's energy use and efficiency [35, 36] (Fig. 1). Human comfort zones for summer and winter climate conditions are displayed in Fig. 2.

Nowadays, massive numbers of conventional heating, ventilation, and air-conditioning (HVAC) systems, in a variety of sizes, types, and capacities, are utilized throughout the world to meet the heating and cooling demands of buildings and provide good IEQ for inhabitants. For example, the International Institute of Refrigeration (IIR) reported in 2019 that there were approximately 1.6 billion residential and commercial air-conditioning (AC) systems in operation globally [41]. HVAC systems, on average, use over half of the total energy consumed in buildings [42, 43], ranging from 40% to 70% [44], although this percentage differs from one country/region to another as a result of local weather conditions, social and economic situations, and levels of development [5, 41]. For instance, in the United States and China, HVAC systems, respectively, account for around 35% [45] and 65% [46] of the overall energy usage in the country's buildings sector. Moreover, in some countries of the Middle East (e.g., Saudi Arabia) AC systems are responsible

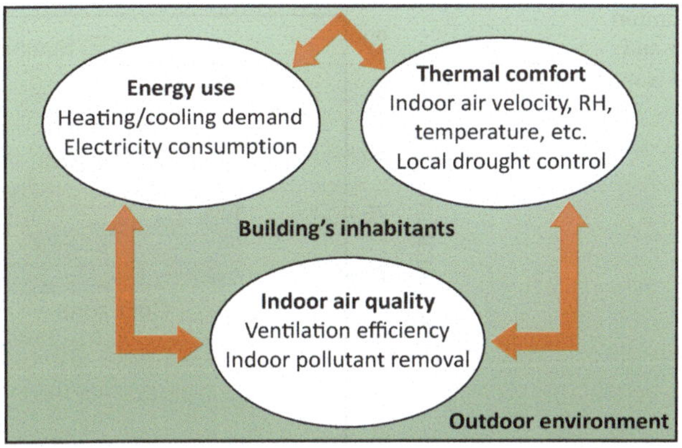

Fig. 1 Correlations among the inhabitants, indoor environment, and building's energy utilization. (Modified from Refs. [37, 38])

for about 70% of the energy utilization in buildings during the hot months of the year [47, 48].

HVAC systems contribute considerably to the global use of electrical energy, generated primarily by fossil-fuel power plants. Based on a report published in 2019 by the IIR [41, 49], the share of AC systems in the consumption of electricity worldwide was greater than 8% at that time. For a variety of reasons—the principal ones being the global increase in population and ambient temperature, the escalating demands for thermal comfort, and the improvement of people's living standards—the demand for these types of systems is continuously and swiftly rising worldwide, leading to increased electricity usage. It is anticipated that, by 2050, electrical energy requirements for space cooling and AC will have tripled [41], reaching about 37% of the overall energy consumption in buildings [49]. The result of this swift, growing, and ongoing demand will be escalating environmental problems and threats (e.g., global warming and climate change), especially via emissions of greenhouse gas (GHG) and pollution of air.

In addition to enhancing the energy efficiency in conventional HVAC systems, finding, developing, and employing sustainable and eco-friendly energy technologies can support efficient, practical, and even cost-effective solutions both for meeting people's thermal comfort needs and diminishing environmental challenges. For example, natural cooling and ventilation systems like wind towers or wind-catchers—called "Baudgeer[1]" or "bâdgir" in Persian and "Malqaf[2]" in Arabic—either alone or integrated with other technologies, can help not only in providing the conditions of thermal comfort and good IAQ for residents throughout the hot/warm months but also in decreasing energy use and costs in buildings, particularly during hours of peak demand and the subsequent environmental footprints.

[1] The word "Baudgeer" is a compound noun in Farsi. *Baud* means *wind* and *geer* means *catcher* [50].
[2] "Malqaf" means *catcher (or grabber) of air*, used in Egypt [50].

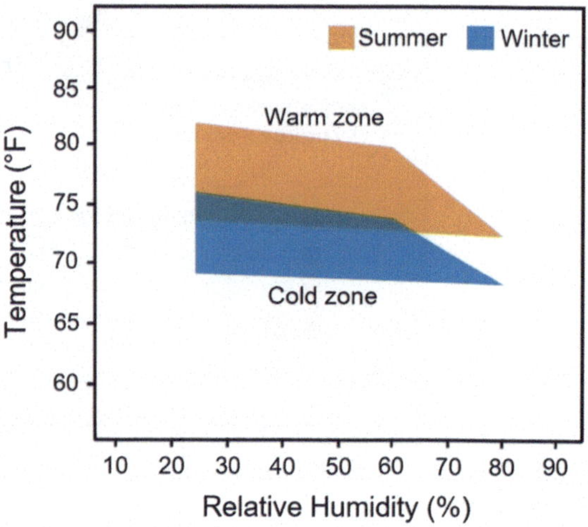

Fig. 2 Human comfort zones for summer and winter. (Modified from Refs. [5, 39, 40])

Wind towers are traditional architectural structures, of different sizes, types, and shapes, mainly made of brick/mud-brick, clay, and plaster, which have been used for centuries to naturally ventilate and passively cool a variety of buildings/dwellings and even to naturally circulate airflow inside public cisterns or underground cold-water reservoirs—called "Aub-anbar[3]" in Persian—in both hot/warm, dry and hot/warm, humid regions [50, 51]. The main purpose of the current chapter is to examine comprehensively a variety of aspects of both traditional/conventional and modern wind towers. This exploration will cover their history, typologies, and components; their ventilation and thermal performance; their role in improving energy efficiency; and their capability of lowering the energy costs of buildings. Also explored are possible combinations with other technologies like renewable energy and energy storage systems to enhance their performance.

2 Traditional/Conventional Wind Towers

Traditional/conventional wind towers were employed in a variety of Middle Eastern countries—including Iran, Iraq, Kuwait, Bahrain, Qatar, Oman, the United Arab Emirates (UAE), Jordan, and Egypt—as well as Pakistan and Afghanistan, as shown in Fig. 3, to provide natural ventilation and pleasant cool air and a thermally comfortable environment inside dwellings during the hot/warm months [50]. Nowadays, they continue to be utilized in their traditional forms in the old parts of various cities and villages. Figure 4 illustrates the traditional/conventional wind towers in certain Middle Eastern countries and Pakistan.

[3] The word "Aub-anbar" is a compound noun in Farsi. *Aub* and *anbar*, respectively, mean *water* and *tank* or *reservoir* [51].

Fig. 3 Distribution of traditional/conventional wind towers in the Middle East, as well as Pakistan and Afghanistan [50, 52]

In a number of Iranian cities, such as Yazd, Meybod, Kashan, and Naein, located in hot, arid regions, builders/architects situated several wind towers next to the domed roofs of large, public cisterns, as shown in Fig. 5, thus encouraging natural airflow over the surface of water stored within the reservoir to retain its cold temperature [51, 54]. In fact, the movement of air over the water's surface (Fig. 6) and heat exchange via convection and evaporation, plus radiative heat exchange between the water surface and the roof, create thermal stratification inside the water reservoir. Consequently, the layers of water near the bottom of the reservoir—the position of the tap—were always cold enough to drink all through the hot/warm months of the year, even when ambient air temperatures reached above 40 °C [55–59].

Although no accurate, comprehensive history of traditional wind towers is available, according to existing historical evidence—such as literary and historical texts, accounts (usually diary entries) by foreign travellers who visited Persia[4] in past times (the first one by Marco Polo (1254–1324), a Venetian merchant, explorer, and writer), historical paintings/pictures, and the remnants of ancient sites/buildings (e.g., Arg-e Bam)—Bahadori and Dehghani-Sanij [50] indicated that Persians first invented and developed wind towers. Based on the historical documents available, their history goes back to the early ninth century, about 1200 years ago [50]. For example, Rudaki Samarqandi (c. 858–940), a Persian poet, regarded as the first great literary genius of the Modern Persian language, refers to the word "Baudghard"—elsewhere termed a "Buadgeer" [60]—in one of his couplets:

> In many places, dwellings/buildings have a *Baudghard* / where inside them, happiness is established, and the residents eat with pleasure.[5]

[4] The old name of Iran.

[5] بسا جاي كاشانه‌ی بادغرد / بدو اندرون شادي و نوش خورد

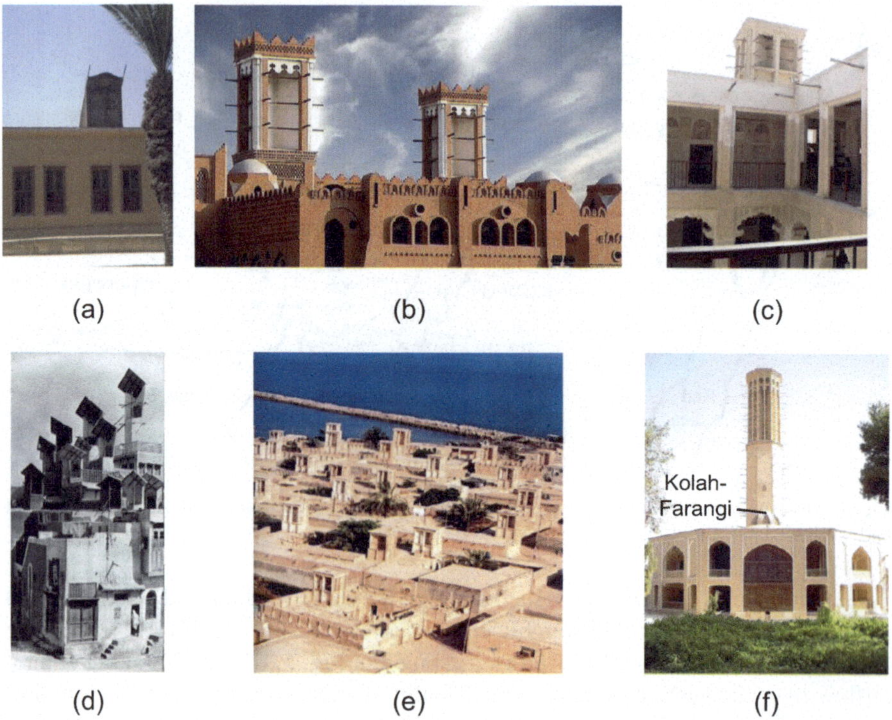

Fig. 4 Traditional/conventional wind towers in (**a**) Cairo, Egypt, (**b**) Doha, Qatar, (**c**) Dubai, UAE, (**d**) Heydar Abad, Pakistan (wind towers are known as "wind scoops" there [50]), plus (**e**) Bandar Lengeh (alongside the Persian Gulf) and (**f**) Yazd, both in Iran [50, 53]

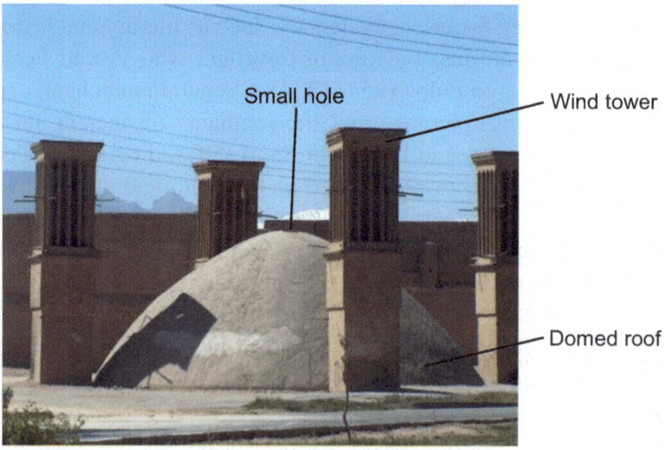

Fig. 5 Aub-anbar of "Rostam Giv" in Yazd City, Yazd, Iran [51]

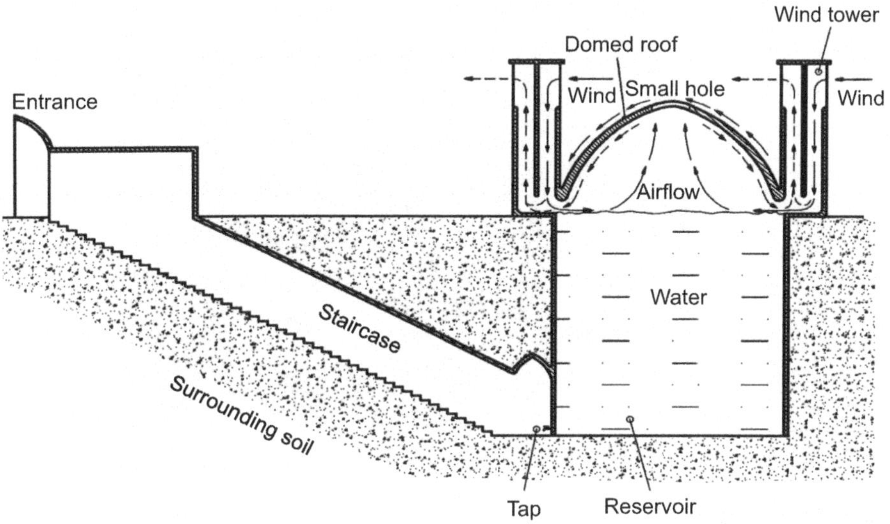

Fig. 6 Schematic illustration of a typical large, public cistern [51, 54, 57]

The above couplet reveals that having a wind tower—or perhaps several wind towers—in buildings was common practice at that time, one valued for creating a comfortable and happy indoor environment for inhabitants. Moreover, Rosenthal [61], Professor of Semitic Languages (1914–2003), in line with the claim made by Bahadori and Dehghani-Sanij [50], based on his extensive study of literary and historical texts, particularly Arabic texts and poems, states that Iran is the origin of the wind tower, which was called "bādhanj[6]" (Arabicized "bādahanjun") in olden times. Keeping its Persian name, it has been exported from Iran to other places such as Egypt over time.

Based on years of research on wind towers, the author of this chapter believes that early wind towers were simple architectural structures, made of clay and mud-brick, most likely one-sided, with an opening to capture the ambient wind and draw it into an inner space. It seems that people who lived in hot, dry regions and on the outskirts of the central desert of Iran learned to build these structures for the first time due to the crucial need for thermal comfort in their habitats on extremely hot summer days. Even today, Meybod, a major desert city located in the Province of Yazd, Iran, still has many one-sided Baudgeers (Fig. 7). Meybod, an ancient city, lying on important old caravan routes, dates back to the pre-Islamic era. This city was one of the most important cities in Iran during its history and has notable architectural monuments (e.g., Narenj, or Narin, Castle and Chapar Khaneh[7]). Meybod

[6] The word "bādhanj" or "bādāhanj" is a compound Persian noun, which is called "Baudahang" in Farsi. Note that in Arabic, there is no the letter *G* (گ) that is used in Farsi. *Baud* means *wind* and *ahang* means *to pull out* [62]. According to the Persian dictionary, *Baudahang* is another name for *Baudgeer* [60].

[7] This Persian term means the "house of a courier," referring to the postal service used during the Achaemenid period (550–330 BC) [63].

Fig. 7 A photo of one-sided Baudgeers in Meybod City, Yazd Province, Iran [50, 52]

Fig. 8 View of (**a**) two-sided, (**b**) six-sided, and (**c**) cylindrical wind towers [50, 52]

or other historical desert cities, such as Yazd, Ardakan, Naein, and Tabas, where one-sided Baudgeers still exist in abundance, are all possible candidate sites for the invention of these structures.

Traditional/conventional wind towers can generally be classified into six types [50, 52, 53]: one-sided (Figs. 4a, d and 6); two-sided (Fig. 8a); four-sided (Figs. 4b, c, e and 5), the most common form; six-sided, or hexahedral (Fig. 8b); eight-sided, or octahedral (Fig. 4f); and cylindrical structure (Fig. 8c). In areas where the direction of the prevailing and pleasant wind is known, builders/architects have built wind towers in a one-sided form facing the wind, but in areas with variable wind directions, they have made other types—especially cylindrical wind towers, the most evolved type, which are able to grab the ambient airflow from all directions. The historical and evolutionary course of traditional wind towers, from one-sided to cylindrical types, reveals that there have also been extensive engineering, technology, and architectural advances over time. In the past, as demonstrated in Fig. 9, the culmination of the architects/builders' masterpiece was the construction of two- or three-storey wind towers of different sizes and forms. These types of wind towers can capture wind from different heights—acknowledging the important fact that with the rising height from the ground surface, wind velocity increases and becomes more consistent because of lowering the surface friction. These wind towers have separate air channels on each storey.

Traditional/conventional wind towers have two main components (Fig. 10), including (1) the *head,* seen in various forms, heights, and decorative styles, with one to several air-openings, and (2) the *column* in different altitudes and shapes, inside which air channels are divided into different forms (e.g., cross-shaped), excluding one-sided wind towers [50].

Traditional/conventional wind towers act in various ways, depending on the velocity of the ambient wind and the time of day, as shown in Fig. 11, described as follows [50, 64]:

1. When the ambient wind velocity is high enough, the wind tower pulls airflow into the inner space of a building, like a fan, due to the pressure difference created between the wind tower's air-opening(s) and the building's windows and door(s). In this case, a part of the ambient airflow enters the interior space of the building—such as the main floor and/or basement—via the wind tower's air-opening(s) and channel(s) that face the wind direction. Then, after distribution and circulation inside, the air exits through the building's windows and door(s); another part of the airflow intake comes out directly through the wind tower's air channel(s) and opening(s) that are positioned opposite the direction of the wind, without having circulated indoors. Note that there are doors or windows between the wind tower's column and the main floor and/or basement. Airflow is controllable by opening and closing them.
2. When the velocity of the ambient wind is zero or very low, depending on the amount of heat or cold stored in the mass/body of the wind tower, the air flows

Fig. 9 View of two-storey wind towers of various sizes and shapes in (**a**) Abarkouh City, and (**b**) Kharānaq village, Ardakan City, Yazd Province, as well as (**c**) Sarhang Abad village, Zavareh district, Ardestan City, Isfahan Province, Iran [50, 52]

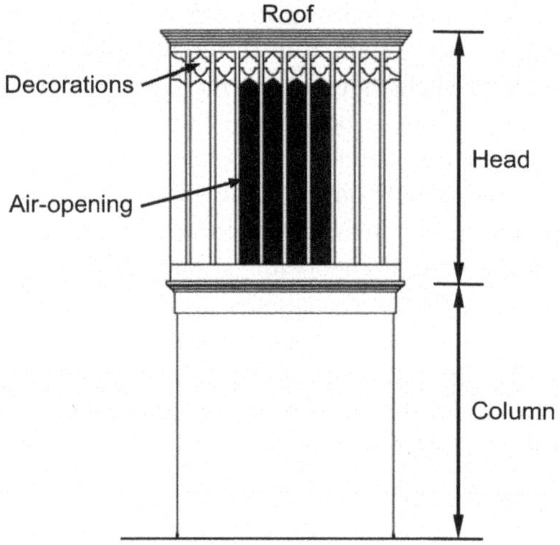

Fig. 10 Traditional/conventional wind tower components [50]

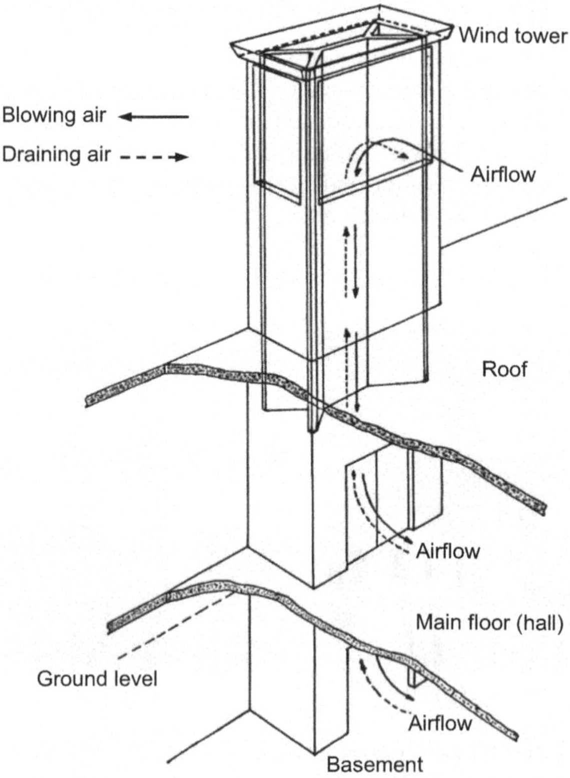

Fig. 11 Operation of a traditional/conventional wind tower based on the existence or absence of wind and the time of day [50, 64]

upwards or downwards naturally in the column, respectively. During the day, the wind tower's body absorbs heat through the sun and its surrounding surfaces, so the absorbed heat causes the air inside the column to warm up. Thus, the air inside flows upwards due to less density. This leads to the flow of air entering the interior space of the building via the windows and door(s) facing the courtyard after cooling as a result of passing through shrubs, plants, and often, a pool of water that has a large surface and shallow depth. In contrast, at night, the mass of the wind tower loses its heat because of the exchange of heat with its surroundings, so the air inside the column cools and flows downwards due to higher density and enters the inner space of the building. It should be noted that the amount of thermal energy (heat or cold) that can be stored in the wind tower's body is limited because of its low mass and specific heat. For this reason, establishing natural airflow due to buoyancy effect into or out of the building via the wind tower is also restricted when there is no wind and is only possible for a few hours. However, hot and dry areas of Iran have relatively constant seasonal and daily patterns of wind.

For example, Mazidi et al. [65, 66], based on their experimental measurements of the main mansion of Dowlat Abad Garden in Yazd City (Fig. 4f), showed that when the velocity of ambient air is higher than ~5 m/s, the wind tower plays the role of drawing air into the building, but once the velocity is less than ~5 m/s, its role is to suck the air from the building like a chimney, which means to pull out the air from the interior space and release it to the environment. In both cases, the wind tower is able to naturally ventilate the inner space. However, when the temperature of the ambient air is high, its thermal performance is not suitable.

In general, the operation of a traditional/conventional wind tower is not fixed during the day and night in the hot/warm months of the year. Its cooling impact and contributions to air circulation in the interior space depend on various factors, such as air temperature fluctuations, wind conditions (velocity and direction), and the intensity of solar radiation [64].

In some old buildings, builders/architects have combined traditional/conventional wind towers with other structures—for instance, domed roofs (Fig. 12), a Kolah-Farangi[8] (Fig. 13), an underground canal (Fig. 14) or a Qanat[9] (Fig. 15), basements, and courtyards—in order to improve ventilation and thermal performance and consequently establish a more pleasant thermal environment for inhabitants during hot summer days. Domed roofs help in reducing heat transfer from the outdoors to the interior spaces of a building, as the dome's height causes rising warm air to accumulate under it, producing cooler conditions at floor level [64, 67, 68]. As illustrated in Figs. 5 and 6, domed roofs have also been employed in large, public cisterns for the reason stated above. Additionally, a small hole was commonly utilized at the tip of the dome to release warm air from inside to the environment.

[8] This is a Persian word that refers to a particular type of architectural structure.

[9] A Qanat (or Karez) is a gently sloping underground canal used to transport water over long distances from an aquifer, or water well to the surface, for drinking and irrigation [51, 69].

Fig. 12 An integration of several wind towers and domed roofs in one of the mansions of Dowlat Abad Garden in Yazd City, Yazd Province, Iran [50]

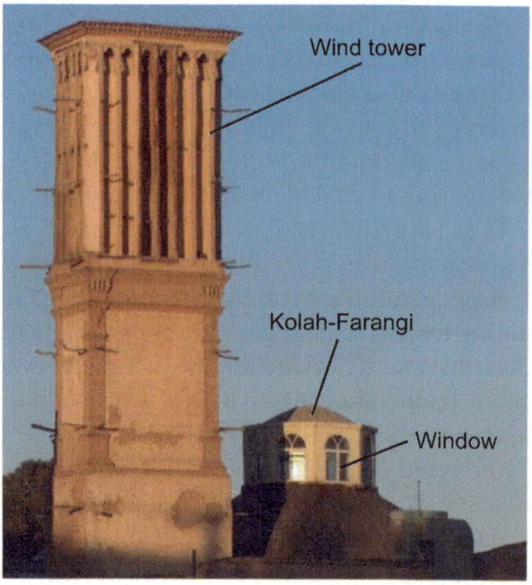

Fig. 13 A combination of a wind tower and a Kolah-Farangi in the Cultural Heritage Building in Yazd City [50]

As demonstrated in Fig. 13, a relatively large Kolah-Farangi that contains a number of openings, or windows in new designs, which can be opened or closed to allow the airflow in and out, was erected upon the main hall (or room) of the mansion and next to a tall wind tower for two purposes [50, 52]: (1) raising the rate of ventilation and (2) providing natural lighting. Figure 4f, showing the main mansion of Dowlat Abad Garden, and Fig. 9a, displaying the Aghazadeh mansion, are other examples of a combination of a Kolah-Farangi and a traditional/conventional wind tower.

In some designs, builders/architects have combined an underground tunnel (Fig. 14) or a Qanat (Fig. 15) and a basement with a traditional/conventional wind tower in order to take advantage of evaporative cooling and improve the quality of

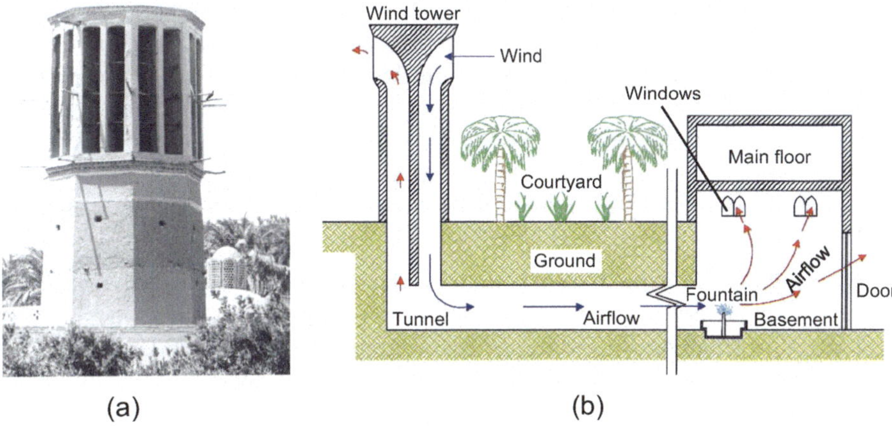

Fig. 14 (**a**) Photo and (**b**) schematic cross-section of an old wind tower, built outside of the building, located in Bam City, Kerman Province, Iran [50, 64]

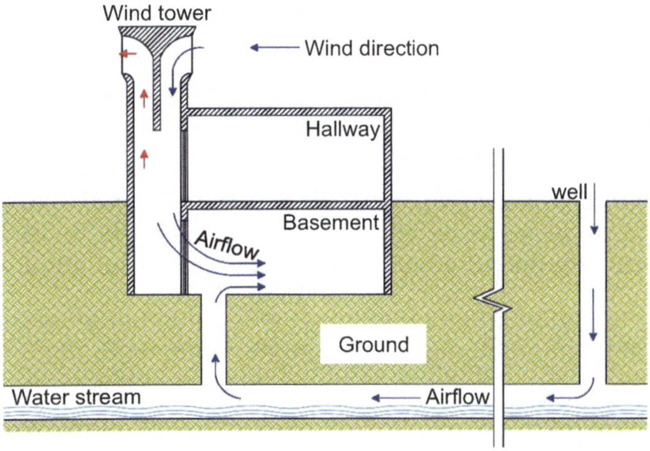

Fig. 15 Schematic cross-section of a combination of a wind tower and a Qanat [50, 64]

air entering the dwellings. In these designs, air that passes over the moist surfaces of the underground channel and basement walls, or the water surface of a Qanat, has increased humidity and reduced temperature, and even dust is removed. In Fig. 14, a small water pool and its fountain further cool the cold airflow out of the underground tunnel evaporatively. In the city of Yazd, a small pool and a fountain can be observed under many wind towers.

The design demonstrated in Fig. 15 is related to the city of Yazd (based on a personal communication with Prof. Bahadori). Many Qanats once supplied its drinking water and irrigated its agricultural fields, but most of them are abandoned today. These Qanats also allowed the intelligent builder/architect to utilize them to cool the air via evaporation. The mixture of two flows of air—one supplied by the wind tower and another passing over the water surface of the Qanat—created a thermally pleasant environment inside the basement. In general, in hot and arid areas, basements were the best place for inhabitants to rest on hot summer afternoons, and wind towers provided thermal comfort and good IAQ for these places. Figure 16 shows an example of a combination of a one-sided wind tower and a basement in Iraq.

Despite the benefits of using traditional/conventional wind towers, they also have some restrictions, such as [50, 52, 53]:

- The head of these wind towers is immobile and cannot rotate to grab the maximum ambient airflow in zones with variable wind directions.
- They allow small birds, insects, and dust to come into inner spaces.
- These exposed structures gradually erode due to natural causes like precipitation, wind, and sun.
- A portion of air entering through wind towers, excluding the one-sided type, exits with no circulation through building interiors.
- Only poor performance is possible in zones with very low wind velocity.
- Installation and maintenance limitations occur, and only a small number of wind towers can be utilized in one building.
- The amount of coolness that can be stored in the mass/body of these wind towers is restricted and may not be sufficient to meet cooling demands on hot days.
- In very still weather, circulation is purely due to density differences in the air, which can prevent adequate flow.

To diminish or even eliminate the restrictions of traditional/conventional wind towers, as well as to enhance their performance and improve the conditions of thermal comfort in dwellings, several new, modern wind tower designs, has been

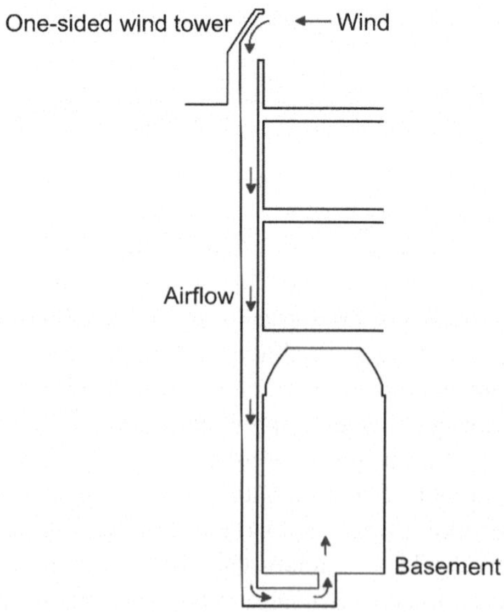

Fig. 16 Schematic illustration of a combination of a one-sided wind tower and a basement in Iraq [50, 70]

proposed [50, 52, 53, 71–87], each with its own benefits and limitations. The next section briefly describes some of them.

3 Innovative, Modern Wind Towers

Various innovative, modern wind tower designs, most of which rely on evaporative cooling techniques, have been offered and developed for natural space cooling and AC. Evaporative cooling means to cool flowing air to, or close to, its dew point temperature by saturating it with water vapour (e.g., by spraying water into the airflow). In other words, in the evaporation process, water absorbs a large quantity of heat from the air, consequently decreasing the temperature of the airflow [38, 88]. This technique is one of the most effective ways to lessen the temperature and increase the RH of the airflow entering the living/working spaces of a building in hot/warm, arid environments. A number of these innovative, new designs have been constructed and implemented, and some of them are at the stage of development and commercialization. Several modern wind towers, both implemented and developing, will be explained below.

Bahadori [89, 90], who is a pioneer in wind tower studies, proposed two innovative wind tower designs, namely "wind towers with wetted surfaces" (Fig. 17) and "wind towers with wetted columns" (Fig. 18). They take advantage of techniques of evaporative cooling to naturally ventilate and passively cool buildings situated in hot, dry regions. Bahadori et al. [84, 85, 91–100] examined the operation and efficiency of these two innovative designs by experimental, numerical, and analytical approaches and compared the results with those of conventional wind towers. In the first design, cellulose pads or straw packs are used in the wind tower's head at its openings, as displayed in Fig. 17. The pads or straw packs are moistened via pouring water on them. A small pump is employed to circulate water in the system. The ambient airflow is evaporatively cooled via passing these moistened pads or straw packs, then enters the building through the column. In this design, when the velocity of the ambient wind is zero or very low, the solar chimney or air heater can help create the necessary airflow within the building. The air inside the solar chimney or air heater moves upwards after absorbing heat, causing the ambient air to enter the interior space after passing through the wetted pads or straw packs and cooling. Additionally, due to the buoyancy effect created in the wind tower's column, the airflow can also move downwards within it. In the second design, thick fabric panels—kept taut by stretching them between two horizontal rods—or clay conduits are mounted in the wind tower's column, as indicated in Fig. 18. At the top of the column, to moisten the panels or conduits, water drops from horizontal pipes pierced with holes to ensure constant wetting. Run-off water gathers in a small pool at the bottom of the tower's column; by means of a small pump, water is then recirculated to the top of the column. The incoming airflow to the column of the tower moves downward over these moistening structures and cools evaporatively; the cooled air then enters the interior spaces. These innovative wind towers have been built in two

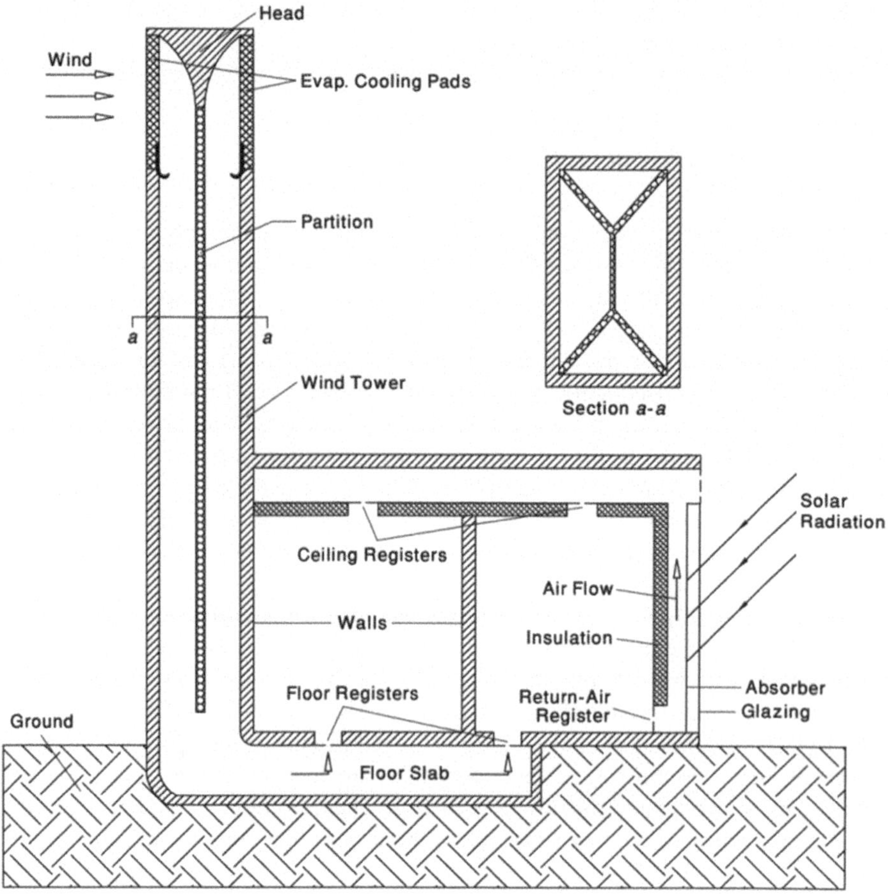

Fig. 17 Schematic cross-section of an innovative wind tower with wetted surfaces, and integrated with a solar chimney or an air heater [71, 90]

different places in Iran to evaluate their operation and efficiency empirically [50]: one on the mosque of Yazd University in Yazd and another at the site of Scientific and Industrial Research Organization near Shahryar in Tehran. The results demonstrate that the innovative wind tower designs have better thermal performance than conventional ones, achieving lower temperatures and higher RHs—as shown in Fig. 19, for example. Moreover, the wind towers with wetted columns are more efficient relative to other wind towers once the velocity of the ambient wind is high sufficient. In contrast, the wind towers with wetted surfaces are more efficient when the ambient wind velocity is low.

A modular wind tower with moistened surfaces (Fig. 20), proposed and constructed by M.R.Khani et al. [71, 72, 100], has been specifically designed to provide low-energy AC in hot, arid zones. To determine the operation and efficiency of this modular design under zero (>0.01 m/s) ambient wind velocity, the parameters of

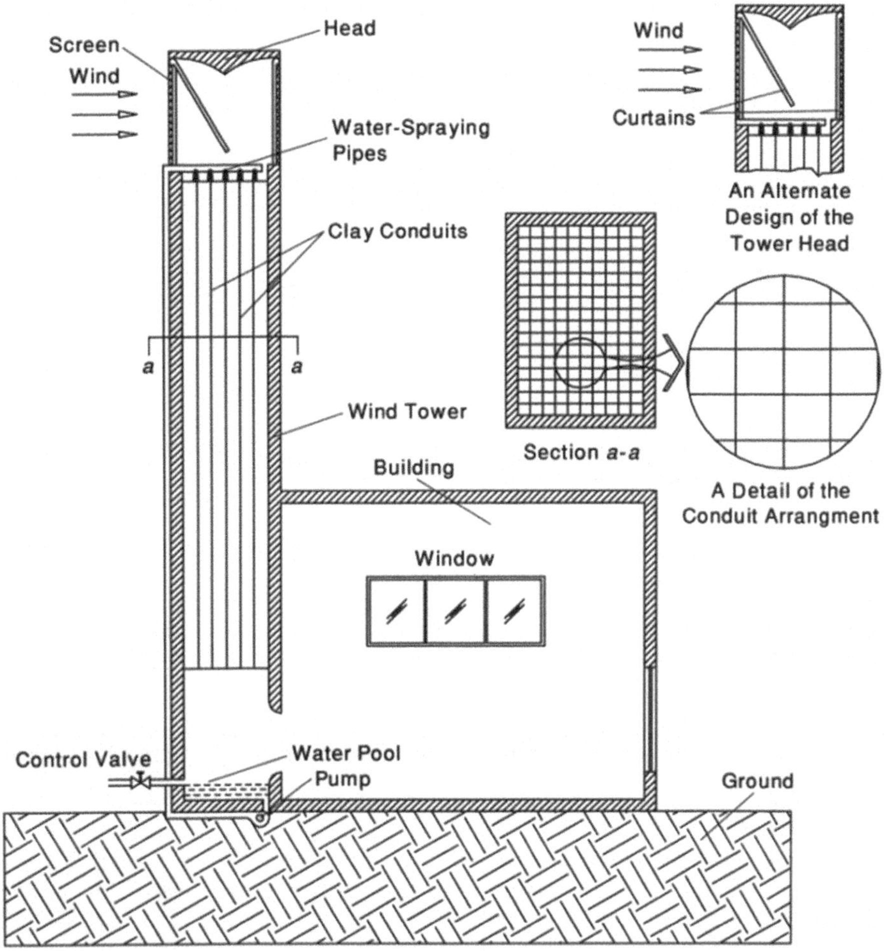

Fig. 18 Schematic cross-section of an innovative wind tower with wetted columns (either clay conduits or thick fabric panels) [71, 89]

airspeed, temperature, and RH were empirically and analytically examined at various times and several points (labelled 1–6 in Fig. 21) for a test building situated in the city of Kerman, Iran. Moistened straw packs were employed at the openings of this tower to evaporatively cool the air entering the column and then flowing to the living room. The highlight and importance of the modular wind tower is the feasibility of industrial production in a factory. Another notable feature is its simple and easy assembly and installation on a building or in an enclosed environment.

The empirical and analytical results obtained for the hottest days of the year, from 21 to 30 Aug. 2013, in Kerman indicate that the modular wind tower was able to substantially lessen the temperature of inflowing air—by an average of nearly 10 °C—and raise its RH—on average, by about 40%—by the time the air entered

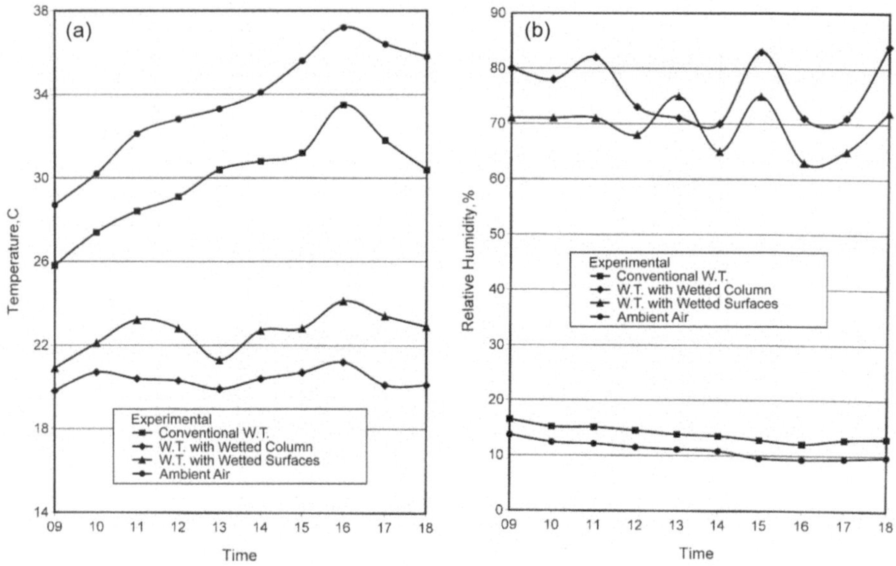

Fig. 19 Variations of (**a**) temperature and (**b**) relative humidity of the air leaving the wind towers (W.T.) and the ambient air at different hours on 10 Sept. 2003, based on experimental measurements on the mosque of the Yazd University in Yazd, Iran [91]

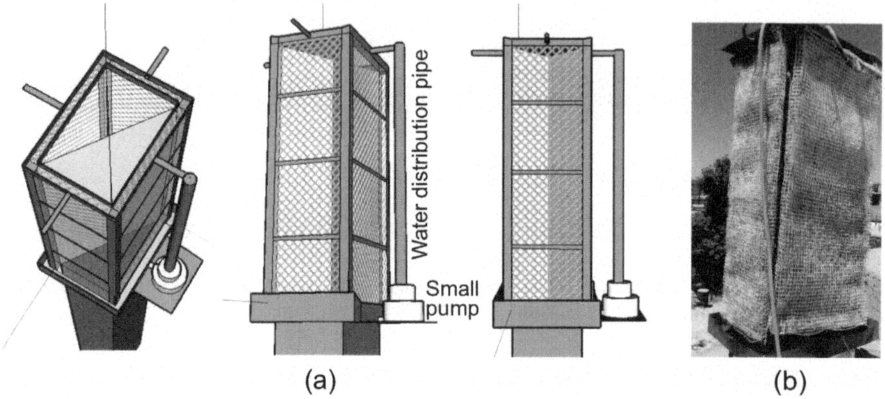

Fig. 20 (**a**) Schematic sketch and (**b**) a photograph of the modular wind tower's head [71, 72, 100]

the living room, even when there was no wind. Moreover, owing to the buoyancy effect, the tower was able to propel the mass flow rate of the ventilated air at an average speed of ~1.8 m/s by the time it entered the room, despite the absence of ambient wind. The thermal performance of this modular wind tower is greatly enhanced in the presence of wind. According to the psychometric chart shown in Fig. 22, when this wind tower was in operation, the living room air met the comfort-zone conditions established for Kerman.

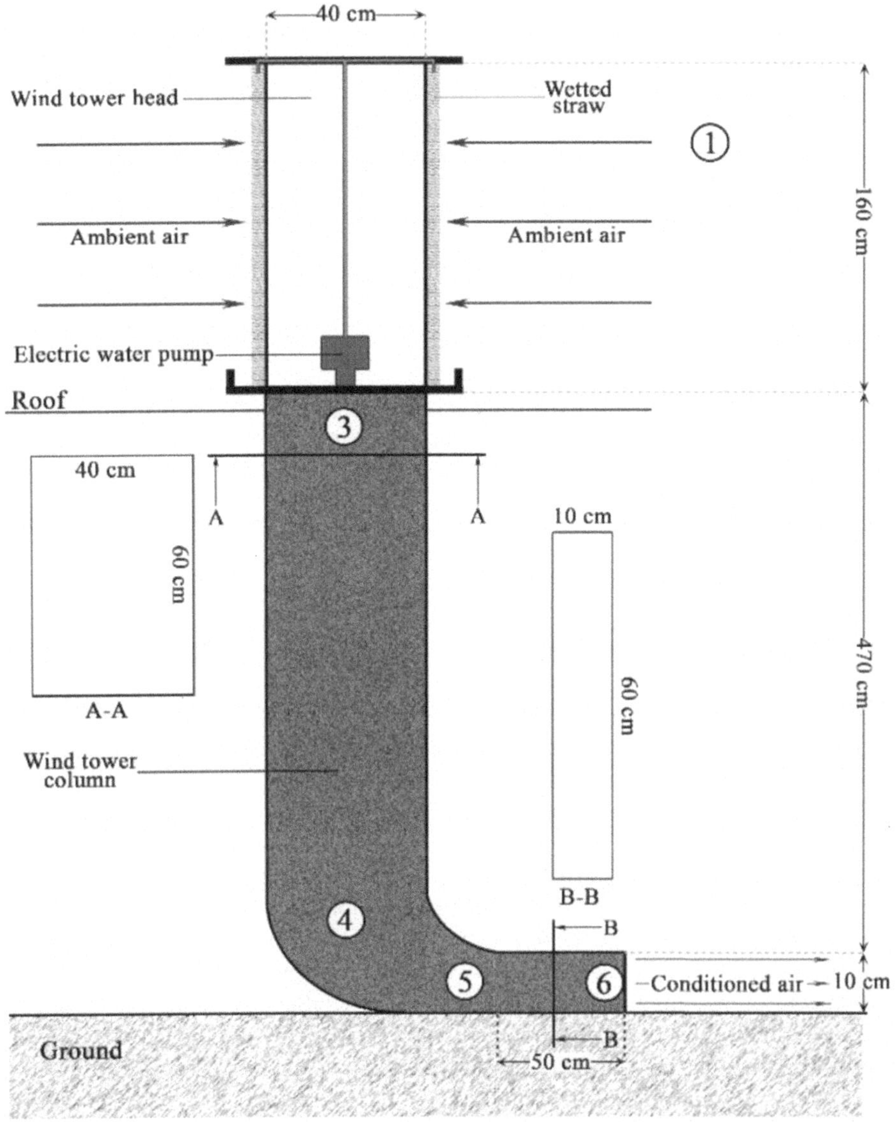

Fig. 21 Schematic cross-section of the modular wind tower. Circled numbers (1–6) show the experimental measurement points [71, 72, 100]

In Fig. 22, the beginning and ending points of the arrow represent the conditions of the ambient air and the air inside the room in which the wind tower was installed, respectively. Sample data presented in this psychometric chart were taken from the experiment done at 1:45 pm on 26 Aug. 2013. The direction of the arrow reveals that the variations in the air temperature and RH are caused by the cooling and humidifying that occur when air passes through the wind tower. The increase in RH and

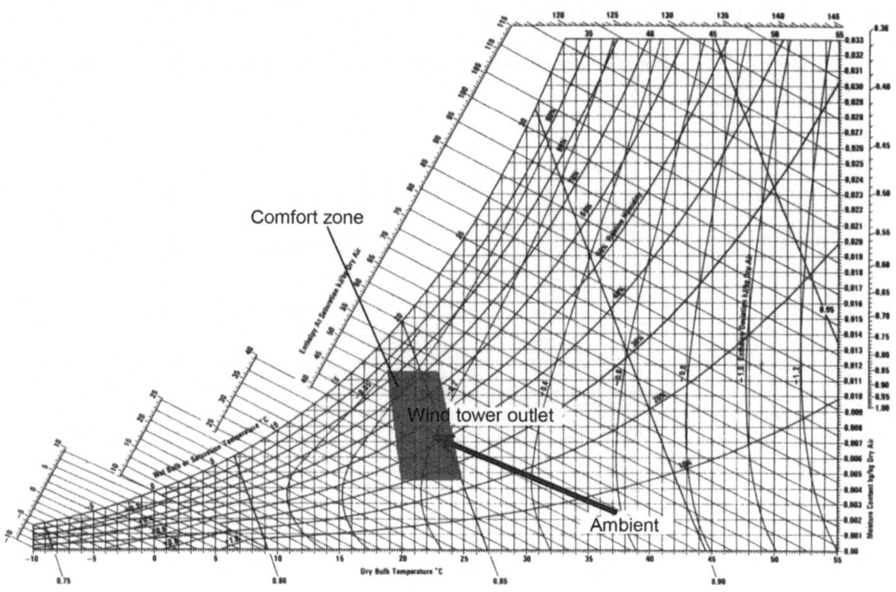

Fig. 22 Air property changes due to modular wind tower with moistened surfaces in a sample experiment, taken at 1:45 pm on 26 Aug. 2013 [71, 72, 100]

decrease in temperature respectively were from 7% to 44.3% and from 37.2 to 22.5 °C between the ambient air and the wind tower's exit opening, confirming the design's effective thermal performance.

To eliminate some the shortcomings of traditional/conventional wind towers, Dehghani-Sanij et al. [52, 53, 101] offered an innovative "wind tower with moving wetted surfaces" design. It consists of a fixed column, a rotating and vertically movable head—which can be adjusted manually or automatically in the direction of a region's maximum ambient wind velocity—a screened air-opening, a moistened pad, and a small pump (if needed), plus two sliding windows installed at the column base to control incoming airflow into a dwelling or confined space (Fig. 23). These windows can be opened or closed manually or automatically. When there is no need to operate the wind tower, its head can be retracted into an empty space between the interior and exterior parts of the column (Fig. 23b, section x-x). Furthermore, the head has an inclined roof that is slightly larger than the outer diameter of the column in order to prevent water leakage into the column during precipitation. To detect an area's maximum wind velocity direction and adjust the air-intake-openings perpendicular to this direction, a wind vane can be utilized. The tower can be made of various materials, for example, galvanized steel, fiberglass, and wood with preservative finish. For use in hot, dry zones and maximized thermal efficiency, a cellulose pad (preferred) or a straw pack can be inserted in the air opening and kept moist by means of a small circulating pump so that incoming air is cooled evaporatively, as illustrated in Fig. 24. In sunny zones, a small solar system can be installed to supply

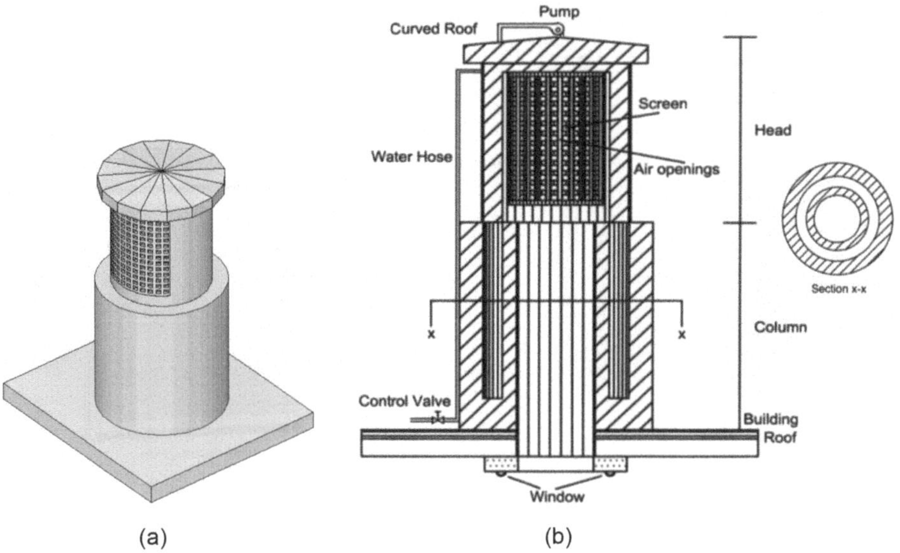

Fig. 23 (**a**) View of the innovative wind tower design and (**b**) a schematic cross-section with details [52, 53, 101]

electricity to the pump. A screen mounted over the air-intake-opening prevents small birds and insects entering the building.

To examine the ventilation performance of this innovative design, a computational fluid dynamics (CFD) simulation of airflow patterns around and inside the wind tower has been done [53]. The parameters of velocity, total pressure, and pressure coefficient distributions around and within the wind tower for three wind speeds of 0.1, 1.5, and 4 m/s were studied. Figure 25 shows the distributions of velocity and streamline evolution, total pressure, and pressure coefficient in the xz-plane for an inlet wind speed of 1.5 m/s, for example. Figure 26 indicates the distribution of velocity and streamlines at three different positions within the column of the wind tower—namely, at the entrance, middle, and exit of the column—for an inlet wind speed of 1.5 m/s. The amount and pattern of the flow of air at the entrance, middle, and exit of the column differ. There are also two rotational streams within the column, although they are not symmetric, nor are the distributions of velocity uniform. Based on the simulation outcomes, the innovative wind tower design by Dehghani-Sanij et al. [52, 53, 101] offers efficient, practical, and cost-effective natural ventilation and passive cooling of buildings/enclosed spaces, especially for windy, hot, and arid areas.

To improve the operation and efficiency of modern wind towers and enhance the conditions of indoor thermal comfort as well as IAQ of dwellings and enclosed spaces, the wind towers can be combined with a variety of technologies such as Trombe walls, solar chimneys, thermosyphons, low-powered or solar fans, Kolah-Farangis, as well as approaches such as renewable energy (more specifically, solar, geothermal, and wind), long- or short-term energy storage systems (for instance,

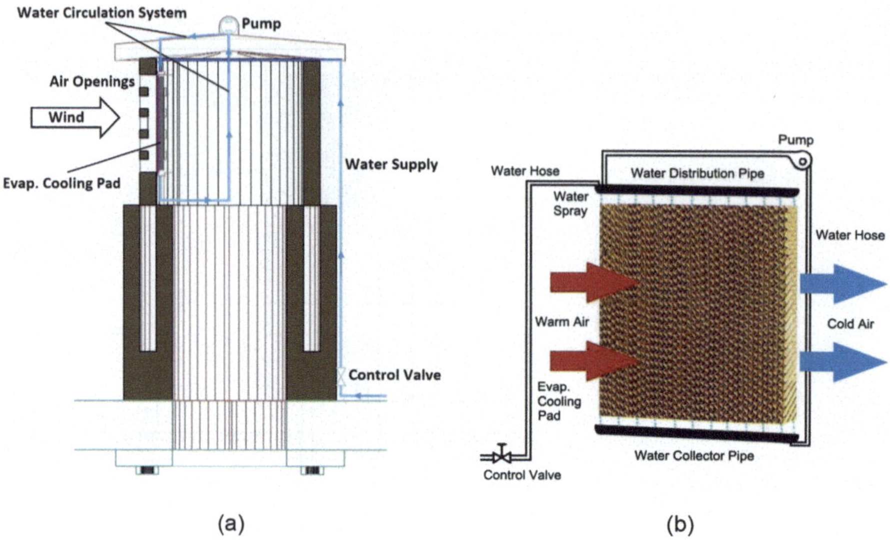

Fig. 24 A schematic depiction of (**a**) location for installing wetted pad in the modern wind tower and (**b**) water distribution system plus air cooling via the wet pad [53, 101]

thermal energy storage), and semi-natural/low-energy methods for ice production [5, 52, 87].

In the book [5], Dehghani-Sanij and Bahadori offer two new hybrid AC systems, one for hot/warm, humid climates (Fig. 27) and another for hot/warm, arid climates (Fig. 28). Both integrate a modern wind tower with the rotating and vertically movable head as suggested by Dehghani-Sanij et al. [52, 53, 101] but with some changes, a cold thermal energy storage (CTES) system and/or a shallow geothermal system (as cold sources), a solar (or low-powered) fan, and an air filter. Note that it is possible to utilize various cold sources, depending on their accessibility, local climate conditions, and geographical features [5]. In Fig. 27, the inlet fresh air is cooled after entering the wind tower's column and passing over pipes mounted inside the column and containing cold water. The cooled fresh air, after purification from dust, other air pollutants, and pathogens by a passive filter, enters the building, providing natural ventilation and cooling of the inner spaces. After distribution and circulation indoors, the air is directed out via the window(s). Small pumps move the cold water provided by the CTES and/or shallow geothermal systems to a heat exchanger where the cold thermal energy is transferred to back the water that circulates inside the pipes within the wind tower's column. The damp proof and mould-resistant materials used for the internal surfaces of the wind tower's column prevent damage from the water that forms due to condensation on the external surfaces of the pipes. This water and any taken from the local water distribution system gather in a small pool at the end of the column and can be reused in the hybrid AC system or utilized for other purposes (e.g., irrigation of green spaces and gardens) after filtration by an appropriate filter.

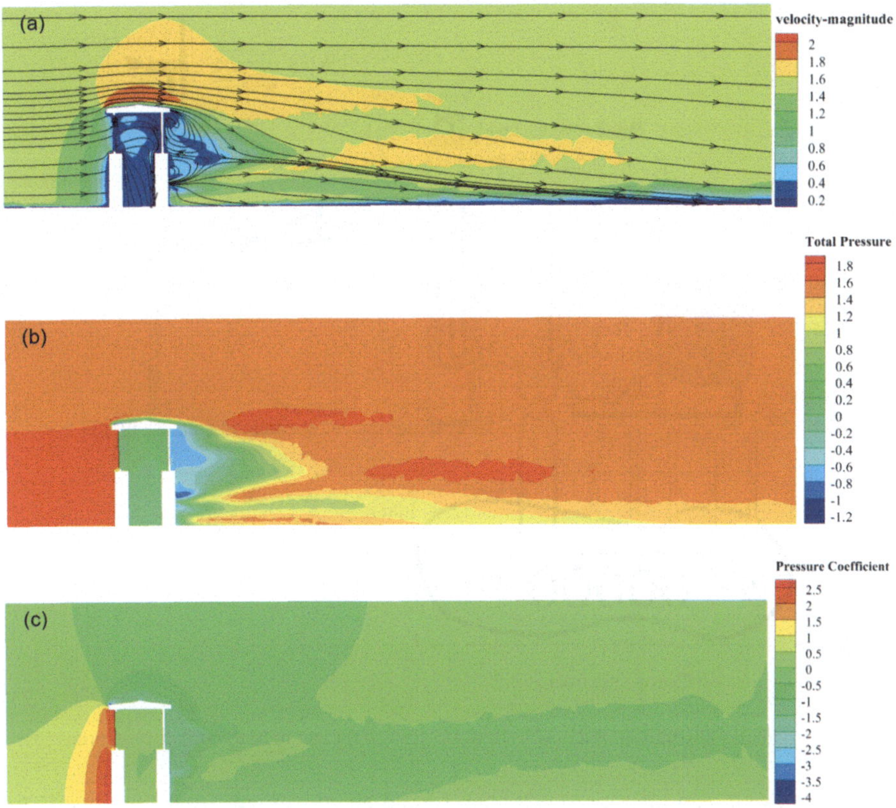

Fig. 25 Distributions of (**a**) velocity and streamline evolution, (**b**) total pressure (in kPa), and (**c**) pressure coefficient in the xz-plane for inlet wind speed of 1.5 m/s [53]

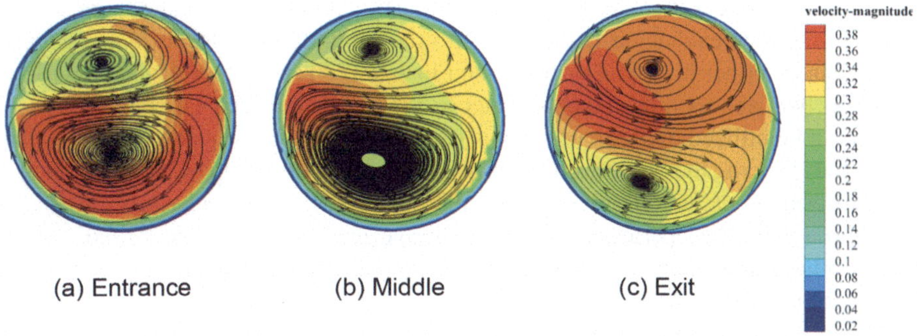

Fig. 26 Velocity distribution and streamlines at the (**a**) entrance, (**b**) middle, and (**c**) exit of the modern wind tower's column for inlet wind speed of 1.5 m/s [53]

When the velocity of the ambient wind is insufficient or under calm conditions, a low-powered or solar fan, installed either inside the wind tower's column or at the

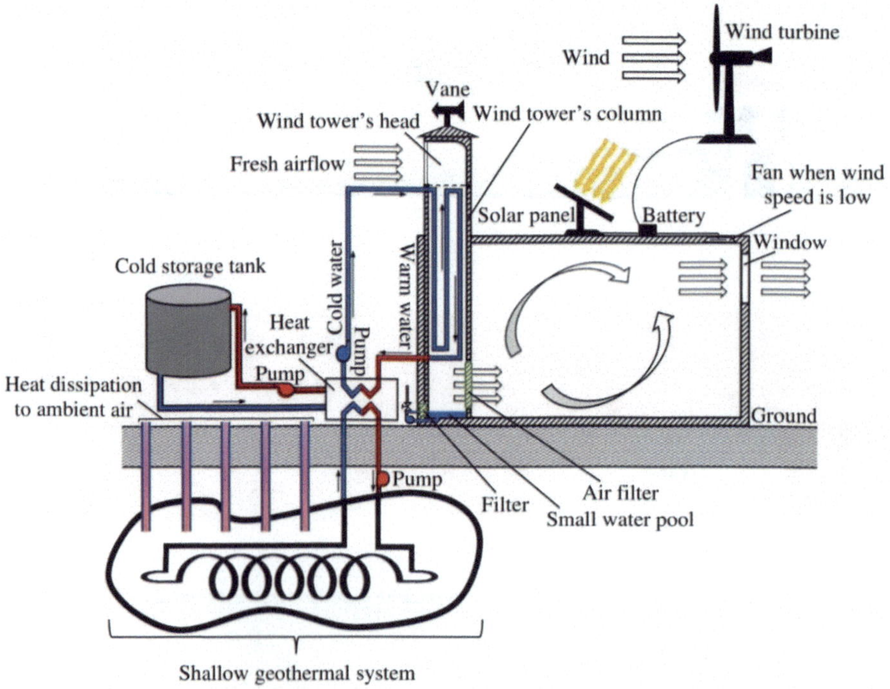

Fig. 27 Schematic illustration of a new hybrid AC system for use in hot/warm, humid areas [5]

place of air exit, can operate to establish forced air movement from the outside to the building in order to sustain the system. To supply the electrical energy needs of all electrical equipment used in the hybrid AC system, such as the fan and small circulating pumps, a small solar or wind energy system can be employed.

In the shallow geothermal system, as demonstrated in Figs. 27 and 28, the warm water in the heat exchanger is directed into the ground via a closed-loop system, which can be either a vertical or horizontal installation, sized appropriately, to dissipate its heat and then resend cold water to the heat exchanger. Passive thermosiphons can be utilized to dissipate surplus ground heat into the outside, especially at night, thus keeping the ground as cool as possible. This action can be called recharging the cooling potential of the geothermal repository, so as not to lose its effectiveness throughout the operation. Different storage mediums, such as chilled water, ice, or other phase change materials (PCM) can be applied in the CTES system to store cooling energy [5].

The proposed hybrid AC system for use in hot/warm, dry regions consists of moistened thick fabric panels or clay conduits, installed within the wind tower's column (similar to the new design offered by Bahadori [89]), instead of using a closed-loop pipe system to take advantage of evaporative cooling to diminish and enhance, respectively, the temperature and RH of the hot, dry fresh air before admitting it to flow inside the building. Furthermore, a moistened cellulose pad or straw

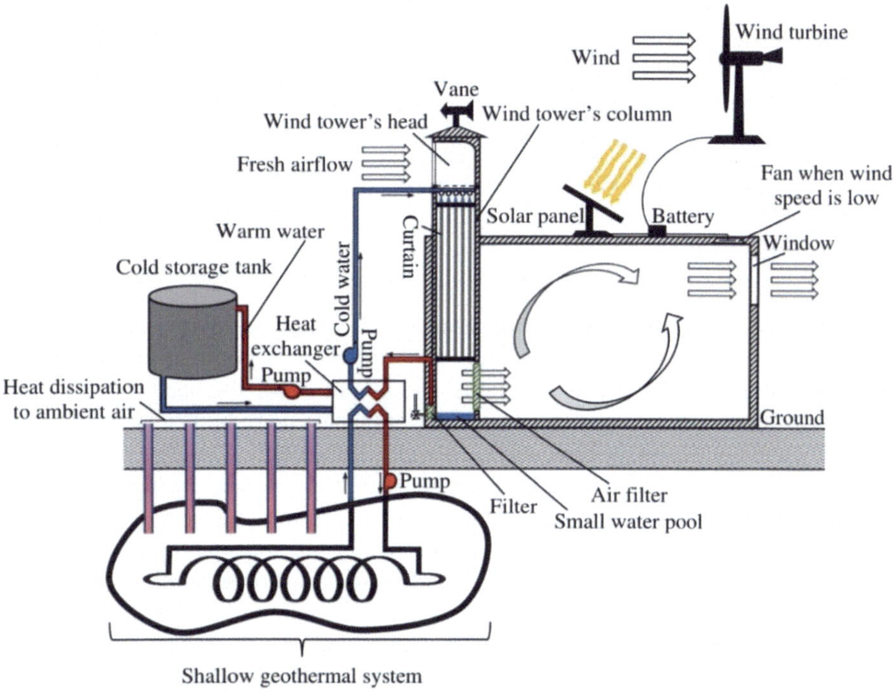

Fig. 28 Schematic illustration of a new hybrid AC system for use in hot/warm, arid areas [5]

pack can be mounted at the air-opening of the tower's head to evaporatively cool the flow of air via pouring cold water on the pad or straw. As shown in Fig. 28, a number of thick fabric panels, installed inside the column, are wetted by spraying cold water on them in order to cool the fresh airflow that enters the interior spaces after passing via a passive filter. Excess water is gathered, filtered, and reused by recirculating it in the heat exchanger and then returned to the wind tower by a small pump.

With the development of these proposed hybrid AC technologies and using them as a central (or district) thermal system, the cooling demands of a bunch of buildings and even a small community can be met during the hot/warm months. These hybrid, integrated AC systems can help improve the buildings' energy effectiveness, optimize total energy use, and manage energy demands, particularly during peak times, and reduce associated energy costs. A smart control system is required to run and manage the hybrid, integrated HVAC systems, especially for those proposed hybrid AC technologies.

In the end, it is necessary to mention that in conventional AC systems, there is the potential for pathogen transmission (e.g., COVID-19 and legionnaire's disease). They do not provide adequate fresh air turnaround in dwellings/buildings, as the cold air is recirculated indoors to save energy, sometimes with filtering or disinfection. To remove or minimize these health problems, in addition to modification or even redesign of the conventional AC systems, the modern wind towers, individually or integrated with other technologies/approaches, can be employed, since they are capable of providing the cooled fresh air naturally and exchanging the internal air frequently. If needed, an ultraviolet light fixture can be installed inside the column, or at its air outlet, of a modern wind tower in order to destroy COVID-19 and other viruses.

4 Conclusions

As a result of the rise in global ambient air temperatures (by over 2 °C above the pre-industrial era's level [102]), in addition to the world's population growth, the improvement in people's living standards, and other factors, the need for using AC systems (both mobile and stationary) of different types, sizes, and capacities is dramatically and continuously growing worldwide. The electrical energy required for most of AC systems, which includes huge quantities of power consumption, is supplied from fossil fuel-based power plants, thus leading to pose local, regional, and global environmental challenges and threats such as air, water and soil pollutions, climate change, and depletion of fossil fuel resources. For example, these challenges and threats can result in the phenomenon of "environmental or climate refugees." It is estimated that by 2050, the negative impacts of global warming and climate change will compel around 150–200 million individuals to leave their dwellings and immigrate to other places (either in their own countries or to other countries totally), causing widely political, social, and economic problems and perturbations [5, 103, 104]. Another notable point is that more than one billion people at the global level currently lack access to electrical energy, mainly living in rural or economically disadvantaged regions (e.g., in Sub-Saharan Africa and India), that suffer from the absence of having appropriate air-conditioners [5, 105]. Hence, in addition to essential alterations in all sectors that emit carbon globally, especially in the AC sector, finding, developing, and applying new, green, efficient, practical, sustainable, and environmentally friendly strategies are vital to attain a cleaner environment, sustainable habitats, and well-being for all people around the world.

In the AC sector, in addition to improving the energy efficiency in all conventional AC systems and adding sufficient fresh air to the flow of air that circulates inside the buildings, the use of natural cooling and ventilation systems like wind towers, particularly innovative and modern ones, either individually or combined with other technologies, can aid not only in providing the conditions of indoor thermal comfort and good IAQ for inhabitants in the warmer months (specifically for rural, poor, and economically vulnerable jurisdictions that suffer from energy poverty) but also in diminishing energy consumption, its related costs in buildings, especially at times of on-peak demand, plus the subsequent environmental footprints. Innovative and modern wind towers, as stand-alone or hybrid, integrated AC systems, can be part of efficient, eco-friendly, and sustainable architectural designs in new, green buildings, which today are called "net zero energy buildings."

References

1. ASHRAE Guideline 10P, Interactions Affecting the Achievement of Acceptable Indoor Environments, Second Public Review, The American Society of Heating, Refrigerating and Air-Conditioning Engineers (ASHRAE), Inc., Atlanta, GA, USA; 2010.
2. Salvador, M., Grieu, S. Methodology for the design of energy production and storage systems in buildings: minimization of the energy impact on the electricity grid, Energy and Buildings, Vol. 47, 2012, 659–673.
3. Klepeis, N.E., Nelson, W.C., Ott, W.R., Robinson, J.P., Tsang, A.M., Switzer, P., Behar, J.V., Hern, S.C., Engelmann, W.H. The National Human Activity Pattern Survey (NHAPS): A Resource for Assessing Exposure to Environmental Pollutants, Journal of Exposure Science & Environmental Epidemiology, Vol. 11, No. 3, 2001, 231–252.
4. Liquna, G., Yanqunb, G. Study on building materials and indoor pollution, Procedia Engineering, Vol. 21, 2011, 789–794.
5. Dehghani-Sanij, A.R., Bahadori, M.N. Ice-houses: Energy, Architecture and Sustainability, Elsevier, Imprint by Academic Press, UK; 2021.
6. Ghodrati, N., Samari, M., Shafiei, M.W.M. Green buildings impacts on occupants' health and productivity, Journal of Applied Sciences Research, Vol. 8, No. 8, 2012, 4235–4241.
7. Evans, G.W., Stecker, R. Motivational consequences of environmental stress, Journal of Environmental Psychology, Vol. 24, No. 2, 2004, 143–165.
8. Boyce, P.R. The impact of light in buildings on human health, Indoor and Built Environment, Vol. 19, No. 1, 2010, 8–20.
9. Ni, M.Y., Yao, X.I., Cheung, F., Wu, J.T., Schooling, C.M., Pang, H., Leung, G.M. Determinants of physical, mental and social well-being: a longitudinal environment-wide association study, International Journal of Epidemiology, Vol. 49, No. 2, 2020, 380–389.
10. Awada, M., Becerik-Gerber, B., Hoque, S., O'Neill, Z., Pedrielli, G., Wen, J., Wu, T. Ten questions concerning occupant health in buildings during normal operations and extreme events including the COVID-19 pandemic, Building and Environment, Vol. 188, 107480, 2021.
11. Loomis, D., Grosse, Y., Lauby-Secretan, B., El Ghissassi, F., Bouvard, V., Benbrahim-Tallaa, L., Guha, N., Baan, R., Mattock, H., Straif, K. The carcinogenicity of outdoor air pollution, Lancet Oncology, Vol. 14, Dec. 2013, 1262–1263.
12. Davis, A. Home Environmental Health Risks, The Online Journal of Issues in Nursing, Vol. 12, No. 2, May 2007.
13. Sofuoglu, S.C., Aslan, G., Inal, F., Sofuoglu, A. An assessment of indoor air concentrations and health risks of volatile organic compounds in three primary schools, International Journal of Hygiene and Environmental Health, Vol. 214, No. 1, 2011, 36–46.
14. Evans, G.W. The built environment and mental health, Journal of Urban Health, Vol. 80, No. 4, 2003, 536–555.
15. Al horr, Y., Arif, M., Katafygiotou, M., Mazroei, A., Kaushik, A., Elsarrag, E. Impact of indoor environmental quality on occupant well-being and comfort: A review of the literature, International Journal of Sustainable Built Environment, Vol. 5, 2016, 1–11.
16. Nimlyat, P.S., Kandar, M.Z., Sediadi, E. Multitrait-multimethod analysis of subjective and objective methods of indoor environmental quality assessment in buildings, Building Simulation, Vol. 11, No. 2, 2018, 347–358.
17. De Giuli, V., Da Pos, O., De Carli, M., Indoor environmental quality and pupil perception in Italian primary schools, Building and Environment, Vol. 56, 2012, 335–345.
18. Saraiva, T.S., De Almeida, M., Bragança, L., Barbosa, M.T. Environmental comfort indicators for school buildings in sustainability assessment tools, Sustainability, Vol. 10, No. 6, 1849, 2018.
19. Pitarma, R., Marques, G., Ferreira, B.R. Monitoring indoor air quality for enhanced occupational health, Journal of Medical Systems, Vol. 41, No. 23, 2017, 1–8.

20. Moreno-Rangel, A., Sharpe, T., Musau, F., McGill, G. Field evaluation of a low-cost indoor air quality monitor to quantify exposure to pollutants in residential environments, Journal of Sensors and Sensor Systems, Vol. 7, No. 1, 2018, 373–388.
21. Cincinelli, A., Martellini, T. Indoor air quality and health, International Journal of Environmental Research and Public Health, Vol. 14, 1286, 2017.
22. Moreno-Rangel, A., Baek, J., Roh, T., Xu, X., Carrillo, G. Assessing Impact of Household Intervention on Indoor Air Quality and Health of Children with Asthma in the US-Mexico Border: A Pilot Study, Journal of Environmental and Public Health, Vol. 2020, Article ID 6042146, 2020.
23. ISO Standard 7730:1994, Moderate Thermal Environments—Determination of the PMV and PPD Indices and Specification of the Conditions for Thermal Comfort, International Organization for Standardization (ISO), Geneva, Switzerland; 1994.
24. ASHRAE Standard 55-1992, Thermal environmental conditions for human occupancy, The American Society of Heating, Refrigerating, and Air Conditioning Engineers (ASHRAE), Inc., Atlanta, GA, USA; 1992.
25. Li, J. The science of clothing comfort, Textile Progress, Vol. 31, No. 1–2, 2001, 1–135.
26. Huang, T., Li, J., Xie, Y., Niu, J., Mak, C.M. Simultaneous environmental parameter monitoring and human subject survey regarding outdoor thermal comfort and its modelling, Building and Environment, Vol. 125, 2017, 502–514.
27. Butera, F.M. Chapter 3—Principles of thermal comfort, Renewable and Sustainable Energy Reviews, Vol. 2, No. 1–2, 1998, 39–66.
28. Ji, W., Luo, M., Cao, B., Zhu, Y., Geng, Y., Lin, B. A new method to study human metabolic rate changes and thermal comfort in physical exercise by CO_2 measurement in an airtight chamber, Energy and Buildings, Vol. 177, 2018, 402–412.
29. Quang, T.N., He, C., Knibbs, L.D., Dear, R., Morawska, L. Co-optimisation of indoor environmental quality and energy consumption within urban office buildings, Energy and Buildings, Vol. 85, 2014, 225–234.
30. Nicol, J.F., Humphreys, M.A. Adaptive thermal comfort and sustainable thermal standards for buildings, Energy and Buildings, Vol. 34, No. 6, 2002, 563–572.
31. Smolander, J. Effect of cold exposure on older humans, International Journal of Sports Medicine, Vol. 23, No. 2, 2002, 86–92.
32. Frontczak, M., Wargocki, P. Literature survey on how different factors influence human comfort in indoor environments, Building and Environment, Vol. 46, No. 4, 2011, 922–937.
33. Lovins, A. Air-conditioning comfort: Behavioral and cultural issues, U.S. Department of Energy, Office of Scientific and Technical Information, Oak Ridge, TN, USA; 1992.
34. de Dear, R.J., Brager, G.S. Developing an adaptive model of thermal comfort and preference, ASHRAE Transactions, Vol. 104, No. 1, 1998, 145–67.
35. Katafygiotou, M.C., Serghides, D.K. Indoor comfort and energy performance of buildings in relation to occupants' satisfaction: Investigation in secondary schools of Cyprus, Advances in Building Energy Research, Vol. 8, No. 2, 2014, 216–240.
36. Catalina, T., Iordache, V. IEQ assessment on schools in the design stage, Building and Environment, Vol. 49, 2012, 129–140.
37. Li, Y. Thermal performance analysis of a PCM combined solar chimney system for natural ventilation and heating/cooling, PhD Thesis, Coventry University, Coventry, UK, Sept. 2013.
38. Jomehzadeh, F., Nejat, P., Calautit, J.K., Yusof, M.B.M., Zaki, S.A., Hughes, B.R., Yazid, M.N. A review on windcatcher for passive cooling and natural ventilation in buildings, Part 1: Indoor air quality and thermal comfort assessment, Renewable and Sustainable Energy Reviews, Vol. 70, 2017, 736–756.
39. ISO Standard 7730:2005, Ergonomics of the Thermal Environment—Analytical Determination and Interpretation of Thermal Comfort Using Calculation of the PMV And PPD Indices and Local Thermal Comfort Criteria, International Organization for Standardization (ISO), Geneva, Switzerland; 2005.

40. Boduch, M., Fincher, W. Standards of Human Comfort: Relative and Absolute, Seminar in Sustainable Architecture, The University of Texas at Austin, TX, USA; 2009.
41. Dupont, J.L., Domanski, P., Lebrun, P., Ziegler, F. The Role of Refrigeration in the Global Economy, 38th Note on Refrigeration Technologies, The International Institute of Refrigeration (IIR), Paris, France, June 2019.
42. Homod, R.Z., Togun, H., Abd, H.J., Sahari, K.S. A novel hybrid modelling structure fabricated by using Takagi-Sugeno fuzzy to forecast HVAC systems energy demand in real-time for Basra city, Sustainable Cities and Society, Vol. 56, 2020, 102091.
43. Chan, H.Y., Riffat, S.B., Zhu, J. Review of Passive Solar Heating and Cooling Technologies, Renewable and Sustainable Energy Reviews, Vol. 14, 2010, 781–789.
44. Son, J., Kim, H. Sensorless air flow control in an HVAC system through deep learning, Applied Sciences, Vol. 9, 2019, 3293.
45. Liu, X., Hewings, G.J.D., Wang, S. Evaluation on the impacts of the implementation of civil building energy efficiency standards on Chinese economic system and environment, Energy and Buildings, Vol. 41, 2009, 1084–1090.
46. Hart, R., Selkowitz, S., Curcija, C. Thermal performance and potential annual energy impact of retrofit thin-glass triple-pane glazing in US residential buildings, Building Simulation, Vol. 12, No. 1, 2019, 79–86.
47. Enteria, N., Mizutani, K., The role of thermally activated desiccant cooling technologies in the issue of energy and environment, Renewable and Sustainable Energy Reviews, Vol. 15, 2011, 2095–2122.
48. Reda, A.M., Ali, A.H.H., Morsy, M.G., Taha, I.S. Design optimization of a residential scale solar driven adsorption cooling system in upper Egypt based, Energy and Buildings, Vol. 130, 2016, 843–856.
49. Birol, F. The future of cooling: opportunities for energy-efficient air conditioning, Technology Report, The International Energy Agency (IEA), Paris, France; May 2018.
50. Bahadori, M.N., Dehghani-Sanij, A.R. Wind Towers: Architecture, Climate and Sustainability, Edited by Sayigh, A., Springer International Publishing, USA; 2014.
51. Dehghani-Sanij, A.R. Cisterns: Sustainable Development, Architecture and Energy, Edited by Sayigh, A., River Publishers, Denmark; 2016.
52. Dehghani-Sanij, A.R., Soltani, M., Raahemifar, K. A New Design of Wind Tower for Passive Ventilation in Buildings to Reduce Energy Consumption in Windy Regions, Renewable and Sustainable Energy Reviews, Vol. 42, 2015, 182–195.
53. Soltani, M., Dehghani-Sanij, A., Sayadnia, A., Kashkooli, F.M., Gharali, K., Mahbaz, S., Dusseault, M.B. Investigation of Airflow Patterns in a New Design of Wind Tower with a Wetted Surface, Energies, Vol. 11, No. 5, 2018, 1100.
54. Dehghan, A.A., Dehghani, A.R., Experimental and Theoretical Investigation of Thermal Performance of Underground Cold-Water Reservoirs, International Journal of Thermal Sciences, Vol. 50, 2011, 816–824.
55. Razavi, M., Dehghani-Sanij, A.R., Khani, M.R., Dehghani, M.R., Comparing Meshless Local Petrov-Galerkin and Artificial Neural Networks Methods for Modeling Heat Transfer in Cisterns, Renewable and Sustainable Energy Reviews, Vol. 43, 2015, 521–529.
56. Ameri-Siahoui, H.R., Dehghani, A.R., Razavi, M., Khani, M.R. Investigation of Thermal Stratification in Cisterns Using Analytical and Artificial Neural Networks Methods, Journal of Energy Conversion and Management, Vol. 52, No. 1, 2011, 505–511.
57. Arefmanesh, A., Dehghan, A.A., Dehghani, A.R., Thermal Characteristics of an Underground Cold Water Reservoir: Analytical and Experimental Studies, Applied Thermal Engineering, Vol. 29, No. 14–15, 2009, 3261–3265.
58. Madoliat, R., Razavi, M., Dehghani, A.R. Modeling of heat transfer in cisterns using artificial neural networks, Journal of Thermophysics and Heat Transfer, Vol. 23, No. 2, 2009, 411–416.
59. Razavi, M., Dehghani, A.R., Khanmohammadi, M. Simulation of thermal stratification in cisterns using artificial neural networks, Journal of Energy, Heat and Mass Transfer, Vol. 31, 2009, 201–210.

60. Dehkhoda, A.A. Dictionary, Vol. II, 2nd ed., University of Tehran Press, Tehran, Iran; 1998. (in Farsi)
61. Rosenthal, F. Poetry and Architecture: the Badhanj, Journal of Arabic Literature, Vol. 8, No. 1, 1977, 1–19.
62. King, D.A. Architecture and astronomy: The ventilators of medieval Cairo and their secrets, Journal of the American Oriental Society, Vol. 104, No. 1, 1984, 97–133.
63. Barghandan, M. Iran's new social media-friendly approach, Turkish Policy Quarterly, Vol. 14, No. 1, 2015, 137–145.
64. Bahadori, M.N. Passive Cooling Systems in Iranian Architecture, Scientific American, Vol. 238, No. 2, 1978, 144–155.
65. Mazidi, M., Dehghani, A.R., Aghanajafi, C. Wind towers' role in natural air conditioning and passive cooling of buildings in hot, arid regions, WSEAS Transactions on Fluid Mechanics, Vol. 1, No. 11, 2006, 959–966.
66. Mazidi, M., Dehghani, A.R., Aghanajafi, C. The Study of the Airflow in Wind Towers for the Old Buildings Air Conditioning, In: Proceedings of the 4th WSEAS International Conference on Fluid Mechanics, Gold Coast, Queensland, Australia, pp. 76–81, Jan. 17–19, 2007.
67. Bahadori, M.N. Natural Cooling in Hot Arid Regions, In: Solar Energy Application in Buildings, edited by A.A.M. Sayigh, Academic Press, New York, pp. 195–225, 1977.
68. Faghih, A.K., Bahadori, M.N. Thermal performance evaluation of domed roofs, Energy and Buildings, Vol. 43, No. 6, 2011, 1254–1263.
69. Semsar Yazdi, A.A. Labbaf Khaneiki, M. Qanat Knowledge: Construction and Maintenance, Springer, Dordrecht, The Netherlands; 2017.
70. Warren, J., Fethi, I. Traditional Houses in Baghdad, The Coach Publishing House Ltd., Horsham, England; 1982.
71. M.R.Khani, S., Bahadori, M.N., Dehghani-Sanij, A.R., Nourbakhsh, A. Performance Evaluation of a Modular Design of Wind Tower with Wetted Surfaces, Energies, Vol. 10, No. 845, 2017, 1–22.
72. M.R.Khani, S., Bahadori, M.N., Dehghani-Sanij, A.R. Experimental Investigation of a Modular Wind Tower in Hot and Dry Regions, Energy for Sustainable Development, Vol. 39, 2017, 21–28.
73. Goudarzi, N., Zhu, W.D., Bahari, H. Wind Energy Conversion: The Potential of a Novel Ducted Turbine, for Residential and Commercial Applications, In: Proceedings of the ASME International Mechanical Engineering Congress & Exposition IMECE, San Diego, CA, USA, 2013.
74. Elzaidabi, A.A.M. Low Energy, Wind Catcher Assisted Indirect - Evaporative Cooling System for Building Applications, PhD Thesis, University of Nottingham, Nottingham, UK, Sep. 2008.
75. Pearlmutter, D., Erell, E., Etzion, Y. A multi-stage down-draft evaporative cool tower for semi-enclosed spaces: Experiments with a water spraying system, Solar Energy, Vol. 82, 2008, 430–440.
76. Pearlmutter, D., Erell, E., Etzion, Y., Meir, I.A., Di, H. Refining the use of evaporation in an experimental down-draft cool tower, Energy and Buildings, Vol. 23, 1996, 191–197.
77. Erell, E., Pearlmutter, D., Etzion, Y. A multi-stage down-draft evaporative cool tower for semi-enclosed spaces: Aerodynamic performance, Solar Energy, Vol. 82, 2008, 420–429.
78. Soutullo, S., Olmedo, R., Sanchez, M.N., Heras, M.R. Thermal conditioning for urban outdoor spaces through the use of evaporative wind towers, Building and Environment, Vol. 46, 2011, 2520–2528.
79. Soutullo, S., Sanjuan, C., Heras, M.R. Energy performance evaluation of an evaporative wind tower, Solar Energy, Vol. 86, 2012, 1396–1410.
80. Soutullo, S., Sanchez, M.N., Olmedo, R., Heras, M.R. Theoretical model to estimate the thermal performance of an evaporative wind tower placed in an open space, Renewable Energy, Vol. 36, 2011, 3023–3030.

81. Issa, R.J., Chang, B., Performance Prediction of a Multi-Stage Wind Tower for Indoor Cooling, Journal of Thermal Science, Vol. 21, No. 4, 2012, 327–335.
82. Ji, Zh., Su, Y., Khan, N. Performance Evaluation and Energy Saving Potential of Wind Catcher Natural Ventilation Systems in China, International Journal of Architecture, Engineering and Construction, Vol. 1, No. 2, 2012, 84–95.
83. Javid, A.H., Modernizing the Traditional Ventilation Principles of Yazd to Minimize the Energy Consumption for Hot Climate, M.Sc. Thesis, Department of Architecture, University of Sheffield, Sheffield, UK, Sept. 2007.
84. Pakzad, A.R. Analysis of Two New Designs of Wind Towers, M.Sc. Thesis, School of Mechanical Engineering, Sharif University of Technology, Tehran, Iran, Sept. 1997.
85. Dehghani, A.R., An Experimental Evaluation of Two New Designs of Wind Towers and a Comparison with a Conventional Wind Tower, B.Sc. Thesis, Yazd University, Yazd, Iran, 2001.
86. Hughes, B.R., Calautit, J.K., Ghani, S.A. The development of commercial wind towers for natural ventilation: A review, Applied Energy, Vol. 92, 2012, 606–627.
87. Zhang, H., Yang, D., Tam, V.W., Tao, Y., Zhang, G., Setunge, S., Shi, L. A critical review of combined natural ventilation techniques in sustainable buildings, Renewable and Sustainable Energy Reviews, Vol. 141, 2021, 110795.
88. Santamouris, M., Kolokotsa, D. Passive cooling dissipation techniques for buildings and other structures: the state of the art, Energy and Buildings, Vol. 57, 2013, 74–94.
89. Bahadori, M.N. An improved design of wind towers for natural ventilation and passive cooling, Solar Energy, Vol. 36, No. 2, 1985, 119–129.
90. Bahadori, M.N. A passive cooling/heating system for hot, arid regions, In: Proceedings of the 13th National Passive Solar Conference, Cambridge, MA, USA, pp. 364–367, June 20–24, 1988.
91. Bahadori, M.N., Mazidi, M., Dehghani, A.R. Experimental investigation of new designs of wind towers, Renewable Energy, Vol. 33, 2008, 2273–2281.
92. Bahadori, M.N., Pakzad, A.R. Performance Evaluation of New Designs of Wind Towers, Paper No. FEDSM 2002–31247, In: Proceedings of the ASME FEDSM02, ASME 2002 Fluids Engineering Division Summer Meeting, Montreal, QC, Canada, July 14–18, 2002.
93. Karakasanis, C., Bahadori, M.N., Vickery, B.J. Evaluation of air flow rates in employing wind towers, Solar Energy, Vol. 37, 1986, 363–374.
94. Bahadori, M.N. Pressure coefficients to evaluate air flow patterns in wind towers, In: Proceedings of the International Passive and Hybrid Cooling Conference, Miami Beach, FL, USA, Nov. 6–16, 1981; American Section of the International Solar Energy Society: Newark, DE, USA, 1981; pp. 206–210.
95. Bahadori, M.N. Viability of wind towers in achieving summer comfort in the hot, arid regions of the Middle East, In: Proceedings of the Third World Renewable Energy Congress, Reading, UK, pp. 879–892, Sept. 11–16, 1994.
96. Amoli, F.J. An Analytical Study of Two Types of Modern Wind Towers, B.Sc. Thesis, School of Mechanical Engineering, Sharif University of Technology, Tehran, Iran, 1990.
97. Memarie, G.H. An Analysis of the Performance and Comparison of Two Modern Designs of Wind Towers, B.Sc. Thesis, School of Mechanical Engineering, Sharif University of Technology, Tehran, Iran, 1991.
98. Mosavi Davar, S. An Empirical Study and Performance Analysis of Wind Towers and Their Feasibility for Use in the Buildings, B.Sc. Thesis, Department of Mechanical Engineering, Faculty of Polytechnic, University of Guilan, Rasht, Guilan, Iran, 2002.
99. Mazidi, M. Assessment of the Efficiency of the Wind Tower of Yazd University Mosque through Analytical and Empirical Methods, M.Sc. Thesis, Department of Power-Plant, Shahid Abbaspour Power and Water University of Technology, Tehran, Iran, 2005.
100. M.R.Khani, S. Industrial Production of Wind Towers to Replace Evaporative Coolers, M.Sc. Thesis, School of Mechanical Engineering, University of Tehran, Tehran, Iran, 2014.
101. Dehghani-Sanij, A.R., Soltani, M. Wind tower with moving wetted surfaces, Industrial Property General Office of Iran, No. 92228, Tehran, Iran, 2017.

102. Dobrescu, E.M. Reducing the polluting emissions, a source for diminishing the climate changes, Revista Romana de Economie, Vol. 28, No. 2, 2009, 1–15.
103. Merone, L., Tait, P. 'Climate refugees': Is it time to legally acknowledge those displaced by climate disruption?, Australian and New Zealand Journal of Public Health, Vol. 42, No. 6, 2018, 508–509.
104. Myers, N. Environmental refugees: a growing phenomenon of the 21st century, Philosophical Transactions of the Royal Society of London, Series B: Biological Sciences, Vol. 357, No. 1420, 2002, 609–613.
105. Kaygusuz, K. Energy for sustainable development: A case of developing countries, Renewable and Sustainable Energy Reviews, Vol. 16, No. 2, 2012, 1116–1126.

Alireza Dehghani-Sanij is a PhD graduate from Memorial University of Newfoundland, and his discipline is Mechanical Engineering (Thermo-Fluids). Alireza, currently a Senior Research Fellow at the University of Waterloo, Canada, studies energy issues such as energy sources, storage, and conversion; stand-alone and hybrid energy systems, specifically in geothermal contexts; and passive, natural cooling systems (e.g., wind towers). In two decades of academic and industrial experience, he has published more than 40 papers in scientific journals and presented more than 30 papers at conferences. He has also collaborated with other researchers to publish several books in both English and Farsi, such as "Ice-houses: Energy, Architecture and Sustainability," published by Elsevier in 2021, and "Wind Towers: Architecture, Climate and Sustainability," published by Springer in 2014.

Thermal Comfort and Climatic Potential of Ventilative Cooling in Italian Climates

Giacomo Chiesa, Francesca Fasano, and Paolo Grasso

1 Introduction

Energy consumption for space cooling is characterized by a continuously rising trend [1]. In parallel, the number of installed domestic cooling units is significantly growing, as confirmed by International Energy Agency (IEA) documents and by yearly reports of sector-specific companies [2–4]. The global penetration of the air conditioning market is, in fact, quickly increasing, especially in Asia, but also in the Americas and in Europe [5]. This trend is connected to several causes, including climate change, urban heat islands, comfort culture, and building design choices that are inconsistent with respect to local climate [1, 6]. It is hence evident that alternative solutions for cooling, when environmental conditions are favourable, may be adopted and developed in order to reduce cooling energy needs and consequent GHGs (greenhouse gas emissions) [7]. Among low-energy alternatives, ventilative cooling (VC) is a valuable technique to reduce energy needs and consumption in buildings supporting free-running and/or fan-assisted ventilation for space cooling. This technique was demonstrated to be very effective in reducing overheating risks, but also to guarantee thermal and IAQ (indoor air quality) comfort in buildings during the summer season. However, the ventilative cooling potential occurs when external air temperatures are below comfort thresholds; therefore, its applicability is local and time-specific and is connected to local climate/weather conditions [8, 9]. As underlined for the majority of passive and low-energy cooling solutions, the non-homogeneous specific local potential has limited the current applications of VC with respect to passive heating technologies, that is sunspaces. Nevertheless, a consistent need to support the diffusion of VC solutions is evident, given this approach is not sufficiently covered and valorised by current regulations – see, for

G. Chiesa (✉) · F. Fasano · P. Grasso
Department of Architecture and Design, Politecnico di Torino, Turin, Italy
e-mail: giacomo.chiesa@polito.it

A. Sayigh (ed.), *Achieving Building Comfort by Natural Means*, Innovative Renewable Energy, https://doi.org/10.1007/978-3-031-04714-5_18

example, the recent analysis reported in [10] – even in those climates in which it may support thermal comfort without cooling system activations during the majority of hours [11, 12]. This requires the development of methodologies to calculate the potential of ventilative cooling – see, for example, the results of the IEA EBC ANNEX 62 on Ventilative Cooling and other references [13, 14] supporting the widespread of VC.

1.1 Objectives and Organisation

This chapter aims at investigating, on one hand, climatic key performance indicators (KPIs) for ventilative cooling (VC) (extended summer season from May to September) and, on the other hand, the impact of the same low-energy cooling technology on thermal comfort variations in dynamically simulated building spaces. Additionally, the chapter aims at comparing the two analyses to help define and choose early design indicators to estimate the ventilative cooling potential in terms of thermal comfort and building cooling energy need reduction. The methodology is applied to a large sample of locations, covering the whole Italian territory, to map the thermal comfort/climatic potential of this low-energy dissipation cooling technique.

The chapter is composed of the following sections: Section 2 describes the proposed methodology and introduces the considered KPIs, the assumed reference building, and the climate dataset; Sect. 3 is devoted to reporting the results of the analyses, including climate and comfort KPI distribution at the Italian level. Furthermore, Sect. 3 reports the comparison between the two approaches (the climatic one and the building simulated one); and finally, Sect. 4 reports chapter conclusions.

2 Methodology

As mentioned, the chapter describes, calculates, and compares the climate and comfort/energy effects of VC. This analysis is applied to a sample territory (Italy) to study the local distribution of KPIs. Furthermore, potential correlations between climate-based and building-simulation-based indices are also discussed. The adopted KPIs and their calculation methodology – including the defined simulated sample building and the locations selected for the analysis – are reported in this section. Section 2 is structured in sub-sections that are devoted to defining the assumed key performance indicators (KPIs) considering climate-based analyses (Sect. 2.1) and building-comfort analyses based on dynamic energy simulations (Sect. 2.2). Furthermore, additional sub-sections are included, describing the assumed sample building (Sect. 2.3) and the considered set of locations (Sect. 2.4).

2.1 Climate-Comfort KPIs

Several climate-related KPIs are reported in literature to estimate the local cooling needs and the local potential of ventilative cooling solutions.

Among them, one of the most popular KPIs defining the local climatic cooling needs is the Cooling degree hours/days (CDH/CDD), which was demonstrated to be linearly correlated to the local intensity of building cooling needs – see, for example, [5, 15]. Furthermore, this KPI is also compared to overheating risks during the free-running mode. CDH and CDD integrate hourly differences in temperature between environmental values and a building base temperature over which cooling or overheating is expected. The assumed cooling base temperature may differ according to the purposes of the analysis, ranging from 15.5 °C [16] to 26 °C [17]. Furthermore, the calculation may be assessed for the whole summer period or for specific calculation times. For the purpose of this chapter, the following expression is adopted for the CDH index:

$$\mathrm{CDH} = \sum m_h \cdot \left(\vartheta_{\mathrm{e},h} - \vartheta_{\mathrm{b}}\right) \quad \begin{cases} m_h = 1 \leftarrow \vartheta_{\mathrm{e},h} > \vartheta_{\mathrm{b}} \\ m_h = 0 \leftarrow \vartheta_{\mathrm{e},h} \le \vartheta_{\mathrm{b}} \end{cases} \tag{1}$$

where:

the variable m_h is a climate cooling activation check and ϑ_{b} is the cooling base temperature, ranging for the purpose of this chapter in the domain {18 °C; 21; 24; 26; 28}, while $\vartheta_{\mathrm{e},h}$ is the hourly dry-bulb environmental temperature.

Additional calculation approaches are suggested in literature, such as the approach described in EN ISO 15927-6:2008. European Statistical Office (EUROSTAT), for example, supports the calculation of CDD on the basis of two thresholds and considering daily average temperatures. The first threshold defines for which days the CDD needs to be calculated, while the second, lower one is used as base calculation temperature – see Eq. (2) [18].

$$\mathrm{CDD}_{eu} = \sum m_d \cdot \left(\vartheta_{\mathrm{e},d} - 21\right) \quad \begin{cases} m_d = 1 \leftarrow \vartheta_{\mathrm{e},d} \ge 24^\circ\mathrm{C} \\ m_d = 0 \leftarrow \vartheta_{\mathrm{e},d} < 24^\circ\mathrm{C} \end{cases} \tag{2}$$

where:

$\vartheta_{\mathrm{e},d}$ is the environmental daily average temperature, and 24 °C and 21 °C are the first and second mentioned thresholds, respectively.

Degree days/hours are also at the basis of different KPIs that were developed to climatically analyse the potential impact of low-energy technologies. For example, a series of indices are able to define the residual amount of the CDH ($\mathrm{CDH}_{\mathrm{res}}$, percentage or absolute) by calculating the theoretical impact of different heat gain dissipation technologies, that is evaporative cooling and ground-coupled earth-to-air exchangers – see, for example [19–21].

Nevertheless, for this work, we calculated the climate cooling potential (CCP) and index that was firstly defined by Artman et al. [22] and supported in several publications, for example [12, 13]. This index is based on the internal-external air temperature difference and is expressed in Kelvin per hour [K/h], suggesting the night cooling potential of ventilative cooling. The calculation methodology is expected to work during night hours (from 19:00 to 7:00), considering an office space conditioned only during the daytime in which internal temperature is following a sinusoidal variation around 24.5 °C of ±2.5 °C (max temperature is set at 19:00). These internal temperature ranges were suggested on the basis of European Committee for Standardization (CEN) thermal comfort categories – for example in EN 16798-4:2019, the comfort category III for offices suggests temperature ranges between 22 and 27 °C in internally conditioned spaces [23]. CCP is generally considered only during night hours to define the night-time mean CCP value in a given period. Hence, the daily average (CCP_d) may be defined by dividing the CCP by the number of days in the considered period. Nevertheless, it is also possible to calculate the cumulative distribution of CCP.

In this chapter, the cumulative CCP is assumed to directly compare results with the cumulative cooling energy savings due to ventilative cooling activation in a conditioned space – see Eq. (3). Nonetheless, this analysis is performed for the whole summer period, considering that ventilative cooling may be also performed during the daytime when external conditions are favourable (controlled ventilation). The CCP_d is also calculated.

$$\mathrm{CCP} = \sum m_h \cdot \left(\vartheta_{\mathrm{b},h} - \vartheta_{\mathrm{e},h}\right) \quad \begin{cases} m_h = 1 \leftarrow \vartheta_{\mathrm{b},h} - \vartheta_{\mathrm{e},h} \geq 2°\mathrm{C} \\ m_h = 0 \leftarrow \vartheta_{\mathrm{b},h} - \vartheta_{\mathrm{e},h} < 2°\mathrm{C} \end{cases} \tag{3}$$

where:

$$\vartheta_{\mathrm{b},h} = \left(24.5 + 2.5 \cdot \cos\left(\frac{2\pi\left(h - 19\right)}{24}\right)\right) [°\mathrm{C}] \tag{4}$$

and h is the number of daily hours [1–24]. A critical difference in temperature between the base and environmental temperature of 2 °C is assumed, even if higher differences may be considered for a precautionary approach.

In addition to DH-DD indices and based on the same principle, it is possible to define a KPI to calculate the climate heat gain dissipation potential of ventilative cooling. By knowing the airflow rate and the thermal properties of air it is possible to calculate the cooling dissipation Q_{ach} [Wh] by considering the difference in temperature between internal and external air and adopting the well-known expression [9, 24, 25]:

$$Q_{ach} = \sum m_h \cdot ACH \bullet Vol \cdot \rho_{air} \cdot c_{air} \cdot \left(\vartheta_{set} - \vartheta_{e,h}\right) \quad \begin{cases} m_h = 1 \leftarrow \vartheta_{e,h} \leq \left(\vartheta_{set} - 2°C\right) \\ m_h = 0 \leftarrow \vartheta_{e,h} > \left(\vartheta_{set} - 2°C\right) \end{cases} \tag{5}$$

where:

the variable *ACH* is the number of air changes per hour, which is assumed for this paper in the domain {0;2.5;5}; Vol is the assumed net volume [m^3] of the building; air density and heat capacity are assumed to be, respectively, 1.2 [kg/m^3] and 0.28 [W/kg°C], while the set point temperature was set to be equal to the simulated cooling set point (26 °C). Finally, 2 °C was the assumed minimal difference in temperature from the set point, to consider a sufficient activation of ventilative cooling. The latter value generally ranges between 2 and 3 °K – see also the critical temperature description defined in the IEA EBC Annex 62 documents [13].

2.2 Building-Comfort KPIs

Focusing on building thermal comfort indices, two approaches are followed in line with EU standard 16798-1:2019. The first approach refers to mechanically cooled building configurations. In these cases, we assumed the thermal comfort model described in ISO 7730, which is based on the PMV (predicted mean vote) and PPD (predicted percentage of dissatisfied) indices. This approach, in line with the Fanger comfort theory [26, 27], is based on six parameters: four environmental parameters (air temperature, humidity, mean radiant temperature, and relative air velocity) and two personal parameters (metabolic rate and clothing level). Considering statistical correlations based on personal expected thermal sensations, the PMV index classifies user thermal sensation on a 7-point scale, ranging from −3 (very cold) to +3 (very hot), in which values around 0 are the most comfortable (neutral). Different comfort categories are suggested in the EU standard according to PMV ranges around the neutral point – that is category I (±0.2 PMV); II (±0.5 PMV); and III (±0.7 PMV).

The calculation procedure for the PMV and PPD was implemented in line with the one described in both EN and ISO standards by developing a specific Python code. The clothing level is assumed to be 0.7 clo, considering underwear, shirt, trousers, socks, and shoes (ISO 9920) – an intermediate balance between hot central hours of summer and September/May evening – while the metabolic rate was set to 1.2 met, corresponding to standing relaxed or sitting activity conditions – a typical value for residential living spaces (EN 16798-1). In addition to these basic indices, cumulative versions are available. For this study, the cumulative value of PPD whenever hourly PMV is higher than category III is calculated according to the following rule:

$$\mathrm{PPD}_{\mathrm{exceedance}} = \sum m_h \cdot \mathrm{PPD}_h \qquad \begin{cases} m_h = 1 \leftarrow \mathrm{PMV}_h \geq 0.7 \\ m_h = 0 \leftarrow \mathrm{PMV}_h < 0.7 \end{cases} \tag{6}$$

where:

PMV_h is the hourly computed value for predicted mean vote and PPD_h the corresponding hourly predicted percentage of dissatisfied.

The second thermal comfort approach is assumed for free-running building configurations. For the purpose of this work, it is considered that a building is working in a free-running mode in the summer when cooling systems are not present or are turned off. This definition is in line with past and current research projects, that is IEA Annex 62 and H2020 EU-co-funded project E-DYCE [28]. In these cases, the adaptive thermal comfort theory is assumed – see the works of Fergus and Humphreys [29, 30]. The calculation was performed in line with the above-mentioned EN standard, assuming the upper comfort category II as a reference. Additionally, cumulative values of hourly distances from the central comfort line and from upper thresholds of different comfort categories are calculated according to the following rules:

$$\mathrm{dist}_0 = \begin{cases} \sum m_h \cdot \mathrm{dist}_{0,mid,h} \quad \text{if } 10°\mathrm{C} < \vartheta_{rm,h} < 33°\mathrm{C} \quad \text{where } \begin{cases} m_h = 1 \leftarrow \mathrm{dist}_{0,mid,h} > 0 \\ m_h = 0 \leftarrow \mathrm{dist}_{0,mid,h} \leq 0 \end{cases} \\ \sum m_h \cdot \mathrm{dist}_{0,\mathrm{up},h} \quad \text{if } \vartheta_{rm,h} \geq 33°\mathrm{C} \quad \text{where } \begin{cases} m_h = 1 \leftarrow \mathrm{dist}_{0,\mathrm{up},h} > 0 \\ m_h = 0 \leftarrow \mathrm{dist}_{0,\mathrm{up},h} \leq 0 \end{cases} \end{cases} \tag{7}$$

$$\mathrm{dist}_3 = \begin{cases} \sum m_h \cdot \mathrm{dist}_{3,mid,h} \quad \text{if } 10°C < \vartheta_{rm,h} < 33°\mathrm{C} \quad \text{where } \begin{cases} m_h = 1 \leftarrow \mathrm{dist}_{3,mid,h} > 0 \\ m_h = 0 \leftarrow \mathrm{dist}_{3,mid,h} \leq 0 \end{cases} \\ \sum m_h \cdot \mathrm{dist}_{3,\mathrm{up},h} \quad \text{if } \vartheta_{rm,h} \geq 33°\mathrm{C} \quad \text{where } \begin{cases} m_h = 1 \leftarrow \mathrm{dist}_{3,\mathrm{up},h} > 0 \\ m_h = 0 \leftarrow \mathrm{dist}_{3,\mathrm{up},h} \leq 0 \end{cases} \end{cases} \tag{8}$$

where:

- $\mathrm{dist}_{0,\,\mathrm{mid},\,h} = \vartheta_{\mathrm{op,\,i},\,h} - (0.33 \cdot \vartheta_{\mathrm{rm},\,h} + 18.8)$ is the distance from the central comfort line whenever the running mean is in the (10,33) °C range;
- $\mathrm{dist}_{0,\,\mathrm{up},\,h} = \vartheta_{\mathrm{op,\,i},\,h} - (0.33 \cdot 33 + 18.8)$ is the distance from the central comfort line whenever the running mean is in the [33,x] °C range;
- $\mathrm{dist}_{3,\,\mathrm{mid},\,h} = \vartheta_{\mathrm{op,\,i},\,h} - (0.33 \cdot \vartheta_{\mathrm{rm},\,h} + 18.8 + 3)$ is the distance from the upper threshold of comfort category II whenever the running mean is in the (10,33) °C range;

– $dist_{3,\,up,\,h} = \vartheta_{op,\,i,\,h} - (0.33 \cdot 33 + 18.8 + 3)$ is the distance from the upper threshold of comfort category II whenever the running mean is in the [33,x] °C range;

and $\vartheta_{op,\,i,\,h}$ is the hourly indoor operative temperature, while $\vartheta_{rm,\,h}$ is the hourly updated running mean temperature calculated in line with EN 16798-1:2019. Different comfort categories are also considered assuming limits mentioned in the standard.

The cumulative intensity of discomfort for free-running buildings may be adapted to a simpler indicator: the cooling internal degree hours (CIDH). This indicator is the building-correlated counterpart of the CDH indicator – see the definition reported in Pellegrino et al. [31]. For this research, it was calculated as below:

$$\mathrm{CIDH} = \sum m_h \cdot \left(\vartheta_{i,h} - \vartheta_c\right) \qquad \begin{cases} m_h = 1 \leftarrow \vartheta_{i,h} > \vartheta_b \\ m_h = 0 \leftarrow \vartheta_{i,h} \le \vartheta_b \end{cases} \tag{9}$$

where:

the variable m_h is the climate cooling activation check and ϑ_c the cooling comfort temperature, assumed in this chapter not only as a fixed reference set point (i.e. 26 °C) but ranging in the domain {18 °C; 20; 22; 24; 26; 28} to verify correlations with climate CDH, while $\vartheta_{i,\,h}$ is the hourly indoor air temperature.

2.3 Building Sample Definition

A sample residential building unit was adopted to test the proposed methodology. The considered unit is a typical flat of a multi-storey building. Its internal organisation and definition are in line with suggested residential building typologies included in well-known architectural technical manuals – see, for example [32, 33] – and in line with the methodological approach followed in previous climate-correlated studies [34]. The considered building is composed of two units per floor, while a single unit is simulated for this chapter. The simulated spaces are considered to be at an intermediate floor with an upper floor and lower floor working at the same

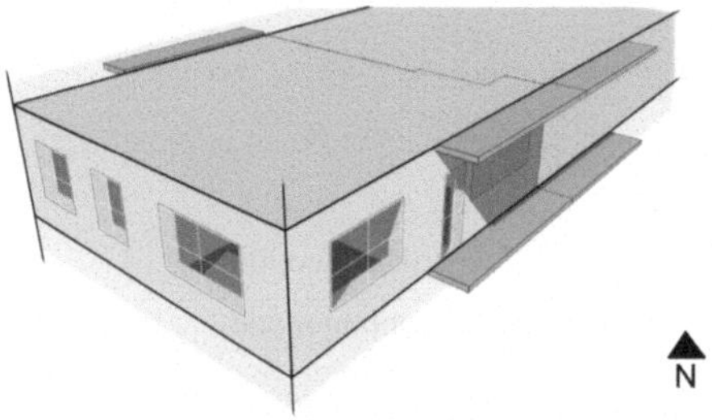

Fig. 1 The considered sample residential building unit

temperature (adiabatic). Similarly, the simulated unit is touching a specular one: Confining walls are also assumed to be adiabatic. Upper-floor balconies are included to consider shading effects – see Fig. 1.

Dynamic hourly energy and thermal simulations are performed by adopting EnergyPlus [35]. Different simulations were performed for each location to compare building scenarios. We investigated, on one hand, the free-running building mode and, on the other hand, the mechanically cooled building mode, assuming EnergyPlus ideal loads and retrieving net energy needs. Five configurations are considered assuming different levels of ventilative cooling air changes per hour (ACH) in order to test the impact of ventilative cooling (VC) – indifferently natural or fan-driven – on cooling energy needs (mechanical cooling mode) and on thermal comfort indices (free-running cooling mode). Table 1 describes the adopted set of simulations.

Ventilative cooling is defined by adopting the scheduled natural ventilation approach in EnergyPlus, supporting VC activation by controlling temperature thresholds. For the free-running mode, external temperature control rules assume that there is no maximum outdoor temperature above which the ventilation is shut-off since there is no other cooling system in the building, although an activation difference in temperature between internal and external values of 2 °C is defined in line with climate KPIs described above. Differently, for the mechanical cooling mode, a control on maximum outdoor temperature (26 °C) is also added to avoid the risk to introduce external airflows in the building at a temperature higher than the cooling set point, which is set to 26 °C, with a consequent rise of cooling loads. Cooling scheduling is activated, when control conditions are appropriate, for the whole extended summer period. Internal gains and occupancy profiles are assumed in line with the ones suggested by EnergyPlus references – see also the description in [34]. Envelope definition is the same in both free-running and mechanically cooled modes. The U-value of walls was set to 0.287 [W/m^{2}K] (10 cm of insulation), while the U-value for windows was 1.499 [W/m^{2}K] (Double glass, Low-E with argon). Simulations are controlled by using Python coding.

Table 1 The adopted simulation configurations (VC = ventilative cooling)

Free-running mode	ACH	Mechanical-cooling mode	ACH
1a. Reference case	0	1b. Reference case	0
2a. low VC	2.5	2b. low VC	2.5
3a. VC	5.0	3b. VC	5.0

2.4 Climate Data and Locations

For this chapter, the above-described KPIs are applied to the whole Italian territory, assuming a calculation point for each municipality. The Italian municipality database was based on the ISTAT (national statistics agency) data (31/12/2017 version) and includes 7978 data points.

Typical meteorological years (TMY) for each point are retrieved by the Meteonorm tool [36] assuming reference periods 1991–2010 and 2000–09 for irradiation and for temperatures, respectively. For all locations, the entire set of simulations was performed, including the calculation of simulated and climate-related KPIs. This large dataset (47,868 simulations) allows comparison analyses among different KPIs. A comparison between building comfort-related indices and external-climatic KPIs is performed to verify the possibility to predict the ventilative cooling potential on building comfort and energy needs without running dynamic building energy simulations.

The analysis is performed for an extended summer period ranging from May to September (included). This assumption will allow us to better evaluate the ventilative cooling potential to cover the summer overheating also in the hottest locations, in which a reduced summer period (e.g. June to August) will not be representative of the whole summer season. This assumption is in line with previous analyses [20].

KPI devoted maps are produced by using a geospatial Python plotting library, while the Italian borders are retrieved by a GeoJSON file provided by Openpolis [37].

3 Results

Three main categories of results are presented. On one hand, KPI results – both climatic (Sect. 3.1) and building-related ones (Sect. 3.2) – are illustrated. On the other hand, mutual correlations between climate and simulated building KPIs are analysed (Sect. 3.3) to verify consistencies and the ability of climate indices to represent preliminary building trends.

3.1 Climate-Based KPIs

The distribution of climatic KPIs over the Italian territory is illustrated, with the aim to represent local specific behaviours.

In line with the description in Sect. 2, the CDH indicator represents the cumulative hourly differences between the environmental air temperature and an assumed base reference temperature, while CDD considers a daily cumulative cooling severity. Figure 2a shows the CDH distribution for the whole dataset, considering 18 °C as base temperature, while Fig. 2b illustrates the CDD distribution. Both KPIs

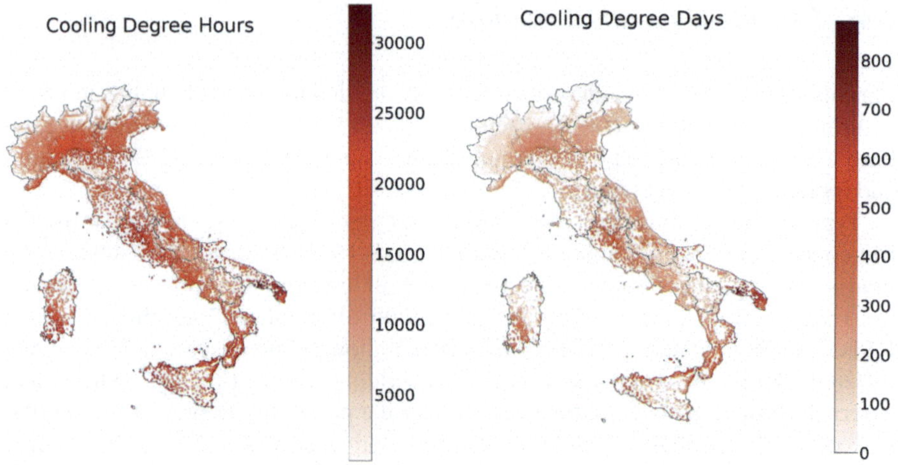

Fig. 2 Distribution of (**a** – left) CDH_{18} and (**b** – right) CDD on the Italian territory

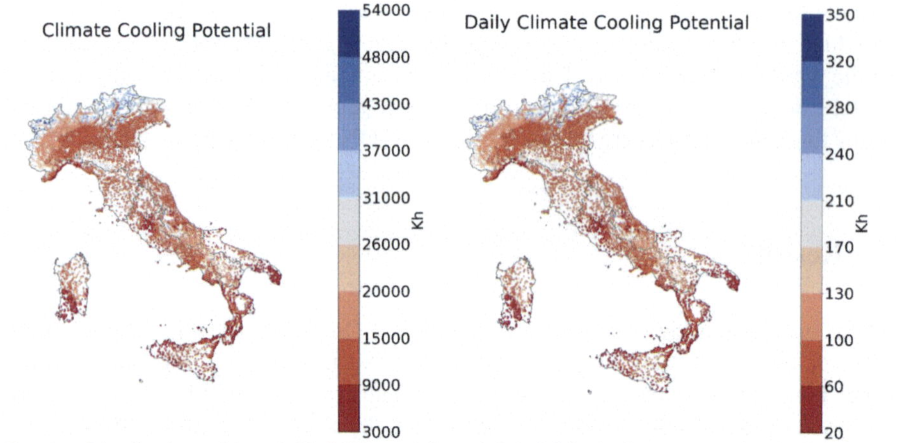

Fig. 3 Distribution of (**a** – left) CCP and (**b** – right) CCP_d indicators on the Italian territory

largely depend on the specific local climate conditions. In the northern mountain zones, lower values of both CDH and CDD are retrieved, given the number of overheating hours or days is very limited. Only for a few hours, the TMY environmental temperatures are above the assumed thresholds, while in the central and southern areas these KPIs reached higher values due to the warmer climate conditions. It could be interesting to notice that locations nearer to Rome and Milan, which are the two most populated Italian cities – see ISTAT databases [38], show higher CDH and CDD values. This phenomenon is in line with the expected impact of urban heat island conditions. In these sites, higher cooling needs or higher discomfort conditions are expected. Furthermore, the high concentration of building units underlines

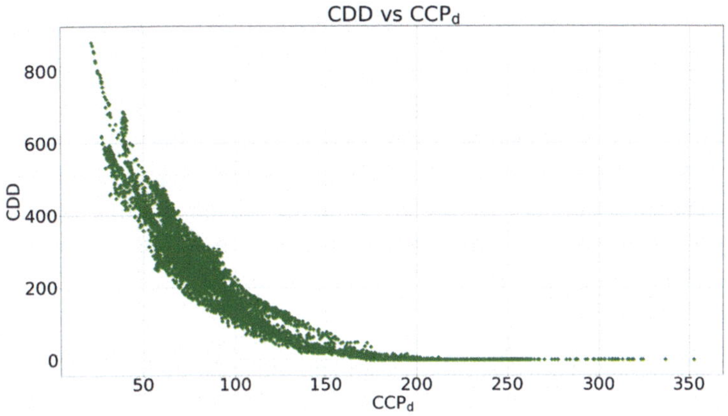

Fig. 4 CCD vs. CCP_d comparison

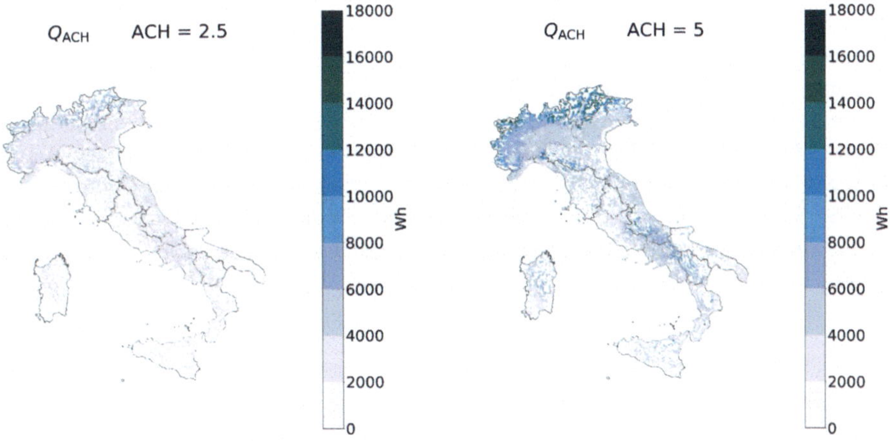

Fig. 5 Distribution of Q_{ACH} indicator on the Italian territory

that the adoption of low-energy cooling technologies in these locations may have an impact on a high number of end-users.

Focusing on the climate ventilative cooling KPIs, Fig. 3 reports the Italian distribution of (a) CCP and (b) CCPd values. These two maps show that the zones in Italy with the higher climate cooling potential are located in the north of the peninsula, due to the colder climate, which allows a higher natural dissipative cooling potential by adopting air as a heat sink. For most summer hours, the environmental air temperature in such zones is low enough to provide climate comfort conditions. Instead, the southern Italian locations are affected by lower ventilative cooling potentials due to higher external temperatures. This condition limits in both time and intensity the possibility to adopt ventilative passive cooling methodologies to achieve a climatic indoor comfort when internal temperatures rise above the comfort threshold.

Figure 4 underlines that CDD is correlated with CCP_d by following a decreasing exponential trend. In fact, a high CDD means that the cooling-correlated energy demand of a virtual building is high when the daily external temperature is above the assumed base threshold, prompting people to turn on mechanical cooling systems when present. Nevertheless, CCP_d values retrieved for hotter days are also expected to be low, since the high environmental air temperature limits the applicability of ventilative cooling. Oppositely, a higher climate cooling potential for a specific day means that, at least at night, environmental temperatures are lower, allowing ventilative cooling activations and consequently a lower virtual climatic cooling demand.

Figure 5 shows the cooling dissipation potential (Q_{ach}) – see Eq. (5) – for the two assumed ACH values. The ACH values in climate analyses are based on the assumed sample building volume to allow KPI comparison. The analysis shows that with an ACH equal to 2.5, the cooling dissipation potential by ventilative cooling is limited in most locations. Differently, considering an ACH equal to 5, several Italian zones – in particular the mountain ones in the north and the one at the centre of the Peninsula – are characterized by an evident natural ventilation dissipative potential able to virtually reduce space cooling needs.

Furthermore, Fig. 6 compares the two calculated climatic ventilative cooling potential KPIs. This figure shows that Q_{ach} and CCP_d are almost linearly correlated. The CCPd may in fact be translated to a dissipation potential by considering an airflow – see, for example, [39]. Although, this result underlines that both indices may be adopted to describe the ventilative cooling potential of a location, even if Q_{ach} is directly reflecting the dissipative potential for a given ACH, while CCP_d focuses on climate ventilative cooling availability.

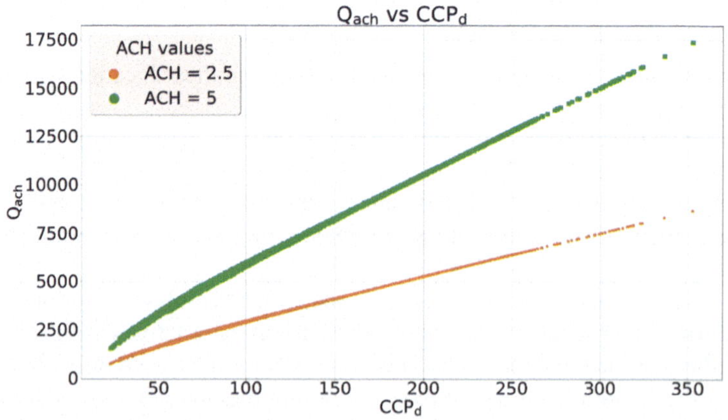

Fig. 6 Q_{ach} vs. CCP_d comparison

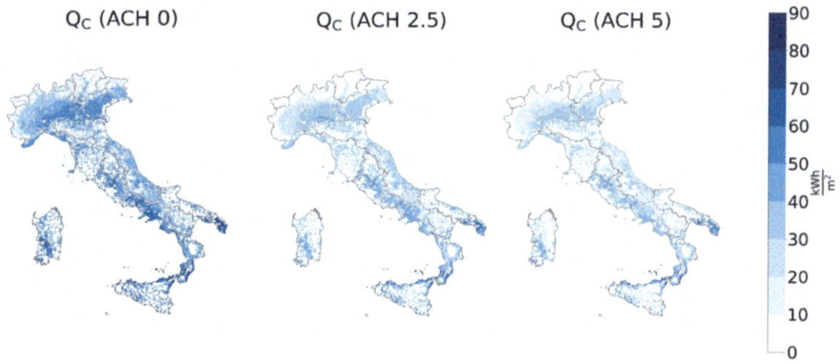

Fig. 7 Building simulation results under mechanical cooling mode. Cooling need distribution in the Italian territory (**a**) without ventilative cooling and with ventilative cooling at different ACH – (**b**) 2.5 and (**c**) 5

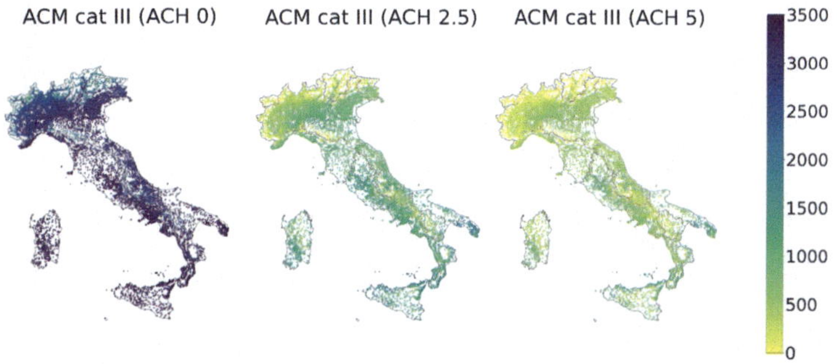

Fig. 8 No. of hours over ACM Cat. II for (**a**) the case without ventilative cooling, (**b**) with a ventilative cooling ACH of 2.5, and (**c**) an ACH of 5

3.2 Building-Based Comfort KPIs

Section 3.2 reports the building simulation results focusing on both mechanical-cooling mode and free-running building mode. Considering the mechanical-cooling mode, Fig. 7 shows that cooling needs in Italy are correlated to different climate conditions – see Fig. 7a. This location-dependent distribution of cooling needs is even more evident when ventilative cooling is considered – see Fig. 7b, c. Nevertheless, the adoption of ventilative cooling strategies can drastically reduce the cooling needs in the majority of locations of the Peninsula. In several cases, ventilative cooling is more than halving the local energy cooling need. The difference between the case with an ACH of 2.5 and the one with an ACH of 5 is not as large as the difference between ACH 2.5 and no ACH, suggesting that even with limited airflow rates, ventilative cooling may strongly reduce building cooling needs in Italy.

Focusing on the free-running mode, Fig. 8 shows the distribution of the number of discomfort hours (above adaptive comfort mode – ACM – upper category II)

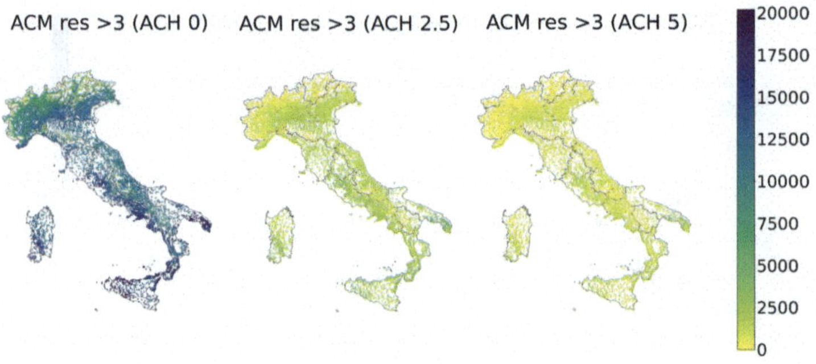

Fig. 9 Cumulative distances from ACM Cat. II upper threshold, varying ventilation ACH

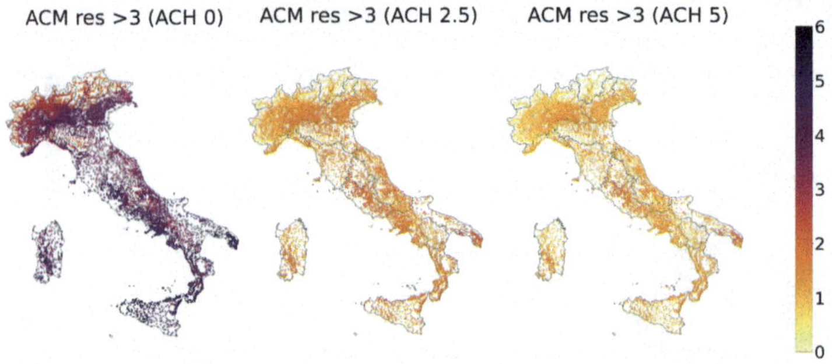

Fig. 10 Average distances from ACM Cat. III lower threshold, varying ventilation ACH

under different ACHs – (a) without ventilative cooling; (b) with ACH set to 2.5; and (c) with an ACH of 5. These maps clearly underline that ventilative cooling may strongly reduce the number of discomfort hours in free-running buildings in almost all Italian locations. In this building mode, a difference between ACH values is underlined especially in southern coastal locations and in the Po Valley. Residual discomfort hours are evident even with a higher ACH in the Eastern part of Apulia regions and in some Tyrrhenian sites, which are also the ones that show higher residual cooling needs in the mechanical cooling mode.

To consider not only the time distribution of discomfort hours but also the discomfort intensity, Fig. 9 reports the cumulative distance from the adaptive comfort model category II upper limit. Comfort category II is assumed considering general suggestions for residential spatial units, assuming in this analysis that discomfort is composed of hours in which building simulated conditions are in both summer Cat. III and Cat. IV – see EN 16798-1. This figure is, on one hand, confirming the above results, while, on the other hand, it shows that even in those locations in which residual discomfort hours are underlined, the intensity of the discomfort may be drastically reduced by adopting ventilative cooling solutions.

Figure 10 shows the same results reported in Fig. 9, but averaged on the basis of the number of discomfort hours shown in Fig. 8. These specific maps are useful to

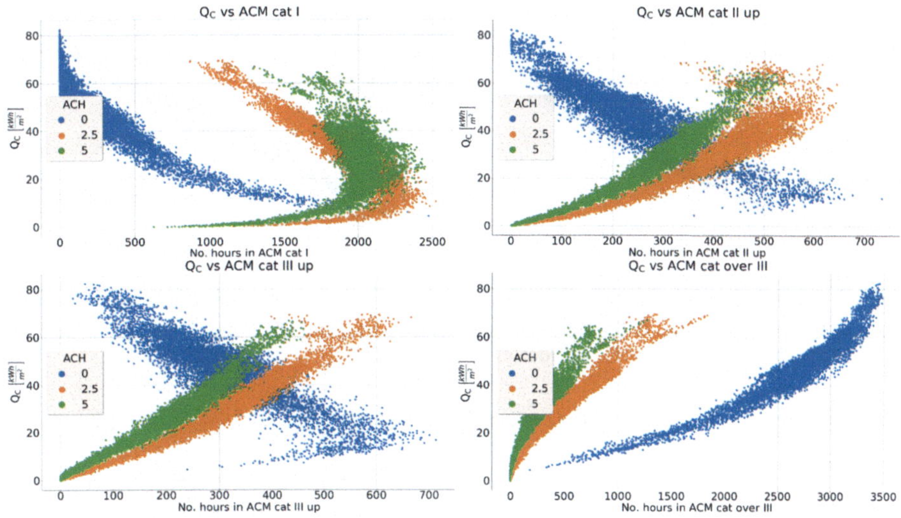

Fig. 11 Cooling consumption (simulated cases B – see Table 1) vs. No. of hours in ACM category (**a**) I, (**b**) II, (**c**) III, and (**d**) IV (simulated cases A – see Table 1) considering different ACHs

give consistency to the discomfort intensity expressed in Kelvin per hour, a synthetic value able to clearly represent discomfort consistency. The aforementioned ability of ventilative cooling to reduce discomfort is immediately evident here, the discomfort intensity shifting in the hottest locations from about 5 [K/h] to about 2 [K/h], a more tolerable value, very near to the upper comfort boundary of adaptive comfort category III (moderate expectations). The distance between the upper limit of Cat. II and III is in fact 1 °C.

Furthermore, a series of correlation analyses have been performed by plotting the cooling energy needs of buildings under mechanical cooling mode as a function of the number of hours for different adaptive comfort categories, considering the simulation results of buildings under free-running cooling mode with the same ACH – see Table 1. These graphs are useful to underline potential virtual correlations between cooling needs and free-running discomfort values. Figure 11 shows the cooling needs of mechanically cooled cases in relation to the number of free-running hours recorded in (a) adaptive comfort category I, (b) Cat. II, (c) Cat. III, and (d) Cat. IV, where the latter includes all discomfort hours outside the moderate comfort category, and the others encompassing, respectively, the hours below the Cat. I upper limit, between Cat. I and Cat. II upper limits, and the hours between Cat. II and III upper limits to avoid hour repetitions (theoretically, Cat. II also includes all hours in Cat. I). These graphs illustrate that in the first three cases – the ones representing different comfort expectations – there is an inverse relationship (ACH = 0) between the number of hours in the comfort category (free-running mode) and cooling needs (mechanical-cooling mode). Other than the last category (discomfort for all expectation levels), a direct correlation is shown for all ACHs, although when VC is considered, a reduction in hours in this category is strongly underlined in all

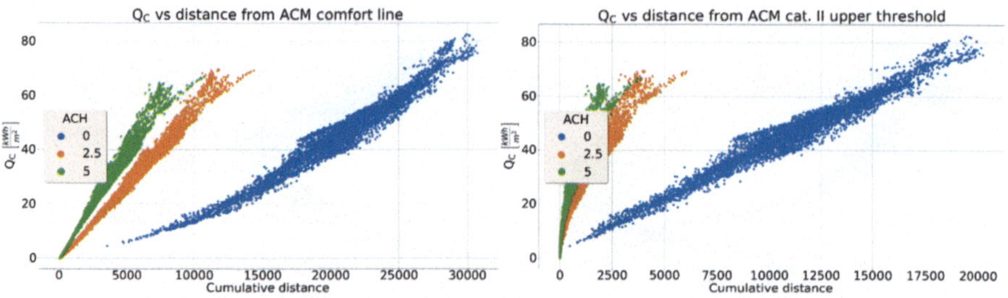

Fig. 12 Cooling consumption (simulated cases B – see Table 1) vs. the cumulative distance from the adaptive thermal comfort (**a**) running mean temperature and (**b**) upper limit of category II (simulated cases A – see Table 1) considering different ACHs

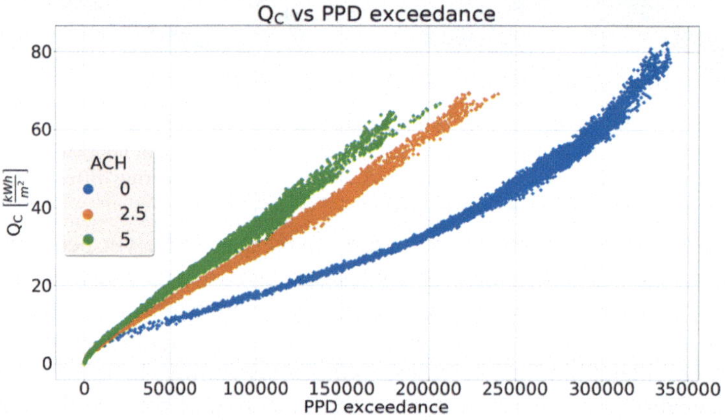

Fig. 13 Cooling consumption (mechanical-cooling mode) vs. PPD exceedance (free-running cooling mode), varying ACH

cases – as mentioned in the previous analyses. ACH activation is impacting the number of hours moving from higher to lower adaptive comfort categories. The difference between ACH activated and ACH = 0 cases is linked to the fact that mechanical cooling works on a fixed set point (26 °C) while adaptive comfort does not. For this reason, hours turned to comfort by ventilative cooling in free-running mode have a corresponding mechanical cooling need for the same ACH value, and the higher the comfort category, the higher the slope of the regression line of VC cases – see, especially, the change among Fig. 11b–d. Nevertheless, it is expected that this outcome will be more evident when discomfort intensity is assumed instead of comfort hours, while direct correlations will be underlined for all cases.

Figure 12 plots the simulated building energy needs as a function of the adaptive comfort cumulative distance from (a) the running mean temperature and (b) the upper limits of comfort Cat. II considering free-running corresponding cases. These graphs show that a correlation is evident between the two building functioning modes. The envisaged slope of regression lines is higher when ACH is active. A slight difference is underlined for ACH 5 with respect to 2.5, but a very evident benefit is retrieved between ventilated cases with respect to the reference case (ACH = 0) supporting the high potential of VC in the Italian territory.

Additionally, cooling needs are compared to the PPD exceedance calculated for the corresponding free-running building cases. This analysis is forcing the above results when Fanger-correlated comfort indices are assumed – which is a VC non-favourable model. Figure 13 shows that cooling needs and free-running PPD exceedance define evident correlation paths for all ACH cases. Hence, the positive impact of VC is underlined even when the adaptive comfort model is not assumed.

3.3 Comparisons Between Climate and Building KPIs

In this section, the results of Sects. 3.1 and 3.2 are compared. Firstly, the building energy need is compared to the climate CDH and CDD. Figure 14 shows the comparison regression lines for (a) Qc vs. CDH considering different CDH base temperatures, and (b) Qc vs. CDD. In all cases, the three levels of ACH are also plotted. A clear correlation between climate and building indices is underlined and the regression model confidence is growing at higher ACHs, given that in these cases the extra effects of building gains are reduced by dissipation. Table 2 reports the retrieved coefficients of determinations (R^2) for different polynomial degrees for the CDH case considering the different base temperatures. In line with other works [34], lower base temperatures better represent, as climate KPIs, the correlated building cooling energy need, especially when VC is not activated. This outcome is directly correlated to the meaning of base temperature in the summer, cooling needs in a building being connected not only to external air temperature, but also to solar and internal gains. Hence, the adoption of a CDH_{18} (or lower) is suggested when this climate KPI is used to climatically analyse the local energy needs. Similarly, Table 3

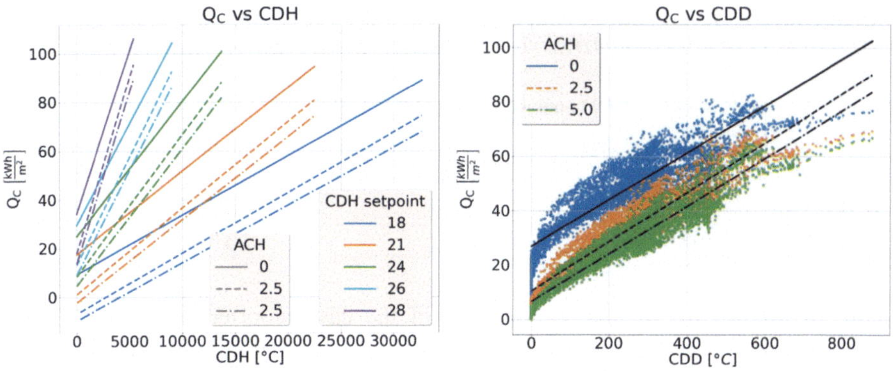

Fig. 14 Building cooling energy needs vs. (**a**) CDH – for different base temperatures, and (**b**) CDD considering different ACH values

Table 2 R^2 results of Qc vs. CDH for different base temperatures – see also Fig. 14a

	Base = 18 ° C	21 ° C	24 ° C	26 ° C	28 ° C
ACH = 0.0	0.893	0.844	0.749	0.656	0.537
ACH = 2.5	0.930	0.904	0.828	0.745	0.629
ACH = 5.0	0.930	0.915	0.850	0.772	0.661

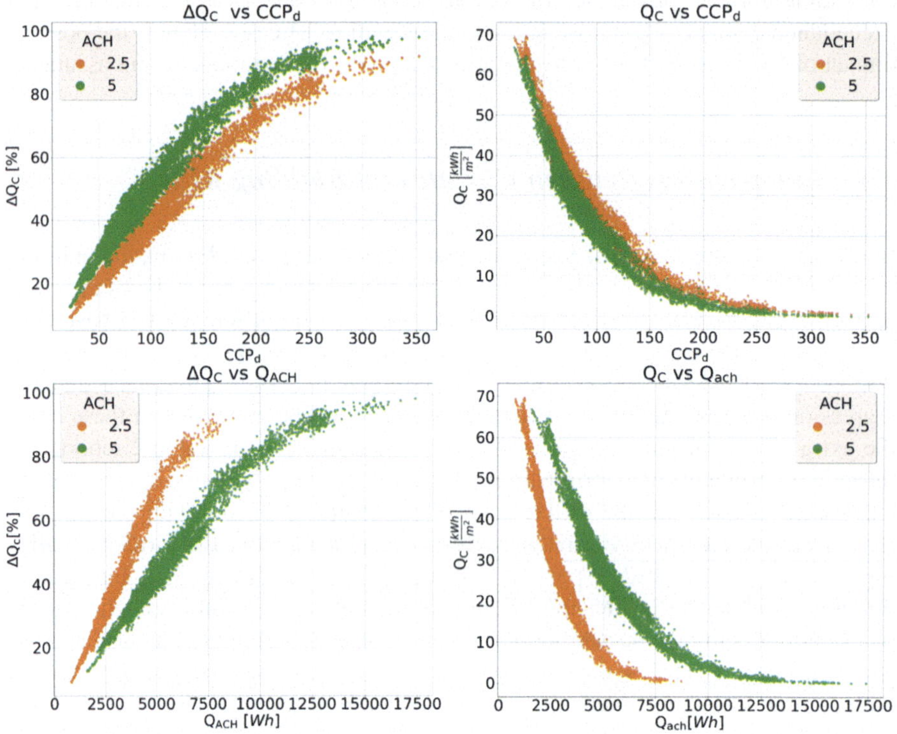

Fig. 15 The difference (percentage of the reference case) in building cooling needs between the reference case (ACH = 0) and the other ACH cases vs. (**a**) CCP_d and (**c**) Q_{ach}. Building cooling energy needs vs. (**b**) CCP_d and (**d**) Q_{ach} considering different ACH values for cooling needs

Table 3 R^2 results of Qc vs. CDD analysis – see also Fig. 14b

	Polynomial degree 1	Polynomial degree 2	Polynomial degree 3
ACH = 0.0	0.805	0.850	0.862
ACH = 2.5	0.893	0.912	0.916
ACH = 5.0	0.918	0.929	0.930

shows R^2 values for the regression lines of the case reported in Fig. 14b. The retrieved values, performed for different polynomial degrees, show that CDD also has a very good potential in representing (climatically) the building cooling energy needs. Even in this case, higher ACHs will increase the correlation. Nevertheless, the CDH_{18} looks to have a slightly higher correlation potential being based on hourly values as opposed to average daily ones.

Furthermore, the other considered climate KPIs are compared with the building simulation results. Figures 15a, b show the difference in building cooling needs between the reference case (ACH=0) and the other ACH configurations (mechanical cooling mode), respectively, expressed as a percentage of the reference case and plotted as a function of the climate CCP_d indicator. Graph (b) underlines that CCP_d and building cooling needs are strongly inversely correlated and that when an ACH of 5 is assumed for the VC building configuration, a slightly higher dissipative potential is shown. The percentage of coverage of the original building cooling needs with VC – see Fig. 15b – ranges from 9.5% to 93.5% for an ACH of 2.5 and from about 12.9% to 98.5% for an ACH of 5.

A similar result is also shown for the Q_{ach} indicator. Figures 15c, d report the same analyses for the latter KPI. Unlike CCP_d, Q_{ach} better represents differences between ACH values since it includes the airflow in its calculation process, while the other indicator is only based on cumulative temperature differences. Nevertheless, both may be adopted to predict, since early design (climate analyses), the local VC potential.

Climate vs. building analyses are also performed by comparing climate indices and the adaptive thermal discomfort defined for building cases in the free-running mode. Firstly, following the same approach as above, CDD, CDH_{18}, and CDH_{26} are compared with the cumulative intensity of adaptive discomfort. Two reference limits are assumed: a theoretical distance from the running mean temperature – here called ">0" – also including comfort conditions until upper Cat. I, but able to represent all cases; and distances from the upper limit of comfort Cat. II (positive values only). Tables 4 and 5 report the R^2 values of polynomial regression lines plotting the adaptive comfort cumulative discomforts as a function of the local CDD and CDH (both CDH_{18} and CDH_{26}), respectively. For higher polynomial degrees, a slightly higher coefficient of determination is reached in both cases, even if these differences are very limited, especially when buildings are set at a high ACH value, that is 5. Similarly, Fig. 16 shows the same behaviour described above by using a graphical representation. As underlined in both tables, the case without VC (ACH = 0) is highly statistically dispersed, while stronger correlations are underlined in VC cases. Both indices are able to represent the expected discomfort intensity in the free-running building mode. Also in this case, a low CDH base temperature (18 °C) shows a higher correlation with respect to adaptive cumulative discomfort values when compared to a higher CDH base temperature (26 °C).

Table 4 R^2 results (adaptive cumulative discomfort vs. CDD)

	Polynomial degree 1		Polynomial degree 2		Polynomial degree 3	
	>0	>3	>0	>3	>0	>3
ACH = 0.0	0.711	0.734	0.765	0.768	0.787	0.783
ACH = 2.5	0.916	0.930	0.926	0.940	0.930	0.940
ACH = 5.0	0.961	0.926	0.961	0.968	0.963	0.968

Table 5 R^2 results (adaptive cumulative discomfort vs. CDH)

	Polynomial degree 1		Polynomial degree 2		Polynomial degree 3	
	>0	>3	>0	>3	>0	>3
	CDH_{18}					
ACH = 0.0	0.811	0.805	0.822	0.806	0.825	0.807
ACH = 2.5	0.932	0.852	0.940	0.942	0.940	0.943
ACH = 5.0	0.934	0.803	0.967	0.962	0.967	0.965
	CDH_{26}					
ACH = 0.0	0.567	0.584	0.647	0.645	0.693	0.679
ACH = 2.5	0.790	0.848	0.823	0.848	0.838	0.850
ACH = 5.0	0.857	0.883	0.869	0.894	0.877	0.895

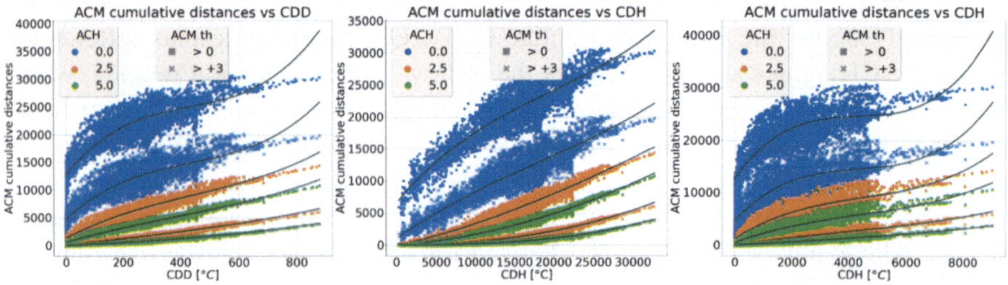

Fig. 16 Building free-running mode. Adaptive thermal comfort cumulative distances from the running mean temperature and upper limit of Cat. II vs. (**a**) CDD, (**b**) CDH_{18}, and (**c**) CDH_{26}

Furthermore, CCP (cumulative) values are compared to adaptive thermal comfort hours in free-running building mode considering the effect of VC. This comparison is based on changes due to VC in the number of hours for comfort categories with respect to the hour distribution of the reference case (ACH = 0). Figure 17a plots the number of hours for different adaptive comfort categories as a function of the local CCP when ACH=0. This figure underlines that, at low CCP values, a high number of hours are included in Cat. IV, while a higher number of hours in Cat. I is underlined for higher CCP values. Differently, the other graphs in Fig. 17 show differences in the number of hours for each category between the reference case and the VC ones. Figure 17b shows an almost linear correlation between the increase (negative delta) in hours below the comfort Cat. I and local CCP values when VC is considered. Focusing on Cat. I, Fig. 17c underlines that at low CCP, an increase in hours in this category is underlined for VC cases, while a reduction in the original hours in this range is shown at a higher CCP, suggesting that VC is moving them below the lower limit. The same trend is evident in all the following graphs, which show that VC is supporting a reduction in space temperatures with respect to the reference case (ACH = 0), especially for locations with a high CCP. Nevertheless, Fig. 17f shows that a reduction in hours above the upper limit of Cat. III is evident for all sites.

Considering the other VC climate indicator, a strict inverse proportionality is underlined in Fig. 18 between local Q_{ach} and the corresponding cumulative distance

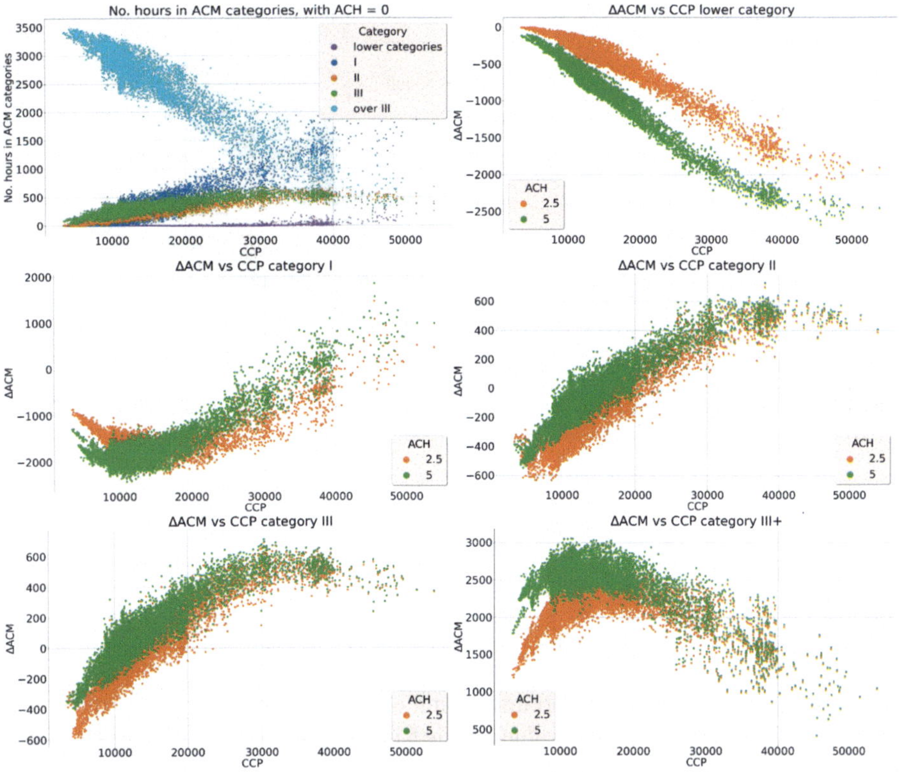

Fig. 17 Free-running building hours for adaptive comfort categories vs. CCP, considering (**a**) the simple number of hours in the reference case with ACH = 0; differences in the number of hours (case ACH_0 – case $ACH_{2.5}$ and ACH_5) by adaptive comfort category: (**a**) below the lower limit of Cat. I; (**c**) in Cat. I; (**d**) in Cat. II; (**e**) in Cat. III; and (**f**) above the upper limit of Cat. III

Fig. 18 Building adaptive comfort cumulative distances from the upper limit of Cat. II plotted as a function of the local Q_{ach}

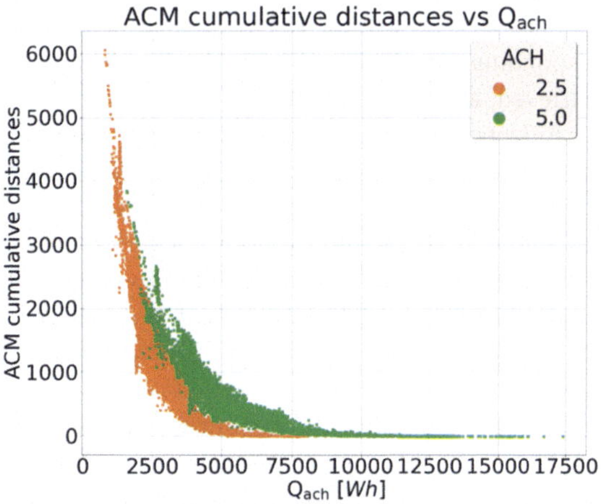

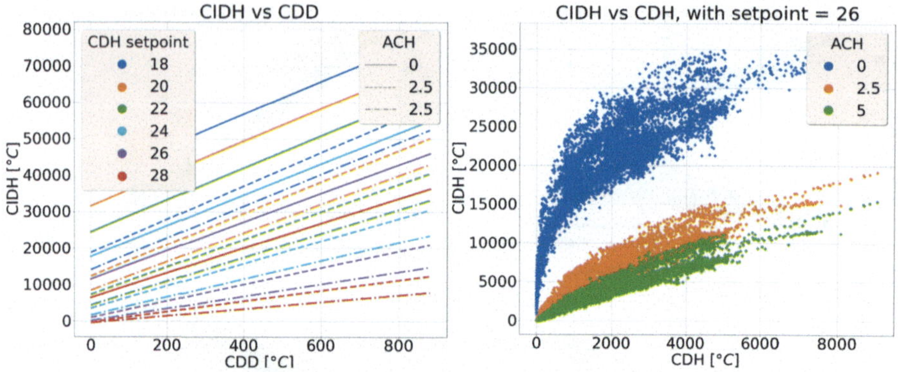

Fig. 19 Building CIDH plotted as a function of local (**a**) CDD and (**b**) CDH. In (**a**), different CIDH base temperatures are assumed, while in (**b**), both KPIs have a base temperature of 26 °C

Table 6 R^2 results ($CIDH_{base}$ vs. CDD)

	Base = 18 ° C	20 ° C	22 ° C	24 ° C	26 ° C	28 ° C
ACH = 0.0	0.778	0.779	0.787	0.801	0.820	0.843
ACH = 2.5	0.851	0.873	0.904	0.940	0.970	0.971
ACH = 5.0	0.878	0.907	0.943	0.974	0.987	0.961

from the adaptive thermal comfort upper limit of Cat. II (considering only the hours above this limit). A higher ACH value, that is 5 instead of 2.5, increases the local VC potential.

Finally, for the free-running building cooling mode, the CIDH is also assumed, considering that in mechanical cooling mode this indicator is functional to the selected set point temperature, thus losing relevance. Figure 19 compares local climate (a) CDD and (b) CDH vs. building CIDH. In graph (a), different CIDH base temperatures are compared with CDD, showing good correlations especially for higher bases – see Table 6. Differently, in graph (b), $CIDH_{26}$ is compared to the corresponding environmental CDH_{26} for different ACH values. High R^2 values are retrieved, as shown in Table 7.

A deeper analysis was performed by comparing the coefficient of determinations for different CDH and CIDH base temperatures (free-running building mode). Results are plotted in Fig. 20a–c for the three considered ACH values {0;2.5;5}, respectively. As mentioned above, a very high correlation is underlined in all cases, especially for CDH_{18}, in line with the results of the mechanical-cooling building mode. For lower CDH bases, medium-to-high CIDH temperatures show a higher R^2, while by increasing the ACH the optimal combination moves to lower CIDH base temperatures. This latter result is due to the increase in the correlation between indoor and environmental conditions, thanks to greater airflow exchanges.

Table 7 R^2 results (CIDH vs. CDH – base temperature 26 for both)

	Polynomial degree 1	Polynomial degree 2	Polynomial degree 3
ACH = 0.0	0.820	0.866	0.883
ACH = 2.5	0.970	0.972	0.973
ACH = 5.0	0.987	0.988	0.988

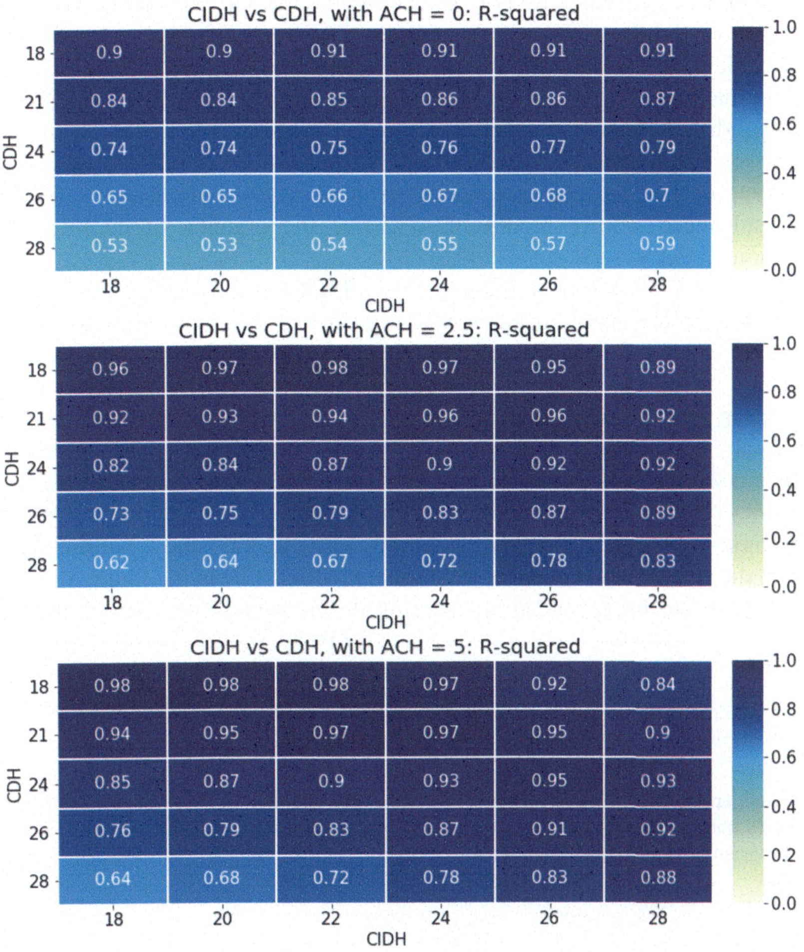

Fig. 20 R^2 values combining different ACH and CIDH base temperatures for: (**a**) ACH = 0; (**b**) ACH = 2.5; and (**c**) ACH = 5 in free-running building mode

4 Conclusions

The chapter analyses the ventilative cooling potential in Italy by performing climate-based and dynamic building energy simulation analyses. The research defines a series of KPIs, both climate and building-simulation based, to represent VC potentialities. A correlation is underlined between climate and building-simulation retrieved KPIs. This supports the possibility to analyse, since the early design stages, such as building programming – that is when a specific building shape is not yet finalized – the VC potential of a location and the local need for cooling in terms of climate intensity. This possibility is very important to support, in the earliest stage possible, the identification of correct bioclimatic and low-energy solutions to reduce space cooling energy needs and take advantage of the local natural/low-energy cooling heat sink potential.

Additionally, the chapter underlines a direct correlation between mechanically cooled building indicators and free-running building ones, which is an important outcome toward the possibility to correlate the performances of these two types of cooling approaches considering potential evaluations and optimisations, for example for advanced control modes or to define the urgency of cooling system installation.

Finally, results have shown that in almost all Italian locations VC may considerably reduce cooling energy needs when coupled with mechanically cooled spaces and turn a large number of discomfort hours to comfort hours in free-running buildings.

Considering the limitations of this study, the presented work is based on a single TMY database, while in another ongoing research, a larger set of climate databases will be considered including past, present, and future TMY models to verify the consistency of the VC potential under climate changes and under specific perturbation phenomena, such as heatwaves. Additionally, in this study, building data is based on the simulation of a single residential model, while in the future, this analysis is expected to be expanded considering different building configurations, for example referencing the Typology Approach for Building Stock Energy Assessment (TABULA) building database and different envelope parameters. Finally, these indicators will be compared in real building operational conditions to verify the obtained results.

Acknowledgements This work includes preliminary results by WP3 (T3.2) of the PRELUDE project [40] that has received funding from the European Union Horizon 2020 research and innovation programme under Grant Agreement N°958345.

References

1. M. Santamouris, ed., Advances in Passive Cooling, Earthscan, London, 2007.
2. Daikin Industries, Annual Report 2019. Fiscal Year Ended March 31, 2019, 2019. https://www.daikin.com/investor/data/report/daikin_jar19.pdf.
3. The Future of Cooling – Analysis, IEA. (n.d.). https://www.iea.org/reports/the-future-of-cooling (accessed June 4, 2020).
4. IEA Electricity Information 2020, (n.d.). http://wds.iea.org/wds/pdf/ele_documentation.pdf.

5. M. Santamouris, ed., Cooling energy solutions for buildings and cities, World Scientific, New Jersey, 2019.
6. G. Chiesa, M. Grosso, D. Pearlmutter, S. Ray, Editorial. Advances in adaptive comfort modelling and passive/hybrid cooling of buildings, Energy and Buildings. 148 (2017) 211–217.
7. P. Holzer, T. Psomas, eds., Ventilative cooling sourcebook, IEA EBC Annex 62, Aalborg University, Aalborg, 2018.
8. B. Givoni, Man, climate, and architecture, Elsevier, Amsterdam, New York, 1969.
9. M. Grosso, ed., Il raffrescamento passivo degli edifici, IV, Maggioli, Sant'Arcangelo di Romagna, 2017.
10. C. Plesner, M.Z. Pomianowski, Ventilative Cooling in Standards, Legislation and Compliance tools, in: Innovations in Ventilative Cooling, SPRINGER, Cham, 2021: p. 26pp.
11. F. Flourentzou, J. Bonvin, COOLINGVENT – Refroidissement par ventilation pour les bâtiments à basse consommation, rénovés ou neufs. IEA ECBCS Annex 62 on Ventilative cooling, OFEN, Geneva, Switzerland, 2017.
12. G. Chiesa, M. Kolokotroni, P. Heiselberg, eds., INNOVATIONS IN VENTILATIVE COOLING., SPRINGER, S.l., 2021.
13. M. Kolokotroni, P. Heiselberg, IEA EBC Annex 62 – Ventilative Cooling State-of-the-Art Review, Aalborg University, Aalborg, 2015. https://venticool.eu/wp-content/uploads/2013/09/SOTAR-Annex-62-FINAL.pdf.
14. P. Heiselberg, IEA EBC Annex 62 – Ventilative Cooling Design Guide, Aalborg University, Aalborg, 2018.
15. M. Santamouris, D. Asimakopolous, eds., Passive Cooling of Buildings, James and James, London, 1996.
16. CISBE, Degree-days: Theory and application, The Chartered Institution of Building Services Engineers, London, 2006.
17. H. Xuan, B. Ford, Climatic applicability of downdraught cooling in China, Architect. Sci. Rev. 55 (2012) 273–286.
18. EUROSTAT, Energy statistics – cooling and heating degree days (nrg_chdd), (2019). https://ec.europa.eu/eurostat/cache/metadata/en/nrg_chdd_esms.htm.
19. G. Chiesa, Calculating the geo-climatic potential of different low-energy cooling techniques, Build. Simul. 12 (2019) 157–168. https://doi.org/10.1007/s12273-018-0481-5.
20. G. Chiesa, N. Huberman, D. Pearlmutter, Geo-climatic potential of direct evaporative cooling in the Mediterranean Region: a comparison of key performance indicators, Building and Environment. 151 (2019) 318–337.
21. G. Chiesa, A. Zajch, Geo-climatic applicability of earth-to-air heat exchangers in North America, Energy and Buildings. 202 (2019) 109332.
22. N. Artmann, H. Manz, P. Heiselberg, Climatic potential for passive cooling of buildings by night-time ventilation in Europe, Applied Energy. 84 (2007) 187–201. https://doi.org/10.1016/j.apenergy.2006.05.004.
23. EN 16798-1:2019, Energy performance of buildings. Ventilation for buildings. Indoor environmental input parameters for design and assessment of energy performance of buildings addressing indoor air quality, thermal environment, lighting and acoustics. Module M1-6, CEN, 2019.
24. G. Chiesa, Climatic potential maps of ventilative cooling techniques in Italian climates including resilience to climate changes, IOP Conf. Ser.: Mater. Sci. Eng. 609 (2019) 032039. https://doi.org/https://doi.org/10.1088/1757-899X/609/3/032039.
25. F. Allard, ed., Natural ventilation in buildings: a design handbook – EC, ALTENER Prorgamme, James and James (Science Publishers) Ltd, London, 1998.
26. P. Fanger, Thermal comfort-analysis and applications in environmental engineering. PhD, Technical University of Denmark, 1970.
27. J. van Hoof, Forty years of Fanger's model of thermal comfort: comfort for all?, Indoor Air. 18 (2008) 182–201. https://doi.org/10.1111/j.1600-0668.2007.00516.x.

28. EDYCE, D1.2 Operational dynamic Energy Perfomance Certificate (EPC) specifications, PoliTO, 2021. https://edyce.eu/reports-and-results/.
29. M.A. Humphreys, F. Nicol, S. Roaf, Adaptive thermal comfort: foundations and analysis., ROUTLEDGE, S.l., 2020.
30. F. Nicol, M. Humphreys, Derivation of the adaptive equations for thermal comfort in free-running buildings in European standard EN15251, Building and Environment. 45 (2010) 11–17. https://doi.org/10.1016/j.buildenv.2008.12.013.
31. M. Pellegrino, M. Simonetti, G. Chiesa, Reducing thermal discomfort and energy consumption of Indian residential buildings: Model validation by in-field measurements and simulation of low-cost interventions, Energy and Buildings. 113 (2016) 145–158. https://doi.org/10.1016/j.enbuild.2015.12.015.
32. E. Neufert, G.M. Di Giuda, V. Villa, A. Gottfried, P. Piantanida, Enciclopedia pratica per progettare e costruire: manuale a uso di progettisti, costruttori, docenti e studenti : fondamenti, norme e prescrizioni per progettare, costruire, dimensionare e distribuire a misura d'uomo, U. Hoepli, Milano, 2013.
33. B. Zevi, L. Zevi, C. Mancosu, Il nuovissimo manuale dell'architetto: [nozioni generali di progettazione, prestazioni degli organismi edilizi, esercizio professionale, progettazione ..., Mancosu, Roma, 2019.
34. G. Chiesa, M. Palme, Assessing climate change and urban heat island vulnerabilities in a built environment, TECHNE – Journal of Technology for Architecture and Environment. (2018) 237–245 Pages. https://doi.org/10.13128/TECHNE-22086.
35. DOE, NREL, EnergyPlus™, DOE, BTO, NREL, 2020. https://energyplus.net/.
36. Meteotest, METEONORM, METEOTEST Genossenschaft, Bern, 2015. meteonorm.com.
37. openpolis, GeoJSON, (2018). https://github.com/openpolis/geojson-italy.
38. ISTAT, Principali statistiche geografiche sui Comuni, (2021). https://www.istat.it/it/archivio/156224.
39. P. Heiselberg, Ventilative Cooling Principles, Potential and Barriers, in: Innovations in Ventilative Cooling, Springer International Publishing, Cham, 2021: p. 22. https://doi.org/10.1007/978-3-030-72385-9.
40. PRELUDE, PRELUDE (prescient building operation utilizing real time data for energy dynamic optimisation) project, (2021). https://prelude-project.eu/.

Giacomo Chiesa Giacomo Chiesa is an associate professor in Architectural Technology at Politecnico di Torino, Department of Architecture and Design. He holds a PhD in Technological Innovation for the Built Environment. His research focuses on bioclimatic and climatic design, passive solutions for heating and cooling buildings, ICT and IT integration in the urban and architectural contexts, intelligent buildings, and smart cities. He leads a research group that pursues a performance-driven methodological approach for design and operative optimisation/management issues correlated to the built environment at the different scales (components, buildings, urban contexts). He is a member of the POLITO's Didactical Board in Architecture and the Didactical Board in Electronic, Telecommunication and Physics Engineering. He is currently the local PI of two EU H2020 Innovation Action-funded projects and, during his 15 years of career, has authored more than 120 scientific publications.

Francesca Fasano Francesca Fasano holds a bachelor's degree in Physical Engineering and a master's degree with Laude in Telecommunication Engineering (ICT for Smart Society) at Politecnico di Torino; she is currently a research fellow at POLITO, Department of Architecture and Design, supporting ICT and IT integrated issues for the built environment. Additionally, to python coding and tool development actions, she keeps building energy management and optimisation, building monitoring and advanced building simulation, including climate and climate change analyses.

Paolo Grasso Paolo Grasso holds a bachelor's degree in Telecommunication Engineering and a master's degree in Telecommunication Engineering (ICT for Smart Society) at Politecnico di Torino. He is a research fellow at the same university at the Department of Architecture and Design. He supports python coding, forecasting building behaviours, and building energy monitoring and management. He works in the ICT and IoT integration in building stocks, including IT platform development and server facility management.

Energy Retrofit of Traditional Buildings in a Warm-Humid Urban Climate

Natali Collado Baldoquin, Dania González Couret, and Luis Alberto Rueda Guzmán

1 Introduction

Buildings are responsible for approximately 40% of energy consumption in Europe [1], which is even higher in warm-humid climates [2], with a high proportion of cooling demand [3]. It could be, for example, 30% of the total electricity produced in Singapore and Malaysia, [4]. According to Aldossary et al. [5] as well as Oropeza-Pérez and Ostergaad [6], artificial conditioning of buildings represents between 50% and 86% of the total energy consumption in warm and humid climates.

Electricity generation is very polluting in Cuba [7], mainly based on imported fuel, and the residential sector is responsible for 60% of the consumed electric power [8]. This demand has been increasing by 66% with respect to the former decade [9].

In spite of the government's interest in reducing energy consumption in the country, the role that buildings play in it has not been enough recognized or promoted [10]. Precedent researches show that national regulations about energy efficiency and thermal environment are not sufficient, and high energy architecture doesn't reach expected international standards to reduce environmental impact [11].

Housing renting rooms for tourists in Cuba is part of the residential buildings with higher energy demand, because of the major use of artificial conditioning, among other reasons. In 2018, the Ministry of Tourism confirmed that 26% of the available rooms for tourists were located in the residential sector [12], and 23% of them, in the "Zona de Valor histórico-cultural de El Vedado" (Zone of historical and cultural value in Vedado) (ZVHCV) [13], recognized territory in Havana because of its historical and cultural values, centrality and network of services and facilities, especially the cultural ones, the reason why it was assumed as the study area in this research.

N. C. Baldoquin · D. G. Couret (✉) · L. A. R. Guzmán
Technological University of Havana, Havana, Cuba

A. Sayigh (ed.), *Achieving Building Comfort by Natural Means*, Innovative Renewable Energy, https://doi.org/10.1007/978-3-031-04714-5_19

The 22% of housing that rent rooms for tourists in El Vedado are low detached buildings with tall ceilings and then highly exposed to the solar radiation, with a great thermal mass, and built during the first three decades of the twentieth century. This architectural type with potentialities for energy retrofit can be found in other zones of Cuba and has been selected as study object in the present research.

2 Materials and Methods

The research was carried out in four steps. During the first one, using theoretical methods like documentary analysis and historic – logic and analysis – synthesis, a discussion of the main related issues was developed. The key scientific and methodologic conflicts and problems not solved in this field were identified, since most of the precedent researches related to energy renovation and reduction of the energy consumed by air conditioning have been mainly advanced in developed regions with climatic and economic conditions different from the Cuban ones. In the same way, previous results related to buildings in warm-humid climates correspond, mostly to study objects different from the one in this work, focused on isolated and tall buildings with the predominance of glass envelope, or housing with low thermal mass. The methods used to evaluate the impact of constructive actions and their relationship with the design variables are presented in the theoretical framework.

The second stage was aimed at characterizing the identified study object for energy retrofit in order to define the theoretical model taken as basis for the thermo-energy evaluations by dynamic simulations. For that, buildings with rooms for rent in El Vedado were identified and georeferenced. After that, and taking into account file information, and reality observation by a field work, these buildings were classified from the energy point of view, based on the variables identified in the theoretical framework. About 22 study cases were selected in which the owners wanted to collaborate on the research, located in the "zona de baja intensidad de El Vedado" (low-intensity Vedado zone); the area was selected because, according to the urban regulations, its morphology won't suffer great future transformations. These housings were characterized and preliminarily evaluated with respect to the identified variables in order to establish the parametric model for simulations. As a result, two extreme situations and other two intermit context conditions were identified, departing from the height and distance from the neighbour buildings; a volumetric and spatial scheme characteristic of the more recurrent distribution with central, circulation corridor were made, and two envelope conditions were defined: one for the initial situation with the original materials and finishing from the early twentieth century and the other for the modified condition, according to the transformations made by the users with the worst possible thermal and energy situation.

In a third step, the initial thermal environment was evaluated for the selected cases, in order to identify the higher impact parameters. The evaluation was developed by two monitoring campaigns for temperature and relative humidity in interior spaces and a referent exterior one (eight buildings, first in naturally ventilated rooms, and 2 for the second campaign in bedrooms with artificial conditioning). To evaluate by dynamic simulations, the main impute data were defined according to the ranks recommended by specialized literature and consulted experts. The behaviour of outdoor temperature was modified to reflect the urban heat island effect, and values of the thermal properties for the envelope materials and infiltration and temperature of the soil surface were selected according to the study object. The thermal environment was simulated in an iterative way with each combination of the input data values until the differences between the temperature monitored in a real case and the one simulated in this build model complemented the statistical indicators recommended for this kind of research (ASHRAE). This comparison between the thermal reality measured and the one simulated was repeated with another study case and, when the statistical requirements were complimented again, the main input data were considered validated. The energy impact of independent parameters and variables as well as invariants in the demand for air conditioning were determined by dynamic simulations with the validates theoretical model.

Design recommendations for the energy retrofit were defined in the last stage, organized according to their energy and economic impacts, depending on the orientation and the building context. For that, the thermal energy impact of the main retrofit interventions was determined, according to the theoretical framework and the dynamic simulations. The economic viability of the solutions was evaluated according to the benefits achieved during the useful life and the impact of the initial investment for a five-year period. The discussion about the thermal, energy and economic impact of each design parameter on the evaluated interventions constitutes the scientific basis for the final proposed recommendations.

3 Results

3.1 Theoretical Framework

3.1.1 Energy Retrofit

Energy retrofit is referred to the interventions on equipment, systems and parts of the building in order to improve its energy behaviour [14], besides the inclusion of sensors and meters [15], which can be classified as actives and passives [3]. The last ones constitute the study object in the present research. It is also considered as a tool to protect the buildings, since it is a guarantee for their use instead of being abandoned and demolished. The solutions should reduce consumption and be technically viable, without causing collateral structural humidity damage [16], improve building aesthetic [15–19] and consider reversibility, authenticity and transformability levels [6, 16, 18, 20, 21].

3.1.2 Methods for the Evaluation of Energy Impacts

The main methods to evaluate the thermal energy transformation in existing buildings may be grouped as: environmental and microclimatic monitoring, in-situ measurements of the envelope and simulation of the thermal energy behaviour [16]. It is usual to calibrate the simulation models, comparing the environmental parameter measured and simulated [16, 21–25]. Monitored temperature and humidity were used in this research to evaluate the departing conditions of the study cases and to validate the parameters assumed for the simulation, which allows to unify the starting situation to compare the determinant parameters, which in real life are presented in a wide variation of situations, as well as to foresee the impact of possible interventions without expensive experiments in the lab or real scale. However, the standardized input data sometimes are not appropriate to be used, and many of the available software can't represent the physical phenomena present in ancient buildings as the thermal vertical stratification and the humidity transfer through the envelope [16].

The resources designed for this purpose differ, mainly because of the mathematical models used to determine the physical phenomena influencing the energy consumption and the indoor environment [26]. According to Mazzeo et al. [27, 28], EnergyPlus is one of the most used tools to develop precise, flexible and high execution speed dynamic simulations, validated as trustworthy in several researches [27, 29, 30]. This calculation engine has been used in this work, added to its graphic interphase OpenStudio because of its strengths and actualized twice a year for being a free open code software [31].

3.1.3 Strategies and Actions for Energy Retrofit in a Warm and Humid Climate

Precedent research has identified design variables influencing the thermal environment and the reduction of energy for artificial conditioning [10, 19, 32–41], based on which four variables have been assumed: Context, Volume, Space and Envelope. Volume and Space have been considered as a fixed and unitary invariant, as a part of the study object, which won't be transformed by the retrofit interventions. Context is the independent variable influencing the behaviour of the Envelope, taken as the dependent variable, constituted by the roof, walls, windows and shading devices, which could be modified by the energy retrofit (Fig. 1).

Passive design strategies have been widely discussed to guarantee a better indoor thermal environment and to reduce energy in buildings [6, 42–49], but publications referring to warm-humid climates are scarce. Referents about energy retrofit of housing with the characteristics of the study object in this kind of climate were not found in the consulted bibliography. Most of the precedent studies deal with interventions for no residential functions [3, 50–62], such as offices and hotels, generally tall buildings, and other ones are focused on general recommendations from the discussion of the state of the art. More than half of the authors evaluate isolated

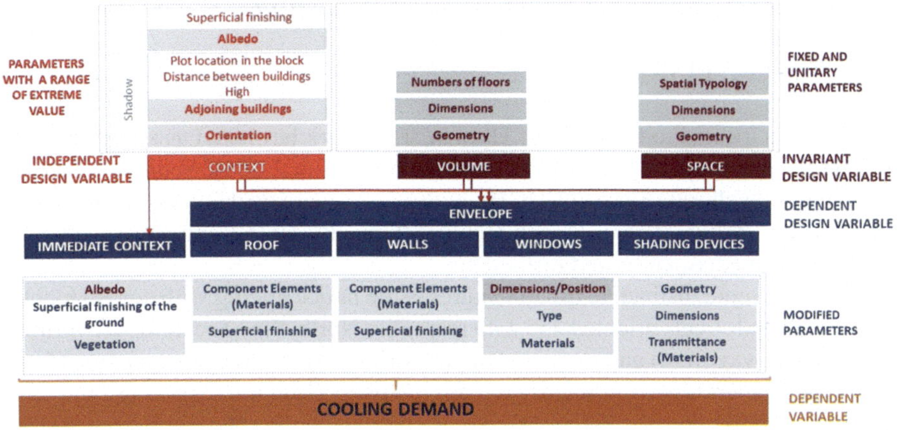

Fig. 1 Variables and parameters for characterizing the study cases

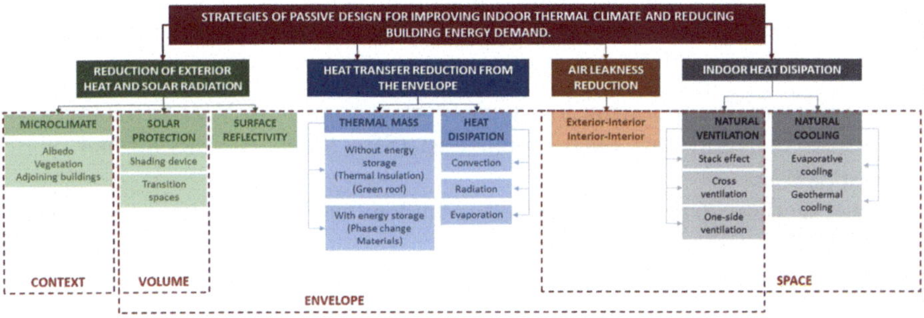

Fig. 2 Strategies of passive design for improving indoor thermal climate and reducing building energy demand

building without the influence of an urban context or in lab tests [3, 25, 51–53, 55, 58–60, 62–79], and a minority are deferred to the impact of energy retrofit on housing with light roof and walls, built using local materials [57, 67, 71, 72, 75].

Strategies to improve indoor thermal environment and reduce energy demand of buildings applicable to warm and humid climates may be classified into four groups [44, 80]: reduction of heat and exterior solar radiation, thermal conduction through the envelope, exfiltration and dissipation of indoor heat (Fig. 2).

Natural ventilation has not been a study object in the present research, since it is known that people use to close the windows to avoid noise and insects or because of privacy and security. Besides that, wind is very variable, mainly in urban contexts [19]. Taking into account that other strategies like evaporative cooling are not effective in warm-humid climates, the research has been concentrated on passive strategies related to reflexion and reduction of incident solar radiation as well as diminishing heat transfer.

3.2 Theoretical Model for Dynamic Simulations

3.2.1 Context: El Vedado

El Vedado is located to the north of Havana. This work has taken as study area the 505 Ha occupied by the "Zona de Valor históńrico-cultural de El Vedado" (zone of historical and cultural value) (ZVHCV), with a great touristic demand because of its historical and cultural values, which includes three city centres and a wide network of theatres, cinemas, clubs and restaurants.

One of its more distinctive urban features is the lattice tracing with square blocks 100 m long and streets oriented 45° with respect to North. Street sections are formed from the car way (8 m) and the wooded pedestrian ways (4 m), plus gardens (5 m) and private galleries (4 m). Blocks were originally divided into 12 plots (15 m width and 50 m depth).

The study object of the present research was limited to the "low intensity zone" (ZIU-I) [81], where most of the original components are preserved in a uniform and well-defined context, and important future transformations are not foreseen. As a result of the context characterization according to its possible influence on the indoor environment and the energy consumption, four context types were defined.

- **Orientation:** The following four typical orientations in El Vedado tracing were considered: North East (NE), South East (SE), South West (SW) and North West (NW).
- **Street section:** 26 m width, considering gardens, car and pedestrian ways, established by the Urban Regulations [81].
- **Adjoining:** Two limit and two intermit situations are considered:

  Context with low solar protection (CSP-L). All adjoining buildings have the same height as the analyzed case (5,6 m), and then, they don't cast shadow over the roof. Separation between buildings is 8 m to the rear, as well as 1,4 m and 8,0 m to the side (Fig. 3).
  Context with high solar protection (CSP-H). All adjoining buildings are 11 m tall (maximum permitted height in the ZIU-I), and separation between buildings is constant (Fig. 4).
  Context with mid solar protection (CSP-M). It represents two intermit situations with respect to the adjoining buildings, at 5,6 m and 11,0 m (Fig. 5).

- **Vegetation:** It doesn't have currently a great impact on reducing solar radiation on the roof, so it was decided not to take into account the trees as shadow elements, considering the most critical insolation condition for the simulations.
- **Albedo:** The number 0,2 was used as the initial value, generally employed for urban zones [82, 83] and to determine the impact of the modified situations, 0,8 was defined for white and polished surfaces [84].
- **Reflexion coefficient of the adjoining walls**: The number 0,6 was used as the initial value, corresponding to clear surfaces [84], and to evaluate modified situations, 0,8 for white and polished surfaces and 0,15 for dark finishing [84].

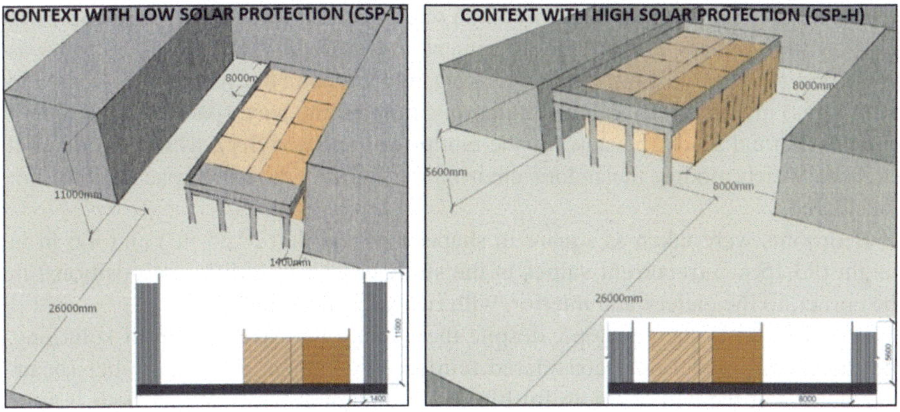

Figs. 3 and 4 Schemes of the extreme boundary situation for the immediate context

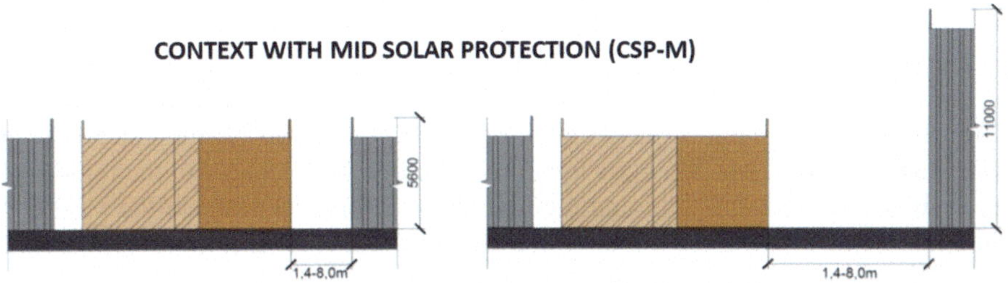

Fig. 5 Schemes of the intermediate boundary situation for the mid-solar protection context

3.2.2 Architectural Typology, Volume, Space, Envelope

From the study cases, three housing types were identified, according to their initial conception, the interior circulation (central or sided), the spaces distribution and relationship, the ceiling height and the date of construction, which determines building materials and techniques.

In order to construct the theoretical model, the central corridor housing type was taken as a reference, since it corresponds to 73% of the cases, and its conditions with respect to the immediate context, volume and space propitiate the major energy consumption. These houses have wide bedrooms lightly transformed, adjoining shadows are less effective due to the volume and separation between buildings in the context is wider, allowing more incident solar radiation. Additionally, they are the currently less modified housing, so recommendations for energy retrofit may have the same impact for most of them.

Two scenarios were taken into account for some parameters: the initial situation (Ini. Sit.), related to the buildings preserving the main original features, and the modified housing (Mod. Sit.), for those elements transformed by the inhabitants, generating major consumes than those estimated to the initial situation.

The theoretical model is composed of a rectangular volume (23 m long and 11 m wide), with a gallery across the entire facade (4 m depth) (Figs. 6 and 7). It was simplified to 4 main spans (without taking into account the gallery, services and bathrooms) in order to facilitate simulations. In the same way, spaces with different internal thermal loads (because of the equipment and activity), as bathrooms and kitchens which modify the indoor environment in the adjoining spaces, were not considered.

Bedrooms were taken as square in shape, 4,5 m wide (20,25 m^2) and 4,6 m in height (93.15 m^3) (recurrent values in the study cases), in 3 different positions: in the corner, to the gallery and interior, with two window models.

With respect to the envelope, despite there being two structural roof solutions, the theoretical model only considered reinforced concrete, since the other (beam and slab) is at the end of its useful life and the thermal resistance of both is very similar (0,10 m^2K/W – 0,11 m^2K/W), according to slab solutions studied in Belgium [85]. For the traditional waterproofing system, 0,20 m^2K/W was assumed [86], the reflectivity of the ceramic tile in the exterior surface finishing was considered 0,3 [87], and 0,1 to the modified situation with the asphaltic layer [84].

The main material in the walls is solid brick, 30 cm for the structural ones and 15 cm in the rest of them. The departing situation considered two variants for the "French" windows: with 3 glass lights, one on top and two on the sides (1,6 m wide), and with only one top glass light (1,4 m wide), both located in the middle of the wall at 1 m from the floor and total height of 2,6 m. To evaluate the modified situation, a full simple glass window was assumed (1,6 × 2,6 m).

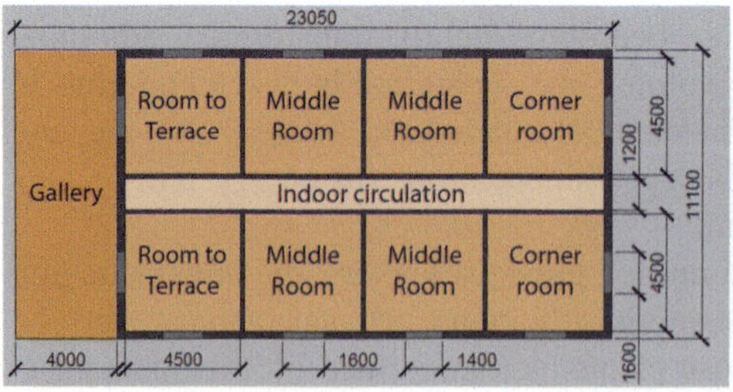

Figs. 6 and 7 Planimetric and volumetric schemes of the theoretical model

Fig. 8 Location and reference image of the monitored study cases

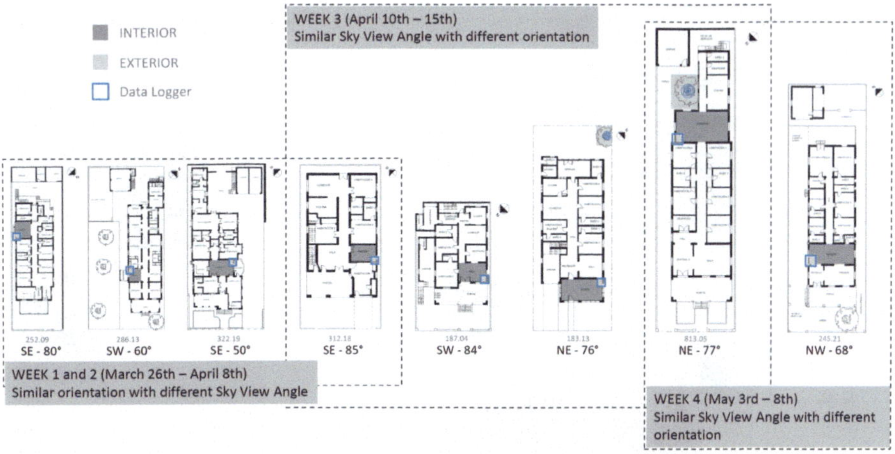

Fig. 9 Thermal sensors position in the monitored spaces at the stage I. It is indicated for each space the exterior wall orientation (SE, SW, NE) and the sky view angle (50–85°)

4 Discussion

4.1 Monitoring and Simulation. Calibration and Validation

4.1.1 Monitoring

In order to evaluate the impact of the design variables on indoor thermal conditions, two monitoring steps (temperature and relative humidity) were carried out in central corridor housing. The first one aimed to compare the effect of the main design variables, simultaneously measuring not acclimatized spaces, which generally correspond to the living and the dining rooms, in thermal environments not modified by air conditioning. Measurements were carried out in 8 interior spaces[1] located in 8 houses, during 1 or 2 weeks in March, April and May, 2019 (Figs. 8 and 9).

The experiences and results achieved during the first step showed that, in order to precise the influence of the design variables, measurements should be carried out in spaces with less internal loads, which was developed during the second monitoring step in bedrooms, which usually are closed when they are not rented, and so less factors influence the thermal environment (persons, equipment, ventilation). Measurements also allowed for precise temperatures and schedule when the spaces were in use with air conditioning. During this step, seven bedrooms were monitored[2] in two houses for 3 months (from October to December 2019 (Figs. 10 and 11).

Calibration of simulation parameter from temperature measurements in real conditions has been treated in former researches [21, 22]. For this work, the information about the thermal environment of one housing monitored during the second step

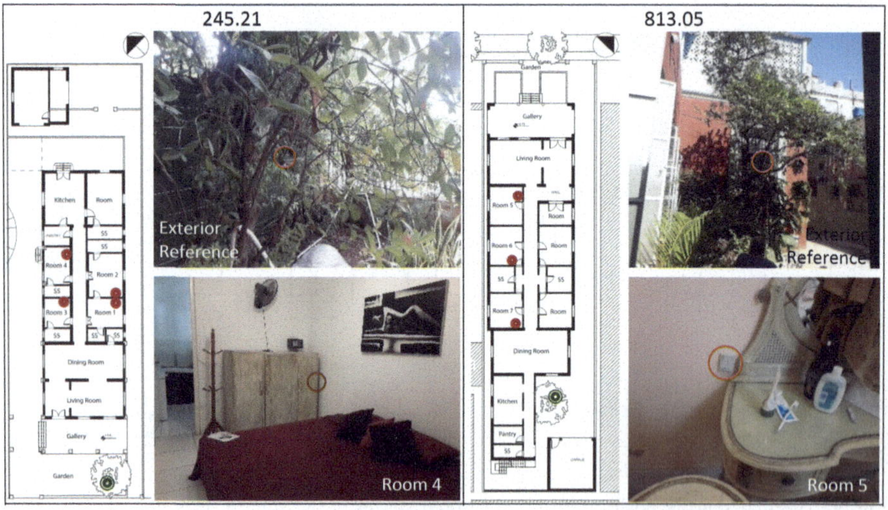

Figs. 10 and 11 Reference plan and images of monitored spaces in 2 study cases

[1] Las mediciones se realizaron con sensores a 1.5 m de altura y ubicados lejos de fuentes de calor.

[2] Los equipos de medición se ubicaron a 1 m de altura entre la cama y la mesa de noche en 5 dormitorios, en las otras 2 habitaciones se ubicaron a 1,5 m en otra pared interior, para evitar el calor de electrodomésticos.

was also considered (code 245.21) to adjust the input data for the simulation and to use the information from another housing (code 813.05) with similar envelope but different volume and context to validate the fixed conditions for the simulation. Then, information taken in building 245.21 was used to calibrate and those taken from building 813.05 for validation.

The study cases were selected (Fig. 8) according to the amount of available data loggers (HOBO) for temperature and relative humidity; diverse conditions of the design variables in order to compare results (similar orientations to different sky view angle (SVA) and analogous SVA for diverse orientations (Fig. 9); as well as the disposition of the owners to collaborate with the monitoring process, allowing to collect the needed information (touristic occupation, schedules and type of interaction with the spaces).

Temperature and relative humidity were registered every 15 minutes, and this information was processed to evaluate the indoor environment from mean hourly values. A sensibility analysis among sensors had been previously realized, monitoring for one week a space with the same conditions. Measurements showed maximum differences of 0,1 for the mean and only 0,02 for the standard deviation, so they were considered as calibrated between them to measure simultaneously serval spaces.

One sensor was also located in a reference outdoor space, inside a meteorological cabinet under a tree. In order to evaluate the behaviour of temperature in high ceiling spaces, measurements were taken at different levels in rooms 2 and 4 of case

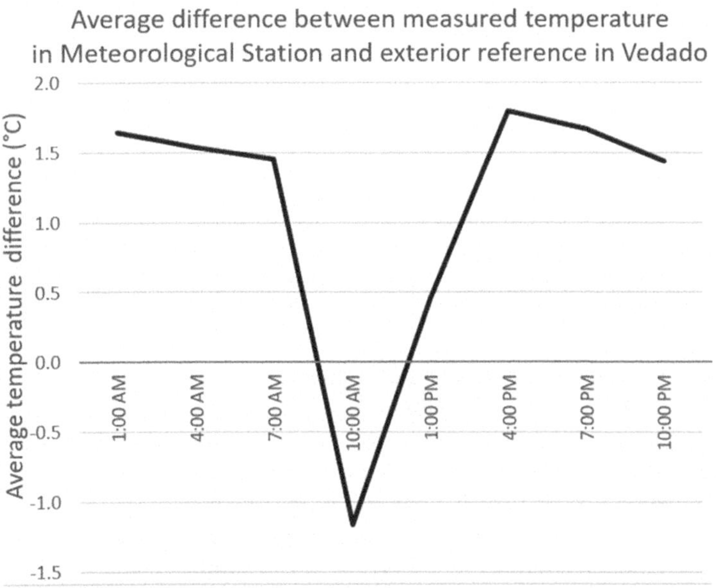

Fig. 12 Average difference between measured temperature in Meteorological Station and exterior reference in Vedado during October, November and December of 2019

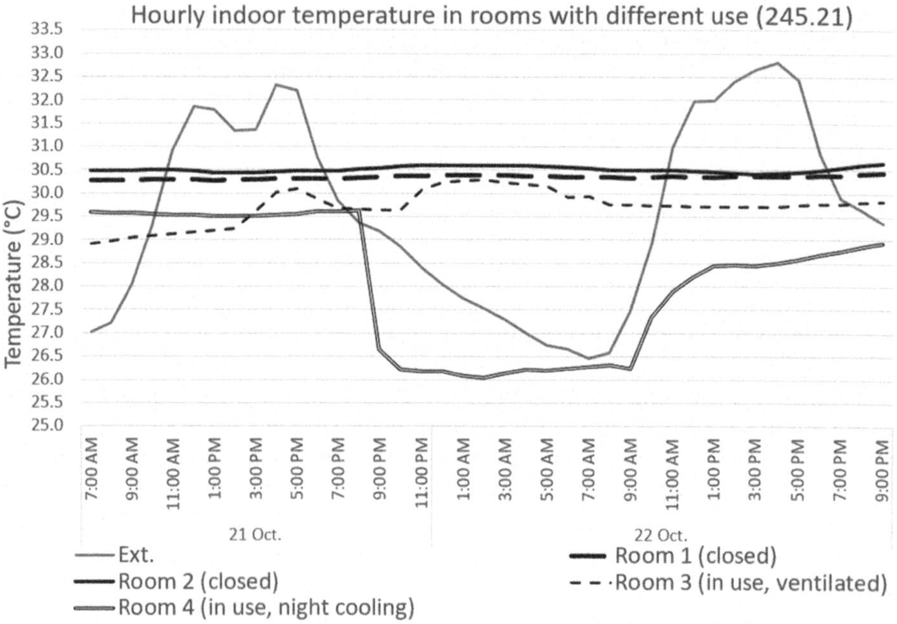

Fig. 13 Impact of natural ventilation and cooling system use in indoor temperature

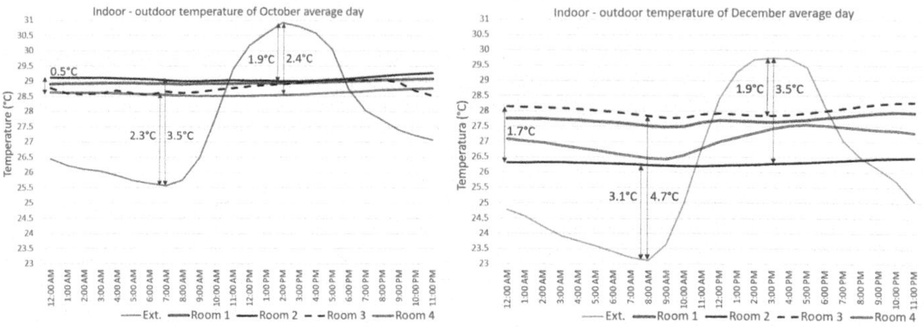

Figs. 14 and 15 Main differences between indoor-outdoor temperature during an average day of October and December in study case 245.21

245.21. The owner registered day and night measurement of the power meter during a month, information needed to calibrate electric consume in the simulation model.

Values registered from the urban context were compared with those from the meteorological station (InsMet) (MEC) during the analyzed periods. The climatological simulation files were also compared to the real ones.

The average difference between the temperature values in MEC and those measured in the outdoor reference points (Fig. 12) is similar to the ones defined by Nieves and Prilipko [88], who, since then, defined variations between +1,5 °C and

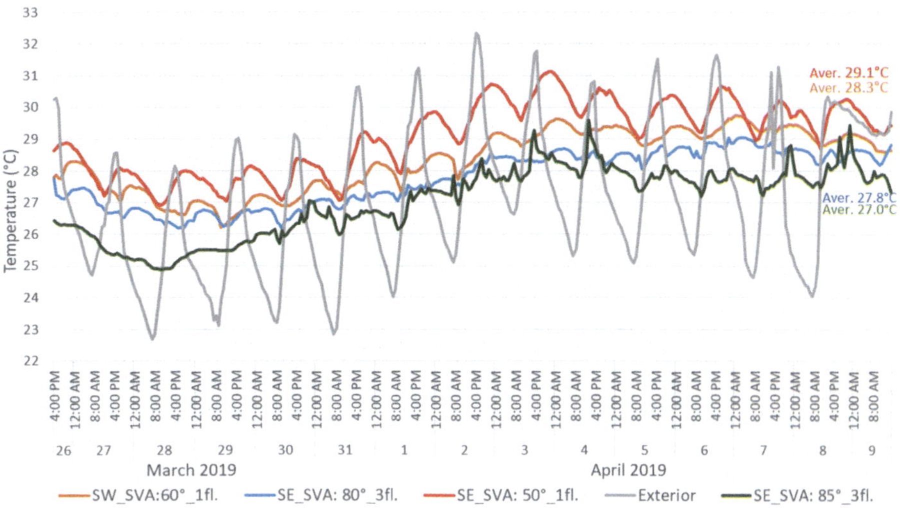

Fig. 16 Impact of the surrounding shadow in the indoor thermal behaviour of naturally ventilated spaces with different orientation (SE/SW), diverse adjacent building high (1–3 floors) and sky view angle (50–85°). The average temperature is indicated (Aver. °C)

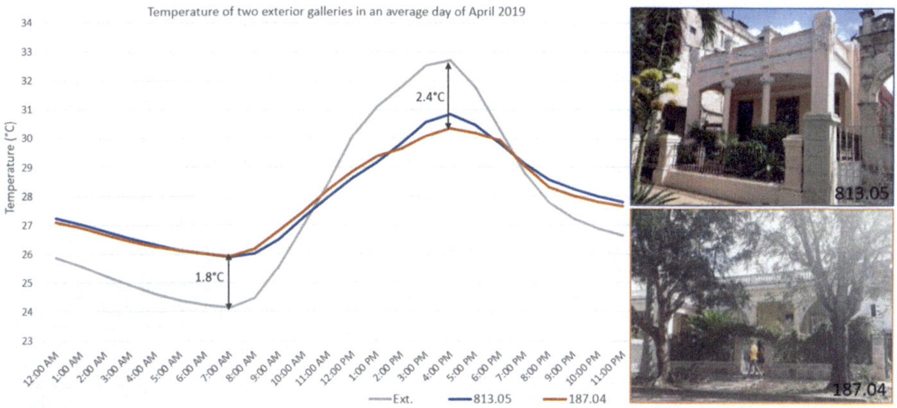

Fig. 17 Temperature of two exterior galleries with Northeast orientation on an average day of April (10–15 2019)

−1 °C as annual average in this urban area. The monitored outdoor environment shows the urban heat island effect.

During the first monitoring step, it was hard to identify the causes of the daily temperature fluctuation, mainly due to the role of some factors, such as the indoor natural ventilation and the human activity in those spaces. In the second measurements campaign, it was possible to establish the moments when the bedrooms were rented, almost always with night air conditioning, as well as the periods when they were closed and occasionally opened for cleaning and maintenance. It allowed to

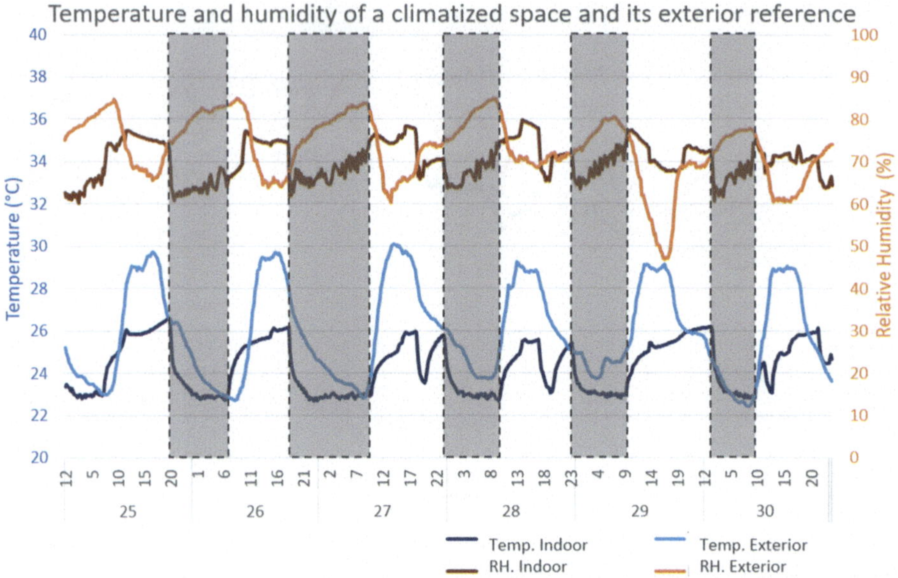

Fig. 18 Hourly behaviour of Temperature and relative humidity of a climatized space (room 2, 245.21) and its exterior reference. The period using cooling systems is indicated

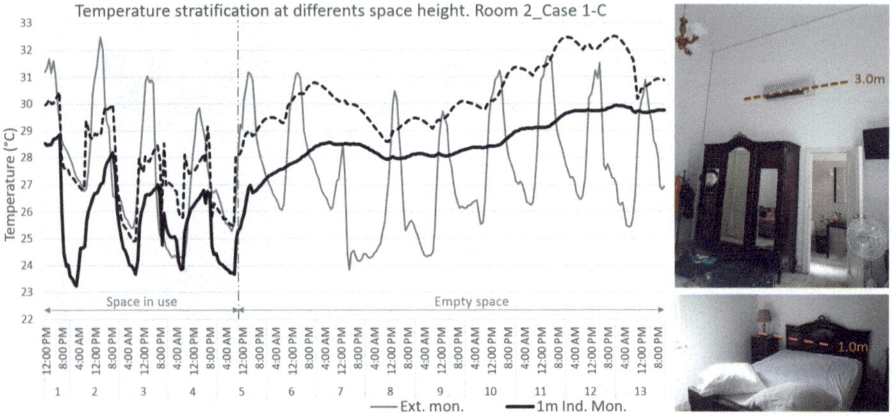

Fig. 19 Thermal behaviour at different heights (1 m and 3 m) during 13 days of October 2019 in room 2 of study case 245.21

demonstrate the important influence of some factors alien to the building, such as human activities, equipment and lighting.

In used rooms with nocturnal air conditioning, night temperatures are, generally, lower than the outdoor reference. When bedrooms are not in use, but closed, they maintain a constant temperature, higher to the outdoor mean, varying only when the windows are open or closed. In Fig. 13, the closed spaces (Room 1 and 2) maintain a constant temperature over 30 degrees, while in room 3, indoor temperature is more variable, due to the windows opening.

The high thermal inertia of this type of buildings is evident, due to the heavy materials in their envelope (roofs and walls) (Figs. 13, 14 and 15). The shadow from the context generates higher reductions in the indoor temperature than the orientation of the space (Fig. 16). In cases with the same orientation (North West), the surrounding vegetation diminishes the maximum temperature value at 2.4 °C, on average (Fig. 17).

Room temperature, when acclimatized, varies between 23 and 26 °C. Measurements also evidenced that artificially conditioned spaces reach indoor temperatures similar or higher than outdoors (Fig. 18).

In order to evaluate the influence of the high ceilings characteristic of these housing on the indoor environment, the temperature was measured at a different height (1,0 m and 3,0 m) in room 2 of the study case 245.21 (Fig. 19). As a result, a stratification was evidenced, both, in the closed disused room and in the period it was acclimatized.

At the bed level, temperature remained 2,5 °C less than 2 m above in the acclimatized space at night (average), while, when the room was closed, the difference was approximately 1,9 °C. It was also possible to observe that variation of the air temperature is higher in upper strata, reaching major values, because of the thermal influence of the roof.

The touristic occupation of a house has a considerable influence on energy consumption. The average consumes of 16 kWh/day quadruples when the owner has clients, as during the analyzed period, with three occupied rooms out of four. The identified causes are, first of all, the use of air conditioning at night and the washing machines, electric water heater, and cooking equipment during the day.[3] The impact of the temperature differences (indoor – outdoor) on the night electricity consume, mainly related to air conditioning, is evident.

4.1.2 Simulation

In order to achieve trustable results, the importance of the input data quality and the way in which technologies and the real complex phenomena are simplified are recognized [89]. The most used parameter that determines the simulated energy consumption and thermal indoor environment are the meteorological variables, building material and elements, internal thermal loads (persons, equipment and lamps), ventilation and air infiltration, as well as the configuration of the used acclimatization system.

The methodology followed to define these parameters was developed in 4 steps, repeated in an iterative way until recommended statistical values for the comparison

---

[3] Los consumos de energía son los que registra el metro contador eléctrico de la vivienda recogidos por la dueña de la casa en dos momentos en el día, generalmente a las 9 am y 9 pm durante todo el mes de octubre.

between simulation and monitoring were complimented in case 245.21. The steps repeated iteratively were:

- Definition of the admissible range of values to be used for each parameter, from the bibliography, historical files, architectural survey and interviews to owners.
- Analysis of the impact on the indoor environment generated by the variation of these parameters, based on EnergyPlus simulation.
- Calibration of the simulation parameters, from the comparison between the simulated and monitored temperature values in four rooms of case 245.21. Values were adjusted in an iterative way to compliment the recommended statistical indicators ASHRAE Guideline 14/2002: mean bias error (MBE), coefficient of variation of the root mean square error (CV (RMSE), root mean square error (RMSE) and mean absolute errors (MAE).
- Validation of the simulation parameters. The adjusted input data, calibrated with case 245.21, were used in case 813.05 to validate its reliability, comparing the simulated and monitored temperature, based on the recommended statistical indicator ASHRAE Guideline 14/2002: MBE, CV (RMSE), RMSE y MAE.

Climatic data defined by the Department of Energy of the United States of America for a typical meteorological year (TMY) in Casablanca were taken as basis for being precise enough to estimate the energy performance of buildings in a long term [26] and the one available with most meteorological variables defined by hourly values. Comparing mean temperature measured at MEC and the one at TMY, a similar thermal behaviour, is appreciable, despite the increment of mean monthly temperatures during the last years at MEC being evident (Fig. 20).

Despite the recognized effect of the urban heat island with increment up to 200% in the energy demand, mainly in residential building [90, 91], the energy impact of buildings in these contexts is still estimated in practice based on TMY coming from meteorological stations [26, 90–93]. The comparison between data from MEC and

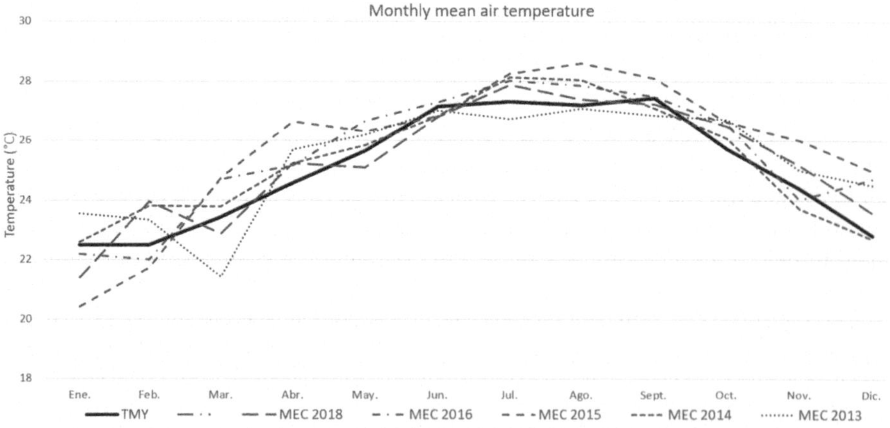

Fig. 20 Mean monthly temperature of the typical meteorological year (TMY) and measured value at Meteorological Station of Casablanca (MEC)

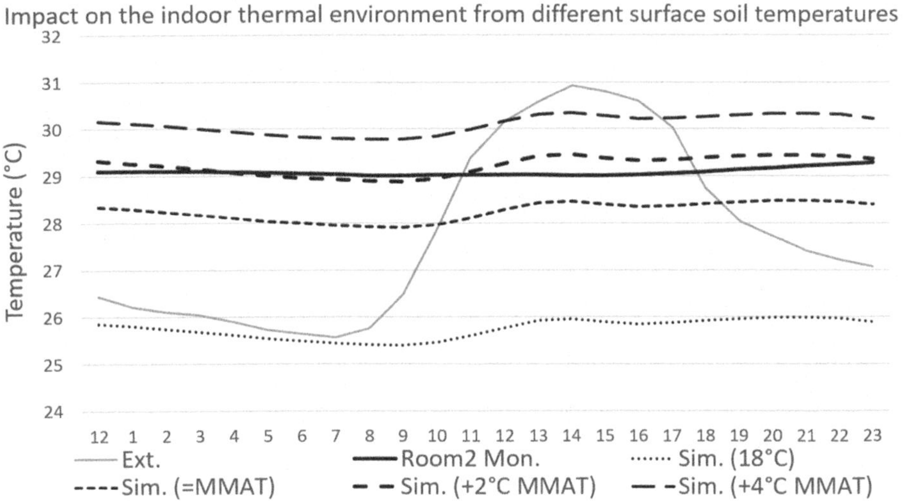

Fig. 21 Indoor thermal environment in a simulated room during an average day of October, with different surface soil temperature: with 18 °C, same as mean monthly air temperature (=MMAT) and with 2 °C and 4 °C above the mean monthly air temperature (+2 °C MMAT) (+4 °C MMAT)

monitored temperature in exterior spaces at El Vedado showed the impact of this phenomena in the study area.

Previous researches show that it is also necessary to adjust the superficial soil temperature assumed by the software due to its impact on the indoor thermal environment [94]. The predefined value (18 °C) is too low, since, according to [95], the mean annual temperatures of soil in the tropics are between 2 and 4 °C higher than the air temperature. Most of the studies to estimate the soil temperature are realized with agro ecologic and forestall aims, and they confirm that, despite its relationship to the air temperature, it also depends on the type of soil, its coverage, the water content, humidity, wind velocity and solar radiation [96, 97]. In urban contexts, the surface temperature is also affected by the own indoor building temperature and its context, mainly when artificial conditioning is used [98]. This parameter has a considerable impact on indoor temperature simulation. Although it couldn't be estimated from Kiva calculation tool, because of the lack of departing information about the local soil, a value 2 °C higher than the mean monthly air temperature was assumed. This value is, within the recommended ones, with which it is possible to obtain simulated interior thermal behaviours more similar to those monitored (Fig. 21).

They were not found any previous research in Cuba about buildings of the same age that the study cases, which considered the thermal properties of the building materials required by the software. Initial simulations with the data available in the software library offered great differences between the simulated indoor environment (with notable variation between day and night) and the one monitored (with more uniform temperatures) [99]. With respect to this, Webb [16] indicated that it

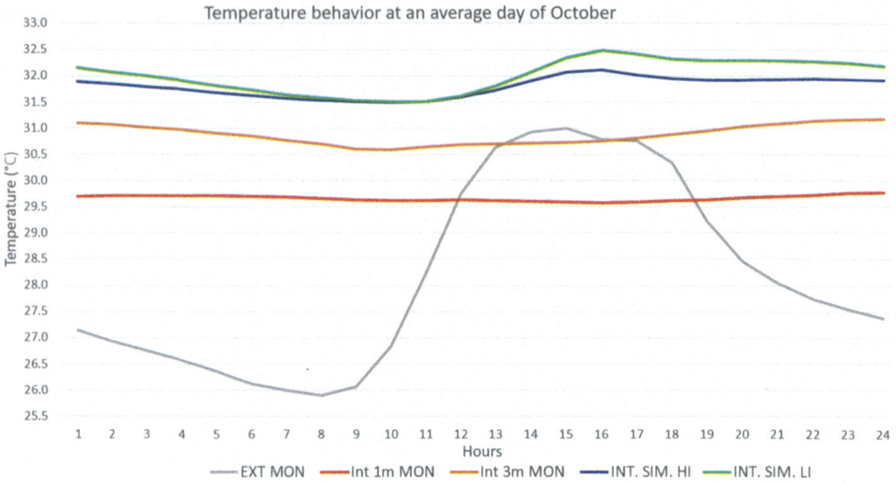

Fig. 22 Temperature on an average day of October in a study case. It is indicated the monitored values (exterior and interior at 1 and 3 m high) and the simulated with material with high thermal inertia (HI) and low (LI)

could not be appropriate to use the material properties defined for modern buildings to simulate traditional ones, due to the differences in construction methods, the heterogeneity of row materials and the use of local components. In the analyzed study cases, one possible cause identified to explain the difference in the thermal behaviour is that real materials have much higher inertia than the defined ones.

Based on a bibliographic study about thermal properties of building materials commonly used in buildings similar to the study cases, according to typology and time in diverse places, the simulation process was repeated, selecting for each building element the properties generating the major thermal inertia. Then, a similar behaviour to the one monitored at 3 m height was achieved (Fig. 21), since the simulated air temperature is an average value of the entire space, while the recorded values may have variations according to the temperature stratification and the position of the sensor [100], which is accentuated in this spaces with high ceilings.

According to the values used in previous researches and their impact on study case 245.21, for this work, it was assumed 0,3 air exchanges per hour (CAH), recurrent in residential spaces [101], which maintain an indoor temperature difference similar to the monitored one (Fig. 22).

Since it is not an objective of this work to evaluate natural ventilation as a strategy to evacuate heat or for passive cooling, the European Standard EN 15242: 2006 [102] was used to define a uniform value of the indoor airflow due to ventilation. For that, first of all, wind velocities of TMY and MEC were compared, evidencing that, except the maximum velocity, the rest of the indicators are similar.

The formula for unilateral ventilation recommended by the referred standards was applied, considering two windows areas (1,5 m^2 for 50% of openings and 3,2 m^2 to full permeable area) for a window height of 2 m. Ranges of temperature differences (indoors–outdoors) between 0–5 °C were used, as well as wind velocities at the meteorological station between 0 and 4 m/s. If the windows are totally opened, the airflow is between 0,16 and 0,4 m^3/s, and if it is assumed that the windows are generally opened in a partial way, the values are from 0,08 to 0,19 m^3/s. It was evident that despite energy consumption doesn't diminish considerably, opening the windows in the morning may reduce indoor temperature up to 1,4 °C, compared to a closed room.

For this work, an airflow of 0,15 m^3/s was assumed, within the previously determined range (0,08–0,40 m^3/s). The value corresponds to a wind velocity of 3 m/s at the meteorological station, near to the mean and the average, and an indoor–outdoor temperature difference of 3 °C, with a window partially opened. Ventilation is used in the model between 9 am and 12 m, when windows are usually opened to clean the room if it is rented by tourists.

A maximum of 2 persons per room was assumed, and the metabolic load was established from the predefined values [101], considering the daily routine declared by the owners in the interviews. From the same reference it was taken 3,9 W/m^2 for the contribution of artificial lighting (20% as a radiant fraction and 20% visible for fluorescent lamps) and 6,7 W/m^2 as the load from the electric equipment, values recommended for mean and large apartments. The schedule for each energy component was also determined according to the information declared in the interviews.

The analysis of these variables in a study case demonstrates that not considering them in the energy evaluation of buildings may represent a variation between 4% and16% of the demand for air conditioning.

In order to estimate the energy demand for acclimatization, the model *"Ideal Loads Air System,"* provided by EnergyPlus, previously used in former researches [10] and considered as a variable airflow system which supplies it up to satisfy the desired thermal conditions, was assumed [103]. A thermostat temperature of 24 °C was used, as recommended for most of the indoor acclimatized spaces in Cuba [104], which is within the frequent temperature range in acclimatized rooms, according to the monitoring results. A schedule for acclimatization was assumed between 9 pm and 9 am, for being, according to the owners, frequent in rented rooms.

4.1.3 Calibration and Validation

The calibration process to readjust the input data was made by comparing the monitored and simulated data in case 245.21. In order to eliminate the impact generated by the variation of no controlled parameters (activity, windows opening, equipment use), the influence of persons, natural ventilation and daylighting was supressed as input data in simulations for calibration, since these rooms were not in use. Infiltration was adjusted from 0,3 air changes per hour (ACH), formerly defined, to

Table 1 Comparison of statistical indicators between simulation and monitoring

Indicator	Maximum permissible indicator[a]	Calibration (for study case 245.21)		
		Room 1 (1104 values)	Room 2 (1008 values)	Room 3 (1128 values)
MBE (mean bias error)	±10% / ±5%	−0,5%	−1,2%	−1,0%
CV (RMSE) (coefficient of variation of root mean square error)	30% / 10–15%	2,3%	2,0%	2,4%
RMSE (root mean square error)	0,5–0,78 °C	0,6 °C	0,5 °C	0,6 °C
MAE (mean absolute errors)	5–10%	1,7%	1,6%	1,9%

[a]Los primeros indicadores MBE y CV (RMSE) son los recomendados por De Backe [22] primero y luego los utilizados por Ascione et al. [21]. Los de RMSE y MAE son los definidos por De Backe [22]

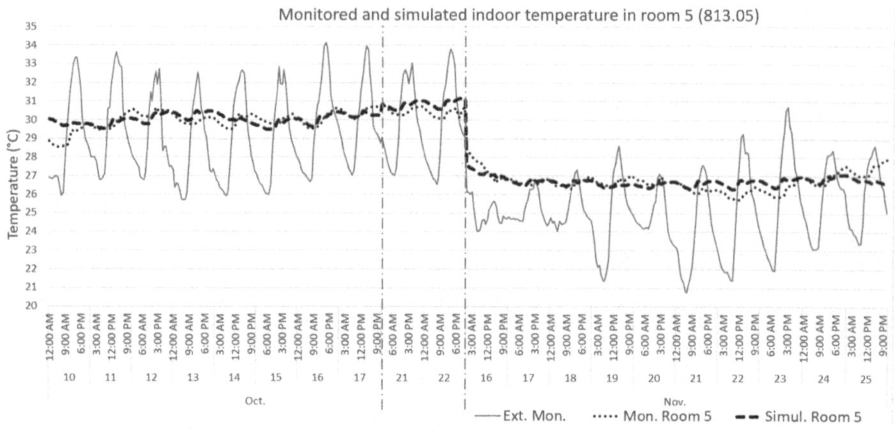

Fig. 23 Thermal Behaviour of a simulated and monitored room during 20 days of October

0,1 ACH,[4] and the equipment schedule was maintained in 10% of use, considering that they, even turned off, maintain an electric consume [105].

It was finally considered that the values are reliable enough when they satisfy the statistical indicators recommended for this kind of researches ASHRAE Guideline 14/2002, discussed by Ascione et al. [21] and De Backe [22] (Table 1).

The same parameters, previously calibrated, are used to simulate the other study case (813.05). The comparison between the simulated and monitored temperatures allowed to validate the input data, since the statistical indicators also complement the recommended ones (Fig. 23 and Table 2).

[4] Se considera que en habitaciones cerradas; el valor de infiltración es menor pues no se abren las aberturas. Se asume 0,1 CAH pues era el que generaba una oscilación de temperaturas similar a los espacios analizados.

Table 2 Comparison of statistical indicators between simulation and monitoring

Indicator	Maximum permissible indicator[a]	Validation (for study case 813.05) Room 4 (480 values)
MBE (mean bias error)	±10% / ±5%	−0,3%
CV (RMSE) (coefficient of variation of root mean square error)	30% / 10–15%	1,5%
RMSE (root mean square error)	0,5–0,78 °C	0,4 °C
MAE (mean absolute errors)	5–10%	1,1%

[a]Los primeros indicadores MBE y CV (RMSE) son los recomendados por De Backe [22] primero y luego los utilizados por Ascione et al. [21]. Los de RMSE y MAE son los definidos por De Backe [22]

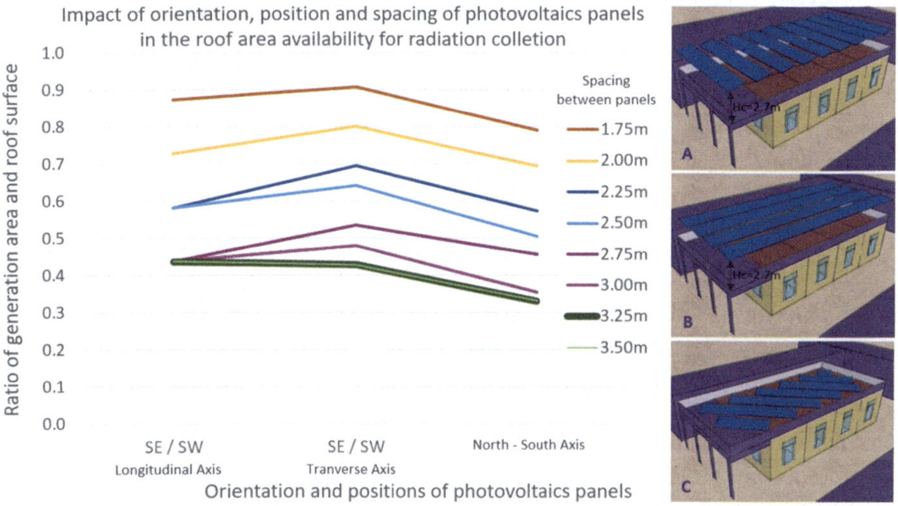

Fig. 24 Ratio of generation area and roof surface for different spacing between photovoltaics panels (1,75–3,5 m) and position (along longitudinal, transverse and North-South Axis

The uncertainty about the real use of the rooms is always a parameter that can modify the values, because of the interaction between the users and the space in a long period of time.

4.1.4 Design Recommendations

The difference about the soil reflexion coefficient (CRs = 0,2 and 0,8) generates savings up to 17 kWh/m^2/year in low protection contexts (CSP-L), between 6 and 14 kWh/m^2/year in those with mid protection (CSP-M) and 3 kWh/m^2/year for the highly protected ones (CSP-H). It is evident that major reductions in energy consume are obtained when the buildings are more separated, and in walls oriented to the South (SE/SW) when the context offers low or mid protection.

The evaluations indicate that, for any orientation and context type, it is recommended to use not reflecting pavements, and preferable with vegetation. Retrofit actions should start in those exterior spaces with buildings more separated (any orientation), and later, in the ones located to the South (SE/SW).

Interventions related to the roof surface reduce 21–25% of consume, with energy savings between 50–66 kWh/m²/year, compared to the traditional waterproof system (Ini. Sit.) and between 58–77 kWh/m²/year with respect to an asphalt layer (Mod. Sit.).

The thermal transmittance of the shading element has a major impact than the height of a double roof. However, the distance between the shading element and the roof has greater influence in contexts of low protection (CSP-L) and when the rooms are oriented to the South East or South West. It is also possible to conclude that variants receiving similar solar radiation may differ in energy saving, depending on the incidence time.

The double roof made of photovoltaic panels nationally produced DCM 250 (990 x 1650 mm) reduce more solar radiation when they are oriented to South East, along the longitudinal or the transversal roof axis. More panels can be located according to the transversal axis, followed by the longitudinal one and less amount oriented South (Fig. 24). The shadow from the parapet affects more the panels located on the longitudinal axis.

Of the evaluated angles for the PV panels (15°, 20°, 25° y 30°), more annual solar radiation is received at 15°, as recommended by Stolik Novygrod [106] in order to avoid shadow projection. Distances shorter than 2,50 m between panels are not recommended, since differences major than 1 W/m² are evidenced on average incident solar radiation, exponentially increased with wider distances. This coincides with what was recommended by Sarmiento Sera [107], who indicated that panels should be separated 1,5 times the module longitude (for panel DCM 250 separation would be 2475 m).

Panels oriented to South East present more generation than the ones oriented South, since the roof area is better used. Differences between both orientations are up to 21kWh/m²/year in low protection contexts, depending on the panels' position. Solutions located 2,7 m over the roof favour a major reduction of energy for air conditioning and generate more electricity, because extra PV panels can be located, since the parapet shadow doesn't affect, there is no space for circulation needed and shadow projected by the context is less (Fig. 25).

The use of an insulation layer with thermal resistance of 2,15 m²°C/W and 4,41 m²°C/W generates a consume reduction of 14–19%. The evaluated solutions diminish energy consumption and indoor temperature, at the same time. However, when thermal insulation is added to the roof, the amount of hour with temperature lower than 26° decrease, as well as hours with values over 29 °C. The most energy-saving solution with higher energy reduction, which is the interior insulation with thermal resistance of 4,41 m²°C/W (Aisl. Int. R = 4,41), only increment indoor temperature during the cool months, like January, while the balance in July and annually indicates that it improves indoor thermal conditions.

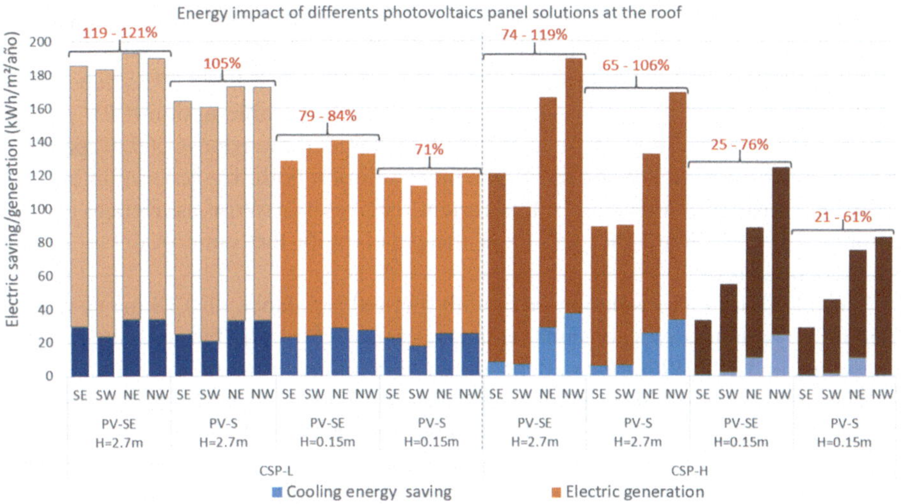

Fig. 25 Energy impact by generation and cooling saving of different photovoltaic panel solutions at the roof. It's indicated the electric coverage by photovoltaic generation

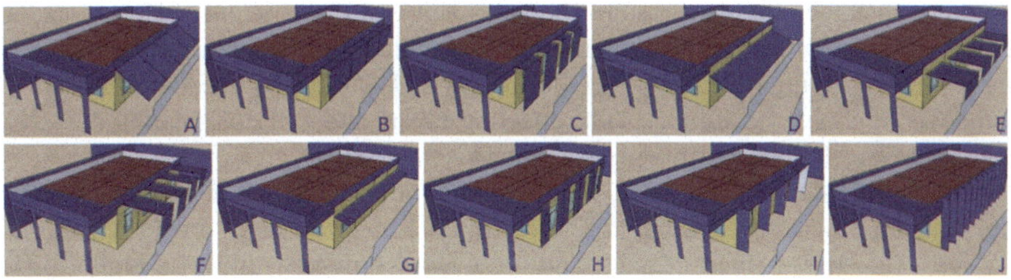

Fig. 26 Evaluated solutions of shading device for the walls

Increasing reflectivity by painting external walls reduces up to 7% of the consume in a low protection context, with annual savings of 13–46 kWh/m²/year and around 2% in highly protected contexts, which means 5–15 kWh/m²/year, depending on the wall orientation and the initial surface colour.

The use of clear surface finishing is recommended. Major impacts are evidenced on the sunny walls in contexts with low protection (CSP-L) and rooms oriented to the South (SE/SW). When near buildings project more shadow on the external walls, as in highly protected contexts, it is more important to guarantee reflectivity in North orientation, because of the effect of the indirect solar radiation.

The use of shading devices on external walls may reduce consume up to 10% in rooms oriented to SE and SW (around 25 kWh/m²/year) and by 5% when they are oriented to NE and NW (up to 15 kWh/m²/year) in a context with low protection. In context with intermit protection, shading devices may reduce consume between 5 and 18 kWh/m²/year, depending on the orientation, and in those with high protection, with short eaves, it diminishes up to 2 kWh/m²/year (Fig. 26).

Simulations show that in low protection contexts, which only reduce around 1% of the incident radiation, PV panels could also be used. In context with intermit protection, shadow could reduce incident solar radiation in 14 and 23%, depending on the PV panel position and orientation. The most favourable option if PV panels will be used is to the South West, at 15°, with the larger dimensions(1,65 m) perpendicular to the wall, this is the most beneficial solution both, by generation and by reducing consumption. However, this variant receives 193 W/m^2 as average a year, compared to 203 W/m^2 that arrives at the panels with the same orientation, but located on the roof.

The addition of an insulating layer with thermal resistance of 0,18 m^2°C/W and 2,54 m^2°C/W generates a consume reduction of 0,4–3,5%, which represents savings between 1 and 10 kWh/m^2/year, depending on the orientation and the context.

Thermal insulation in walls also generates increments of indoor temperature in no acclimatized spaces, the same as when it is used on the roof. Horus with temperatures over 29° increase between 1 y 2%. It is, then, a no recommended variant to be used in the study cases, since consume reduction should not worsen the thermal environment without air conditioning, in whose case, it would incentive its use.

The traditional "French" window is the best solution, only surpassed by simple glass windows with solar protection, since the glass portions of the French windows are not protected against the sun. The solar heat gain coefficient (SHGC) of the glass is the parameter with major impact in reducing consume, which can be increased by adding reflectant layers to the glass. Then, if it is needed to use glass windows, it should be simple, with reflecting layers and exterior shading devices, and not to use double or triple glass windows.

Simulations demonstrated that simple glass windows with a double exterior window may diminish consume up to 17 kWh/m^2/year in contexts of low protection, 9 kWh/m^2/year in the intermit ones, and up to 2 kWh/m^2/year in those highly protected. The shadow angle used to optimize the context protection when designing louvers should be the top window point, which means a difference of kWh/m^2/year with respect to the bottom one. Dimensions and inclination of louvers are not relevant in energy direct solar radiation, but receive more indirect one. The same happens comparing horizontal and inclined positions.

Other shading devices for windows, like awnings, vertical devices and eaves may generate similar savings, despite they don't provide shadow to the total window area (Fig. 27), both, in the glass windows (VV) and in the French one (VF) (Fig. 28).

5 Conclusions

- The research allows to demonstrate the energy impact and the economic viability of the retrofit actions proposed for historic buildings with high thermal mass in urban contexts with warm and humid climates, which has not been addressed in former researches and it is applicable to this architectural type in urban zones with similar conditions.

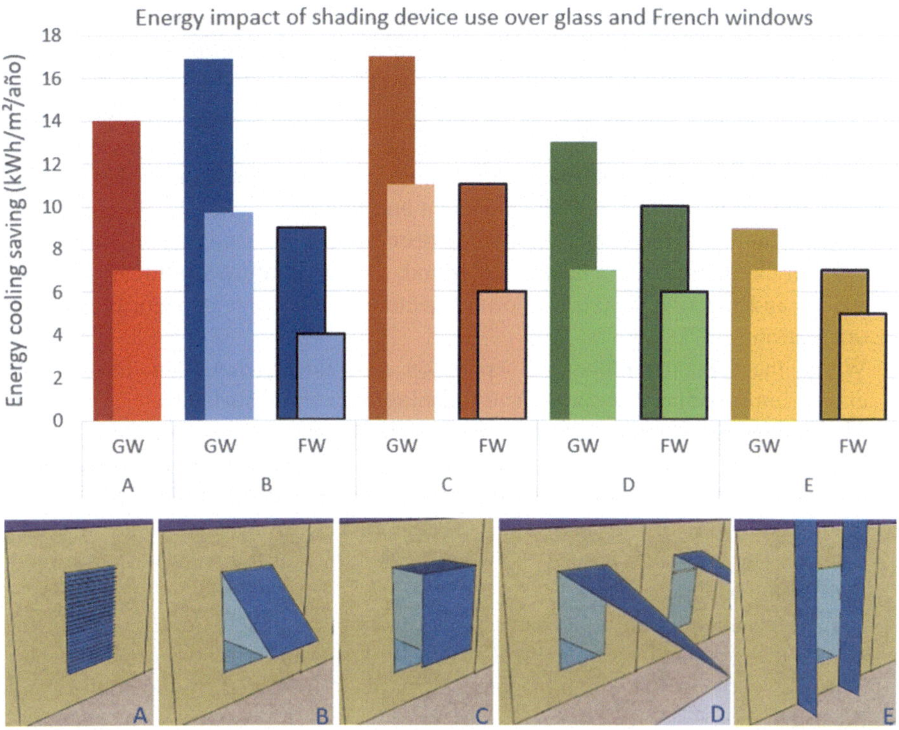

Fig. 27 Energy impact of shading device use on glass window (GW) and French window (FW)

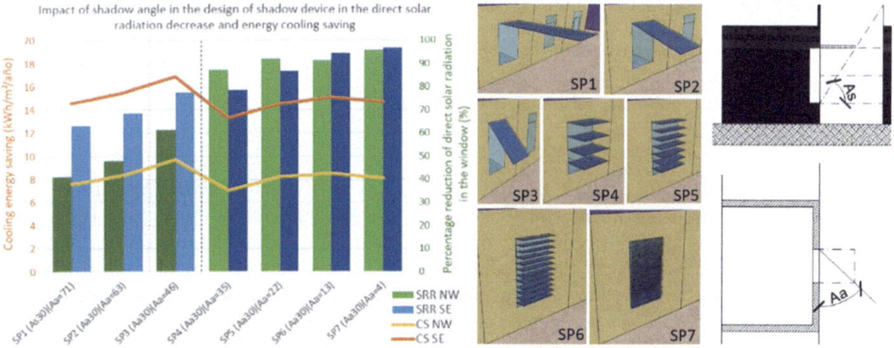

Fig. 28 Impact on the direct solar radiation reduction (SRR) and cooling saving (CS) because of the shadow angle in the design of shadow device

- The importance to adjust the values of the input parameter to the software for dynamic simulation, from the monitoring results, is demonstrated in order to bring them closer to the real behaviour in the specific conditions of each climate, context and study object.

- For urban contexts in warm and humid climates, it has been demonstrated that it is not convenient to apply solutions about which effectiveness there is a widespread belief, such as low transfer glasses and thermal insulation in the envelope.
- Within the input parameters for simulations which have a relevant impact but have been little discussed before, there are the adjustment of the typical meteorological year, the soil surface temperature and the definition of the thermal properties of building materials in ancient buildings.
- The evaluation of the initial thermal environment in the study cases allowed to confirm the impact of the heat urban island, the temperature stratification in tall spaces and the role of the shadow projected by the context according to the orientation.
- Within the context parameters, separation and building height generate more energy impacts than orientation and finishing of external surfaces.
- The shadow projected by different parametrized context on the roof of the theoretical model demonstrates that between 19 and 75% of its area receive more than 90% of the solar radiation with respect to an isolated building, which constitutes a potential to be used as renewable energy, even in this kind of urban context. The adjoining reduces between 11 and 98% of the solar radiation incident on the walls.
- The interventions with major energy and economic impacts are those realized, first on the roof and later on the walls. The third priority (windows or pavements) depends on the initial situation, the solar protection offered by the context and the orientation.
- Design recommendations simulated with the theoretical model may generate annual energy savings up to 193 kWh/m2, which could represent benefits at 10 years valuated up to 7260 CUP/m2 in each retrofitted building.

References

1. PARLAMENTO EUROPEO Y EL CONSEJO DE LA UNIÓN EUROPEA. DIRECTIVA 2010/31/UE relativa a la eficiencia energética de los edificios (refundición) [Internet]. 2010/31/UE May 19, 2010. Available from: https://eur-lex.europa.eu/eli/dir/2010/31/oj/spa
2. Nandapala K, Halwatura R. Design of a durable roof slab insulation system for tropical climatic conditions. Hussain RR, editor. Cogent Eng [Internet]. 2016 Jun 22 [cited 2020 Apr 28];3(1). Available from: https://www.cogentoa.com/article/10.1080/23311916.2016.1196526
3. Sun X, Gou Z, Lau SS-Y. Cost-effectiveness of active and passive design strategies for existing building retrofits in tropical climate: Case study of a zero energy building. J Clean Prod. 2018 May;183:35–45.
4. Yang J, Mohan Kumar D llamathy, Pyrgou A, Chong A, Santamouris M, Kolokotsa D, et al. Green and cool roofs' urban heat island mitigation potential in tropical climate. Sol Energy. 2018 Oct;173:597–609.
5. Aldossary NA, Rezgui Y, Kwan A. Domestic energy consumption patterns in a hot and humid climate: A multiple-case study analysis. Appl Energy. 2014 Feb;114:353–65.
6. Oropeza-Perez I, Østergaard PA. Active and passive cooling methods for dwellings: A review. Renew Sustain Energy Rev. 2018 Feb;82:531–44.

7. Oficina Nacional de Estadística e Información. ONEI. Electricidad en Cuba. Indicadores seleccionados. Enero-Diciembre 2018 [Internet]. La Habana, Cuba; 2019 May. Available from: http://www.onei.gob.cu/sites/default/files/prodconselectdici18.pdf
8. Segunda comparecencia especial del Presidente Miguel Díaz Canel [Internet]. Mesa Redonda. La Habana; 2019. Available from: http://www.cubadebate.cu/noticias/2019/09/12/cuba-no-esta-paralizada-contamos-con-una-estrategia
9. Panfil M, Whittle D, Silverman-Roati K. What's next for Cuba's electricity sector? Electr J. 2017 Oct;30(8):38–44.
10. de la Paz Pérez GA. Influencia del contexto urbano y la orientación en la protección solar de los vanos y en la demanda energética de los edificios en Cuba [Doctorado]. [La Habana]: Universidad Tecnológica de La Habana José Antonio Echeverría, CUJAE; 2019.
11. Collado Baldoquin N, Rueda Guzmán LA, González Couret D. Hoteles de consumo energético casi nulo. Potencialidades y restricciones para Cuba. Eco Sol. 2019;69:34–44.
12. Turismo cubano por buen camino [Internet]. Juventud Rebelde. 2019. Available from: http://www.juventudrebelde.cu/cuba/2019-01-02/turismo-cubano-por-buen-camino
13. Collado Baldoquin N, Rueda Guzmán LA, González Couret D, Lemus Mesa KM. Urban tourism – The impact of tourist accommodation in residential buildngs at Vedado, Cuba. Rev Interam Ambiente Tur. 2020;16, no.1:35–48.
14. ASHRAE Terminology [Internet]. [cited 2020 Mar 20]. Available from: https://xp20.ashrae.org/terminology/index.php?term=retrofit&submit=Search
15. Rabani M, Madessa HB, Nord N. A state-of-art review of retrofit interventions in buildings towards nearly zero energy level. Energy Procedia. 2017 Oct;134:317–26.
16. Webb AL. Energy retrofits in historic and traditional buildings: A review of problems and methods. Renew Sustain Energy Rev. 2017 Sep;77:748–59.
17. Moran P, O'Connell J, Goggins J. Sustainable energy efficiency retrofits as residenial buildings move towards nearly zero energy building (NZEB) standards. Energy Build. 2020 Mar;211:109816.
18. Almeida M, Ferreira M. Ten questions concerning cost-effective energy and carbon emissions optimization in building renovation. Build Environ. 2018 Oct;143:15–23.
19. Silvero F, Montelpare S, Rodrigues F, Spacone E, Varum H. Energy retrofit solutions for heritage buildings located in hot-humid climates. Procedia Struct Integr. 2018;11:52–9.
20. Lidelöw S, Örn T, Luciani A, Rizzo A. Energy-efficiency measures for heritage buildings: A literature review. Sustain Cities Soc. 2019 Feb;45:231–42.
21. Ascione F, Bianco N, Iovane T, Mauro GM, Napolitano DF, Ruggiano A, et al. A real industrial building: Modeling, calibration and Pareto optimization of energy retrofit. J Build Eng. 2020 May;29:101186.
22. De Backe L. Preventive Conservation of Works of Art in Historic Buildings:Assessment and Modelling [Internet]. [Ghent]: Ugent University; 2018. Available from: https://biblio.ugent.be/publication/8560690/file/8560691
23. Shittu E, Stojceska V, Gratton P, Kolokotroni M. Environmental impact of cool roof paint: case-study of house retrofit in two hot islands. Energy Build. 2020 Mar;110007.
24. Kolokotroni M, Shittu E, Santos T, Ramowski L, Mollard A, Rowe K, et al. Cool roofs: High tech low cost solution for energy efficiency and thermal comfort in low rise low income houses in high solar radiation countries. Energy Build. 2018 Oct;176:58–70.
25. Nematchoua MK, Vanona Noelson JC, Saadi I, Kenfack H, Andrianaharinjaka A-ZFR, Ngoumdoum DF, et al. Application of phase change materials, thermal insulation, and external shading for thermal comfort improvement and cooling energy demand reduction in an office building under different coastal tropical climates. Sol Energy. 2020;207:458–70.
26. Hensen JLM, Lamberts R. Building Performance Simulation for Design and Operation. Oxford, UK: Spon Press; 2011.
27. Mazzeo D, Matera N, Cornaro C, Oliveti G, Romagnoni P, De Santoli L. EnergyPlus, IDA ICE and TRNSYS predictive simulation accuracy for building thermal behaviour evaluation

by using an experimental campaign in solar test boxes with and without a PCM module. Energy Build. 2020 Jan;109812.
28. Mazzeo D, Matera N, Cornaro C, Oliveti G, Romagnoni P, De Santoli L. EnergyPlus, IDA ICE and TRNSYS predictive simulation accuracy for building thermal behaviour evaluation by using an experimental campaign in solar test boxes with and without a PCM module. Energy Build. 2020 Apr;212:109812.
29. Kamal R, Moloney F, Wickramaratne C, Narasimhan A, Goswami DY. Strategic control and cost optimization of thermal energy storage in buildings using EnergyPlus. Appl Energy. 2019 Jul;246:77–90.
30. Cetin KS, Fathollahzadeh MH, Kunwar N, Do H, Tabares-Velasco PC. Development and validation of an HVAC on/off controller in EnergyPlus for energy simulation of residential and small commercial buildings. Energy Build. 2019 Jan;183:467–83.
31. EnergyPlus | EnergyPlus [Internet]. [cited 2020 Feb 10]. Available from: https://energyplus.net/
32. Bano F, Sehgal V. Finding the gaps and methodology of passive features of building envelope optimization and its requirement for office buildings in India. Therm Sci Eng Prog. 2019 Mar;9:66–93.
33. Rueda Guzmán LA. Influencia del diseño en la carga térmica y el consumo de energía en haitaciones de hoteles con “destino sol y playa”en Cuba [Doctorado]. [La Habana]: Instituto Superior Politécnico José Antonio Echeverría, CUJAE; 2003.
34. González Couret D, Rueda Guzmán LA, González Milián N, Rodríguez García E, Llovet Salazar M. Evaluación cualitativa de la influencia del diseño arquitectónico en el ambiente interior. Arquit Urban. 2015;XXXVI(3):53–66.
35. TABULA Project Team. Typology Approach for building stock energy assessment. Main results of the TABULA project [Internet]. 2012 Oct p. 43. Report No.: Final Project Report. Available from: https://ec.europa.eu/energy/intelligent/projects/sites/iee-projects/files/projects/documents/tabula_final_report_en.pdf
36. Sánchez Martínez O. Usos activos y pasivos del sol y el viento en la vivienda urbana en La Habana [Maestría]. [La Habana]: Instituto Superior Politécnico José Antonio Echeverría, CUJAE; 2015.
37. Martins TA de L, Adolphe L, Bastos LEG, Martins MA de L. Sensitivity analysis of urban morphology factors regarding solar energy potential of buildings in a Brazilian tropical context. Sol Energy. 2016 Nov;137:11–24.
38. Tardioli G, Kerrigan R, Oates M, O’Donnell J, Finn DP. Identification of representative buildings and building groups in urban datasets using a novel pre-processing, classification, clustering and predictive modelling approach. Build Environ. 2018;140:90–106.
39. Hong Y, Ezeh CI, Hong S-H, Peng Z, Tang Y. Correlation between building characteristics and associated energy consumption: Prototyping low-rise office buildings in Shanghai. Energy Build. 2020;217:1–14.
40. Muringathuparambil RJ, Musango JK, Brent AC, Currie P. Developing building typologies to examine energy efficiency in representative low cost buildings in Cape Town townships. Sustain Cities Soc. 2017;33:1–17.
41. Triana MA, Lamberts R, Sassi P. Characterisation of representative building typologies for social housing projects in Brazil and its energy performance. Energy Policy. 2015;87:524–41.
42. Al-Obaidi KM, Ismail M, Abdul Rahman AM. Passive cooling techniques through reflective and radiative roofs in tropical houses in Southeast Asia: A literature review. Front Archit Res. 2014 Sep;3(3):283–97.
43. Prieto A, Knaack U, Auer T, Klein T. Passive cooling & climate responsive façade design Exploring the limits of passive cooling strategies to improve the performance of commercial buildings in warm climates. Energy Build. 2018;175:30–47.
44. Bhamare DK, Rathod MK, Banerjee J. Passive cooling techniques for building and their applicability in different climatic zones—The state of art. Energy Build. 2019 Sep;198:467–90.

45. Givoni B. Indoor temperature reduction by passive cooling systems. Sol Energy. 2011;85:1692–726.
46. Kaboré M, Bozonnet E, Salagnac P, Abadie M. Indexes for passive building design in urban context –indoor and outdoor cooling potentials. Energy Build. 2018;173:315–25.
47. Lechner N. Heating, Cooling, Lighting: Sustainable Design Methods for Architects. 4th Edition. New Jersey, Canada: Wiley; 2015. 731 p.
48. Mohammad Yusoff WF, Mohamed MF. Building Energy Efficiency in Hot and Humid Climate. In: Encyclopedia of Sustainable Technologies [Internet]. Elsevier; 2017 [cited 2020 Feb 12]. p. 159–68. Available from: https://linkinghub.elsevier.com/retrieve/pii/B9780124095489101915
49. Nguyen A-T, Reiter S. A climate analysis tool for passive heating and cooling strategies in hot humid climate based on Typical Meteorological Year data sets. Energy Build. 2014 Jan;68:756–63.
50. Li S, Lu Y, Kua HW, Chang R. The economics of green buildings: A life cycle cost analysis of non-residential buildings in tropic climates. J Clean Prod. 2020 Apr;252:119771.
51. Alhuwayil WK, Abdul Mujeebu M, Algarny AMM. Impact of external shading strategy on energy performance of multi-story hotel building in hot-humid climate. Energy. 2019 Feb;169:1166–74.
52. Al-Masrani SM, Al-Obaidi KM. Dynamic shading systems: A review of design parameters, platforms and evaluation strategies. Autom Constr. 2019 Jun;102:195–216.
53. Al Touma A, Ouahrani D. Evaporatively-cooled façade integrated with photovoltaic thermal panel applied in hot and humid climates. Energy. 2019 Apr;172:409–22.
54. Shin M, Baltazar J-C, Haberl JS, Frazier E, Lynn B. Evaluation of the energy performance of a net zero energy building in a hot and humid climate. Energy Build. 2019 Dec;204:109531.
55. de Abreu-Harbich LV, Chaves VLA, Brandstetter MCGO. Evaluation of strategies that improve the thermal comfort and energy saving of a classroom of an institutional building in a tropical climate. Build Environ. 2018 May;135:257–68.
56. Dutta A, Samanta A. Reducing cooling load of buildings in the tropical climate through window glazing: A model to model comparison. J Build Eng. 2018 Jan;15:318–27.
57. Yew MC, Yew MK, Saw LH, Ng TC, Chen KP, Rajkumar D, et al. Experimental analysis on the active and passive cool roof systems for industrial buildings in Malaysia. J Build Eng. 2018 Sep;19:134–41.
58. Dutta A, Samanta A, Neogi S. Influence of orientation and the impact of external window shading on building thermal performance in tropical climate. Energy Build. 2017 Mar;139:680–9.
59. Wan Mohd Nazi WI, Wang Y, Chen H, Zhang X, Roskilly AP. Passive Cooling Using Phase Change Material and Insulation for High-rise Office Building in Tropical Climate. In: Energy Procedia. Cardiff, UK: Elsevier; 2017. p. 2295–302.
60. Bano F, Kamal AM. Examining the Role of Building Envelope for Energy Efficiency in Office Buildings in India. Archit Res. 2016;6(5):107–15.
61. Dewiyana E, Ibrahim N, Hajar NH. The Green Aspects of Adaptive Reuse of Hotel Penaga. Procedia – Soc Behav Sci. 2016 Jun;222:631–43.
62. Lau AKK, Salleh E, Lim CH, Sulaiman MY. Potential of shading devices and glazing configurations on cooling energy savings for high-rise office buildings in hot-humid climates: The case of Malaysia. Int J Sustain Built Environ. 2016 Dec;5(2):387–99.
63. Somasundaram S, Chong A, Wei Z, Thangavelu SR. Energy saving potential of low-e coating based retrofit double glazing for tropical climate. Energy Build. 2020 Jan;206:109570.
64. Daemei AB, Eghbali SR, Khotbehsara EM. Bioclimatic design strategies: A guideline to enhance human thermal comfort in Cfa climate zones. J Build Eng. 2019 Sep;25:100758.
65. Dehwah AHA, Asif M. Assessment of net energy contribution to buildings by rooftop photovoltaic systems in hot-humid climates. Renew Energy. 2019 Feb;131:1288–99.

66. Halawa E, Ghaffarianhoseini A, Ghaffarianhoseini A, Trombley J, Hassan N, Baig M, et al. A review on energy conscious designs of building façades in hot and humid climates: Lessons for (and from) Kuala Lumpur and Darwin. Renew Sustain Energy Rev. 2018 Feb;82:2147–61.
67. Ekoe A Akata AM, Njomo D, Agrawal B. Assessment of Building Integrated Photovoltaic (BIPV) for sustainable energy performance in tropical regions of Cameroon. Renew Sustain Energy Rev. 2017 Dec;80:1138–52.
68. Bui VP, Liu HZ, Low YY, Tang T, Zhu Q, Shah KW, et al. Evaluation of building glass performance metrics for the tropical climate. Energy Build. 2017 Dec;157:195–203.
69. Chen X, Yang H, Wang Y. Parametric study of passive design strategies for high-rise residential buildings in hot and humid climates: miscellaneous impact factors. Renew Sustain Energy Rev. 2017 Mar;69:442–60.
70. Cyrille Vincelas FF, Ghislain T, Robert T. Influence of the types of fuel and building material on energy savings into building in tropical region of Cameroon. Appl Therm Eng. 2017 Jul;122:806–19.
71. Hashemi A, Khatami N. Effects of Solar Shading on Thermal Comfort in Low-income Tropical Housing. Energy Procedia. 2017 Mar;111:235–44.
72. Hashemi A. Effects of thermal insulation on thermal comfort in low-income tropical housing. Energy Procedia. 2017 Oct;134:815–24.
73. Aste N, Adhikari RS, Del Pero C, Leonforte F, Timis I. Sustainable Building Design in Kenya. Energy Procedia. 2017 May;105:2803–10.
74. Kameni Nematchoua M, Ricciardi P, Reiter S, Yvon A. A comparative study on optimum insulation thickness of walls and energy savings in equatorial and tropical climate. Int J Sustain Built Environ. 2017 Jun;6(1):170–82.
75. Thateenaranon P, Amornkitbamrung M, Hirunlabh J, Khedari J, Waewsak J. Full-scale field investigation of a bio-climatic house under Thailand tropical climate. Build Environ. 2017 Dec;126:54–67.
76. Tong S, Li H. Life-cycle cost analysis of roofing technologies in tropical areas. Energy Build. 2017 Sep;151:283–92.
77. Lei J, Yang J, Yang E-H. Energy performance of building envelopes integrated with phase change materials for cooling load reduction in tropical Singapore. Appl Energy. 2016 Jan;162:207–17.
78. Mirrahimi S, Mohamed MF, Haw LC, Ibrahim NLN, Yusoff WFM, Aflaki A. The effect of building envelope on the thermal comfort and energy saving for high-rise buildings in hot–humid climate. Renew Sustain Energy Rev. 2016 Jan;53:1508–19.
79. Zhang Z, Tong S, Yu H. Life Cycle Analysis of Cool Roof in Tropical Areas. Procedia Eng. 2016;169:392–9.
80. Omar AI, David D, Vergnault E, Virgone J, Idriss AI. A new set of indicators to evaluate the bioclimatic performance of air conditioned buildings in a hot and humid climate. J Build Eng. 2020 Sep;31:101350.
81. Colectivo de Autores. Regulaciones urbanísticas Ciudad de La Habana. El Vedado, Municipio Plaza de la Revolución. Edificiones Boloña & Ediciones Unión. Colección Arquitectura y Ciudad. La Habana; 2007.
82. Jandaghian Z, Berardi U. Analysis of the cooling effects of higher albedo surfaces during heat waves coupling the Weather Research and Forecasting model with building energy models. Energy Build. 2020 Jan;207:109627.
83. Falasca S, Ciancio V, Salata F, Golasi I, Rosso F, Curci G. High albedo materials to counteract heat waves in cities: An assessment of meteorology, buildings energy needs and pedestrian thermal comfort. Build Environ. 2019 Oct;163:106242.
84. Solar radiation absorbed by various materials [Internet]. The Engineering ToolBox. Available from: https://www.engineeringtoolbox.com/
85. Belgisch Staatsblad. Bijlage B. Warmteweerstand (R-waarde) van niet-homogene bouwmaterialen. Monit Belge. 2007 Nov 13;323:57240.

86. Alemany Barrera A, Díaz Quintero G. Cubiertas para el trópico húmedo. Manual de consideraciones climáticas. Tomo I: Cubiertas planas pesadas. Facultad de Arquitectura. Instituto Superior Politécnico "José Antonio Echeverría." La Habana; 1988.
87. Urban B, Roth K. Guidelines for selecting cool roofs. U.S. Department of Energy Building Technologies Program and Oak Ridge National Laboratory; 2010 p. 23. Report No.: V. 1.2.
88. Nieves ME, Prilipko G. Urbanismo y régimen térmico. AU. 1988;3:8–12.
89. Al-janabi A, Kavgic M, Mohammadzadeh A, Azzouz A. Comparison of EnergyPlus and IES to model a complex university building using three scenarios: Free-floating, ideal air load system, and detailed. J Build Eng. 2019 Mar;22:262–80.
90. Yang X, Peng LLH, Jiang Z, Chen Y, Yao L, He Y, et al. Impact of urban heat island on energy demand in buildings: Local climate zones in Nanjing. Appl Energy. 2020 Feb;260:114279.
91. Palme M, Inostroza L, Villacreses G, Lobato-Cordero A, Carrasco C. From urban climate to energy consumption. Enhancing building performance simulation by including the urban heat island effect. Energy Build. 2017 Jun;145:107–20.
92. Bourikas L, James PAB, Bahaj AS, Jentsch MF, Shen T, Chow DHC, et al. Transforming typical hourly simulation weather data files to represent urban locations by using a 3D urban unit representation with micro-climate simulations. Future Cities Environ. 2016 Jul 8;2(0):7.
93. Yang X, Yao L, Peng LLH, Jiang Z, Jin T, Zhao L. Evaluation of a diagnostic equation for the daily maximum urban heat island intensity and its application to building energy simulations. Energy Build. 2019 Jun;193:160–73.
94. Sigalingging RC, Chow D, Sharples S. Modelling the impact of ground temperature and ground insulation on cooling energy use in a tropical house constructed to the Passivhaus Standard. IOP Conf Ser Earth Environ Sci. 2019 Oct 11;329:012010.
95. Buol SW. TROPICAL SOILS | Humid Tropical. In: Encyclopedia of Soils in the Environment [Internet]. Elsevier; 2005 [cited 2020 Jan 30]. p. 187–95. Available from: https://linkinghub.elsevier.com/retrieve/pii/B0123485304005117
96. Dolschak K, Gartner K, Berger TW. A new approach to predict soil temperature under vegetated surfaces. Model Earth Syst Environ. 2015 Dec;1(4):32.
97. P. Sharma, M. K. Shukla, T. W. Sammis. Predicting Soil Temperature Using Air Temperature and Soil, Crop, and Meteorological Parameters for Three Specialty Crops in Southern New Mexico. Appl Eng Agric. 2010;26(1):47–58.
98. Group – Location – Climate – Weather File Access: Input Output Reference — EnergyPlus 8.4 [Internet]. bigladder software. [cited 2020 Jan 31]. Available from: https://bigladdersoftware.com/epx/docs/8-4/input-output-reference/group-location-climate-weather-file-access.html#sitegroundtemperaturebuildingsurface
99. Lemus Mesa KM. Recomendaciones de diseño para reducir la temperatura del aire interior en alojamientos para el turismo cultural en El Vedado [Tesis Diploma]. [La Habana, Cuba]: Universidad Tecnológica de La Habana José Antonio Echeverría, CUJAE; 2019.
100. Guo H, Ferrara M, Coleman J, Loyola M, Meggers F. Simulation and measurement of air temperatures and mean radiant temperatures in a radiantly heated indoor space. Energy. 2020 Feb;193:116369.
101. ASHRAE. 2017 ASHRAE handbook: Fundamentals : SI edition. Atlanta, GA: ASHRAE, 2017, ©2017; 2017.
102. Ventilation for buildings – Calculation methods for the determination of air flow rates in buildings including infiltration. European Standard 2006.
103. Ideal Loads Air System: Engineering Reference — EnergyPlus 8.0 [Internet]. [cited 2020 Feb 10]. Available from: https://bigladdersoftware.com/epx/docs/8-0/engineering-reference/page-092.html
104. Oficina Nacional de Normalización. Edificaciones-Requisitos de diseño para la eficiencia energética- Parte 3: Sistemas y equipamiento de calefacción, ventilación y aire acondicionado. NC 220-3: 2009 2009 p. 35.
105. Fariñas Acosta L. Calentadores solares de agua: apuesta por la calidad de vida y el ahorro [Internet]. Granma. 2019 [cited 2020 Nov 13]. Available from: http://www.granma.cu/

cuba/2019-11-06/calentadores-solares-de-agua-apuesta-por-la-calidad-de-vida-y-el-ahorro-06-11-2019-23-11-16
106. Stolik Novygrod D. Energía fotovoltaica para Cuba. Editorial CubaSolar. La Habana, Cuba; 2019. 555 p.
107. Sarmiento Sera A. Orientación, inclinación y espaciamiento de los módulo PV. Conferencia de electiva de pregrado "Fuentes Renovables para la Arquitectura" presented at; 2018; La Habana, Cuba.

Prof Natalí Collado Baldoquin Natalí Collado Baldoquin Architect (2012), Master in Science (2015), PhD, (2021). She is Associate Professor at the Faculty of Architecture, in the Technological University of Havana "José Antonio Echeverría" CUJAE. During the last years, her research work has been focused on sustainable architecture, bioclimatic design, buildings energy retrofit, as well as monitoring and simulation of the thermal built environment. Results have been published in several scientific journals, such as the *Interamerican Journal of Environment & Tourism*, RIAT ("The impact of tourist accommodation in residential buildings at Vedado, Cuba, 2020"); *Journal of Construction Engineering*, RIC ("Challenges for a definition of Nearly Zero Energy Buildings, 2019"), *Eco-Solar* ("Proposals to improve the habitat quality in Las Terrazas community" and "Nearly zero energy hotels. Potentialities and restrictions for Cuba") and *Architecture and Urbanism*, AU ("Sustainable development in the community Las Terrazas. First results of an international project"). Some papers have also been presented at international congresses, such as the International Congress Built Environment and Sustainable Development, Havana 2018, International Conference Sustainable Development of Built Environment, London 2018, and International Conference on Renewable Energy, Energy Saving and Education, Havana 2017.

Prof Dania González Couret Dania González Couret Architect (1979), PhD (1994), Doctor in Sciences (2007), Full Professor (2001) and Emeritus Professor (2022) at the Faculty of Architecture, in the Technological University of Havana "José Antonio Echeverría" CUJAE. She is a member of the Cuban Academy of Science, the National Technical Committee for Standardization on Bioclimatic Design; member of the Leading Board of Cubasolar and Arc.Peace International; president of the International Congress Built Environment and Sustainable Development, and the National Court of Scientific Degrees in Architecture and Urbanism. During the last 40 years, her research work has been focused on issues such as sustainable built environment, bioclimatic design, passive low energy architecture and renewable energy. She is the author of three books published in Havana, and some chapters on books published by Elsevier (where she is Reviewer) and Springer, as well as a contributor to books edited by Un Habitat. Her articles and papers have been published in several journals and proceedings, such as *Renewable Energy*; *World Renewable Energy Congress* (member of the Steering Committee); *Passive Low Energy Architecture*; *Teaching in Architecture*; *Rebuilding the European City*; *Architecture, Energy and Environment*, among others.

Prof Luis Alberto Rueda Guzmán Luis Alberto Rueda Guzmán Architect (1995), PhD (2003), and Associate Professor at the Faculty of Architecture, in the Technological University of Havana "José Antonio Echeverría" CUJAE. During the last 20 years, his research work has been focused on evaluating the influence of architectural design on the indoor thermal environment and energy consumption of buildings, as well as on tourist facilities and ecotourism. He coordinates the international joint research project with the University of Gent in Belgium "Renewable energy and bioclimatic architecture improving sustainability and development in ecotouristic settlement: Las Terrazas". His last articles have been published in the *Interamerican Journal of Environment & Tourism*, RIAT ("The impact of tourist accommodation in residential buildings at Vedado, Cuba, 2020"); *Journal of Construction Engineering*, RIC ("Challenges for a definition of Nearly Zero Energy Buildings, 2019"), *Eco-Solar* ("Proposals to improve the habitat quality in Las Terrazas community" and "Nearly zero energy hotels. Potentialities and restrictions for Cuba") and *Architecture and Urbanism* ("Sustainable development in the community Las Terrazas. First results of an international project"). Some papers have been presented at the International Conference Built Environment and Sustainable Development, Havana 2018, International Conference Sustainable Development of Built Environment, London 2018 and International Conference on Renewable Energy, Energy Saving and Education, Havana 2017. He has held different academic and professional responsibilities in recent years as Dean of the Architecture Faculty in Havana from 2012 to 2021 and currently vice-rector of the Technological University of Havana (CUJAE). He has also presided over the National Architecture Career Commission since 2015 and has been a member of the National Commission of Monuments since 2012.

Correction to: The Role of Shading, Natural Ventilation, Daylighting, and Comfort in Enhancing Indoor Environmental Quality and Liveability in the Age of COVID-19

Mohsen Aboulnaga and Maryam Elsharkawy

Correction to:
Chapter 8 in: A. Sayigh (ed.), *Achieving Building Comfort by Natural Means*, Innovative Renewable Energy, https://doi.org/10.1007/978-3-031-04714-5_8

Correction to Chapter 8 and Front matter in: Author name

The name of the author Maryam Elsharkawy was unfortunately published with an error. The initially published version has now been corrected.

The correction has been updated in the Front matter of the book.

The updated original version of this chapter can be found at https://doi.org/10.1007/978-3-031-04714-5_8

A. Sayigh (ed.), *Achieving Building Comfort by Natural Means*, Innovative Renewable Energy, https://doi.org/10.1007/978-3-031-04714-5_20

Correction to: Thermal and Visual Adaptive Comfort: Field Studies in Portugal

L. Matias, A. Santos, and M. Correia Guedes

Correction to: Chapter 11 in: A. Sayigh (ed.), *Achieving Building Comfort by Natural Means*, Innovative Renewable Energy, https://doi.org/10.1007/978-3-031-04714-5_11

The photos of Dr. António Santos and Dr. Luis Matias were interchanged in the Author Biography section and also the order of presentation of the authors' CVs was incorrect. The initially published version has now been corrected.

The updated original version of this chapter can be found at https://doi.org/10.1007/978-3-031-04714-5_11

A. Sayigh (ed.), *Achieving Building Comfort by Natural Means*, Innovative Renewable Energy, https://doi.org/10.1007/978-3-031-04714-5_21

Conclusions

In this book, several approaches and experiences that address the important issue of comfort in well-designed buildings achieved with minimal use of external energy are presented.

The international ASHRAE Standard 55-2004 describes comfort standard requirements in buildings and the thermal environmental conditions for human occupancy. It combines indoor thermal environmental factors and personal factors to achieve acceptable comfort for majority of the occupants within the buildings. The standard allows the psychometric comfort charts to be applied to spaces where the occupants have activity levels resulting in metabolic rates between 1.0 and 1.3 and with wearing clothing it provide between 0.5 and 1.0 of thermal insulation. The comfort zone is based on the predictive mean vote (PMV) values between -0.5 and +0.5, as empirically mentioned by Fanger in the introduction of this book.

Prof Zeiler in Chap. 1 states that to achieve comfort, one has to get optimal combination between natural means and other introduced efficient systems, that is, the right balance between passive and active systems.

The concept of "living" microbial systems as a strategy for modulating building comfort is a new field of research, explained well by Prof. Armstrong in Chap. 2. Even with respect to the fundamental science that characterizes human/microbial relations, where findings are still emerging. Despite their novelty, the role of microbial technologies in buildings is of strategic importance in an ecologically stressed world, where gardened microbial systems in buildings will stop waste streams being tipped into the environment, reduce consumption and provide probiotic living spaces with regular impacts on health and well-being. As we enter an age of increasing zoonotic epidemics and pandemics, their capacity to act as external immune systems by eliminating pathogens from waste streams is of particular interest.

One of the major controlling elements in building design is the use of appropriate thermal mass. For thermal mass to be effective, it must be closely coordinated with suitable passive design strategies including orientation, shading of glazing, proper ventilation and appropriate level of insulation. The use of thermal mass may not be

A. Sayigh (ed.), *Achieving Building Comfort by Natural Means*, Innovative Renewable Energy, https://doi.org/10.1007/978-3-031-04714-5

appropriate if a location has poor air quality and high levels of noise pollution (characteristic of many city centres). This was the conclusion of Dr. Azerbijani and Prof. Thaddeus in Chap. 3. They maintained that it must be possible to access the real impacts of living spaces both on the planet, in terms of energy efficiency and conscious use of resources, as well as on people, in terms of comfort and well-being, in order to support nature-based regeneration processes.

In Chap. 4, the authors described how to enhance the comfort in Urban Isles of the Mediterranean Region. This was a specific study related to Cyprus.

Most of the work analysis was based on Fanger formula (PMV). "It was noticed that the albedo increase has created a negative impact on comfort, if the replaced material has less than 0.4 difference in albedo value, since the decrease in air temperature is counteracted by the increase in mean radiant temperature. Also, it is critical for the awnings to be retractable to allow solar access in winter, because shading affects in a noticeably negative way, the outdoor and possibly indoor comfort in winter, which is expected to impact additionally the energy consumption of the surrounding buildings.

Mr. Tony Book in Chap. 5 approaches the comfort issue by using a smart system in monitoring energy consumption. For example, in a standard home, the system is saving about 6 tons of CO_2 per year. If it was feasible to similarly equip half the homes in the United Kingdom similarly over the next 10–15 years, then the savings (of CO_2) would be over 30 million tons per year.

Prof Balocco in Chap. 6 states the importance of lighting. Light makes possible to see more clearly the architectural, historical and social value of buildings and their environment. This fact implies a different light design aimed at the optimal combination between quantity and quality of natural and artificial lighting. It ensures quality of vision and perception, guaranteeing protection and preventive conservation. Nowadays, the LED technology, both for spectrum and correlated colour temperature tuning, allows adding sophisticated functions to standard lighting solutions for energy saving and for using smart lighting technologies based on the human-centric lighting concept.

Dr. Amina and Dr. Baba in Chap. 7 address how to stay cool under the blazing hot sun in Nigeria. Cultural and religious factors play a substantive role in building construction decisions, more so than bio-climatic factors.

In order to stay cool, the builder's choice of using earth as a building material is for its thermal qualities and abundance.

In addition, cooling in traditional Hausa is achieved naturally by building the walls with two or more air gaps in between, placing windows in such a way as to harness the south and northerly winds plus limiting the size of any openings to reduce the flow of hot, dusty air into the cool interiors.

Prof. Aboulnaga and Dr. ElSharkawy in Chap. 8 discuss the role of shading, natural ventilation, daylighting and comfort in enhancing indoor environmental quality and liveability in the age of COVID-19. Twenty-six buildings are presented and assessed in terms of shading strategies, ventilation, daylighting and sunlight as well as thermal comfort. In addition, four buildings were studied regarding their natural comfort and space requirement during the Covid-19 restriction. Among them was the new Grand Museum in Giza, which satisfies all green sustainable criteria.

Chapter 9 investigates the use of lighting in the ancient Minoan buildings, which shows that light has been used continuously since ancient times to achieve comfort and sustainability. In Minoan architecture, light wells were used primarily for illumination. The integration of the known principles of light transmission and the benefits of proper orientation in the domestic architecture of the classical period, as well as the careful design of the Greek temples with regards to the daylight and intensifying religious experience.

Dr. Adam and Dr. Ghafar in Chap. 10 related in Malaysia, research heat gain or loss resulting through building roofs. Based on the evolution of roof design. Designers may choose appropriate roof geometry not only based on architectural style and local identity but also on building thermal performance, which results in reducing heat gain and energy consumption in buildings.

In Chap. 11, Prof Guedes focuses on an interdisciplinary approach in considering specific Portuguese weather heat gain characteristics and is supported by extensive objective measurements and subjective data obtained through questionnaires regarding occupants' response and reaction to thermal and luminous indoor environments.

However, a closer look at objective/subjective relationships reveals that it is possible to attain satisfaction "differently" and actually in a more sustainable way. The Portuguese temperate climate, with predominantly non-overcast skies and traditional adaptability, bears out that premise.

In Chap. 12, Dr. Trombadore and Dr. Calcagno maintain that architects and builders must address the critical requirements of the need to adopt ecological rethink in their architectural models towards a man–nature harmony. They must address the concept of predictive capacity of digital twins in integrated design processes, envisioning healthy and sustainable built environments.

Prof. Al-Sallal in Chap. 13 looks at comfort in museums and galleries in Canada. His recommendation is that there should be a balance between visual acceptance, lighting intensities and space security; also avoiding glare and too much direct sunlight coupled with moderate use of artificial lighting is preferable to achieve harmony.

In Chap. 14, Prof. Kasinath has reviewed five houses in rural Andhra Pradesh, India which utilize different construction materials for heat gain. The authors discovered that the south-west wall gained more heat than other sides and therefore recommended it to be shaded. Regarding the roofing, it was found that reed thatch roof performs best and is recommended as compared with RCC slab when used in hot and humid climate, which resulted in 50% heat loss.

In Chap. 15, Prof. Carolina Ganem Karlen asserted that buildings should be designed

to operate in a flexible way and must adapt to their occupants' requirements and comfort. Prof. Karlen followed Schweiker's principles in establishing the relationship between paradigms of comfort parameters and resilience. This will lead to several opportunities to scrutinise building design and practices together with research on thermal comfort related to shading strategies, ventilation, daylighting and sunlight gain.

Dr. AlKubaisy in Chap. 16 has analysed in depth how to stay cool in the arid zones of the Arab countries and recommends the use of rectangular-shaped buildings, long in north-south orientation with gardens either at the front or at the back of the buildings. The buildings must also use thicker walls than normal, utilizing the concept of thermal mass as diode to reduce much solar gain. The use of glazing should be avoided, especially on the south-facing. Buildings preferably should have small windows. The use of wind tower and north-facing "Mulqaf – Hawa" are recommended.

In Chap. 17, Dr. Alireza Dehghani-Sanij discusses the well-documented tradition of Iranians to achieve building comfort in dry-hot regions by combining natural ventilation with radiative cooling from man-made tunnels to the ground connected with water wells. This is known as the wind tower or badger concept. Again the importance of using shading devices, thermal mass and ventilation coupled with night radiation to the black sky at night encouraged the people to sleep on the roofs of open buildings at night for cooling.

Dr. Chiesa in Chap. 18 discusses the major elements of comfort in hot climate regions by the use of ventilative cooling (VC), which has been around for many years and has a history of success. A correlation is found to exist between climate and the building system, that is the design and construction of the building as indicated in key performance indicators (KPI). It is thus possible to analyse at an early stage of the building process the VC potential of a specific location and its need for cooling in terms of climate intensity.

Chapter 19, Prof Dania Couret, Many research investigations as in Chap. 4, were carried out at various universities and building establishments which demonstrated the energy impact and the economic viability of the retrofit actions proposed for historic buildings with high thermal mass in urban contexts in warm and humid climates. This has not been addressed in former research, and it is applicable now to Cuba. For urban contexts in warm and humid climates, the suitability to apply solutions leading to effective comfort has been demonstrated, such as low transfer glasses and thermal insulation in the buildings' envelopes.

Achieving comfort by natural means in a building is always possible if architects and builders avoid the use of excessive electrical energy and rethink their design to use proper thermal mass, ventilation, shading and daylighting and in creation of green effect with appropriate window opening and building orientation in line with the regional location and climate.

Index

A. Sayigh (ed.), *Achieving Building Comfort by Natural Means*, Innovative Renewable Energy, https://doi.org/10.1007/978-3-031-04714-5

C

The manufacturer's authorised representative in the EU is Springer Nature Customer Service Centre GmbH, Europaplatz 3, 69115 Heidelberg, Germany. If you have any concerns regarding our products, please contact ProductSafety@springernature.com

Printed and bound by CPI Group (UK) Ltd, Croydon, CR0 4YY
01/10/2026
02226185-0010